FLORA ZAMBESIACA

Flora terrarum Zambesii aquis conjunctarum

VOLUME FIVE: PART ONE

FLORA ZAMBESIACA

MOZAMBIQUE

MALAWI, ZAMBIA, ZIMBABWE

BOTSWANA

VOLUME FIVE: PART ONE

Edited by
E. LAUNERT

on behalf of the Editorial Board:

G. Ll. LUCAS
Royal Botanic Gardens, Kew

E. LAUNERT
British Museum (Natural History)

M. L. GONÇALVES
Centro de Botânica, Instituto de Investigação Científica Tropical, Lisboa

Published by the Managing Committee on behalf of
the contributors to Flora Zambesiaca
1989

Typeset at the Royal Botanic Gardens, Kew, by
Pam Arnold, Christine Beard, Brenda Carey
and Helen O'Brien

Printed in Great Britain by
Whitstable Litho Printers Ltd.,
Whitstable, Kent

ISBN 0 9507682 4 3

CONTENTS

EDITOR'S NOTE
Special recognition must be given to the editorial assistance of Mr G.V. Pope, Krukoff Curator of African Botany.

LIST OF FAMILIES INCLUDED IN VOLUME V, PART 1

94.- Rubiaceae (Rubioideae)

LIST OF NEW NAMES AND TAXA PUBLISHED IN THIS WORK

Agathisanthemum bojeri subsp. *angolense* (Bremek.) Verdc., comb. et stat. nov.

Agathisanthemum bojeri subsp. *angolense* var. *linearifolia* Verdc., var. nov.

Anthospermum rigidum subsp. *pumilum* (Sond.) Puff, stat. nov.

Conostomium gazense Verdc., sp. nov.

Kohautia caespitosa subsp. *brachyloba* (Sond.) D. Mantell, comb. et stat. nov.

Kohautia microflora D. Mantell, sp. nov.

Kohautia subverticillata (K. Schum.) D. Mantell, comb. nov.

Oldenlandia subgen. Phymatesta Verdc., subgen. nov.

Oldenlandia machingensis Verdc., sp. nov.

Oldenlandia robinsonii Verdc., sp. nov.

Oldenlandia verrucitesta Verdc., sp. nov.

Pentas zanzibarica var. *haroniensis* Verdc., var. nov.

Psychotria amboniana subsp. *mosambicensis* (Petit) Verdc., comb. nov.

Psychotria capensis subsp. *riparia* (K. Schum. & K. Krause) Verdc., comb. nov.

Psychotria capensis subsp. *riparia* var. *puberula* (Petit) Verdc., comb. nov.

Psychotria diversinodula (Verdc.) Verdc., comb. nov.

Psychotria pumila var. *subumbellata* (Petit) Verdc., comb. nov.

Psychotria mwinilungae Verdc., sp. nov.

Psychotria peduncularis var. *angustibracteata* Verdc., var. nov.

Psychotria peduncularis var. *rufonyassana* Verdc., var. nov.

Spermacoce bisepala Verdc., sp. nov.

Spermacoce princeae var. *mwinilungae* Verdc., var. nov.

Spermacoce samfya Verdc., sp. nov.

Spermacoce schlechteri K. Schum. ex Verdc., sp. nov.

Triainolepis sancta Verdc., sp. nov.

94. RUBIACEAE

By B. Verdcourt*

Small to large trees, shrubs or less often annual or perennial herbs or woody or herbaceous climbers, sometimes spiny; tissues in many tribes containing abundant rhaphides. Leaves opposite or verticillate, decussate, almost always entire, very rarely (not in Africa) palmatifid, toothed or finely denticulate, always obviously stipulate (save in some Rubieae, where the stipules may be considered foliar or almost absent according to interpretation), the stipules interpetiolar or intrapetiolar, entire or, particularly in herbaceous genera, variously divided into lobes or fimbriae, often tipped or separated by mucilaginous hairs known as colleters** and often with colleters inside the base; the leaves in certain tribes sometimes contain small bacterial nodules. Flowers rarely solitary, mostly in various terminal or axillary inflorescences, all basically cymose but variously aggregated into panicles, etc., occasionally in globose heads to an extent that the ovaries are adnate; bracts vestigial to well developed, even conspicuous; flowers usually hermaphrodite, rarely unisexual, regular or nearly so (except in *Posoqueria* (America)) or corolla tube rarely curved, homostylous or quite often heterostylous with 2 or rarely 3 forms (long-styled (dolichostylous), short-styled (brachystylous) or equal-styled (isostylous)). Calyx*** gamosepalous, the tube mostly adnate to the ovary, (3)4–5(8)-toothed or lobed, sometimes only minutely so, with open, valvate, imbricate or contorted aestivation, 1 or several lobes sometimes slightly to very considerably enlarged to form a leafy often coloured lamina. Corolla small to large and showy, gamopetalous, rotate to salver-shaped or funnel-shaped, the tube often very long, (3)4–5(11)-lobed, the lobes mostly contorted or valvate, sometimes valvate-induplicate, rarely imbricate or quincuncial. Stamens usually as many as the corolla-lobes and alternate with them, epipetalous; anthers basi-or dorsifixed, introrse, the thecae rarely multilocellate transversely. Pollen various, mostly simple, isopolar and 3-colporate, but sometimes porate, the number of colpi or pores varying from 2–25, globose, ovoid or discoid, sometimes (in some *Gardenieae*) in tetrads or rarely polyads. Disc often present, 2-lobed or tubular. Ovary inferior, rarely half-inferior or (in *Gaertnera*) superior, syncarpous of 2–5 or more carpels, but predominantly of 2 and therefore predominantly 2-locular, but 3–5 or even 12 or more (e.g. in *Urophylleae* due to supplementary incomplete partitions); placentation axile or (in some *Gardenieae*) parietal; ovules 1–many per locule, often embedded in fleshy placentas, erect, basal or horizontal, anatropous; style simple, usually long and narrow, the "stigma" either cylindrical, clavate, or otherwise modified to form a "receptaculum pollinis" or divided into 2–many linear, spathulate or clavate lobes, the actual stigmatic surface sometimes confined to certain areas, e.g. the inner faces of the lobes. Fruit small to quite large (0.2–20 cm.), a capsule, berry or drupe or indehiscent or woody, occasionally (e.g. in *Nauclea* and *Morinda*) united to form syncarps, (1)2–many-seeded, if capsules then loculicidal or septicidal or opening by a beak. Seeds small to rather large, sometimes winged; testa cells in some tribes with very distinct pits; albumen present (save in *Guettardeae*); embryo straight or rarely curved, the radicle mostly longer than the cotyledons.****

A large family of about 500 genera and 6000 species, predominantly tropical and adapted to moist environments (except the *Rubieae* which are predominantly temperate or even arctic in distribution). Older descriptions of the family differ in some respects owing to the inclusion of elements now generally excluded, e.g. somewhat zygomorphic

* *Kohautia* by D. Mantell (Vienna); *Otiophora, Paederia, Anthospermum, Rubia* & *Galium* by C. Puff (Vienna).

** See Lersten, Colleter morphology in *Pavetta, Neorosea* and *Tricalysia* (Rubiaceae) and its relationship to the bacterial leaf nodule symbiosis, in Journ. Linn. Soc. Bot., **69**: 125–136 (1974).

*** For terminology of parts see p. 4.

**** Excellent more extensive accounts of morphology and biology of the family will be found in Hallé, Fl. Gabon **12**, Rubiacées: 7–22 (1966) and White, F.F.N.R.: 383–388 (1962).

flowers occur in *Henriquesiaceae*, often considered as a tribe (*Henriquesieae*) of the Rubiaceae. In the flora area there are 96 genera and about 450 species.

The classification of this family into tribes is not really very difficult, but the basic arrangement of these tribes into subfamilies is a matter of contention. The classical division is into two groups, one with the ovary-locules containing one ovule and the other with them containing 2 or more. This is still followed by many authors, e.g. Hallé in Fl. Gabon 12 (1966), and is a partially convenient method, being subject to no more exceptions than most classifications of this type. There is no doubt, however, that this is artificial and the classification proposed by Bremekamp and myself* is more natural, though perhaps less practical in some respects, since it is partly based on the presence or absence of rhaphides. Since neither ovules nor rhaphides are easy to see, use has also been made of other characters in the key, as was first done by Hutchinson in the first edition of F.W.T.A. and later followed by Keay and Hepper in the second. A conspectus of the classification followed in this present account is given below, it differs little from that proposed by me in 1958, except for some modifications adopted from Bremekamp's latest paper and recent work by E. Robbrecht.

Various pollination mechanisms exist in the family from simple protandry, where the pollen is dispersed before the stigmatic surfaces are revealed, to more sophisticated arrangements such as heterostyly, where there are 2 or even 3 types of flowers. In complete cases the style is well included and the anthers well exserted in one sort of flower and exactly the opposite in the other sort, but in some tribes the differences are far less marked. In many homostylous species (i.e. those not showing heterostyly) the stigma and the anthers are well isolated. The ixoroid pollination mechanism, mentioned quite frequently in the text, is where the usually cylindrical or clavate style-head (which could easily be taken for the stigma) acts as a pollen receptacle; it is in close contact with the anthers and pollen is deposited on it before the flower opens. After the flower opens the style elongates and the pollen is ready for insect transfer; later the stigma lobes divide to reveal the true stigmatic surfaces and are ready for pollen from other flowers. Unisexuality is rare in the Flora Zambesiaca area, but occurs in *Anthospermum*, possibly derived from extreme cases of heterostyly. Most species of Rubiaceae are insect-pollinated, many having white flowers strongly scented at night, being obviously pollinated by moths. *Anthospermum*, with its long styles and dangling anthers is clearly wind-pollinated and a few of the larger flowered *Vanguerieae* may even be bird-pollinated. Doubtless, despite the special mechanisms, quite a good deal of self-pollination exists. There is a wealth of material here for studies to be made in the field by biologists on the spot.

Synopsis of subfamilies and tribes

Subfamily **Rubioideae**. Rhaphides present in the tissues (except in *Urophylleae* where these are replaced by thicker styloid crystals). Trees, shrubs or very often herbs. Corolla lobes valvate (with some very rare exceptions none of which occur in Africa). Indumentum of stem, foliage and inflorescence, if present, usually of clearly septate hairs (see TAB. **1**, figs. 8,9), but short undivided hairs are common in some genera. Complete heterostyly is very frequent but can be present or absent in species of the same genus. Ovules solitary to numerous in 2–many locules, erect or pendulous or attached in the middle. Fruit dry or succulent or dehiscent. Stipule sheath very often divided into fimbriae. Style usually divided into 2 linear lobes (or more in the case of multilocular ovaries), bearing the stigmatic papillae on their inner surfaces. Seeds with albumen: testa cells rarely pitted. The first two, fourth, fifth and sixth tribes of this subfamily all contain aluminium accumulators to a rather marked degree. *Genera 1–37.*

Tribe **Psychotrieae** Dumort., Anal. Fam. des Plantes: 33 (1829). Mostly shrubby, rarely herbs or trees. Ovules solitary in 2–8 locules, erect. Fruit usually succulent and containing 2–8 pyrenes, more rarely dry and separating into cocci. Pollen grains porate. *Genera 1–4.*

Tribe **Gaertnereae** Bremek. ex Darwin in Taxon **25**: 601 (1976). Trees or shrubs. Ovary superior (the only tribe thus), 2-locular with one erect solitary ovule in each locule. Fruit more or less fleshy, indehiscent with 2 pyrenes. Pollen grains 3-colpate. *Genus 5.*

* See Verdcourt in Bull. Jard. Bot. Brux. **28**: 209–290 (1958) & Bremekamp in Acta. Bot. Neerl. **15**: 1–33 (1966).

Tribe **Morindeae** Miq., Fl. Ind. Bat, **2**: 241 (1857) [as subtribe in DC., Prodr. **4**: 342 (1830)]. Trees or shrubs. Ovary 2–12-locular; ovules solitary in the locules, affixed at or below the middle. Fruit baccate or drupaceous, 2–12 locular or containing up to 12 pyrenes, often forming a fleshy syncarp. Pollen grains porate. *Genera 6–7.*

Tribe **Triainolepideae** Bremek. in Proc. K. Nederl. Akad. Wetensch., Ser. C, **59**: 3 (1956). Shrubs with 3–7 fimbriate stipules. Ovary 2–10-locular; locules with 2(3) collateral ovules inserted at the base of the septum. Fruits with 1 multilocular pyrene, each locule 1-seeded. *Genus 8.*

Tribe **Urophylleae** Bremek. ex Verdc. in Bull. Jard. Bot. Brux. **28**: 281 (1958). Shrubs or small trees, rarely herbaceous. Ovary 2–many-locular, with numerous ovules in each locule. Fruit a berry, 2–many-locular, many-seeded. Pollen grains colpate, with large apertures. Owing to the lack of true rhaphides Bremekamp has erected a subfamily *Urophylloideae* and has also placed the African members in a separate tribe *Pauridiantheae*. *Genus 9.*

Tribe **Craterispermeae** Verdc. in Bull. Jard. Bot. Brux. **28**: 281 (1958). Shrubs or trees, with rather marked yellow-green foliage on drying, nearly all being marked aluminium accumulators. Ovary 2-locular, with solitary pendulous ovules. Fruit a berry, 1–2-locular. Pollen grains porate. *Genus 10.*

Tribe **Knoxieae** Hook.f. in Benth. & Hook., Gen. Pl. **2**: 9 (1873). Herbs or small subshrubs. Ovary 1–5-locular; ovules solitary, pendulous. Fruit either succulent with pyrenes, dry and indehiscent or dividing into indehiscent cocci. Flower sometimes very bright blue. *Genera 11–12.*

Tribe **Paederieae** DC., Prod **4**: 343; 470 (1830). Usually foetid climbers. Ovary 2–5-locular; ovules solitary, erect. Fruit 2–coccous with thin epicarp, or capsular. *Genus 13.*

Tribe **Hamelieae** Kunth, Nova Gen. **3**: 413 (1820), as *Hameliaceae*, emend. DC., Prod. **4**: 342; 438 (1830). Shrubs. Corolla lobes imbricate. Ovary 5-locular; ovules numerous in each locule. Anthers long. Disk fleshy. Fruit a berry. Only known cultivated in Africa, see p. 6.

Tribe **Hedyotideae** Cham. & Schlecht. ex DC., Prodr. **4**: 342; 401 (1830). Herbs or shrubs. Ovary 2(rarely 3–4)-locular; ovules usually numerous, rarely few or even solitary in each locule, affixed to the base of the septum. Fruit a capsule (save in a few dubiously placed genera not occurring in Africa). *Genera 14–27.*

Tribe **Anthospermeae** Cham. & Schlecht. ex DC., Prodr. **4**: 343; 578 (1830). Herbs or shrubs. Flowers often unisexual. Ovary 1–4-locular; ovules solitary in each locule, erect. Fruit dry, dividing into cocci or capsular. *Genera 28–30.* (The genus *Otiophora* is excluded by C. Puff).

Tribe **Spermacoceae** A. Rich., ex Dumort, Anal. Fam. des Plantes: 33 (1829). Herbs. Ovary 2(rarely 3–4)-locular; ovules solitary in the locules, affixed to the middle of the septum. Fruit dry, capsular or dividing into dehiscent or indehiscent cocci. Pollen grains pluricolpate. *Genera 31–35.*

Tribe **Rubieae**. Herbs. Stipules (with very few exceptions) foliar, there being a whorl of similar leaves and stipules at each node. Stem polygonal with collenchymatous ribs. Ovary 2-locular; ovules attached to the septum. Fruit fleshy, often didymous, or more or less dry. *Genera 36–37.*

Subfamily **Cinchonoideae** Rafinesque in Ann. Gén. Sci. Phys. **6**: 81 (1820) emend. K. Schum. in Engl. & Prantl., Pflanzenf. **4**(4): 16 (1891) & re-emend. Verdc. in Bull. Jard. Bot. Brux. **28**: 280 (1958). Rhaphides absent from the tissues. Trees, shrubs or rarely herbs. Corolla lobes valvate, contorted or imbricate. Indumentum of stem, foliage and inflorescence, if present, of thick-walled non-septate or incompletely septate hairs. Complete heterostyly rare, but limited heterostyly present in a few tribes. Ovules solitary or numerous in 1–many locules, attached to the septum or pendulous or embedded in 1–several placentas. Fruit dry or succulent, dehiscent or indehiscent. Stipules usually undivided. Style with stigma fusiform, capitate or divided into linear lobes. Seeds with albumen; testa cells in some tribes with very conspicuous pits. The tribes in this subfamily will be listed in part 2.

Subfamily **Antirheoideae** Rafinesque in Ann. Gén. Sci. Phys. **6**: 86 (1820). Rhaphides absent from the tissues. Trees or shrubs. Corolla lobes valvate or imbricate. Indumentum consisting of hairs which are not truly septate. Limited heterostyly occurs. Ovary 2–many-locular, with a solitary pendulous ovule in each locule. Fruits drupaceous or with woody putamen, with 2 pyrenes or rarely dicoccous. Seeds with little or no albumen; testa not sculptured or irregularly reticulate. Pollen grains porate. One tribe *Guettardeae* (dealt with in part 2).

Notes on the keys

The family Rubiaceae contains so many genera and species, many of which resemble each other even when not closely related, that it is impossible to make a usable key which does not involve looking at small difficult characters. If, after a preliminary run through the key it is found that the macroscopic characters will not suffice then it is best to examine the plant in detail and make a list of its essential characters. The aestivation of the corolla lobes in bud is a very important character indeed and the main types are shown in TAB. 1, figs. 1–5. This is fairly easy to see in fresh material under an ordinary hand-lens but when only herbarium material is available a bud should be boiled up. The arrangement of the ovules in the ovary is also a crucial character, whether there are one or several to many in each of how many locules and if solitary whether they are attached to the base, apex or middle of the locule. This is admittedly a difficult character to see and dissection under a binocular dissecting microscope is often necessary. A transverse cut across the ovary with a razor blade will often clearly show the number of locules, and in the case of solitary ovules whether thay are attached at the base or apex can be ascertained by seeing which half of the cut ovule drops out easily. A good deal can be learnt by cutting gradually into the side of the ovary until the ovule or placentas are revealed. Care must be taken not to mistake a placenta covered with minute ovules for a single ovule; this mistake has been made even by professional botanists and several species have been redescribed in the wrong genus and tribe as a result. The presence or absence of heterostyly can be a very useful character and is worth noting in the field where a population is available. It is of course difficult or impossible to ascertain from a single herbarium specimen. If flowers showing long exserted anthers and included styles are present then the species is very likely heterostylous but the reverse is by no means true. Other microscopic characters are very useful and not difficult to see if adequate equipment is available. The presence of rhaphides restricts the choice of genera considerably. These are easily seen under a microscope and they break up into very numerous fine needle-shaped crystals if teased out under water; they are particularly easily seen if a polarising microscope is available*, but in many cases, e.g. those between the pyrenes of *Psychotria* fruits, they are clearly visible to the naked eye and they break down into fluffy crystals even when seen dry under a × 10 hand-lens. In other cases rhaphides can be seen in leaf and other tissues with the naked eye (TAB. **1**). Pollen grains vary very considerably in shape in the family and are valuable characters. Even the gross morphology easily visible at low powers of a compound microscope can restrict one's search for the identity of a difficult specimen. The following main classes may be distinguished so far as the Tropical African tribes are concerned.

1) Colpate grains. 3–4(5)-colpate. The usual kind of grain found in the family.
2) Pluricolpate grains. 5–25-colpate. These are divisible into two types:
   A) disk-shaped grains, usually large and with numerous short colpi on the rim — restricted to *Spermacoceae.*
   B) ellipsoidal grains with fewer longer colpi extending over the body — *Spermacoceae* and *Rubieae.*
3) Porate grains. 3–4 porate. Of common occurrence in the family particularly in *Mussaendeae, Hamelieae, Gardenieae, Guettardeae, Vanguerieae, Morindeae, Psychotrieae,* and *Craterispermeae.*
4) Tetrahedral tetrads of 3-porate grains. Restricted to the *Gardenieae.*

Examples of these types are shown in TAB. 1, figs. 10–16.

The authors differ in their interpretation of the calyx. Verdcourt uses the term calyx tube in the same sense of Hooker (Genera Plantarum) being the tissue adnate to the ovary; the free part of the calyx is termed the limb comprised of the limb tube and the lobes, but either or both can be absent. Puff and Mantell use the term ovary for all tissue below the disk and corolla and restrict the term calyx to that above, pointing out that the term inferior ovary is incompatible with the adnate tissue being termed calycine.

Until the family is completed the general artificial key given in F.T.E.A., Rubiaceae 1: 9–23 (1976) will be found useful and covers most genera; a few generic concepts have been altered since that time and some names changed.

* The simplest student's microscope can be used by placing pieces of polaroid in the condenser and on top of the eyepiece.

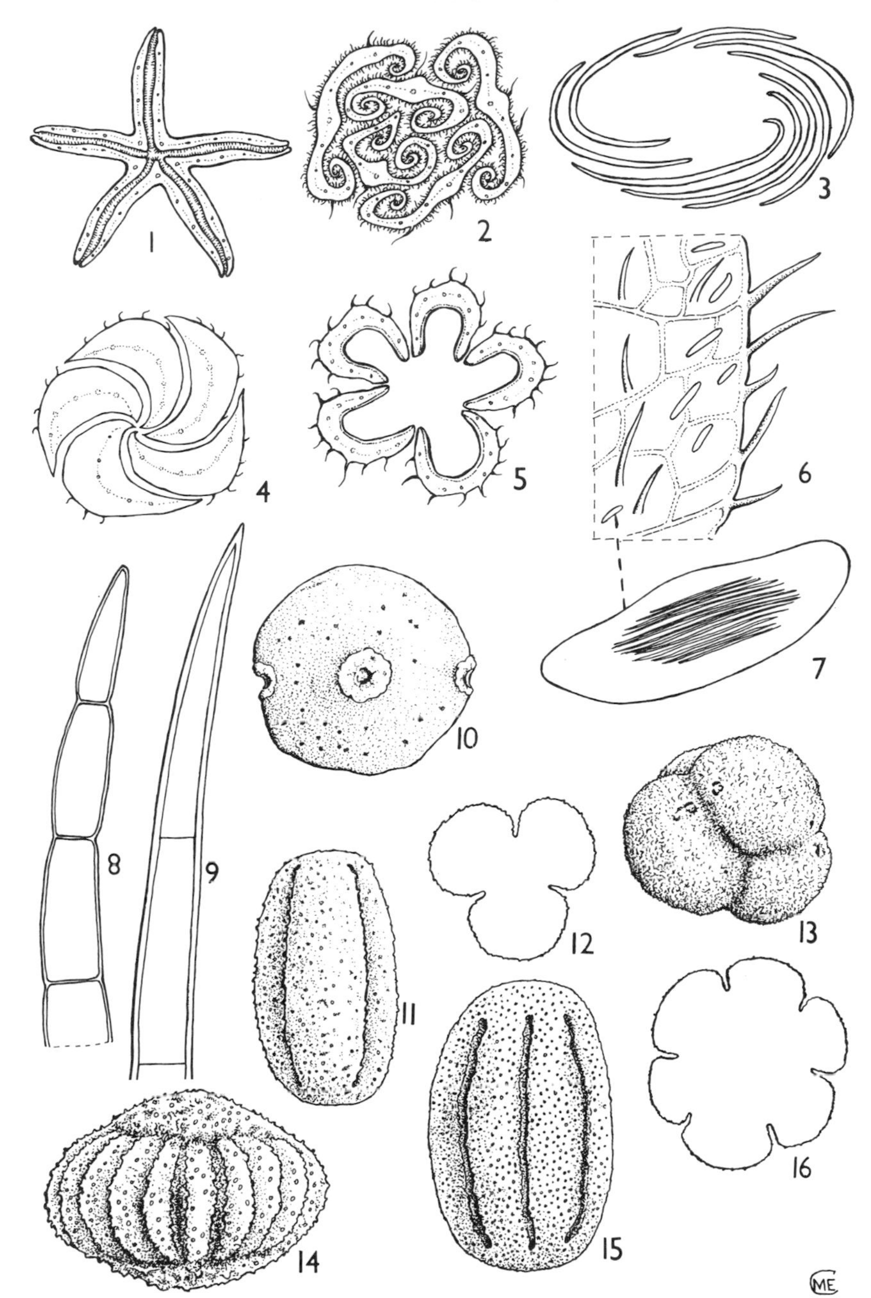

Tab. 1. CRITICAL FEATURES IN RUBIACEAE. — Transverse section of flower bud to show aestivation. 1, reduplicate-valvate, MUSSAENDA ARCUATA (× 6); 2, imbricate, HEINSIA CRINITA subsp. PARVIFLORA (× 14); 3, contorted to the left, GARDENIA JOVIS-TONANTIS (× 4); 4, contorted to the right; 5, valvate, slightly induplicate, PENTAS LANCEOLATA (× 12). —Rhaphides and hairs. 6, part of lower leaf surface of PENTAS LANCEOLATA, showing rhaphides (× 40); 7, cell with rhaphides from same (× 360); 8, true multicellular hair of PENTAS LANCEOLATA (× 80); 9, false multicellular hair of VIRECTARIA MULTIFLORA (× 80). —Pollen grains. 10, porate, GUETTARDA SPECIOSA (× 650); 11, 12, tricolpate, PENTAS BUSSEI (× 1300); 13, tetrad of 3-porate grains, GARDENIA JOVIS-TONANTIS (× 450); 14, pluricolpate, disk-shaped, SPERMACOCE DIBRACHIATA (× 650); 15, 16, pluricolpate, ellipsoidal, GALIUM STENOPHYLLUM (× 2000).

Key to the subfamilies

1. Rhaphides present (except where replaced by thicker styloid crystals in *Urophylleae*); corolla-lobes valvate (save in *Hamelia*, cultivated); flowers mostly, but not always heterostylous, ixoroid pollination mechanism never present; plants herbaceous or woody subfam. *Rubioideae*
– Rhaphides absent; corolla lobes contorted, valvate or sometimes imbricate; ixoroid pollination mechanism frequently present, but reduced heterostyly or rarely heterostyly can occur in combination with valvate aestivation; plants woody, only occasionally herbaceous (see below). - - - - - - - - - - - - - - - - - 2
2. Fruit not as below, if woody then separate pyrenes present; seeds with albumen; corolla not exactly as below - - - - - - - - - - subfam. *Cinchonoideae* (part 2)
– Fruit a large subglobose drupe, with ± lobed woody 4–9-locular putamen and fibrous outer layers; seeds without albumen; corolla white, salver-shaped, velvety outside, the tube c. 2.5 cm. long with 4–9 lobes - - - - subfam. *Antirheoideae* (*Guettarda speciosa* L.) (part 2)

Key to the tribes in Rubioideae

1. Ovary superior; shrub; stipules connate and vaginate forming a sheath; ovules solitary in each locule - - - - - - - - - - - - - - - - **2. Gaertnereae**
– Ovary inferior - - - - - - - - - - - - - - - - - 2
2. Ovules 2 or more in each locule - - - - - - - - - - - - - - 3
– Ovule solitary in each locule - - - - - - - - - - - - - - 5
3. Style divided into 6–8 filiform lobes at the apex; ovary with 4–10 locules, each containing 2 (rarely 3) collateral erect ovules, but only one developing into a seed in each locule; fruit a drupe with woody 4–10-locular putamen - - - - - - - - **4. Triainolepideae**
– Style 2-lobed or rarely with 3–4(5) lobes but if so other characters not agreeing 4
4. Fruit a berry - - - - - - - - - - - - - - - - **5. Urophylleae**
– Fruit a capsule or if indehiscent then calyx characteristic (see TAB. 19) **9. Hedyotideae**
5. Ovules pendulous from near apex of locule - - - - - - - - - - 6
– Ovules erect or ± attached by middle to the septum - - - - - - - - - 7
6. Trees with axillary inflorescences - - - - - - - - **6. Craterispermeae**
– Herbs or small shrublets with mostly terminal but sometimes axillary inflorescences; flowers often bright blue - - - - - - - - - - - - - - - - **7. Knoxieae**
7. Leaves and leaf-like stipules in whorls of 4–8; erect, prostrate, or climbing herbs often with hooked hairs rendering the plant adhesive; flowers rotate; fruit dry or berry-like - - - - - - - - - - - - - - - - - **12. Rubieae**
– Leaves opposite or if in whorls of 3 or more then with other characters quite different 8
8. Calyx tubes confluent; fruit united into a fleshy mass; 1 calyx lobe produced into a coloured lamina in one species; trees or shrubs - - - - - - **3. Morindeae** (in part)
– Calyx tubes not confluent; fruits not forming a fleshy mass - - - - - - 9
9. Usually foetid smelling climbers; fruits flattened, the outer pericarp falling off to expose 2 compressed winged pyrenes supported by long filiform stalks **8. Paederieae**
– Plants not evil-smelling; fruit not of this characteristic structure - - - - - - 10
10. Fruits fleshy, indehiscent; trees or shrubs or in a few cases small forest floor herbs; flowers heterostylous - - - - - - - - - - - - - - - - - 11
– Fruits dry, usually dehiscent; herbs to shrubs - - - - - - - - - - 12
11. Flowers terminal; ovary 2-locular or if rarely 3-locular then inflorescence with a very conspicuous involucre - - - - - - - - - - - - - **1. Psychotrieae**
– Flowers axillary; ovary 4–12-locular, not involucrate - - - - - **3. Morindeae**
12. Ovules erect, attached to the base of the ovary locules; flowers unisexual or hermaphrodite; pollen 3-colporate - - - - - - - - - - - - **10. Anthospermeae**
– Ovules attached to the septum of the ovary; flowers hermaphrodite, mostly in globose clusters at the internodes; pollen pluricolpate - - - - - - - - **11. Spermacoceae**

Artificial key to genera of Rubioideae

1. Corolla lobes imbricate in bud; anthers mostly included, only the tips exserted, about ¾ the length of the corolla; ovary 5-locular; flowers in terminal branched cymes; corolla cylindrical, orange-red, the lobes very short; style filiform, with narrowly fusiform undivided stigma - - - - - - - - - - - - - - - - - - **Hamelia** (cult.)
– Corolla lobes valvate in bud and other characters not found combined - - - 2
2. Nodes with whorls of 4–8 leaves and similar leaf-like stipules; herbs and herbaceous climbers with rotate corollas (save introduced *Sherardia*); ovules solitary in each locule; fruits globose, indehiscent; plants often adhesive due to prickles and harsh hairs (*Rubieae*) 3
– Nodes with leaves opposite or, if in whorls of 3–5, then plants either shrubs or trees, corolla not rotate but stipules much smaller - - - - - - - - - - - - - - 5

3. Calyx with 4–6 distinct teeth; flowers in small terminal involucrate heads; corolla-tube cylindrical - - - - - - - - - - - - - - **Sherardia** (introduced)
– Calyx limb reduced to rim; inflorescence not involucrate; corolla rotate - - - - 4
4. Leaf blades ovate to lanceolate, the petiole very well developed; corolla usually 5-merous - - - - - - - - - - - - - - - - **36. Rubia**
– Leaf blades mostly narrow, linear or lanceolate or if wider and elliptic, then petiole very short; corolla usually 4-merous - - - - - - - - - - - - **37. Galium**
5. Climbing herbs or lianes - - - - - - - - - - - - - - - - 6
– Erect, decumbent or procumbent herbs, shrubs or trees - - - - - - - - 8
6. Cultivated climbing plant with tubular scarlet and yellow corollas 1.7 cm. long; calyx teeth long, lanceolate; fruit capsular and seeds numerous, narrowly winged **Manettia** (cult.)
– Not as above; only 2 seeds per fruit - - - - - - - - - - - - 7
7. Evil-smelling plants; fruits flattened, the outer pericarp falling off to expose two compressed winged pyrenes supported by long filiform stalks - - - - - - **13. Paederia**
– Plants not evil-smelling; fruit not of this characteristic structure but fleshy, ovoid with 2 semiglobose pyrenes - - - - - - - - - - **1. Psychotria ealaensis**
8. Leaf blades with distinct bacterial nodules present, sometimes restricted to the midrib (not to be confused with various fungal spots, insect galls, discolourations etc. — if in doubt ignore this couplet since it is merely a short cut) - - - - - - - **1. Psychotria** (in part)
– Leaf blades without distinct bacterial nodules - - - - - - - - - - 9
9. Calyces joined together and fruit a compound infructescence of fused drupes; corolla-lobes valvate; ovaries with 1 ovule in each locule - - - - - - - - **7. Morinda**
– Calyces separate; drupes not joined to form a compound infructescence - - - - 10
10. Style divided into 6–8 filiform lobes; flowers heterostylous; ovary with 4–10 locules, each containing 2 (rarely 3) collateral erect ovules, but only 1 develops into a seed in each locule, the undeveloped ovule being stuck to the seed; shrub 1.8–6 m. tall; corolla white, with tube 1 cm. long, woolly tomentose outside; drupes red, with woody putamen; mainly littoral or in coastal bushland - - - - - - - - - - - - - - **8. Triainolepis**
– Style 2-fid or if several lobes then without above characters combined - - - - 11
11. Ovules 2 or more in each locule - - - - - - - - - - - - - 12
– Ovules solitary in each locule - - - - - - - - - - - - - 27
12. Calyx limb spreading, eccentric, entire or shallowly lobed, venose and accrescent, up to 2.8 cm. wide in fruit; fruit dry and tardily dehiscent or not; corolla tube narrowly tubular, 1.2–3(3.5) cm. long - - - - - - - - - - - - - - **17. Carphalea**
– Calyx limb not as above; usually with 4–5 distinct lobes or teeth - - - - - - 13
13. Fruit a capsule or splitting into 2 cocci; herbs to small shrubs - - - - - - 14
– Fruit a berry; trees or shrubs - - - - - - - - - - - **9. Pauridiantha**
14. Small shrubs or shrubby herbs with winged seeds; somewhat similar to native *Pentas* - - - - - - - - - - - - - - **Bouvardia** (cult.)
– Herbs or small shrubs; seeds not winged - - - - - - - - - - - 15
15. Flowers mostly 4-merous; leaf blades frequently narrow and uninerved or with lateral nerves obscure - - - - - - - - - - - - - - - - - - 16
– Flowers mostly 5-merous; leaf blades usually broad and with very obvious lateral and tertiary venation (but not always) - - - - - - - - - - - - - - 24
16. Anthers and stigmas included, the latter always overtopped by the former; corolla tube narrowly cylindrical - - - - - - - - - - - - - - **18. Kohautia**
– Anthers and/or stigmas exserted or if both included then anthers overtopped by the stigma; corolla tube cylindrical or funnel-shaped - - - - - - - - - - - 17
17. Corolla tube cylindrical, at least usually 1 cm. long; anthers included and style exserted; flowers not heterostylous - - - - - - - - - - - - **19. Conostomium**
– Corolla tube cylindrical or funnel-shaped, always less than 2 cm. long, sometimes with included anthers and exserted styles but then usually heterostylous - - - - - - - 18
18. Flowers solitary or fascicled at numerous nodes forming long interrupted spike-like inflorescences; leaves subulate or filiform, in one species in whorls of 3–4 but appearing verticillate with a greater number due to very abbreviated shoots **24. Manostachya**
– Not as above, if leaves as narrow then inflorescence not an extensive spike 19
19. Capsule opening both septicidally and loculicidally - - - - - - - - - 20
– Capsule opening loculicidally - - - - - - - - - - - - 22
20. Beak of capsule as long as or longer than the rest of the capsule - - - - - 21
– Beak of the capsule shorter than the rest of the capsule **22. Agathisanthemum**
21. Fragile annual herb with capsule of very characteristic shape, emarginate at the base, the beak much exceeding the rest of the capsule (TAB. 23, fig. 7); stipules triangular with 2 lobes - - - - - - - - - - - - - **20. Mitrasacmopsis**
– A more robust subshrubby herb; beak as long as the rest of the capsule; stipules 3–7-fimbriate - - - - - - - - - - - - - **21. Hedythyrsus**
22. Rush-like plant with linear or filiform leaves; stipule sheath tubular, truncate or with 2 minute teeth; seeds dorsiventrally flattened - - - - - - - - **23. Amphiasma**
– Plant not rush-like even if leaves linear; if seeds dorsiventrally flattened then leaves not linear - - - - - - - - - - - - - - - - - - 23

23. Capsule with thick woody wall and a solid beak, tardily dehiscent - - **26. Lelya**
– Capsule with horny wall, with or without a beak but never solid, early dehiscent - - - - - - - - - - - - - - - - - - **27. Oldenlandia**
24. Leaves uninerved, mostly fairly small; decumbent herb of wet places with small white or blue flowers ± 5 mm. long in very lax elongated axillary cymes - - - - - **25. Pentodon**
– Leaves larger, pinnately nerved; flowers mostly terminal or if axillary then 1–2.5 cm. long - - - - - - - - - - - - - - - - - - 25
25. Flowers solitary, axillary or pseudoterminal; woody plant mainly of rock crevices, with small revolute leaves - - - - - - - - - - - - - - - **16. Batopedina**
– Flowers in more extensive terminal and axillary inflorescences; plant of different habit usually with much longer leaves - - - - - - - - - - - - - - 26
26. Flowering inflorescences capitate or lax much branched complicated cymes, not elongating into simple spikes in fruit, although individual branches sometimes become spicate; fruit globose or triangular, less often ovoid-oblong - - - - - - - - - - **14. Pentas**
– Flowering inflorescences capitate, later elongating into a long simple "spike", rarely with axillary spikes from the upper axils and frequently with solitary flowers at the lower nodes; fruits oblong - - - - - - - - - - - - - - - **15. Otomeria**
27. Ovule pendulous from or near the apex of the locule - - - - - - - - - 28
– Ovule erect from or near the base or attached towards the middle of the septum 30
28. Herbs with terminal or axillary inflorescences - - - - - - - - - - 29
– Trees with axillary inflorescences - - - - - - - - **10. Craterispermum**
29. Slender annual, with 4-merous flowers; corolla tube 3 mm. long; fruit breaking off to leave a small cup, which is the persistent woody flanged pedicel - - - - **11. Paraknoxia**
– Perennials with mainly 5-merous flowers; corolla-tube exceeding 3 mm.; fruit not leaving a cup-like organ when falling; corolla often bright blue - - - **12. Pentanisia***
30. Ovule erect from the base of the locule - - - - - - - - - - - - 31
– Ovule attached ± near the middle of the septum - - - - - - - - - - 41
31. Herbs or subshrubs with dry capsular, indehiscent or succulent (cult.) fruits 32
– Shrubs, trees or subshrubs or forest floor herbs (*Geophila*) with succulent fruits; flowers mostly heterostylous - - - - - - - - - - - - - - - 36
32. Inflorescence terminal, very lax and extensive with filiform pedicels elongate, up to 2 cm. long - - - - - - - - - - - - - - - - - **30. Galopina**
– Inflorescence much less lax, pedicels much shorter - - - - - - - - - 33
33. Stamens inserted near the base of the corolla but filaments sometimes lightly adnate to tube (cult.) - - - - - - - - - - - - - - - - - - - 34
– Stamens inserted near the throat of the corolla (cult.) - - - - - - - - - 35
34. Flower unisexual; style divided to base into two lobes; stipules entire or toothed - - - - - - - - - - - - - - - - - **Coprosma** (cult.)
– Flowers hermaphrodite; style shortly bifid, stipules with 3 or more stiff bristly lobes - - - - - - - - - - - - - - - - - - **Serissa** (cult.)
35. Calyx lobes unequal, 1 or more enlarged, flowers hermaphrodite, not heterostylous, both stamens and style exserted; stigmas smooth; corolla tube exceedingly fine **28. Otiophora**
– Calyx lobes equal, small; flowers usually unisexual or polygamous and partially heterostylous in some plants; stigmas feathery - - - - - - - - - **29. Anthospermum**
36. Flowers in axillary inflorescences; fruits mostly blue, with 4–12 pyrenes; trees or shrubs; seeds with albumen soft and oily - - - - - - - - - - - - **6. Lasianthus**
– Flowers in mostly terminal inflorescences; fruit with 2 (rarely 3) pyrenes - - - - 37
37. Ovary superior - - - - - - - - - - - - - - - - - **5. Gaertnera**
– Ovary inferior - - - - - - - - - - - - - - - - - - 38
38. Pyrenes without well-marked dehiscence; seeds with a red-brown or purplish testa, often with ruminate endosperm; mostly shrubs or small trees, less often subshrubby or herbaceous - - - - - - - - - - - - - - - - - **1. Psychotria****
– Pyrenes with well-marked dehiscence; testa of seed pale; albumen not ruminate 39
39. Herbs of the forest floor, mostly procumbent or straggling; pyrenes opening basally by 2 short dorsal slits - - - - - - - - - - - - - - - - - **2. Geophila**
– Shrubs, lianes or small trees, rarely subshrubby herbs but never procumbent or straggling - - - - - - - - - - - - - - - - - - - 40
40. Pyrenes opening by 2 marginal slits; buds never winged; shrub with leaves mostly developing after the flowers have matured; stems corky - - - - - - - **3. Chazaliella**
– Pyrenes opening by 1 dorsal slit; buds and corolla lobes often with longitudinal narrow wing-like keels (but not present in some species) - - - - - - - - - **4. Chassalia**

* *Chlorochorion* has recently been separated on account of its axillary inflorescences etc. (see page 58).

** Including *Cephaelis* with inflorescence surrounded by an involucre and *Megalopus* with inflorescences borne on exceedingly long peduncles and surrounded by large boat-shaped involucre.

41. Ovary 3-locular; stigma-lobes 3; fruit with 3 cocci - - - - - - **35. Richardia**
- Ovary 2-locular; stigmas 2 or 1, capitate; fruits capsular with 2 valves or with 2 cocci or circumscissile - - - - - - - - - - - - - - - - 42
42. Fruit circumscissile about its middle, the top coming off like a lid; flowers minute, in globose nodal clusters; seeds with a ventral impressed x-like pattern - - **34. Mitracarpus**
- Fruit indehiscent or opening by longitudinal slits or 2-coccous - - - - - - 43
43. Succulent creeping plant of the sea-shore with imbricated leaves joined by quite broad sheathing stipules with very short processes; stems rooting at the nodes; fruits indehiscent - - - - - - - - - - - - - - - **31. Phylohydrax**
- Plant not a littoral succulent, leaves not imbricated, stipules with longer processes, fruits dividing into cocci or capsular - - - - - - - - - - - - - - - 44
44. Capsule opening from base to apex, the calyx limb and jointed valves falling off together, leaving the oblong septum - - - - - - - **33. Spermacoce** subgen. **Arbulocarpus**
- Capsule opening from apex to base or fruit 2-coccous - - - - - - - 45
45. Fruit an obvious capsule - - - - - - - - - - - - - **33. Spermacoce**
- Fruit with 2 indehiscent or ± indehiscent cocci* - - - - - - - - **32. Diodia**

Tribe 1. **PSYCHOTRIEAE**

1. Flowers in axillary inflorescences; fruits mostly blue with 4–12 pyrenes; trees or shrubs - - - - - - - - - - - - - - **6. Lasianthus****
- Flowers in mostly terminal inflorescences; fruit with 2(3) pyrenes; trees shrubs, lianes or subshrubby herbs - - - - - - - - - - - - - - - 2
2. Pyrenes without well-marked dehiscence; seeds with a red-brown or purplish testa and often with a ruminate endosperm; herbs to trees but mostly shrubs **1. Psychotria*****
- Pyrenes with ± well-marked dehiscence; testa of seed pale; endosperm not ruminate 3
3. Herbs of the forest floor, mostly procumbent or straggling; pyrenes opening basally by 2 short dorsal slits: - - - - - - - - - - - - - - - - **2. Geophila**
- Shrubs, lianes or small trees, rarely subshrubby herbs, but never procumbent or straggling; dehiscence of pyrenes not as above - - - - - - - - - - - - - 4
4. Pyrenes opening by 2 marginal slits; buds never winged; shrub with leaves mostly developing after the flowers have matured; stems corky; flowers small, in small capitate inflorescences - - - - - - - - - - - - - - **3. Chazaliella**
- Pyrenes opening by 1 dorsal slit; buds and corolla lobes often (but not always) with longitudinal narrow wing-like keels; inflorescences mostly branched and extensive **4. Chassalia**

1. **PSYCHOTRIA** L.

Psychotria L., Syst. Nat., ed. 10: 929 (1759). —Hiern in F.T.A. **3**: 193 (1877) pro parte. —Petit in Bull. Jard. Bot. Brux. **34**: 1–299 (1964); op. cit. **36**: 65–190 (1966). —Steyerm. in Mem. N.Y. Bot. Gard. **23**: 406 (1972) **** nom conserv.
Mapouria Aubl., Pl. Guian. **1**: 175, t. 67 (1775).
Grumilea Gaertn., Fruct. & Sem. **1**: 138, t.1 fig. 7 (1788). —Hiern in F.T.A. **3**: 215 (1877).
Cephaelis Sw., Prodr. Veg. Ind. Occ.: 45 (1788). —Hiern in F.T.A. **3**: 222 (1877). —Hepper in Kew Bull. **16**: 153 (1962); in F.W.T.A., ed. 2, **2**: 202 (1963) nom. conserv.
Camptopus Hook.f. in Bot. Mag. 95, t. 5755 (1869).
Uragoga Baill. in Adansonia **12**: 323 (1879). —Hutch. & Dalz., F.W.T.A. **2**: 127 (1931).
Megalopus K. Schum. in Engl., Bot. Jahrb. **28**: 490 (1900).
Apomuria Bremek. in Verh. K. Nederl. Akad. Wet., Afd. Natuurk., Ser. 2, **54**: 88 (1963).

Mostly shrubs but less often trees, lianes or subshrubs, some even ± herbaceous. Leaves opposite, petiolate, frequently drying reddish-brown in colour, either with domatia or

* Since confusion is easy *Diodia* species are also keyed out under *Spermacoce*.
** Often included in the *Psychotrieae*, but in this account I have followed Petit in transferring it to *Morindeae*.
*** Including *Cephaelis*, with the inflorescences surrounded by an involucre, and *Megalopus*, with inflorescences borne on exceedingly long peduncles and surrounded by a large boat-shaped involucre.
**** See this reference for extensive extra-African generic synonomy.

with bacterial nodules*, the former mostly in the main nerve-axils but the latter either scattered throughout the lamina or variously restricted, sometimes only a very few at the base of the midrib beneath; less often neither domatia nor bacterial nodules are present; stipules either entire or variously fimbriated, usually deciduous; colleters** present. Flowers mostly small, sessile or pedicellate, 4–5-merous, hermaphrodite, heterostylous, in terminal or axillary (in some American species) capitate or paniculate inflorescences; bracts and bracteoles small or large, variously placed, sometimes one bract subtending 2 or 3 flowers or only 1 flower or absent; often there is an involucre of bracts (sometimes large) and in some groups (particularly those previously considered to be worth recognising as *Cephaelis*) united and surrounding the whole inflorescence. Calyx limb usually a short tube with mostly minute or less often shortly linear or ovate lobes. Corolla tube mostly shortly cylindric, hairy at the throat save in a few species; lobes always valvate, often thickened at the apex, sometimes with little projections which show as small horns in bud. Stamens with long filaments in short-styled flowers and shorter filaments in long-styled flowers, mostly inserted about the middle of the tube. Ovary 2 (rarely 3–4)-locular, each locule containing a single erect ovule; style filiform, long or short with 2 (rarely 3–4) linear stigmas; disk small, surrounding the base of the style. Fruit a drupe, very often red or in bracteate species often blue or black, with 1–2(rarely 3 or 4) pyrenes. Seeds mostly semi-ellipsoid with plane ventral face and convex dorsal face; testa usually reddish-brown; endosperm horny, entire, strongly ruminate or with a variously shaped median ventral fissure.

A very large genus of over 900 described species, although 500 may be a better estimate of the actual number; over 200 occur in tropical Africa. Hiern records *Psychotria obtusifolia* Lam. ex Poir. from the Flora area citing "Mozambique *Forbes*" but the specimen most probably came from Madagascar. Petit has given a detailed revision of the African species. His classification is used here with certain corrections suggested by Steyermark. Petit has placed most of the African species in subgenus *Psychotria*, but since this must of necessity include *P. asiatica* L., the type of the genus which belongs to the group formerly recognised as *Mapouria* Aubl. not occurring in Africa, Steyermark has erected a new subgenus to include all the other species except those in subgenus *Tetramerae* (Hiern) Petit. The classification used in the following account is summarised in the following conspectus.

Subgen. HETEROPSYCHOTRIA

Steyerm. in Mem. N.Y. Bot. Gard. **23**: 484 (1972).

Trees, shrubs, herbs or climbers. Stipules various, rounded, acute, truncate or variously divided into teeth or lobes or erose or pectinate, the scars without ferruginous fibrillate hairs. Leaves without bacterial nodules. Inflorescences various, with small to large bracts and bracteoles or sometimes distinct involucres. Flowers 4- or mostly 5-merous. Seeds ribbed or not; endosperm markedly to scarcely or not ruminate, ventral surface sometimes fissured.

Sect. **FLAVIFLORAE** Petit in Bull. Jard. Bot. Brux. **34**: 41 (1964).

Stipules broadly triangular or ovate-triangular, entire at the apex or rarely bifid for less than 1 mm. Inflorescence paniculate, with minute bracts and bracteoles. Seeds often not ribbed; endosperm mostly deeply ruminate on the ventral face, less deeply or not at all ruminate on the dorsal face. Corolla often yellow. *Species 1–2.*

* A study of these nodules has been made by J.F. Gordon and is available in an unpublished thesis "The nature of and distribution within the plant of the bacteria associated with certain leaf-nodulated species of the families Myrsinaceae and Rubiaceae" 370 pp. (1963). N.R. Lersten and his colleagues have carried out a more exhaustive study and some results may be found in *Journ. Bacteriol.* **94**: 2027–2036 (1967), *Amer. Journ. Bot.*, **55**: 1089–1099 (1968); **59**: 89–96 (1972) & *Internat. Journ. Syst. Bacteriol.* **22**: 117–122 (1972).

** Colleters in *Psychotria* are of three types — standard unbranched, dendroid and intermediate types. Curiously in species from Madagascar and the rest of the world the colleters are remarkably uniform and of the standard type, but in those from Africa all three types are found although only in two groups, namely sect. *Flaviflorae* and subgen. *Tetramerae*, is the presence of non-standard types marked. See also Lersten, Morphology and distribution of colleters and crystals in relation to the taxonomy and bacterial leaf nodule symbiosis of *Psychotria* (Rubiaceae), in Amer. Journ. Bot. **61**: 973–981 (1974).

Sect. **HOLOSTIPULATAE** Petit in Bull. Jard. Bot. Brux. **34**: 70 (1964).
Stipules elliptic, ovate or obovate, often large with rounded or obtuse rarely slightly incised apex or with 2 rounded lobes. Inflorescences paniculate, often large and lax, with minute bracts and bracteoles. Seeds often ribbed, the endosperm always uniformly deeply ruminate. Corolla white. *Species 3–5.*

Sect. **PANICULATAE** Hiern in F.T.A. **3**: 193 (1877).
Stipules variously shaped but always deeply divided into acuminate lobes. Inflorescences paniculate; bracts and bracteoles small. Seeds ribbed or smooth; endosperm always ruminate. Corolla usually white. *Species 6–7.*
Species 8 is of uncertain and isolated position.

Sect. **INVOLUCRATAE** Petit & Verdc. in Kew Bull. **30**: 255 (1975).
(*Cephaelis* sensu auctt. Afr., non L.; *Uragoga* auctt. Afr., non Baill.*)
Stipules mostly large, divided into 2 acuminate lobes. Inflorescences capitate, surrounded by large involucral bracts sometimes united to form an involucral cup, sessile or on short to very long peduncles. Fruits usually blue. The species with very long peduncles and 3–4-locular ovaries have been put in the genus *Camptopus* (*Megalopus*), but despite the striking peduncles there is no real character to separate them from *Psychotria*. *Species 9–10.*

Subgen. TETRAMERAE

(Hiern) Petit in Bull. Jard. Bot. Brux. **34**: 28 (1964).

Mostly shrubs or subshrubs, less often almost herbaceous, rarely small trees. Stipules ovate-triangular, mostly divided into 2 acuminate lobes or in one instance multifid. Leaves with bacterial nodules either dispersed throughout the lamina or sometimes restricted to the midrib (occasionally reduced to only very few at its base) and not easy to see. Inflorescence laxly to densely paniculate or subumbellate; bracts and bracteoles minute or obsolete. Flowers 4- or 5-merous. Seeds never ribbed; endosperm entire or with a fissure on the ventral face which is I-, T-, or V-shaped in section. *Species 11–19.*

Key to species of both subgenera

1. Inflorescences condensed, surrounded by an involucre of conspicuous bracts; fruit usually blue - - - - - - - - - - - - - - - - - - - 2
– Inflorescences condensed to quite lax, not surrounded by an involucre; bracts if present small and scattered - - - - - - - - - - - - - - - - 3
2. Shrub or subshrub mostly over 1 m. with wider leaves (to 13.5 cm.) and longer peduncles mostly 1–9(15) cm. long; involucres ± 1 × 1.5–2 cm. when mature (widespread and variable) - - - - - - - - - - - - - - - 9. *peduncularis*
– Subshrub with simple ± herbaceous stems 25–30 cm. tall; leaves 4.5–7.5 × 0.9–3 cm.; peduncles ± 5 mm. long; involucres about 5 × 13 mm. (Mwinilunga) - - - 10. *mwinilungae*
3. Stems climbing - - - - - - - - - - - - - - - - 6. *ealaensis*
– Stems not climbing - - - - - - - - - - - - - - - - 4
4. Leaves with bacterial nodules dispersed throughout the lamina - - - - - - 5
– Leaves without bacterial nodules or, if present, these restricted to petiole or base of main nerve beneath and often very obscure - - - - - - - - - - - - - - 10
5. Lamina with spot-like, star-like and elongate nodules intermixed 16. *diversinodula*
– Lamina with spot-like nodules only, save sometimes on the midrib beneath - - 6
6. Nodules of two sorts, linear or tuberculiform situated on the midrib or on petiole, sometimes reduced to 1–2 and a few spot-like in the lamina as well - - - - - 17. *heterosticta*
– Nodules spot-like, dispersed in the lamina - - - - - - - - - - - - 7

* There is no doubt that the African species usually placed in *Cephaelis* are not related to the American species referred to the same genus including the type but are more closely related to species always maintained in *Psychotria*. See Schnell in Bull. Jard. Bot. Brux. **30**: 357–73, fig. 48–56 (1960). This section badly needs a thorough revision. I have examined all the "species" of *Cephaelis* described by De Wildemann from Zaire but the elucidation of their correct relationships would be a lengthy task. When this has been done certain changes may be necessary to the treatment I have used here.

7. Cymes mostly subumbelliform, secondary axes of the clusters practically suppressed: very often small subshrubby herbs - - - - - - - - - - - - - - 13. *pumila*
– Cymes not subumbelliform, secondary axes mostly present - - - - - - 8
8. Seeds with an air-space; venation often darker and evident beneath in dry state; apical stem-buds and stems often distinctly corky and ferruginous; mostly a small subshrub under 50 cm. tall - - - - - - - - - - - - - - - - 18. *spithamea*
– Seeds without an air-space; tertiary venation often quite obscure or invisible - - 9
9. Nodules smaller, mostly under 1 mm. diameter; leaves glabrous or more usually pubescent; panicles not usually with branches in form of an X; peduncles not winged (widespread) - - - - - - - - - - - - - - - - - - 15. *kirkii*
– Nodules larger, mostly about 1 mm. diam.; leaves glabrous; panicles usually with 4 branches in form of an X; peduncles ± 2-winged (Mwinilunga) - - - - - - 19. *kikwitensis*
10. Seeds with flat face deeply ruminate; trees or shrubs less often smaller subshrubs; no nodules on the leaves - - - - - - - - - - - - - - - - - - 11
– Seeds with flat face with a single variously shaped median fissure or subglobose; shrubs or subshrubs sometimes almost herbaceous; nodules present or absent - - - - - 17
11. Stipules entire or only very slightly divided (rarely with 2 long cusps in *djumaensis)* 12
– Stipules usually distinctly divided - - - - - - - - - - - - - - 15
12. Stipules triangular, mostly less than 1 cm. long; corolla often yellow; dorsal surface of pyrene not conspicuously ribbed - - - - - - - - - - - - - - - - - 13
– Stipules elliptic or oblong, mostly 0.5–1.7 cm. long; corolla white or pale yellow; dorsal surface of pyrenes conspicuously ribbed and showing in dried fruits - - - - 4. *zombamontana*
13. Small subshrub to shrub, usually with young and old shoots rather dissimilar; young stems and leaf blades beneath often densely velvety or at least pubescent; fruits didymous with distinct median groove, pubescent; leaf blades usually obtuse - - - - - 2. *eminiana*
– Shrub or tree with young and old shoots not very different, often glabrous; fruits ± subglobose, glabrous; leaf blades rounded to ± acuminate - - - - - - - - - - 14
14. Leaves broadly elliptic to obovate, mostly rounded or very shortly acuminate 3–20 × 1.5–12 cm., mostly micropuberulous and/or pubescent on venation beneath; inflorescences usually more extensive, 8–12 cm. wide; peduncle 3.5–11 cm. long; more usually a distinct tree (mostly western Zambia) - - - - - - - - - - - - - - 3. *djumaensis**
– Leaves more narrowly elliptic, oblong or oblanceolate, 3–14(20) × 1–6(10) cm. but usually narrow, glabrous or rarely pubescent; inflorescences mostly narrower, mostly under 8 cm. wide; peduncle 1–6 cm. long, usually quite short; shrub or small tree (mostly eastern Zimbabwe and adjacent Mozambique) - - - - - - - - - - - - - 1. *capensis*
15. Young stems and undersurfaces of leaves very shortly pubescent, usually densely and velvety so; fruit didymous with distinct median groove; pyrenes not ribbed; small subshrub or shrub - - - - - - - - - - - - - - - - - - 2. *eminiana*
– Not as above; shrubs or distinct trees - - - - - - - - - - - - - 16
16. Stipules emarginate with 2 rounded lobes; stems, leaves, inflorescence branches etc, entirely glabrous; dorsal face of pyrenes with obscure obtuse ribs (Zambia N) 5. *succulenta*
– Stipules with triangular-acuminate lobes; stems, leaves etc. glabrous to sparsely or densely pubescent; dorsal face of pyrenes only slightly grooved - - - - - 7. *mahonii*
17. Calyx lobes linear, 2–4 mm. long, exceeding the rest of the calyx 12. *linearisepala*
– Calyx lobes usually shorter and not as long as rest of calyx - - - - - - 18
18. Inflorescences distinctly subumbellate - - - - - - - - - - - 13. *pumila*
– Inflorescences with axes ± developed and not appearing so distinctly umbellate (save some inflorescences in the poorly known sp. 14) - - - - - - - - - - - - 19
19. Seeds subglobose, the ventral surface with a deep Y- or V-shaped fissure and having a spherical air-space between the hard outer layer and the albumen; usually subshrub with several stems to 50 cm., rarely a distinct shrub; leaves sometimes drying yellow-green or white beneath and often with tertiary venation quite evident in dry state; nodules scattered but sometimes very difficult to see - - - - - - - - - - - - - - - - - - 18. *spithamea*
– Seeds without such an air-space - - - - - - - - - - - - - - 20
20. Leaves very narrowly oblong-elliptic to oblanceolate, 2.6–6.5 × 0.5–1.3 cm.; pyrenes distinctly 4–5-ribbed; seeds with flat surface without or with scarcely any fissure; flowers not known (S. Malawi, Ruo Gorge)** - - - - - - - - - - - - - - - *14. sp. A.*
– Leaves relatively wider or quite differently shaped; seeds with ventral surface with a distinct fissure or groove - - - - - - - - - - - - - - - - - 21
21. Seeds subglobose with a deep median fissure which is usually branched within; leaf blades rather distinctly margined; nodules absent or sometimes a few long ones along the nerves - - - - - - - - - - - - - - - - - 8. *butayei*
– Seeds semi-globose with an M-shaped groove or V- or U-shaped fissure; leaf blades not so obviously margined; at least some nodules usually quite evident - - - - - 22

* Petit distinguishes this together with some other species in his key by possessing two holes or slits in the disk but I have not found this reliable. The species in this couplet certainly overlap somewhat but *djumaensis* has a characteristic look.
** A dune plant sp. B will also key near here.

22. Leaves 1.5–8(11) × 0.2–3.3(4) cm., with the linear nodules restricted to the basal part of the midrib; seed with a rather shallow groove or M-shaped impression (coastal Mozambique) - - - - - - - - - - - - - - - - 11. *amboniana*
– Leaves 3.5–19 × 0.5–5.5 cm., with the linear nodules scattered along the midrib and some obscure to distinct laminal ones also; seeds with a more distinct mostly U-shaped fissure (N. Malawi & N. Zambia) - - - - - - - - - - - - - - - - 17. *heterosticta*

1. **Psychotria capensis** (Eckl.) Vatke in Oesterr. Bot. Zeitschr. **25**: 230 (1875). —Schonland, Bot. Surv. S. Afr. Mem. **1**: 96 (1919). —Hutch., Bot. in S. Afr.: 671 (1946). —Petit in Bull. Jard. Bot. Brux. **34**: 41, photo 1A (1964) (full references). —Palgrave, Trees S. Afr.: 899 (1977). Type from S. Africa.
Logania capensis Eckl. in S. Afr. Quart. Journ. **1**: 371 (1830). Type as above.
Grumilea capensis (Eckl.) Sond. in Harv. & Sond., F.C. **3**: 21 (1964). —De Wild. in Bull. Jard. Bot. Brux. **9**: 30 (1923). —Marloth, Fl. S. Afr. **3**, 2: 195, t. 50 B (1932). Type as above.

Shrub or small tree 1–5(20) m. tall with pale finely ridged stems, mostly glabrous but with pubescent foliage inflorescences etc. in one variety. Leaves 3–14(20) × 1–6(10) cm., narrowly obovate to oblanceolate or oblong-elliptic, rounded to acuminate at the apex, narrowly cuneate at the base, rather thin to subcoriaceous or coriaceous; domatia often present in nerve-axils with ± small openings; petiole 0.4–2(4) cm. long; stipules 3–8 mm. long, triangular to ovate-triangular, apiculate to acute at the apex, glabrous or rarely pubescent, soon falling. Flowers in much-branched panicles 2–15 cm. long; peduncle 1–10 cm. long; pedicels 0–2 mm. long to 5–12 mm. in fruit; bracts and bracteoles minute. Calyx tube and limb tube together scarcely 1–1.5 mm. long; lobes 0.3–0.5 mm. long, triangular. Corolla cream or yellow; tube (2)3–5 mm. long; lobes 1.5–3 × 1.3–2 mm., oblong-triangular, ± thickened at the apex. Stamens with filaments 1.5–3.5 mm. long in short-styled flowers and 0.3–1 mm. long in long-styled flowers. Style 2–3 mm. long and 4.5–5 mm. long respectively, the stigmatic arms 0.5–1 mm. long. Fruits red, ± 5–6.5 mm. diam., subglobose; pyrene with abaxial face entire. Seed with adaxial face flat and deeply ruminate, abaxial face convex.

Subsp. **capensis**

Flowers mostly distinctly pedicellate the pedicels usually very distinct in fruit.

Var. **capensis**. —Petit in Bull. Jard. Bot. Brux. **34**: 41 (1964).
Psychotria zambesiana Hiern in F.T.A. **3**: 203 (1877). —K. Schum. in Engl., Pflanzenw. Ost-Afr. **C**: 391 (1895). —De Wild., Pl. Bequaert. **2**: 442 (1924). Type: Mozambique, Kongone mouth of Zambezi, *Kirk* 327 (K, holotype).
Grumilea oblanceolata K. Schum. in Engl., Bot. Jahrb. **28**: 103 (1899). —De Wild. in Bull. Jard. Bot. Brux. **9**: 43 (1923). Type: Mozambique, Morrumbene, *Schlechter* 12101 (B, holotype†; BM).

Foliage, inflorescences etc. glabrous.

Zimbabwe. E: Chimanimani, Haroni-Makurupini Forest, fr. 30.v.1969, *Müller* 1190 (K; SRGH). **Mozambique**. Z: Kongone mouth of R. Zambezi, fl. 25.i.1861, *Kirk* 327 (K). MS: Lusitu-Haroni junction, fr. 25.iv.1973, *Mavi* 1436 (K; LISC; SRGH). GI: near Xai-Xai, fl. 11.xii.1940, *Torre* 2314 (C; FHO; LISC; LMU; MO; PRE). M: Namaacha Mts., fl. 22.xii.1944, *Torre* 6928 (BR; C; COI; LISC).
Also in Swaziland and S. Africa; cultivated at Harare and in Mauritius. Riverine and other evergreen forest; 0?–800 m.

Subsp. **riparia** (K. Schum. & K. Krause) Verdc. comb. nov. Type from Tanzania.
Grumilea riparia K. Schum. & K. Krause in Engl., Bot. Jahrb. **39**: 560 (1907). —De Wild. in Bull. Jard. Bot. Brux. **9**: 55 (1923). —Brenan, T.T.C.L.: 497 (1949). —Dale & Greenway, Kenya Trees & Shrubs: 443 (1961) pro parte.
Grumilea bussei K. Schum. & K. Krause in Engl., Bot. Jahrb. **39**: 562 (1907). —De Wild in Bull. Jard. Bot. Brux. **9**: 30 (1923). —Brenan, T.T.C.L.: 497 (1949). Type from Tanzania.
Psychotria riparia (K. Schum. & K. Krause) Petit in Bull. Jard. Bot. Brux. **34**: 43, photo 1/B (1964); in Distrib. Pl. Afr. 4, map 92 (1972). —Verdc. in F.T.E.A., Rubiaceae 1: 38 (1976).
Grumilea buchananii K. Schum. nom. nud. based on Malawi specimen *Buchanan* 310 (B†; BM; K photo.).

Flowers mostly subsessile but some where elements of inflorescence are reduced to one flower distinctly pedicellate and fruits occasionally pedicellate.

Var. **riparia**

Leaves, stems etc. glabrous or nearly so but sometimes with sparse rather ferruginous pubescence on the inflorescences.

Malawi. S: 11 km. E. of Zomba, Mingoli Farm Estate, fl. 29.xi.1977, *Brummitt et al.* 15227 (K; MAL). **Mozambique**. N: Mutuali, Namuilasse Stream by bridge on Malema Road, fr. 13.iii.1954, *Gomes e Sousa* 4242 (K). Z: Lugela, Mocuba, Namagoa, fl. & fr. *Faulkner* Pre. 188 (BM; K; PRE). MS: Shupanga, fr. vi.1859, *Kirk* (K).

Also in East Africa. Gallery forest, *Brachystegia* woodland and transitions, also coastal dunes; 0–900m.

Some seedlings from Zambia N, Luwingu, *Fanshawe* 8660 may belong here.

Var. **puberula** (Petit) Verdc. comb. nov.

Psychotria riparia var. *puberula* Petit in Bull. Jard. Bot. Brux. **34**: 46 (1964). —Verdc. in F.T.E.A., Rubiaceae 1: 39 (1976). —Gonçalves in Garcia de Orta Sér. Bot. **5**: 205 (1982). Type from Kenya.

Leaves, stems, etc., pubescent.

Mozambique. T: near Zóbuè, fr. 2.viii.1943, *Torre* 5779 (LD; LISC; LMU; M; P). MS: Chimoio, near Braunstein sawmill, fr. 27.i.1948, *Mendonça* 3729 (BR; C; COI; LISC; LMU; MO; PRE; WAG).

Also in Ethiopia, Kenya and Tanzania. By streams, stony banks; alt. ?

In F.T.E.A. I kept *Psychotria riparia* up as a species although fully realising it was scarcely distinct from *Psychotria capensis*. Study of the Flora Zambesiaca material has shown that, although most specimens from southern Africa have pedicellate flowers, nearly all from northern areas (particularly Uganda and Kenya) have sessile or subsessile flowers. There is a good deal of variation in both areas and also overlap in Mozambique and Malawi; occasional Tanzanian specimens have quite marked pedicels. It has therefore not been possible to retain them at specific level.

2. **Psychotria eminiana** (Kuntze) Petit in Bull. Jard. Bot. Brux. **34**: 48, photo. 1/E (1964). —Petit in Distr. Pl. Afr. 4, map 93 (1972). —Verdc. in F.T.E.A., Rubiaceae 1: 41 (1976). Type from Sudan.

Uragoga eminiana Kuntze, Rev. Gen. Pl. **2**: 955 (1891).

Grumilea sulphurea Hiern in F.T.A. **3**: 218 (1877). —De Wild. in Bull. Jard. Bot. Brux. **9**: 56 (1923) non *Psychotria sulphurea* Ruiz & Pavon (1794) nec *Psychotria sulphurea* Seemann (1865–73). Type as for *Psychotria eminiana*.

Subshrub or small tree, with woody or herbaceous stems 0.15–2.4(5) m. tall from a woody rhizome; stems glabrous to densely pubescent, sparsely branched. Leaves petiolate; lamina variable, 3–25 × 2–13 cm., ovate, oblong, obovate or elliptic to almost round or oblate, emarginate to acuminate at the apex, cuneate to rounded at the base, glabrous to scabrid-puberulous or pubescent above, glabrous to densely pubescent beneath, often strongly discolorous, pale beneath, often with a characteristic pattern of dark veinlets beneath on drying, thin to ± coriaceous; nodules absent but domatia present; petiole 0.3–3(3.5) cm. long, glabrous to densely pubescent; stipules triangular or ovate-triangular, 0.4–1.1 cm. long, acute to acuminate at the apex, obscurely 2-toothed or ± awned, glabrous or pubescent, deciduous. Flowers heterostylous, (4)5(6)-merous, in usually trichotomous or much-branched inflorescences, each panicle rather dense and many-flowered; peduncle (1)4–13 cm. long, glabrous to densely pubescent; secondary branches 0.5–4 cm. long; pedicels obsolete or 0.5 mm. long; 2 inferior bracts sometimes developed, others and bracteoles minute. Calyx tube ± 1 cm. long, obconic; limb 0.75–1 mm. long, cupuliform, glabrous or pubescent with unequal triangular lobes 0.5–1.75 mm. long, acuminate. Corolla yellow, cream or greenish yellow, glabrous, puberulous or in a variety velvety-tomentose outside; tube 3.5–5 mm. long; lobes 1.7–2.5 × 1–1.2 mm. thickened at the apex. Stamens with filaments 1.5–2.5 mm. long in short-styled flowers, 0.25 mm. long in long-styled flowers. Style 1–2.5 mm. long in short-styled flowers, 6 mm. long in long-styled flowers; stigma lobes 0.5–1 mm. long. Drupes red, with 2 pyrenes, didymously subglobose, 5–6(7)* × 7–8(10)* mm., glabrous or puberulous or shortly pubescent; pyrenes subglobose, 5 × 5 × 4 mm.; dorsal face entire, not folded, sometimes so coated with rhaphides as to appear silvery. Seeds dark brown, subglobose, 4.5 × 4.5 × 3.5 mm.; ventral face flat, dorsal face not ribbed; endosperm with ventral face deeply and dorsal face slightly ruminate.

Var. **eminiana**. —Petit in Bull. Jard. Bot. Brux. **34**: 48, photo 1/E (1964).

Grumilea flaviflora Hiern, Cat. Afr. Pl. Welw. **1**: 495 (1898). —De Wild. in Bull. Jard. Bot. Brux. **9**: 35 (1923). Type from Angola.

Psychotria dewevrei De Wild. in Ann. Mus. Congo, Bot. Sér. 5, **1**: 208 (1904); in Pl. Bequaert. **2**: 361 (1924). Type from Zaire.

* Measurements in parenthesis are from spirit material.

?*Grumilea ungoniensis* K. Schum. & K. Krause in Engl., Bot. Jahrb. **39**: 563 (1907). —De Wild. in Bull. Jard. Bot. Brux. **9**: 56 (1923). —Brenan, T.T.C.L.: 498 (1949). Type from Tanzania.

Grumilea stolzii K. Krause in Engl., Bot. Jahrb. **57**: 48 (1920). —De Wild., Pl. Bequaert. **2**: 480 (1924). —Brenan, T.T.C.L.: 498 (1949). Type from Tanzania.

Psychotria dalzielii Hutch. in F.W.T.A. **2**: 122, 125 (1931). —Hepper in F.W.T.A. ed. 2, **2**: 201 (1963). Type from N. Nigeria.

Psychotria humilis Hutch., Botanist in S. Afr.: 514 (1946). Type: Zambia, Mwinilunga, Matonchi Farm, *Milne-Redhead* 1029 (K, lectotype) non Hiern.

Mapouria flaviflora (Hiern) Bremek. in Verh. Kon. Nederl. Akad. Wetensch., Natuurk. ser. 2, **54**, 5: 11 (1963).

Psychotria eminiana var. *stolzii* (K. Krause) Petit in Bull. Jard. Bot. Brux. **34**: 51 (1964); in Bull. Jard. Bot. Brux. **42**: 354 (1972); in Distr. Pl. Afr. 4, map 94 (1972). —Verdc. in F.T.E.A., Rubiaceae 1: 42 (1976). —Gonçalves in Garcia de Orta, Sér. Bot., **5**, 2: 203 (1982). Type as above.

Subshrub or small shrub 0.15–2 m. tall with herbaceous or woody shoots from a very woody rootstock; young stems not strikingly different from older ones, together with petioles, peduncles and particularly undersurfaces of the leaves etc. usually distinctly velvety pubescent but sometimes thinly pubescent or even ± glabrous; leaves sometimes almost bullate above. Corolla glabrous to puberulous.

Zambia. N: Mbala Distr., Mwamba road, fl. 10.x.1970, *Sanane* 1366 (K). W: 19.2 km. from Solwezi on road to Mwinilunga, fl. 20.xi.1972, *Strid* 2509 (K; S). C: Luangwa Valley, near Kapampa R., fl. 6.i.1966, *Astle* 4253 (K). E: 6.4 km. E. of Chipata on Malawi road, fl. 12.i.1960, *Wright* 282 (K). **Malawi**. N: Mzimba Distr. Mzuzu, St. John's Hospital, fl. 11.xii.1969, *Pawek* 3084 (K). C: Kasungu Distr., Chimaliro Forest, Phaso road, fl. 10.i.1975, *Pawek* 8883 (K; MAL: MO). S: 15 km. NE. of Mangochi, slopes of Uzuzu Hill, above Chowe, fr. 21.ii.1982, *Brummitt* et al. 16019 (K; MAL). **Mozambique**. N: Marrupa, about 20 km. on road to Nungo, margin of R. Messalo between Montanhas Mirenge and Mucuwango, fr. 20.iii.1981, *Nuvunga* 648 (K; LMU). T: between km. 3–10 Macanga (Furancungo) to Bene (Tembuc), fr. 21.iii.1966, *Pereira* et al. 1904 (LMU).

Also in N. Nigeria, Cameroon, Central African Republic, Burundi, Zaire, Sudan, Tanzania and Angola. Deciduous thicket, *Brachystegia* woodland and derived open bushland, usually in dry burnt ground or stony hills but sometimes riverine; 480–1650 m.

Although I maintained var. *stolzii* in the F.T.E.A. account, examination of the mass of material now available suggests that the habit difference between plants with short ± herbaceous stems and those with longer woodier ones and distinctly more shrubby habit are due to differences in burning; those protected for some years grow quite shrubby. Field observations would soon confirm or otherwise. An immature fruiting specimen with distinctly acuminate leaves and awned stipules (Malawi. N: Rumphi District, 8 km. S. of Livingstonia escarpment road junction, S. of Chiweta, 465 m., fr. 20.iv.1972, *Pawek* 5147 (K)) looks totally different from other material but I am certain it belongs to this species. It occurs along the beach; a study of the population should be made.

Var. **heteroclada** Petit in Bull. Jard. Bot. Brux. **34**: 52 (1964); in Distr. Pl. Afr. 4, map 95 (1972). —Verdc. in F.T.E.A., Rubiaceae 1: 42 (1976). Type: Zambia, Lake Mweru, *Fanshawe* 3917 (BR; K, holotype).

Distinctly shrubby (or small tree?) 1.8–3.6 m. tall, the stems becoming very distinctly woody with age and markedly dissimilar from the young growths, together with petioles, peduncles, undersurfaces of the leaves etc. glabrous to distinctly hairy but leaves glabrous above, usually thinner, never bullate; inflorescences often more branched. Corolla glabrescent or densely velvety tomentose outside.

Zambia. N: Lake Mweru, fl. 11.xi.1957, *Fanshawe* 3917 (K; NDO).

Also in Zaire (Shaba) and Tanzania. Lake-shore fringing forest.

This certainly seems a more distinctive variety but could still stand field study to see how it develops and what the early stages are like.

3. **Psychotria djumaensis** De Wild., Miss. Laurent **1**: 349 (1906). —Petit in Bull. Jard. Bot. Brux. **34**: 72, photo. 2/B (1964) (full synonymy). Type from Zaire.

Shrub or small tree 1.5–9(12) m. tall. Branches glabrous, shortly pubescent or densely puberulous, the internodes often flattened, at first green later often with purplish brown bark. Leaves 3–20 × 1.5–12 cm., elliptic to obovate, acute to rounded or emarginate at the apex, cuneate, rounded or sometimes subcordate at the base, thinly coriaceous, glabrous save for minutely puberulous pubescent or glabrescent venation beneath; domatia present in nerve-axils, with distinct pits; leaves dry a grey-green, leaden green or ± chestnut, never bright green and with a very distinctive micro-reticulation beneath (visible under 25 mm. objective); petiole 0.5–4 cm. long; stipules greenish white, obovate-elliptic, 0.2–1.2 cm. × 1.5–7 mm. with a thicker basal triangular area, obtuse or distinctly cuspidate, entire or rarely with 2 cusps, glabrous, puberulous or quite hairy, soon deciduous and revealing a ridge of rusty hairs. Inflorescences dense, many-flowered, 8–12 cm. wide; peduncle 3.5–11 cm. long; secondary branches well-developed; pedicels 0–1 mm. long and some fruiting ones up to 6 mm. but mostly flowers and fruits subsessile; bracts small; all parts glabrescent to minutely puberulous or tomentose. Calyx tube ± 1 mm. long, obconic; limb ± 1 mm. long, cupular, with very short broadly triangular lobes. Corolla white, glabrous to densely pubescent outside; tube 3.5–6 mm. long; lobes ± 2–3 × 1.2 mm., oblong. Stamens with filaments 2.5–3.5 mm. long in short-styled flowers, 0.5 mm. long in long-styled flowers. Style 2–3 and 4.5–5.5 mm. long respectively, the stigmatic arms 0.5–1 mm. long. Disk rather conspicuous, 1 mm. long, cylindric-conic with wide opening. Fruits orange or red, subglobose, 6–8 × 5–7 mm. (± 10 in life); pyrenes hemispherical; seed similar with both flat and curved faces conspicuously ruminate.

Var. *djumaensis* occurs in Cameroon, Gabon, Congo (Brazzaville), Central African Republic, Zaire, Cabinda and Angola.

Var. **zambesiaca** Petit in Bull. Jard. Bot. Brux. **34**: 76 (1964). Type: Zambia, Samfya, *Fanshawe* 313 (BR, holotype; K).
Grumilea sp. 1, White, F.F.N.R.: 409 (1962).

Leaves rounded, retuse or emarginate at the apex, only occasionally ± acute.

Zambia. B: Kalabo, fl. 14.x.1963, *Fanshawe* 8064 (K; NDO). N: Pansa R., fl. 6.x.1949, *Bullock* 1147 (K). W: Mwinilunga Distr., by R. Matonchi below dam, fl. 21.x.1937, *Milne-Redhead* 2884 (K). **Zimbabwe**. E: Vumba, 'Witchwood', fl. 11.xii.1945, *Wild* 553 (K; SRGH) (rather inadequate specimen but appears to belong here). **Malawi**. C: Nkhota Kota Distr., Kanjamwano R., fr. 17.i.1964, *Salubeni* 204 (K).
Also in Zaire (Shaba) and Angola. Swamp forest, particularly with *Syzygium* by lakes and rivers, riparian woodland; 1050–1620 m.
Nash 167 (Mbala, fl. 21.viii.55) (BM)) shows one apical stipule with 2 distinct cusps; they are so deciduous in this species that the possibility of this being not unusual needs study in the field.

4. **Psychotria zombamontana** (Kuntze) Petit in Bull. Jard. Bot. Brux. **34**: 86 photo. 2/F (1964). —Palgrave, Trees S. Afr.: 900 (1977). Type: Malawi, Mt. Zomba, *Kirk* (K, holotype).
Grumilea kirkii Hiern in F.T.A. **3**: 216 (1877). —K. Schum. in Engl., Pflanzenw. Ost-Afr. **C**: 391 (1895). —De Wild. in Bull. Jard. Bot. Brux. **9**: 40 (1923). —Krause in R.E. & T.C.E. Fries in Notizbl. Bot. Gart. Berl. **10**: 607 (1929). —Goodier & Phipps in Kirkia **1**: 64 (1961) non *Psychotria kirkii*. Type as for *P. zombamontana*.
Uragoga zombamontana Kuntze, Rev. Gen. Pl. **2**: 958 (1891). Type as above
Psychotria meridiano-montana Petit in Bull. Jard. Bot. Brux. **34**: 88 photo 2/G (1964); in Dist. Pl. Afr. 4, map 116 (1972). —Verdc. in F.T.E.A., Rubiaceae 1: 47 (1976). Type from S. Tanzania.
Psychotria meridiano-montana var. *angustifolia* Petit in Bull. Jard. Bot. Brux. **34**: 89 (1964). —Verdc. in F.T.E.A., Rubiaceae 1: 47 (1976). Type from S. Tanzania.
Psychotria meridiano-montana var. *glabra* Petit in Bull. Jard. Bot. Brux. **34**: 90 (1964). Type: Zambia, Nyika Plateau, 8.8 km. SW. of Rest House, *Robson* 347 (BM; K, holotype).
Psychotria goetzei var. *meridiana* Petit in Bull. Jard. Bot. Brux. **34**: 84 (1964) quoad *Buchanan* 948* ?non Petit sensu stricto (see note).

Shrub or tree 1.5–9 m. tall with glabrous branches and smooth grey bark. Leaves 3.5–16.5 × 1–8 cm., elliptic to elliptic-oblanceolate, oblanceolate or very narrowly oblong-elliptic, acute to acuminate at the apex, sometimes the actual apex narrowly rounded, narrowly cuneate at the base, usually glabrous but sometimes with scattered ± bristly hairs on midrib beneath, rather thin to thinly coriaceous; domatia lacking or inconspicuous; petiole 2–2.5 cm. long; stipules 0.5–1.7 cm. long, ovate, acuminate, acute or obtuse, glabrous or ciliate, soon falling. Inflorescences much-branched, 3–12 × 4.5–12 cm.; lower bracts up to 2 mm. long; peduncle 3.5–9.5 cm. long, glabrous; pedicels obsolete or up to 1 mm. long, up to 9 mm. in fruit; inflorescence-axes glabrous or shortly pubescent. Calyx tube 0.8 mm. long, obconic; limb 0.5–0.75 mm. long, cupuliform, subtruncate or with short broad teeth. Corolla pale yellow or white; tube 2.5–4 mm. long; lobes 1–1.5 × 0.8 mm. triangular. Filaments 0.4 mm. long in short-styled flowers, the anthers just exserted, included in long-styled flowers; style 0.6–1.5 mm. long in short-styled flowers, exserted about 1 mm. in long-styled flowers; stigmatic lobes 0.3–0.6 mm. long. Fruits yellow to orange (immature?) to crimson-red, subglobose, 5–6 mm. long, drying very ribbed; pyrenes semi-globose with 4–5 ribs on convex face, the other face flat. Seeds similar, the albumen conspicuously ruminate.

* This specimen is in very young bud and the supposed characters do not show anyway.

Zambia. N: Isoka Distr., Mafingi Mts., W. of Chisenga Rest House, bud 21.xi.1952, *White* 3733 (FHO; K). E: Lundazi Distr., Nyika Plateau, upper slopes of Kangampande Mt., fl. 6.v.1952, *White* 2733 (FHO; K). **Zimbabwe**. E: Mutare Distr., Vumba Mts., fl. 28.xii.1956, *Chase* 6281 (K; SRGH). **Malawi**. N: Nkhata Bay Distr., Viphya Plateau, 40 km. SW. of Mzuzu, fl. & fr. 23.ii.1975, *Pawek* 9107 (K; MAL; MO; SRGH; UC). C: E. side of Dedza Mt., fl. 12.xii.1954, *Adlard* 204 (FHO; K). S: Mt. Zomba, fl. xii.1896, *Whyte* (K). **Mozambique**. N: Serra de Ribáuè, Mepalué, fr. 25.i.1964, *Torre & Paiva* 10230 (LD; LISC; LMU; MO; WAG). Z: Gúruè Mts., Estaçao Pecuaria, buds 27.vii.1979, *de Koning* 7393 (LMU). MS: Macequece, Mt. Vengo, fl. 23.xi.1943, *Torre* 6241 (LISC).

Also in Tanzania and S. Africa (Transvaal). Evergreen forest including montane forest, ravine-forest also on forested rocky kloofs, sometimes at edges and other derived situations; 990–2500 (2900) m.

An abundance of material from Malawi and elsewhere shows that *Psychotria meridiano-montana* cannot be retained although I maintained it in F.T.E.A. Some variants of *Psychotria goetzei* are also very similar. Material with hairy midribs is now known from Malawi. Some specimens with long pedicels are distinctive but there is much seemingly uncorrelated variation.

5. **Psychotria succulenta** (Schweinf. ex Hiern) Petit in Bull. Jard. Bot. Brux. **33**: 382 (1963); **34**: 97, photo 2/L (1964); **42**: 356 (1972); in Distrib. Pl. Afr.4, map 121 (1972). —Verdc. in F.T.E.A., Rubiaceae 1: 57 (1976). Type from Sudan.

Grumilea succulenta Schweinf. ex Hiern in F.T.A. **3**: 216 (1877). —K. Krause in R.E. Fries, Wiss. Ergebn. Schwed. Rhod.-Kongo-Exped. Bot. Erg.: 16 (1921). —Mildbr., Wiss. Ergebn. Deutsch. Zentr.-Afr. Exped. 1907–1908 **2**: 68 (1922). —De Wild. in Bull. Jard. Bot. Brux. 9: 55 (1923). —F. White, F.F.N.R.: 408 (1962). Type as above.

Uragoga succulenta (Hiern) Kuntze, Rev. Gen. Pl. **2**: 962 (1891). Type as above.

Psychotria giorgii De Wild., Pl. Bequaert. **2**: 369 (1924). Type from Zaire.

Shrub or small tree 1.2–10(?12) m. tall, with glabrous stems; trunk warty and knobby, with reddish brown or brown sometimes flaking bark; crown rounded. Leaf lamina 4–25 × (1.1)2–10 cm., elliptic or oblong-elliptic, acute to shortly acuminate at the apex, rounded to cuneate at the base, entirely glabrous, with narrowly hyaline, sometimes revolute margins, coriaceous, often drying the characteristic yellow of an aluminium-accumulating plant or young leaves drying purplish, often ± shining above; nodules absent; domatia present, not margined with hairs; petiole 0.5–2 cm. long, glabrous; stipules 1–1.5 cm. long, obovate, emarginate at the apex, the lobes rounded, glabrous, soon deciduous, few hairs present on nodes inside the stipules at the base. Flowers sweet-scented, fleshy, heterostylous, 5-merous, in much-branched panicles 6–20 cm. long; peduncle 2.5–13 cm. long, glabrous; secondary branches 0.8–1.7 cm. long, glabrous; pedicels obsolete; bracts and bracteoles small with a few cilia. Calyx ± 1 mm. long, rounded-conic (subglobose in life), glabrous; limb 2–2.5 mm. long, cupuliform, glabrous; lobes small or obsolete. Corolla white or ? greenish yellow, glabrous outside; tube 4–6 mm. long; lobes 2–2.5(3) × 1.2(2) mm., oblong-triangular, margined in dry state, thickened and inflexed at the apex. Stamens with filaments 2.5–3 mm. long in short-styled flowers, 0.5–1 mm. long in long-styled flowers. Style 3.5–4 mm. long in short-styled flowers, 6 mm. long in long-styled flowers; stigma lobes 0.5–1 mm. long. Drupes red, with 2 pyrenes, 6–7 mm. in diam., subglobose or ellipsoid, glabrous, only slightly grooved even in dry state; pyrenes depressed, semi-globose, 5.5–6 × 5–5.5 × 2.2–3 mm., the dorsal face scarcely grooved. Seeds dark, 5 × 5 × 2.8 mm., semi-globose, ventral face rugose, dorsal face with 3 basally joined obtuse ribs and rugose; albumen strongly ruminate.

Zambia. N: 22.4 km. from Kawambwa on road to Mporokoso, fl. 16.x.1947, *Brenan & Greenway* 8121 (EA; FHO; K). W: Chingola, fl. 10.x.1954, *Fanshawe* 1619 (K; NDO).

Also in Nigeria, Cameroon, Central African Republic, Sudan, Zaire, Rwanda, Burundi, Uganda, Tanzania & Angola. Riverine and streamside forest, evergreen thicket, also in *Brachystegia* woodland, sometimes on termite mounds, 1170–1650 m.

6. **Psychotria ealaensis** De Wild., Miss. Laurent. **1**: 348 (1906); **2**, t. 94 (1906); in Ann. Mus. Congo, Bot. Sér. 5, **2**: 182 (1907) pro parte. —Th. & H. Dur., Syll. Fl. Congo: 281 (1909) pro parte. —Petit in Bull. Jard. Bot. Brux. **34**: 103, photo. 3/A (1964); in Bull. Jard. Bot. Brux. **42**: 357 (1972); in Distr. Pl. Afr. 5, map 123 (1973). —Verdc. in F.T.E.A., Rubiaceae 1: 57 (1976). Type from Zaire.

Grumilea ealaensis (De Wild.) De Wild. in Bull. Jard. Bot. Brux. **9**: 33 (1923); Pl. Bequaert **2**: 456 (1924). Type as above.

Climbing shrub with stems 6–15 m. long and attaining 3 cm. in diam. near the base, at first fairly densely covered with very short pubescence but later glabrous. Leaf blades 3.5–11 × 1–5.5 cm., elliptic, distinctly acuminate at the apex, cuneate at the base, glabrous save sometimes for some pubescence on the lower part of the main nerve beneath, thin;

nodules absent; domatia present; petiole 0.2–1.5 cm. long, pubescent; stipules 4–6.5 mm. long, ovate-triangular, bilobed at the apex, the lobes 1–2.5 mm. long, pubescent or glabrous, deciduous; nodes with long hairs within the stipules. Flowers heterostylous, (4)5(6)-merous, in much-branched panicles 2.5–10 cm. long; peduncle 1–4 cm. long, glabrous or pubescent; secondary branches 0.2–1 cm. long, pubescent; pedicels obsolete or ± 1 mm. long, very shortly pubescent; main bracts ± 3 mm. long, lobed at the base, clasping the stem; rest small, ± pubescent. Calyx tube ± 1 mm. long, conic, glabrescent; limb 0.75–1 mm. long, cupuliform, glabrous or sparsely covered with very short almost papilla-like hairs; lobes very short, ± 0.5 mm. long, broadly triangular; corolla yellowish, greenish or lilac, glabrous to finely densely tomentose outside; tube 3.25–5.5 mm. long; lobes 2–2.5 × 1–1.5 mm., ovate-triangular. Stamens purple, with filaments 2–2.5 mm. long in short-styled flowers, 0.5–0.75 mm. long in long-styled flowers. Style 1.5–3 mm. long in short-styled flowers and 4–5 mm. long in long-styled flowers; stigma lobes 0.75–1 mm. long. Drupes red, 5–7(10 in living state) mm. long, 7(8) mm. wide, ellipsoid, with 2 pyrenes; pyrenes 5.5 × 4.2 × 2.5 mm., depressed, semi-ellipsoid; ventral face plane, dorsal face obscurely ribbed. Seeds dark blackish-red, of similar shape, 4.5 × 4 × 2 mm.; ventral face rugose, dorsal face irregularly 7-ribbed; albumen ruminate on both faces but particularly between dorsal ribs.

Malawi. N: Nyika Plateau, Mwenembwe Forest, st. 8.vi.1983, *Dowsett-Lemaire* ecol. 124 (K). S: Zomba Plateau, Chingwe's Hole, bud 8.x.1981, *Chapman & Tawakali* 5923 (K; MAL). **Mozambique**. Z: Serra de Milange, fl. 12.xi.1942, *Mendonça* 1421 (LISC; LMU).

Also in Cameroon, Gabon, Zaire, Uganda and Tanzania. Evergreen forest extending to submontane region; 1200–2100 m.

When dealing with this species previously Petit and I considered it rather distinctive but the material from S. Tanzania and Malawi is literally no more than a climbing form of *P. mahonii* and I wondered if it were possible to keep up a climbing taxon distinct from *P. ealaensis* but the Cameroon material is also similar to *P. mahonii*. Intermediate in Mozambique (N & Z) are shrubs with climbing branches. I am reluctant to sink the two together since in Zaire and Uganda *P. ealaensis* is a distinct easily recognisable species and for practical reasons I have maintained it at specific rank. It needs further study throughout its range.

7. **Psychotria mahonii** C.H. Wright in Bull, Misc. Inf. Kew **1906**: 106 (1906). —De Wild., Pl. Bequaert. **2**: 386 (1924). —Petit in Bull. Jard. Bot. Brux. **34**: 210 (1964). —Verdc. in F.T.E.A., Rubiaceae 1: 58 (1976). —Palgrave, Trees S. Afr.: 900 (1977). Type: specimen grown at Kew from material collected in S. Malawi, Likangala stream, *Mahon* 597-1898 (K, holotype).

Grumilea macrantha K. Schum. in Engl., Bot. Jahrb. **28**: 102 (1899). —De Wild. in Bull. Jard. Bot. Brux. **9**: 42 (1923). Type: Malawi, Mt. Malosa ('Matosa'), *Whyte* s.n. (B, holotype†; K, isotype?) non *Psychotria macrantha* Müll. Arg.

Grumilea megistosticta S. Moore in Journ. Linn. Soc., Bot. **38**: 256 (1908). —De Wild. In Bull. Jard. Bot. Brux. **9**: 42 (1923). —Dale & Eggeling, Ind. Trees Uganda, ed. **2**: 346 (1952). Type from Uganda.

Grumilea punicea S. Moore in Journ. Linn, Soc., Bot. **40**: 101 (1911). —De Wild. in Bull. Jard. Bot. Brux. **9**: 45 (1923). —Goodier & Phipps in Kirkia **1**: 64 (1961). Type: Zimbabwe, Chimanimani Mts., *Swynnerton* 563 (BM, holotype; K).

Psychotria ficoidea K. Krause in Mildbr., Wiss. Ergebn. Deutsch. Zentr. Afr. Exped. 1907–1908: 336 (1911). —De Wild., Pl. Bequaert. **2**: 365 (1924). —Robyns, Fl. Parc. Nat. Alb. **2**: 365 (1947). Types from Zaire and Rwanda.

Grumilea bequaertii De Wild. in Bull. Jard. Bot. Brux. **9**: 28 (1923); in Pl. Bequaert. **2**: 453 (1924). —Robyns, Fl. Parc Nat. Alb. **2**: 366 (1947). —Dale & Greenway, Kenya Trees & Shrubs: 443 (1961). Types from Zaire.

Grumilea bequaertii var. *pubescens* Robyns in Bull. Jard. Bot. Brux. **17**: 96 (1943); Fl. Parc Nat. Arb. **2**: 367 (1947). Type from Zaire.

Grumilea sp. —Dale & Eggeling, Ind. Trees of Uganda ed. 2: 346 (1952).

Psychotria megistosticta (S. Moore) Petit in Bull. Jard. Bot. Brux. **34**: 112, t. 3, photo 3/F (1964); in Distr. Pl. Afr. 5, map. 129 (1973). Type as above.

Psychotria megistosticta var. *puberula* in Bull. Jard. Bot. Brux. **34**: 116 (1964). Type from Tanzania.

Psychotria megistosticta var. *punicea* (S. Moore) Petit in Bull. Jard. Bot. Brux. **34**: 117 (1964).

Psychotria robynsiana Petit in Bull. Jard. Bot. Brux. **34**: 120 (1964); in Distr. Pl. Afr. 5, map. 132 (1973). Type from Zaire.

Psychotria robynsiana var. *pauciorinervata* Petit in Bull. Jard. Bot. Brux. **34**: 121 (1964). Type from Kenya.

Psychotria robynsiana var. *glabra* Petit in Bull. Jard. Bot. Brux. **34**: 122 (1964). Type from Uganda.

Psychotria mushiticola Petit in Bull. Jard. Bot. Brux. **34**: 122 (1964). Type from Zaire.

Psychotria mahonii var. *puberula* (Petit) Verdc. in Kew Bull. **30**: 253 (1975); in F.T.E.A., Rubiaceae 1: 60, fig. 3 (1976).
Psychotria mahonii var. *pubescens* (Robyns) Verdc. in Kew Bull. **30**: 354 (1975); in F.T.E.A., Rubiaceae 1: 61 (1975).

Tree 5–15(24) m. tall or shrub 1.5–5 m. tall; stem soon rugose or striate or the bark sometimes corky and transversally fissured, glabrous or densely shortly velvety pubescent or grey to ferruginous hairy on the young shoots. Leaf blades 3–23 × 1.5–10 cm., elliptic, oblong-elliptic or obovate, acuminate at the apex, cuneate to rounded at the base, glabrous or sometimes with a mixture of long multicellular and papilla-like hairs on the venation beneath and sometimes with what look like scattered minute pale scales beneath (actually stomata glistening) or sometimes quite densely ferruginous hairy, surface often rugulose beneath, rather thin to coriaceous; nodules absent but sometimes there are lesions and bumps which closely simulate them; domatia usually very marked, hairy; petiole 0.2–3.5 cm. long, glabrous to pubescent; stipules 0.4–1.7 cm. long, obovate, bilobed at the apex, the lobes 1–5 mm. long, glabrous or margins ciliate and basal parts pubescent, deciduous. Flowers sweet-smelling, heterostylous, 5-merous, in much-branched panicles 4–18 cm. long; peduncle 1.5–8 cm. long, glabrous or densely covered with minute papilla-like hairs or with dense spreading ferruginous hairs; secondary branches 0.4–1.4 cm. long, similarly hairy; pedicels 0–3 mm. long, usually glabrous or similarly velvety-papillose; bracts 3–8 mm. long, triangular to linear (very reduced leaves); other bracts and bracteoles small, usually margined with red hairs. Calyx tube 1 mm. long, obconic, glabrous to densely papillate; limb 1–1.5 mm. long, cupuliform, glabrous to velvety-papillate; lobes 0.5–0.75 mm. long, bluntly triangular or rounded, glabrous to densely velvety-papillate. Corolla white, greenish or yellow, glabrous to densely velvety-papillose or pubescent outside; tube 4–6 mm. long; lobes 2–3.5 × 1.3–1.5 mm. wide, elliptic-oblong, incrassate and inflexed at the apex, densely grey papillose inside. Stamens with filaments 2–4 mm. long in short-styled flowers, 0.5 mm. long in long styled flowers. Style 2–3 mm. long in short-styled flowers, 6–7 mm. long in long-styled flowers; stigma-lobes 1–2 mm. long. Drupes red, with 2-pyrenes, 5–6 mm. in diam., subglobose, scarcely grooved; pyrenes 4 × 4.5 × 2 mm., subglobose, ventral surface plane, dorsal surface slightly grooved. Seeds dark blackish red, subglobose, about the size of the pyrenes, plane ventrally, with narrow deep dorsal fissures but not noticeably grooved; albumen ruminate.

Zambia. N: 57.6 km. from Mbala on road to Nakonde, fl. & fr. 22.x.1947, *Brenan & Greenway* 8184 (BM; EA; FHO; K). W: Mwinilunga Distr., N. of Dobeka Bridge, by river, fl. 8.xi.1937, *Milne-Redhead* 3164 (K). **Zimbabwe**. E: Mutare Distr., Vumba, below Elephant forest, fl. 28.xii.1951, *Chase* 6280 (K; LISC: SRGH). **Malawi**. N: Mzimba Distr., Mzuzu, from Marymount towards Lunyangwa, fl. 11.xii.1970, *Pawek* 4074 (K; MAL). C: Dedza Distr., Mua Livulezi Forest, fr. 31.v.1962, *Adlard* 474 (K; SRGH). S: Zomba, fl. & fr. xi.1915, *Purves* 237 (K). **Mozambique**. N: Joana, fl. 27.x.1862., *Meller* (K). MS: Carvalho's Estate, Himalaya, fl. 28.xi.1966 *Dale* 454 (K; SRGH).

Also in Zaire, Rwanda, Uganda, Kenya and Tanzania. Predominantly in riverine evergreen forest also streamsides and lake-edges but also in *Afrormosia–Burkea* and *Brachystegia–Julbernardia* deciduous woodland; 640–1850 m.

I have not kept up the varieties I maintained in F.T.E.A. and have taken a much wider view of the species than did Petit (as *P. megistosticta*). The large amount of material now available shows the variation to be much more complex. *Blackmore & Brummitt* 1468 (K; MAL) and *Dowsett-Lemaire* 938 (both from Malawi, Mulanje, R. Likhubula) have narrowly oblong-elliptic or oblong-spathulate leaves with petioles to 2.5 cm. and may be a local variant; only fruiting material has been seen. See also note under *Psychotria ealaensis*.

8. **Psychotria butayei** De Wild. in Ann. Mus. Congo Bot., Sér. 5, **2**: 180, t. 47 (1907); Pl. Bequaert. **2**: 346 (1924). —Petit in Bull. Jard. Bot. Brux. **34**: 195, photo. 6/D (1964); in Distr. Pl. Afr. 6, map 171 (1973). —Verdc. in F.T.E.A., Rubiaceae 1: 66 (1976). Type from Zaire.

Shrub or subshrub with simple or branched stems 0.15–1.2(2) m. tall from a woody rhizome or with many short unbranched stems from a woody stock; stems dull purplish, glabrous or pubescent, longitudinally rugose or ribbed. Leaves often missing from the lower nodes; blades 1–11 × 0.5–4.8 cm., ovate to obovate or narrowly elliptic, obtuse to acute at the apex, cuneate at the base, completely glabrous or sometimes pubescent on both sides, thin to thinly coriaceous, sometimes drying blackish- to yellowish-green; domatia absent; nodules absent or sometimes a few long tubercular ones along the main nerve; petiole 0–1 cm. long, glabrous; stipules ovate-triangular, 4–6 mm. long, acuminate or narrowly bifid at the apex, glabrous or pubescent outside but with ferruginous hairs within, deciduous. Flowers heterostylous, 5-merous, in sparsely branched often condensed panicles 2–6 cm. long; peduncle 1–5 cm. long, glabrous or pubescent; secondary branches 0.5–2 cm. long; pedicels 1–2 mm. long, glabrous; bracts and bracteoles minute. Calyx tube 0.7 mm. high, obconical, glabrous; limb 0.7–1 mm. tall, cupuliform, glabrous; lobes 0.5–1 mm. long, ± triangular, glabrous or margined with hairs. Corolla white, glabrous or densely minutely papillate outside; tube 4–6 mm. long; lobes 1.7–3 × 1.5 mm., triangular, thickened at apex. Stamens with filaments 3–3.5 mm. long in short-styled flowers, obsolete in long-styled flowers. Style 3 mm. long in short-styled flowers, 4.5–6 mm. long in long-styled flowers; stigma lobes ± 1 mm. long. Drupes red, transversely ellipsoid, divided into 2 subglobose parts by a median groove, with 2 pyrenes, 5.5–7 × 7.5–10 mm, or sometimes subglobose, ± 6 mm. in diameter and containing only 1 pyrene; pyrenes subglobose, 5–7 × 5–7 × 4–5 mm., both faces entire. Seed subglobose, ventral face deeply ruminate, usually with a deep median fissure which is branched within; dorsal face scarcely or not ruminate.

Var. **butayei**. —Petit in Bull. Jard. Bot. Brux. **34**: 195 (1964).

Branched shrub or subshrub with glabrous stems and leaves.

Zambia. B: Zambezi (Balovale), fl. 13.x.1952, *White* 3480 (FHO: K). N: Mbala, Itembwe Gap and Gorge, fl. 7.i.1968, *Richards* 22880 (K). W: Mwinilunga Dist., about 5 km. SE. of Angola border and similar distance SW. of Mujileshi R., fl. 6.xi.1962, *Lewis* 6152 (K; MO). S: Machili, fr. 24.ii.1961, *Fanshawe* 6317 (K; NDO).

Also in Zaire, Tanzania and Angola. woodland and derived bushland (*Marquesia*, *Baikiaea*, *Cryptosepalum–Brachystegia–Guibourtia*) sometimes in rocky places and often on Kalahari Sand, also plateau grassland and occasionally in riverine evergreen forest, 1200–1740 m.

Several of the now fairly numerous specimens available from Zambia are indistinguishable from var. *simplex* Petit in Bull. Jard. Bot. Brux. **34**: 196 (1964) and although I retained this in F.T.E.A. I have not formally recognised it here.

Var. **glabra** (Good) Petit in Bull. Jard. Bot. Brux. **34**: 197 (1964). —Verdc. in F.T.E.A., Rubiaceae 1: 67 (1976). Type from Angola.

Grumilea flaviflora var. *glabra* Good in Journ. Bot. Lond. **64**, Suppl. 2: 31 (1926).

Branched shrub or subshrub with puberulous or pubescent stems and leaves.

Zambia. W: Mwinilunga Distr., about 5 km. from Angola border, Mujileshi R., fl. 6.xi.1962, *Lewis* 6163 (K; MO).

Also in Zaire, Tanzania and Angola. Open riverine woodland on sand; 1140 m.

The pubescent form occurs together with var. *butayei* but apparently rarely. It is unfortunate that the correct name is so inappropriate.

9. **Psychotria peduncularis** (Salisb.) Steyerm. in Mem. N.Y. Bot. Gard. **23**: 546 (1972). —Verdc. in F.T.E.A., Rubiaceae 1: 72 (1976). TAB. **2**. Type a specimen grown from material collected in Sierra Leone.

Cephaelis peduncularis Salisb., Parad. Lond., t. 99 (1808). —Hiern in F.T.A. **3**: 223 (1877). —Hepper in Kew Bull. **16**: 154 (1962); in F.W.T.A. ed. 2, **2**: 204 (1963). Type as above.

A very variable subshrub, shrub or subshrubby herb 0.2–3(4.5) m. tall, the stems branched or unbranched, glabrous or sparsely to densely covered with often ferruginous hairs. Leaf blades 2.5–26.5 × 3.1–13.5 cm., narrowly to broadly elliptic, oblong- to obovate-elliptic or ± lanceolate, acute to abruptly acuminate at the apex, cuneate to ± rounded at the base, glabrous above and beneath or sparsely to densely covered with mostly ferruginous hairs, or tomentose, often thinly coriaceous; petioles 1–6.5 cm. long, pubescent or glabrous; stipules 1–2.2 × 0.8–1.5 cm., obovate-oblong to lanceolate, bilobed at the apex, the lobes 0.5–2 cm. long, acuminate, with a median and 2 lateral lines of basal hairs, glabrous to ciliate on the margins, soon deciduous, the nodes hairy above the scars. Flowers usually ± numerous in involucrate capitate inflorescences; peduncles usually solitary but sometimes 2–3 together, 1–9(15) cm. long, pubescent all round or with 2 lines of pubescence or glabrous; pedicels sometimes white, 2–11 mm. long in fruit; bracts free or joined together to form an entire to lobed involucre, white, green, pale blue-green or yellow-green, 0.6–3 × 0.4–1.5 cm., rounded, glabrous, pubescent or ciliate. Calyx tube 1–3 mm. long, ellipsoid or obconic, limb 1–1.3 mm. long; lobes 0.5–4 mm. long, almost obsolete to attenuate-triangular, pubescent or glabrous. Corolla white; tube 3.5–6.5 mm. long, funnel-shaped, cylindric below; lobes 1–3 × 0.6–1.2 mm., triangular or elliptic-

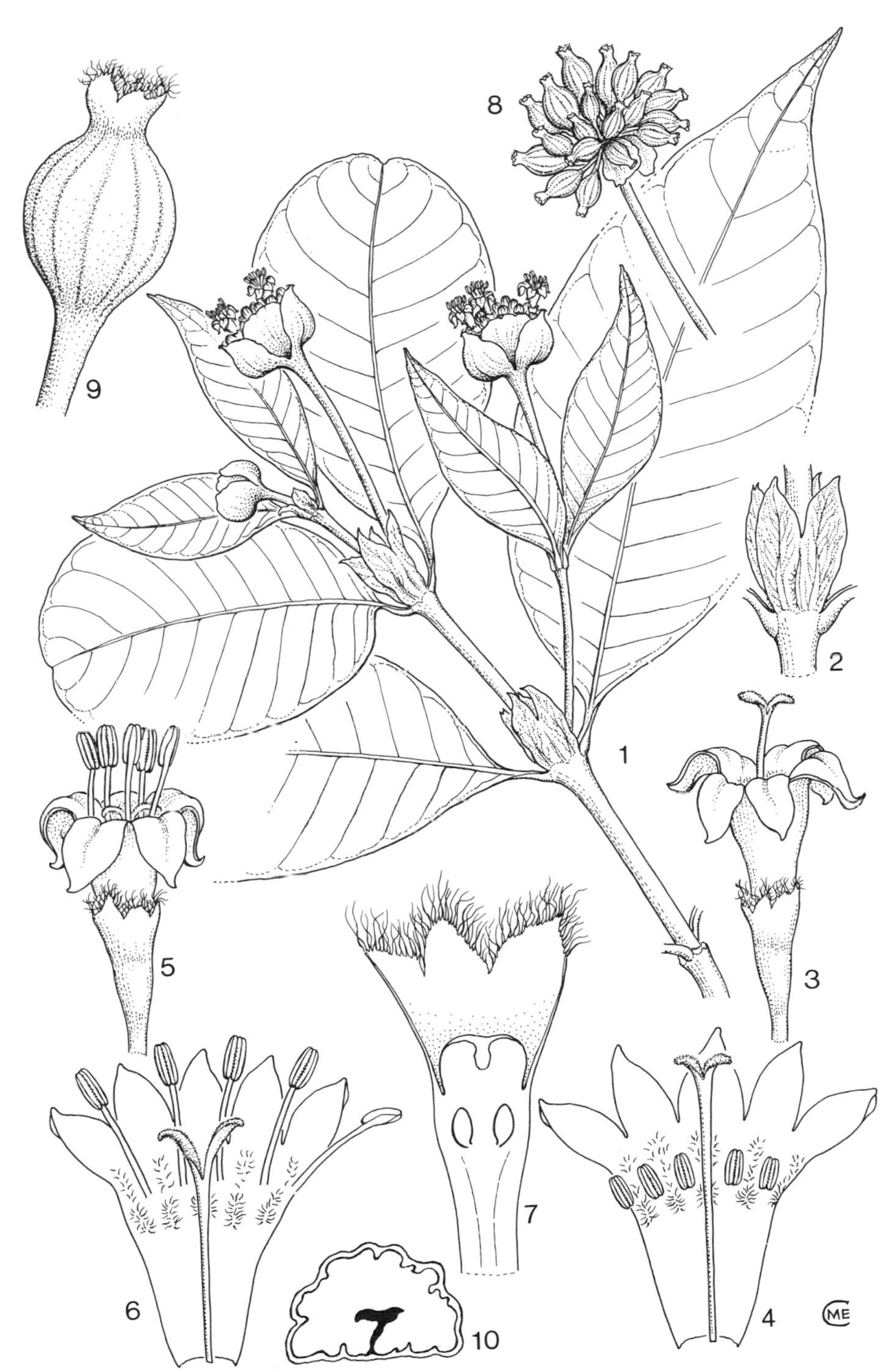

Tab. 2. PSYCHOTRIA PEDUNCULARIS. 1, habit (× $\frac{2}{3}$); 2, stipules (× 1), 1–2 from *Lewis* 6159; 3, long-styled flower (× 4); 4, corolla of long-styled flower opened out (× 4); 5, short-styled flower (× 4); 6, corolla of short-styled flower opened out (× 4); 7, ovary, longitudinal section (× 8), 3–7 from *Milne-Redhead* 2847; 8, infructescence (× 1); 9, fruit (× 4); 10, pyrene, transverse section (× 6), 8–10 from *Brummitt* 10620.

lanceolate, often horned outside at the apex. Anthers just included or exserted in long-styled flowers; filaments exserted (0)1–2 mm. in short-styled flowers. Style 4.3–9 mm. long in long-styled flowers, the stigmas exserted (1)2–3 mm.; 4.5 mm. long in short-styled flowers, the stigmas 1–2.5 mm. long. Fruits blue, blue-black or waxy white, 5.5–8(10 in life) × 3.5–6(9 in life) mm., ellipsoid, grooved in the dry state, crowned or not with the persistent calyx; pyrenes straw-coloured, 4.5–7 × 3.2–5 × 1.5–2 mm., semi-ellipsoid, grooved and convex dorsally, ventrally plane or with 2 narrow grooves. Seeds very dark red, closely conforming in size and ribbing to the pyrenes but not ruminate and with a shallow V-shaped ventral groove.

Var. **nyassana** (Krause) Verdc. in Kew Bull. **30**: 257 (1975); in F.T.E.A., Rubiaceae 1: 74 (1976). Types from Tanzania.

Uragoga nyassana Krause in Engl., Bot. Jahrb. **57**: 50 (1920). —Brenan, T.T.C.L.; 536 (1949). Type as above.

Cephaelis peduncularis sensu F. White, F.F.N.R.: 405 (1962) pro parte non (Salisb.) K. Schum.

Leaf blades more coriaceous than in other varieties, glabrous; venation mostly raised on upper surface in dry state. Stipules with ciliate or glabrous margins. Inflorescence pedunculate, the peduncle 2–5.5 cm. long. Calyx lobes mostly larger and densely margined with hairs up to ± 1 mm. long.

Zambia. N: Mbala Distr., near middle Lunzua Falls, fl. 12.i.1975, *Brummitt & Polhill* 13744 (K). W: Mwinilunga Distr. 4.8 km. SE. of Angola border and 1.8–6.4 km. SW. of Mujileshi R., fl. 6.xi.1962, *Lewis* 6159 (K). **Zimbabwe**. E: Chimanimani, Haroni-Makurupini Forest, fr. 26.v.1969, *Müller* 985 (K; SRGH). **Malawi**. N: 21 km. SW. of Nkhata Bay on road to Chinteche, fr. 11.v.1970, *Brummitt* 10620 (K; MAL). **Mozambique**. MS: Dombo, fr. 31.vii.1945, *Simão* 407 (LISC; LMA).

Also in Tanzania. riverine forest and evergreen seepage patches, also woodland of *Cryptosepalum* or *Brachystegia* etc., lakeshore scrub etc.; 300–1290 (?1400) m.

Many specimens from the Mbala area have the calyx limb ± glabrous.

Var. **rufonyassana** Verdc. var. nov.* Type: Zambia, near Mbala, *Napper* 961 (EA: K, holotype).

Cephaelis peduncularis sensu F. White, F.F.N.R.: 405 (1962) non (Salisb.) K. Schum. sens. str.

Similar to var. *nyassana* but differing in undersurfaces of leaves (particularly the midrib and main venation), and other parts ± ferruginous pubescent.

Zambia. B: Zambezi (Balovale), R. Zambezi near Chavuma, fl. 13.x.1952, *White* 3492 (FHO: K). N: Lake Kashiba, fr. 22.x.1957, *Fanshawe* 3789 (K; NDO). W: Ndola, Lake Ishiku, fl. 18.x.1953, *Fanshawe* 435 (K; NDO). **Malawi**. N: Chitipa Distr., Mafinga Mts., middle slopes below Namitawa summit, fl. 3.iii.1982, *Brummitt* et al. 16275 (K; MAL). S: Mulanje Mts., S. slopes, Ruo Gorge, fr. 7.v.1980, *Blackmore* et al. 1498 (K; MAL).

Also in S. Tanzania.

Evergreen forest (including swamp forest) particularly at streamsides, patches around dambos, riverine etc., also in derived thicket; 1050–1740 m.

Var. **angustibracteata** Verdc. var. nov.**. Type: Zimbabwe, Chirinda Forest, *Chase* 447 (K, holotype).

Similar to var. *nyassana* and var. *rufonyassana* but differing in involucre-bracts being mostly narrower, longer than wide, 7–9 × 1.5–5 mm. Leaves glabrous, or sparsely pubescent on midrib beneath, or densely ferrugineous pubescent beneath. Drupes perhaps juicier with longer pyrenes ± 7.5 mm. long.

Zimbabwe. E: Chipinge Distr., Elisabethville Farm, fr. ii.1962, *Goldsmith* 44/62 (K; LISC; SRGH). **Mozambique**. Z: Serra de Morrumbala, fr. 30.iv.1943, *Torre* 5232 (COI; K; LISC; LMU; MO; SRGH). MS: Gorongosa Mt. Nhandore, fr. 6.v.1964, *Torre & Paiva* 12288 (LISC; LMU; SRGH).

Endemic to Zimbabwe (E) & Mozambique (MS & Z). Riverine (e.g. *Newtonia*) forest; 690–1200 m.

This may well be more distinct than indicated and deserving of specific rank but the available material is not uniform, particularly in indumentum and leaf-shape and size. These bracteate species need detailed revision throughout Africa. Other varieties in Zaire and Uganda have narrow bracts

* **Psychotria peduncularis** var. **rufonyassana** Verdc. var. nov. var. *nyassanae* similis sed foliis subtus praesertim ad nervos rufopubescentibus et a var. *suaveolenti* (Hiern) Verdc. pedunculis maturis longioribus differt.

** **Psychotria peduncularis** var. **angustibracteata** Verdc. var. nov. var. *rufonyassanae* similis sed bracteis involucri angustioribus 7–9 × 1.5–6 mm. differt.

but differ in indumentum. The fruit differences are only hinted at by the herbarium sheets but I suspect something may be involved which will become evident when spirit material of a wide range of material is available. *Obermeyer* Transvaal Mus. 37538 (Mt. Silinda, Chirinda Forest, fl. i.1938) (BM) is very low in habit, ± 12 cm. tall but seems to be this taxon.

10. **Psychotria mwinilungae** Verdc. sp. nov.* Type: Zambia, 8 km. N. of Mwinilunga, West Lunga R., *Brummitt et al.* 14018 (K, holotype).

Rhizomatous subshrub, with ± ascending slender unbranched ± herbaceous flowering stems 25–30 cm. tall, bifariously rufo-pubescent above, ± glabrous beneath; rhizomes slender. Leaves drying grey-green, 4.5–7.5 × 0.9–3 cm., elliptic, oblong-elliptic or ± obovate-elliptic, shortly acuminate at the apex, cuneate at the base, glabrous above and beneath save for red pubescence on midrib, venation drying reddish brown beneath; petiole 3–6 mm. long, densely ciliate with red hairs where blade is decurrent on each side; stipules ± 7 mm. long, divided into 2 acuminate lobes ± 5 mm. long, ± ciliate and base of stipules rufo-pubescent particularly on the main-nerves. Inflorescences with short peduncles ± 5 mm. long, ± 5-flowered, the involucres about 13 × 5 mm. made up of ± ovate bracts up to 14 × 6 mm., glabrous save for ferruginous marginal ciliae, shortly acuminate; pedicels ± 1 mm. long. Calyx ± 1.2 mm. long, the limb distinctly toothed, ± ciliate. Corolla colour not recorded; tube ± 5 mm. long, cylindrical with broadly funnel-shaped throat, 1 mm. wide at base and ± 3 mm. wide at throat, glabrous; lobes c. 2 × 1.2 mm., ovate. Style exserted 2–2.5 mm. in long-styled flowers, the stigmatic lobes ± 0.8 mm. long. Fruit not known.

Zambia. W: 8 km. N. of Mwinilunga, West Lunga R., fl. 23.i.1975, *Brummitt et al.* 14018 (K).
Endemic. Riverine forest; 1300 m.
This comes very close to and is probably conspecific with *Uragoga repens* De Wild. and *Uragoga wellensii* De Wild. (Mem. Inst. Roy. Col. Belg., Sci. Nat. Méd. 8, **4**(4): 167; 184 (1936)) but there are slight indumentum differences; both were described from the Bas-Congo region of Zaire but all the species in this work were described in French shortly after the rules made Latin obligatory.

11. **Psychotria amboniana** K. Schum. in Engl., Pflanzenw. Ost-Afr. **C**: 390 (1895). —S. Moore in Journ. Bot. Lond. **43**: 353 (1905). —De Wild., Pl. Bequaert. **2**: 331 (1924). —Brenan, T.T.C.L.: 523 (1949). —Dale & Greenway, Kenya Trees & Shrubs: 466 (1961). —Verdc. in F.T.E.A., Rubiaceae 1: 77 (1976). Types from Tanzania.
Psychotria albidocalyx K. Schum. in Engl., Pflanzenw. Ost-Afr. **C**: 390 (1895). —De Wild., Pl. Bequaert. **2**: 327 (1924). —Petit in Bull. Jard. Bot. Brux. **36**: 80, t. 5 (1966). Type from Tanzania.
Psychotria albidocalyx var. *angustifolia* S. Moore in Journ. Bot. Lond. **45**: 116 (1907). —De Wild., Pl. Bequaert. **2**: 328 (1924). Type from Kenya.

Shrub 1–3 m. tall; young stems mostly pale, greyish white, often chestnut when older, with grooved slightly peeling epidermis, glabrous or in one variety velvety. Leaf blades 1.5–8(11) × 0.2–3.3(4) cm., oblong-elliptic, narrowly to broadly elliptic, narrowly obovate or even ± linear-lanceolate, acute to shortly acuminate at the apex, very narrowly cuneate at the base, discolorous, glabrous on both surfaces, or in one variety velvety pubescent; nodules few, situated in the basal part of the midrib and domatia often present in the axils of the lateral nerves beneath; petiole 1–3(15) mm. long, passing very gradually into the very narrow base of the lamina; stipules sometimes brown, glabrous, bilobed, the base ovate-triangular 1–1.5 mm. long, the lobes 0.3–0.6(2) mm. long. Flowers heterostylous, 5-merous, in ± many-flowered rather congested glabrous panicles or rarely subumbellate; peduncle 0.7–4 cm. long; secondary peduncles 4–10 mm. long; pedicels 1–4 mm. long; bracts minute. Calyx glabrous; tube 0.5–1 mm. long, campanulate; limb 1.25–2 mm. long, deeply cupuliform, drying pale, with ± triangular teeth 0.25–0.8 mm. long. Corolla white or cream, glabrous outside; tube 4–6 mm. long; lobes 3–4 × 0.9 mm., narrowly oblong, ± appendiculate at the apex. Stamens with filaments 2.5–3.5 mm. long in short-styled

* **Psychotria mwinilungae** *Verdc.* sp. nov. ob habitus fere subherbaceum, pedunculos breves ± 5 mm. longos, involucra parva c. 13 × 5 mm. valde distincta.
Caules floriferi graciles, ± adscendentes, simplices, 25–30 cm. alti, superne bifariam pubescentes. *Folia* elliptica, oblongo-elliptica vel obovato-elliptica, 4.5–7.5 × 0.9–3 cm., apice breviter acuminata, basi cuneata, glabra praeter costam subtus pubescentem. *Stipulae* 7 mm. longae, biacuminatae. *Involucra* brevipedunculata, ± 5-flora, c. 13 × 5 mm. *Corollae tubus* 5 mm. longus, basi 1 mm. latus, apice infundibuliformis, ± 3 mm. lata.

flowers, 0.3–0.4 mm. long in long-styled flowers. Style 2–2.5 mm. long in short-styled flowers, 6–8 mm. long in long-styled flowers; stigma lobes 1–2 mm. long. Drupes red, with 1-2 pyrenes, 4–6 mm. in diam., subglobose or ovoid, glabrous, crowned with the calyx limb; pyrenes 4 × 2.7 × 2.4 mm., ovoid or semiglobose, the dorsal surface 6–7-ribbed. Seeds 4 × 2.7 × 2.4 mm., ± semi-globose or ovoid, grooved on the ventral face; albumen not ruminate.

Subsp. **mosambicensis** (Petit) Verdc. comb. nov. Type: Mozambique, between Costa do Sol and Marracuene, Muntanhane, *Balsinhas* 271 (BM; K, holotype; LMA).
Psychotria albidocalyx var. *mosambicensis* Petit in Bull. Jard. Bot. Brux. **36**: 83 (1966).

Leaves mostly larger with longer petioles up to 1.5 cm. long; calyx and corolla shorter.

Mozambique. GI: Rio das Pedras, fr. vi. 1936, *Gomes e Sousa* 1763 (K). M: Marracuene, Restaurante Costa do Sol, fr. 1.vi.1959, *Barbosa & Lemos* 8545 (K; LISC; LMA).

Endemic. Dune vegetation and forest; ?0–150 m.

Torre & Paiva 9841 (COI; K; LISC; LMU) Mozambique, Cabo Delgado, Macondes, Chomba, between Mueda and Negomano, 3.i.1964, savanna with *Parinari,* 750 m. has wider more ovate leaves than the other varieties and may be distinct.

12. **Psychotria linearisepala** Petit in Bull. Jard. Bot. Brux. **36**: 91 (1966). —Verdc. in F.T.E.A., Rubiaceae 1: 84 (1976). TAB. **3**. Type from Zaire.
Psychotria sp. —Brenan in Mem. N.Y. Bot. Gard. **8**: 453 (1954). —F. White, F.F.N.R.: 418 (1962).

Shrub (or ? small tree) 0.3–2.5(4) m. tall, with slender stems, densely pubescent with whitish or ± ferruginous hairs when young, glabrescent, blackish-or greyish-purple and longitudinally ridged when older; in some specimens the internodes are short and the shoots very nodose and roughened in appearance. Leaf blades 1–9 × 0.5–4.5 cm., elliptic to oblong-elliptic or in some forms almost round, rounded to distinctly acuminate at the apex, narrowly cuneate to rounded at the base, rather markedly discolorous, densely to rather finely puberulous or pubescent on both sides, often rather thickly so beneath in young leaves; nodules few, situated by the lower part of the midrib and visible beneath at its side; petiole 0.2–1.3 cm. long, puberulous to densely pubescent; stipules prominently bilobed, becoming rather thick and somewhat persistent, the triangular base 3–5 mm. long, the lobes acuminate, 1.5–3(4) mm. long, drawn out into fine points when very young but soon lost in older ones, at first ± ferruginous pubescent, later glabrous. Flowers heterostylous, 5–6(7)-merous, in few-flowered condensed subumbellate inflorescences 0.5–2.5 cm. long; peduncle 0.3–2.3 cm. long, rather densely pubescent; pedicels 1–3 mm. long, pubescent; bracts minute. Calyx pale, pubescent; tube 1 mm. long, obconic; limb very short or up to 0.75 mm. long; lobes 2–4 mm. long, linear or narrowly linear-triangular; disk fleshy. Corolla white or greenish cream, pubescent outside; tube 3–5.5 mm. long; lobes 2.5–3 × 0.8 mm., narrowly-oblong. Stamens with filaments 2–2,5 mm. long in short-styled flowers, almost obsolete in long-styled flowers. Style 2.5–4 mm. long in short-styled flowers, 5–5.5 mm. long in long-styled flowers; stigma lobes 0.5–0.75 mm. long. Drupes reddish or scarlet, 3.5–5 mm. in diameter, with 1–2 pyrenes, subglobose, ± grooved between the pyrenes where 2, glabrous or glabrescent, crowned with the persistent calyx. Seeds red-brown, 4 × 3.5 × 2.5 mm., subglobose, the ventral face with a T-shaped median fissure; albumen not ruminate.

Var. **linearisepala**. —Verdc. in F.T.E.A., Rubiaceae 1: 85 (1976) —Gonçalves in Garcia de Orta. Sér. Bot. **5**: 204 (1982).

Leaf blades more oblong-ovate, acuminate, rather less densely pubescent. Young shoots whitish pubescent or often less obviously ferruginous. Shrub over 1 m. tall.

Zambia. N: Mbala Distr., Chilongowelo, fl. 24.xii.1956, *Richards* 7342 (K). W: Ndola, fl. 18.xii.1954, *Fanshawe* 1720 (K; LISC; NDO). C: Serenje Distr., 12.8 km. SE. of Kanona, Kundalila Falls, fl. 17.xii.1967, *Simon & Williamson* 1419 (K; SRGH). **Malawi**. N: 1.6 km. E. of Rumphi, 3.2 km. N. of Chelinda R., fl. 18.i.1976, *Pawek* 10723 (K; MAL; MO; SRGH; UC). C: Ntchisi Forest Reserve, fr. 26.iii.1970, *Brummitt* 9432 (K; MAL). S: Mt. Mulanje Forest Reserve, along Likabula–Lichenya path, fl. 28.xii.1981, *Chapman* 6084 (K; MAL). **Mozambique**. N: Massangulo Mts., fl.xii.1932, *Gomes e Sousa* 1200 (K). T: Macanga, near base of Mt. Furancungo, fr. 17.iii.1966, *Pereira et al.* 1803 (LMU).

Also in S. Zaire and S. Tanzania. *Brachystegia–Uapaca* and other woodland, thicket, evergreen forest remnants, sometimes riverine but often in rocky places; 750–1650 m.

Apparently first collected by McClounie (31) on Mulanje in 1896. A well-marked and easily recognised species. Some specimens particularly the one cited from Mozambique approach var. *subobtusa* Verdc. described from Songea District, Tanzania.

Tab. 3. PSYCHOTRIA LINEARISEPALA. 1, habit (× $\frac{2}{3}$); 2, stipules (× 3); 3, short-styled flower (× 4); 4, corolla of short-styled flower opened out (× 6); 5, long-styled flower (× 4); 6, corolla of long-styled flower opened out (× 6); 7, ovary, longitudinal section (× 16), 1–7 from *Fanshawe* 1720; 8, fruiting branch (× 1); 9, fruit (× 4); 10, pyrene, transverse section (× 6), 8–10 from *Pawek* 14208.

13. **Psychotria pumila** Hiern in F.T.A. **3**: 207 (1877). —K. Schum. in Engl. Pflanzenw. Ost-Afr. **C**: 391 (1895). —De Wild., Pl. Bequaert. **2**: 409 (1924). —Petit in Bull. Jard. Bot. Brux. **36**: 96, t. 6 (1966). —Verdc. in F.T.E.A., Rubiaceae 1: 86 (1976). Type: Mozambique, Morrumballa, *Kirk* (K, holotype).

Subshrub, with erect shoots (2)5–30 cm. tall (rarely taller in intermediates) from ± horizontal underground stems, or shrub 1–5 m. tall; stems shortly densely pubescent, glabrescent or glabrous, older ones longitudinally ridged. Leaf blades 2–14 × 0.5–5.8 cm., narrowly elliptic to elliptic or sometimes ovate or obovate to oblanceolate, rounded to acute or rarely acuminate at the apex, narrowly to rather broadly cuneate at the base, usually markedly discolorous, the undersurface pale or quite whitish, glabrous or finely puberulous to pubescent above, glabrous or with nerves or less often whole surface pubescent beneath, the surfaces often with a minute honeycomb reticulation in the dry state; domatia absent; nodules absent or almost obsolete or rarely dispersed in the lamina; petiole 0.1–1 cm. long, glabrous or shortly pubescent; stipules prominently bilobed, the base 2.5–5 mm. long, ovate, the lobes 1.5–4 mm. long, acuminate, the base somewhat persistent. Flowers heterostylous, 5-merous, in few-many-flowered condensed subumbelliform cymes 1–1.5 cm. in diam., rarely inflorescence branched with a lower whorl of 3 additional cymes; peduncles 0.5–6.5 cm. long, glabrous or puberulous; pedicels 1–5(7) mm. long, ± glabrous; bracts obsolete. Calyx 1 mm. long, turbinate, glabrous or slightly pubescent; limb 1–2 mm long, cupuliform, glabrous or ± pubescent, the lobes unequal, usually short, subtriangular or ovate, up to 0.3 mm. long or practically obsolete. Corolla white, glabrous outside; tube (2.5)5–6.5 mm long; lobes (2.5)3–5 × ± 1 mm. oblong-lanceolate, usually slightly appendaged at the apex. Stamens with filaments 2.5–3 mm. long in short-styled flowers, 0.2 mm. long in long-styled flowers. Style 3–4.5 mm. long in short-styled flowers, 6–7(10) mm. long in long-styled flowers; stigma lobes 0.5–1.5 mm. long. Drupes green turning yellow, orange and finally becoming bright red, with 2 pyrenes, 5–6 mm. in diam., subglobose, glabrous, crowned with the persistent calyx limb; pyrenes pale, 4.2 × 3.6 × 2 mm., semi-ellipsoid, ± 6-ribbed dorsally, ventrally flat, all surfaces rhaphide-packed. Seeds dark red, 3.5 × 3.5 × 2 mm., semi-ellipsoid or semi-globose, with a V-shaped median fissure on the ventral face; albumen not ruminate.

Key to Varieties

1. Small subshrub 2–15(30) cm. tall; foliage glabrous or pubescent - - - var. *pumila*
- Shrubs 1–5 m. tall - - - - - - - - - - - - - - - - - 2
2. Leaves glabrous or with few hairs on midrib beneath; flowering pedicels 3–7 mm. long 3
- Leaves glabrous to densely pubescent; flowering pedicels 1–3 mm. long - - - 4
3. Leaves on young shoots obovate, on old shoots narrowly elliptic to oblanceolate - - - - - - - - - - - - - - var. *subumbellata*
- Leaves ± uniformly elliptic or oblong-elliptic - - - - - - var. *leuconeura*
4. Leaves glabrous to slightly or more densely pubescent beneath - - - var. *buzica*
- Leaves densely velvety pubescent beneath - - - - - - - - - var. *puberula* (see note at end of var. *buzica*)

Var. **pumila**. —Petit in Bull. Jard. Bot. Brux. **36**: 96, t. 6 (1966). —Verdc. in F.T.E.A., 1: 87 (1976). —Gonçalves in Garcia de Orta. Sér., Bot. **5**: 204 (1982). Type as above.

Psychotria brachythamnus K. Schum. & K. Krause in Engl., Bot. Jahrb. **39**: 566 (1907). —De Wild., Pl. Bequaert. **2**: 339 (1924). —Brenan, T.T.C.L.: 522 (1949). Type from S. Tanzania.

Psychotria albidocalyx var. *subumbellata* Petit in Bull. Jard. Bot. Brux. **36**: 83 (1966) pro parte quoad *Brass* 16368.

Psychotria sp. nr. *kirkii* Hiern —Brenan in Mem. N.Y. Bot. Gard. **8** 453 (1954) quoad *Brass* 16368.

Small geophilous plant with stems 2–15(30) cm. tall.

Zambia. N: Mbala to Kasama, fl. 13.ii.1949, *Bullock* 2094 (K). W: just E. of Mwinilunga by R. Lunga, fl. 2.xii.1937, *Milne-Redhead* 3497 (K). C: Lusaka Distr., Great E. Rd., 14 km. E. from Rufunsa, 30.xii.1972, *Kornaś* 2876 (K; KRA). E: Sasare, fl. 8.xii.1958, *Robson* 865 (BM; K; LISC). S: Mumbwa, *Macaulay* 1068 (K). **Zimbabwe**. N: Darwin, Mvurwi Range, Umsengezi R., fl. 22.xii.1952, *Wild* 3936 (K; LISC; SRGH). E: Mutare Distr., Burma Valley, Mangere Farm, fl. 29.xii.1959, *Chase* 7232 (BM; K: SRGH). **Malawi**. N: Chitipa Distr., Misuku Hills, descent from Sokora to Kalenga R., fl. 30.xii.1972, *Pawek* 6252 (K; MAL; MO). C: 23 km. S. of Lilongwe, fr. 19.ii.1970, *Brummitt* 8635 (K; MAL). S: Mulanje, fl. Dec. 1894, *Scott Elliot* 8597 (K). **Mozambique**. N: Marrupa, 20 km. on road towards Lichinga, Naboina, fr. 17.ii.1981, *Nuvunga* 572 (K; LMU). Z: Lugela-Mocuba, Namagoa Estate, fl. i.194?, *Faulkner* Pre 14 (K; PRE). T: Moatize, Zóbuè towards Metengobalama at km 5 from Zóbuè, fl.

10.i.1966, *Correia* 336 (LISC). MS: between Dondo & Muanza, fl. 4.xii.1971, *Pope & Müller* 509 (K; LISC; SRGH). GI: Panda, fr. 25.ii.1955, *Exell et al.* 592 (BR; LISC).

Also in Zaire (Shaba) and S. Tanzania. *Brachystegia–Julbernardia–Uapaca* woodland, particularly in rocky places, plateau woodland, open savanna woodland or occasionally in grassland, also 'Chipya dambo'; 100–1740 m.

Var. **leuconeura** (Schum. & Krause) Petit in Bull. Jard. Bot. Brux. **36**: 99 (1966). —Verdc. in F.T.E.A., Rubiaceae 1: 87 (1976). Type from Tanzania.

Psychotria leuconeura Schum. & Krause in Engl., Bot. Jahrb. **39**: 554 (1907). —De Wild., Pl. Bequaert. **2**: 381 (1924). —Brenan, T.T.C.L.: 525 (1949). Type as above

Shrub 1–5 m. tall, virtually totally glabrous; flowering pedicels 3–7 mm. long.

Zambia. E: Petauke Distr., Nyimba, fl. 11.xii.1958, *Robson* 904 (BM; K; LISC). **Mozambique**. N: Murrupula, 15 km. road to Nampula, Mt. Namuato, fr. 30.i.1968, *Torre & Correia* 17480 (BR; LISC; LMU; SRGH).

Also in S. Tanzania. Undergrowth of riparian fringe; 500–750 m.

The above Zambian specimen described as a straggling shrub 1–2 ft. tall is intermediate with the typical variety. It is not evident if it is due to lack of burning in a sheltered habitat.

Var. **subumbellata** (Petit) Verdc. comb. nov.

Psychotria albidocalyx var. *subumbellata* Petit in Bull. Jard. Bot. Brux. **36**: 83 (1966). Type: Zambia, Kawambwa, *Fanshawe* 3877 (BR, holotype; K, isotype).

Shrub 0.9–1.2 m. tall almost glabrous save for few hairs on midrib beneath but differing from other shrubby varieties in having obovate leaves c. 4 × 2 cm. on young shoots and narrower elliptic or oblanceolate leaves c. 3.5–5 × 0.8 cm. on older shoots; flowering pedicels 3–6 mm. long but mostly c. 4 mm.

Zambia. N: Kawambwa, fl. 10.xi.1957, *Fanshawe* 3877 (K; NDO).

Endemic. Riparian thicket.

Var. **buzica** (S. Moore) Petit in Bull. Jard. Bot. Brux. **36**: 99 (1966). —Verdc. in F.T.E.A., Rubiaceae 1: 88 (1976). —Gonçalves in Garcia de Orta, Sér. Bot. **5**: 204 (1982). Type: Mozambique, Manica e Sofala, Boka, R. Buzi, *Swynnerton* 562a (K, lectotype).

Psychotria buzica S. Moore in Journ. Linn. Soc. Bot. **40**: 100 (1911). —De Wild., Pl. Bequaert. **2**: 346 (1924). Type as above.

Very similar to var. *leuconeura* but sometimes 1.5–3 m. tall (even a 'small tree') and flowering pedicels mostly very short; glabrous to sparsely pubescent or densely so on venation.

Zimbabwe. E: Chipinge Distr., ± 9 km. NE. of Musirizwe/Bwazi, R. confluence, fr. 28.i.1975, *Pope et al.* 1394 (K; SRGH). **Malawi**. S: without locality, *Buchanan* 554 (K). **Mozambique**. N: Marrupa, Estrada towards Nungo 20 km., margin of R. Sezalo between Montanhas Mirenge and Mucuwango, fr. 20.ii.1981, *Nuvunga* 645 (K; LMU). Z: Alto Molócuè, Gilé, km. 10, near Mt Gilé, fl. 21.xii.1967, *Torre & Correia* 16694 (COI; LISC; LMU). MS: Matarara do Lucite, *Gomes Pedro* 4276 (BR).

Also in S. Tanzania. Woodland, open 'forest', thicket, grassland with scattered bushes, sometimes on limestone; 250–300 m.

The pedicels are quite long in fruit. It seems probable that this should not be recognised but the material is scarcely adequate for assessment. Some of the material comes close to var. *puberula* Petit (Type from Tanzania) but this is scarcely worth retaining eg. Mozambique, Beira District, Inhaminga, Corone, fr. 20.iii.1956 *Gomes e Sousa* 4307 (K) and GI: Panda, fr. 25.ii.1955, *Exell et al.* 592 (BM); the lectotype of var. *buzica* itself is slightly but distinctly pubescent. The status of these variants is most unsatisfactory and needs examination in the field to see if var. *pumila* ever gets large in favourable circumstances. The 3 m. tall specimens are clearly distinctly different from the small subshrubs and need further field study.

14. **Psychotria sp. A.**

Slender shrub to 1.5 m. tall; young shoots pale, slightly pubescent; older parts with dark purplish ridged glabrous bark. Leaves 2.5–6.5 × 0.5–1.3 cm., very narrowly oblong-elliptic to oblanceolate, narrowed and acute at the apex, narrowly cuneate at the base, glabrous, the midrib white beneath and with very few sparse elongate bacterial nodules at its base; petiole 0.7–1 cm. long; stipules 2 mm. long, ± triangular, divided at the apex into two subulate lobes 2 mm. long. Inflorescences about 2 cm. diam., known only in fruit; peduncles 3–3.5 cm. long; pedicels 0.3–1 cm long. Corolla not seen. Fruits red, 4.5 × 3.5 mm., glabrous, crowned with calyx limb, ± 1.5 mm. long, tubular with short lobes; pyrenes straw-coloured, 4–5-ribbed, the endosperm with scarcely any furrows of any sort save a vague one on the ventral surface.

Malawi. S: Mulanje, S. slopes, Ruo Gorge near Power station, fr. 7.v.1980, *Blackmore & Brummitt* 1480 (K; MAL).

Endemic. Riverine evergreen forest undergrowth; 900–1350 m.

Known from two specimens both from the Ruo Gorge, too diverse I think to be included in *Psychotria pumila* Hiern. Flowering material is needed before describing it at some level.

15. **Psychotria kirkii** Hiern in F.T.A. **3**: 206 (1877). —K. Schum. in Engl., Pflanzenw. Ost-Afr. **C**: 391 (1895). —De Wild., Pl. Bequaert. **2**: 379 (1924). —Bremek. in Journ. Bot. Lond. **71**: 280 (1933) pro parte. — Garcia in Mem. Junta Invest. Ultramar., Sér. 2, 6: 42 (1959). —Petit in Bull. Jard. Bot. Brux. **36**: 126 (1966). —Verdc. in Kew Bull. **30**: 262 (1975); in F.T.E.A., Rubiaceae 1: 92 (1976). —Gonçalves in Garcia de Orta, Sér. Bot. **5**: 203 (1982). Type: Mozambique, Morrumbala [Maramballa] *Kirk* 9 (K, lectotype).*

Psychotria petroxenos K. Schum. in Engl., Bot. Jahrb. **39**: 557 (1907). —De Wild., Pl. Bequaert. **2**: 403 (1924). —Bremek. in Journ. Bot., Lond. **71**: 270 (1933). —Brenan T.T.C.L. **2**: 523 (1949). Type from S. Tanzania (Lake Malawi).

Shrub or subshrub (0.1)0.2–6 m. tall, rarely somewhat scandent; stems glabrous to densely velvety pubescent, sometimes with yellow-brown hairs, the older stems often becoming glabrescent or glabrous, pale to blackish purple in more shrubby forms. Leaf blades 2–18 × 0.5–9 cm., narrowly to broadly elliptic, elliptic-obovate, ± ovate or oblong-elliptic, rounded to acute or ± acuminate at the apex, cuneate at the base, glabrous or densely velvety pubescent above and beneath or sometimes with only rather sparse long hairs on the midnerve beneath, mostly rather thin but sometimes ± coriaceous, the margins often thin, drying yellow and sometimes very marked, sometimes wavy; domatia absent; nodules numerous, spot-like or rarely linear, scattered in the lamina; petiole 0.1–2.5 cm. long, glabrous or pubescent; stipules usually brown and ± scarious, 2.5–11 mm. long, ovate-triangular, bilobed at the apex, the lobes 0.5–4(6) mm. long, acuminate or filiform, glabrous to densely pubescent. Flowers heterostylous, 5(6)-merous, in panicles or umbels or collections of umbels, 1–14 cm. long, the actual component parts often quite dense; peduncles 0.5–9 cm. long, glabrous to densely pubescent; secondary branches 0.2–2.5 cm. long, similarly hairy; pedicels 1–3(6 in fruit) mm. long, glabrous or hairy; bracts and bracteoles small, filiform or narrowly triangular, up to 4 mm. long. Calyx glabrous or hairy; tube 0.5–1 mm. tall; limb 0.75–1.5 mm. long, cupuliform, glabrous or hairy; lobes 0.2–0.75 mm. long or ± obsolete. Corolla white, cream, greenish or rarely yellow, glabrous outside; tube 2.5–6 mm. long; lobes 2–3(5) × 0.8–1.2 mm. oblong or elliptic. Stamens with filaments 1.5–3 mm. long in short-styled flowers, 0–0.5 mm. long in long-styled flowers. Style 1.5–3.75 mm. long in short-styled flowers, 3.5–6(10) mm. long in long-styled flowers; stigma lobes 0.25–0.75(2) mm. long. Drupes red, with 2 pyrenes, subglobose, 5–7 mm. diam. or didymously subglobose, 7–8 × 4–6 mm., glabrous or sparsely hairy when young; pyrenes 4–4.5 × 3 mm., semi-globose to subglobose, slightly rugulose. Seeds semi-globose to subglobose, similar in size, ventral face with a T-shaped median fissure but rest of albumen not ruminate.

Zambia. N: Mbala Distr., Saisi State Ranch, Saisi Valley, fl. 13.i.1968, *Richards* 22897 (K). W: 10 km. from Chingola on road to Solwezi, fl. 18.i.1975, *Brummitt et al.* 13826 (K). C: Lusaka Distr., Chilanga Fish Farm, fl. 5.v.1963, *Lusaka Nat. Hist. Club* 261 (K). S: Mumbwa, *Macaulay* 1067 (K, pro majore parte). **Zimbabwe**. N: Mtoko Reserve, fr. iv. 1956, *Davies* 1912 (K; SRGH). W: Victoria Falls, fl. i.1910, *Rogers* 5382 (K). C: Beatrice, fl. 23.xii.1924, *Eyles* 4420 (K; SRGH). E: Inyanga, Cheshire, fl. 15.i.1931, *Norlindh & Weimarck* 4407 (K; LD; SRGH). S: Chibi, S. slopes of Nyoni Mts., fr. 20.iv.1967, *Müller* 596 (K; SRGH). **Malawi**. N: Nkhata Bay Distr., Sanga, 16 km. S. of Mzuzu-NB road, fl. 17.xii.1972, *Pawek* 6100 (CAH; K; MAL; MO; UC). S: Zomba, Namasi, fl. xi. & xii. 1899, *Cameron* 71 (K). **Mozambique**. N: Mandimba, fl. 18.xii.1941, *Hornby* 3524 (K). Z: Mocuba, Namagoa, road to Moébede, fl. 31.xii.1947, *Faulkner* Kew 176 (BR; K; PRE; S). T: Moatize towards Metengobalama at km 6 from Zóbuè, edge of R. Mevúzi, *Correia* 344 (LISC). MS: Garuso Distr., Bandula, fr. 28.i.1949, *Chase* 1221 (K; SRGH).

Also in Zaire, S. Ethiopia, Sudan, Uganda, Kenya and Tanzania. *Brachystegia, Colophospermum* and similar woodland, wooded grassland and grassland, also riverine and lakeside thicket, sometimes in rocky places and often on termite mounds; ?60–1350 m.

In my treatment of this species for F.T.E.A. I showed that a half dozen or so species recognised by Petit could not be maintained and reduced them to varietal rank. For the present purpose even this

* There are three pieces on the type sheet, two collected in Dec. 1858 and one on 18 Jan. 1863. Only the former are associated with Rubiaceae no. 9. By citing *Kirk 9*, I assume Petit has chosen a lectotype, actually the central of the three specimens.

seems unacceptable and I am not formally recognising most of these varieties. Most of the material comes close to the concept of var. *kirkii* but some material from S. Malawi and Mozambique e.g. *Faulkner* 176 cited above is named *P. swynnertonii* Bremek. by Petit i.e. var. *swynnertonii* (Bremek.) Verdc.; other glabrous variants from S. Malawi e.g. *Wild* 7696, Mangochi Distr., Monkey Bay I. come very close to var. *nairobiensis* (Bremek.) Verdc. and this is recorded at specific level by Petit from this province. For details of variation in this species and synonymy see F.T.E.A. The differences between *P. kirkii* and *P. spithamea* are dealt with under the latter; despite the fact most material is easily named some specimens are difficult and possibly hybrids. Very likely *P. kirkii* will be combined with *P. punctata* Vatke, unfortunately an earlier name, a littoral species which nevertheless merges with the glabrous variants of *P. kirkii*.

16. **Psychotria diversinodula** (Verdc.) Verdc. comb. et stat nov.* Type from Tanzania.
Psychotria kirkii var. *diversinodula* Verdc. in Kew Bull. **30**: 264 (1975); in F.T.E.A., Rubiaceae 1: 98 (1976). Type as above.

Subshrub 9–25 cm. tall with few slightly branched shoots from a slender woody rhizome; youngest part of stem shortly velvety pubescent, older parts glabrous or grey to rusty-pubescent with dark ridged bark. Leaves variable, 4–14 × 1.1–6.4 cm., narrowly oblong-elliptic to obovate but at least some near apex ± triangular, acute to obtuse at the apex, narrowly to broadly cuneate or ± subtruncate at the base, shortly pubescent, particularly on the veins beneath or sometimes more or less limited to those; bacterial nodules numerous of various shapes, spot-like, substellate or linear; petiole 0.2–2.5 cm. long, pubescent like the young stems; stipules ovate- triangular, the base c. 5 × 5 mm. with two narrow cusps 3–4 mm. long, grey-pubescent, the pair at the junction of new and old parts of stem darkening, thickening and accrescent to 1 cm. but often missing from other nodes on old stems, rusty pubescent. Inflorescences all terminal, many-flowered, congested, 1.5–2 × 1.5 cm., oblong or ovoid; peduncles 1–2.5 cm. long, grey-pubescent; pedicels 1–3 mm. long. Calyx tube about 1 mm. long, the limb reduced to a narrow truncate or undulate rim. Corolla white; tube 1.5–2.5 mm. long, glabrous outside, white-hairy at the throat; lobes 2–2.5 × 1.5 mm., ovate to oblong. Style 4.5 mm. long in long-styled flowers, the stigmatic lobes scarcely 1 mm. long. Fruit scarlet, ± 5 mm. diam.; peduncles up to 4.5 cm. long; pedicels 5 mm. long; seeds c. 4 × 4 × 2.5 mm. with a ± Y-shaped fissure in the ventral-face but rest of albumen not ruminate and with no air-cavity.

Zambia. N: Mbala Distr., Kasulu Farm, fl. 4.i.1969, *Sanane* 386 (K).
Also in S. Tanzania. Shady places in woodland; 1500–1740 m.
Additional material has shown this is distinct enough to be recognised at specific level.

17. **Psychotria heterosticta** Petit in Bull. Jard. Bot. Brux. **36**: 131, photo. 4/F (1966). —Verdc. in F.T.E.A., Rubiaceae 1: 101 (1976). Type from Zaire (Shaba).

Shrub or subshrub 0.2–3 m. tall, or small tree 6–7.5 m. tall, with glabrous, bifariously pubescent or less often completely pubescent stems. Leaf blades 3.5–19 × 0.5–5.5 cm., elliptic, narrowly elliptic, or oblanceolate to narrowly oblong-lanceolate, acute or shortly acuminate at the apex, cuneate at the base, completely glabrous or shortly pubescent on lower part of main nerve beneath, thin to slightly coriaceous; domatia absent; nodules present, usually linear or spot-like along the main nerve beneath and usually but not always some spot-like ones in the lamina; petiole obsolete or 0.5–2 cm. long, glabrous or pubescent; stipules 2.5–6 mm. long, ovate or ovate-triangular, glabrous or pubescent, divided into 2 lobes at the apex 0.5–3 mm. long, eventually deciduous. Flowers heterostylous, 5-merous, usually congested into umbel-like inflorescences, each branch with a ± head of flowers or in panicles 2–8 cm. long; peduncles often 2-ribbed, 1–9 cm. long, glabrous or pubescent; secondary branches 0.4–1.5 cm. long, glabrous or pubescent; pedicels 0.5–3(5 in fruit) mm. long, glabrous or pubescent; bracts and bracteoles minute. Calyx tube 0.75–1 mm. long, ovoid, glabrous; limb 0.5–0.75 mm. long, cupuliform, lobes irregular, very small or rarely 0.5 mm. long. Corolla white or yellowish green, glabrous outside; tube 2.5–4 mm. long; lobes 2–2.5 × 0.8 mm., elliptic-triangular. Stamens with filaments 1.5–2.5 mm. long in short-styled flowers, 0.3 mm. long in long-styled flowers. Style 2–2.5 mm. long in short-styled flowers, 2.5 mm. long in long-styled flowers; stigma lobes 0.5 mm. long. Drupes red, 5–7 × 6–10 mm., subglobose, glabrous,

* **P. diversinodula** (Verdc.) Verdc. affinis *P. kirkii* Hiern et *P. spithameae* S. Moore ab ambabus nodulis heteromorphis linearibus maculiformibus stellatisque intermixtis et a posteriore lacuna propria seminae carenti differt. Suffrutex 9–25 cm. altus.

with 2 pyrenes; pyrenes semi-globose, similar to the seeds. Seeds 4.2 × 5 × 3 mm., semi-globose; ventral face with a V- or U-shaped median fissure but rest of the albumen not ruminate.

Key to infraspecific variants

Leaf blades less distinctly acuminate; lateral nerves 5–11 pairs; petioles 0–5 mm. long - - - - - - - - - - - - - - - - - var. *heterosticta*
Leaf blades more distinctly and more narrowly acuminate; lateral nerves 8–14 pairs; petioles 0.2–2 cm. long - - - - - - - - - - - - - - - - - var. *plurinervata*

Var. **heterosticta**. —Petit in Bull. Jard. Bot. Brux. **36**: 131 (1966). —Verdc. in F.T.E.A., Rubiaceae 1: 101 (1976).

Mostly entirely glabrous; petioles obsolete or up to 5 mm. long, lateral nerves 5–11 pairs; leaf blades mostly narrowly elliptic or narrowly obovate.

Zambia. N: Kawambwa, fl. 30.i.1957, *Fanshawe* 2980 (K; NDO). W: 27 km. W. of Mwinilunga on Matonchi road, W. of Musangila R., fl. 22.i.1955, *Brummitt et al.* 13971 (K).
Also in Zaire, Burundi and Tanzania. Riparian and streamside *Berlinia–Syzygium* forest and evergreen thicket, *Cryptosepalum–Copaifera* woodland on Kalahari Sand; 1350–?1500 m.

Var. **plurinervata** Petit in Bull. Jard. Bot. Brux. **36**: 133 (1966). —Verdc. in F.T.E.A., Rubiaceae 1: 102 (1976). Type from Tanzania.

Mostly entirely glabrous; petioles 0.2–2 cm. long; lateral nerves 8–14 pairs; leaf blades narrowly elliptic to elliptic, more narrowly and distinctly acuminate than in the other varieties.

Malawi. N: Misuku Hills, Wilindi Forest, fl. 12.i.1959, *Robinson* 3189 (K).
Also in Tanzania. Mist-forest edges; 1600–2000 m.

18. **Psychotria spithamea** S. Moore in Journ. Bot., Lond. **48**: 222 (1910). —De Wild., Pl. Bequaert. **2**: 423 (1924). —Petit in Bull. Jard. Bot. Brux. **36**: 134, t. 8 (1966). —Verdc. in F.T.E.A., Rubiaceae 1: 102 (1976). Type: Zambia, Katenina Hills, *Kassner* 2187 (BM, holotype).

Subshrub with several stems 10–50 cm. tall from a mostly creeping woody rhizome; stems usually only leafy at the upper nodes, glabrous to densely pubescent, often becoming covered with soft corky bark. Leaf blades 3–22 × 0.6–5.4 cm., narrowly elliptic to elliptic but variable and sometimes obovate, elliptic-ovate or linear-oblong, rounded to acute at the apex, narrowly cuneate at the base, glabrous to pubescent above and beneath, sometimes densely so on the venation beneath rendering it almost velvety to the touch, mostly distinctly paler beneath and sometimes whitish, thin to ± coriaceous, sometimes drying yellowish-green; domatia absent; nodules mostly numerous, scattered in the lamina and often, at least when dry, not easy to see; petiole 0.3–2 cm. long, glabrous or pubescent; stipules 3–8 mm. long, ovate-triangular, entire at the apex or bifid for almost 1 mm., either thin and soon deciduous or thick and woody and ± persistent. Flowers heterostylous, 5-merous, in usually ± dense slightly branched inflorescences 1.5–5 cm. long; peduncle 0.5–3 cm. long, pubescent or glabrescent often bent ± at right angles; secondary branches 0.3–1.3 cm. long, similarly hairy; pedicels 1–2 mm. long, attaining 5 mm. in fruit, glabrous or pubescent; bracts small, filiform, ± 3 mm. long. Calyx tube ± 1 mm. long, subglobose, glabrous or pubescent; limb 0.75–1.5 mm. tall, cupuliform, usually glabrous, more rarely densely hairy; lobes unequal, 1–2 mm. long, mostly triangular-lanceolate, usually ± ciliate on the margins, or rarely lobes obsolete. Corolla white or cream, glabrous outside; tube 3–5 mm. long; lobes 2.5–3.5 × 0.8 mm., narrowly oblong. Stamens with filaments 1.5 mm. long in short-styled flowers, 0.1–0.2 mm. long in long-styled flowers. Style 3 mm. long in short-styled flowers, 3.5–6.5 mm. long in long-styled flowers; stigma lobes 0.5–1 mm. long. Drupes red, 3.5–6 × 7–10 mm., didymously biglobose, glabrous or pubescent, sometimes with only 1 pyrene due to abortion; pyrenes 5 × 4 × 4 mm. subglobose. Seeds subglobose, ventral surface with a deep Y- or V-shaped fissure and, save for the area of the junction of the 2 pyrenes, having a spherical air space between the hard outer layers and the albumen, which is not otherwise ruminate.

Zambia. N: Mbala to Kasama, fl. 13.xii.1949, *Bullock* 2095 (K). W: Dobeka Plain, fl. 19.xi.1937, *Milne-Redhead* 3321 (K). C: Lusaka East Forest Reserve, near Leopard's Hill road, fl. 11.i.1973, *Kornaś*

2984 (K; KRA; UZL). E: Lundazi-Mzimba km. 6.4, fr. 28.iv.1952, *White* 2496 (FHO; K). S: Livingstone Distr., Situmpa Forest Reserve, Machili, fr. 13.ii.1952, *Fewdays* 24 (FHO; K). **Malawi**. N: Nyika Plateau, fl. 2.i.1976, *Phillips* 853 (K; MO). C: Kasunga-Bua road, fl. 13.i.1959, *Robson & Jackson* 1135 (BM: K; LISC). **Mozambique**. N: near Mandimba, fr. 2.v.1960, *Leach* 9899 (K; SRGH).

Also in Zaire, Burundi, Tanzania and Angola. *Brachystegia–Uapaca–Julbernardia* and similar woodland, *Parinari* thicket, sometimes also in grassland, often in rocky, granite or quartzite areas; 560–1860 m.

Often difficult to tell apart from *P. kirkii*. Short plants with very short petioles, the tertiary venation quite evident and the air-space in the fruits, are unmistakable, also those with leaves drying yellow-green or white beneath but the two need field study to investigate possible characters particularly in areas where both seem to occur e.g. Zambia W., *Jeke* 160 from Malawi, Dedza is described as a 'small tree' but has the seeds of *P. spithamea*; the possibility of hybridisation needs investigation.

19. **Psychotria kikwitensis** De Wild., Pl. Bequaert. **2**: 377 (1924). —Petit in Bull. Jard. Bot. Brux. **36**: 150, photo. 4/1 (1966). Type from Zaire.

Subshrubby herb from woody rhizome or small shrub, 0.3–1 m. tall; stems bicostate and finely ridged in dry state; glabrous save for some hairs around some of the domatia, stipules and sometimes petioles. Leaves (3)7–20 × (1.3)3.5–8 cm., elliptic, oblong or narrowly oblong-elliptic, acuminate at the apex, cuneate at the base, thinly coriaceous; lateral nerves rather prominent in dried state; bacterial spots few to numerous, very evident, ± 1 mm. in diam., dispersed throughout the lamina; domatia usually present, small, ± pilose; petiole 1–3 cm. long; stipules 5–10 mm. long, ovate-triangular, biacuminate at the apex for 2–5 mm., ± glabrous or mostly with ferruginous hairs on margins and keels or all over and ferruginous pilose within. Flowers heterostylous, 5-merous, in branched panicles 2–6 cm. long, the lower inflorescence-nodes with branches in 4's arranged X-wise; peduncle 2.5–5.5 cm. long, ± narrowly 2-winged; bracts small; pedicels up to 1 mm. long, attaining 1–3 mm. in fruit. Calyx tube ± 0.5 mm. long; limb 0.3–0.4 mm. long, ± truncate or shallowly lobed. Corolla probably white; tube 1–2.5 mm. long; lobes 1–1.5 mm. long; filaments in short-styled flowers 1.7 mm. long, in long-styled flowers 0.2 mm. long. Style 1.7 mm. long in short-styled flowers, c. 3.5 mm. long in long-styled flowers; stigma lobes 0.5–0.7 mm. long. Drupes red, 5–7 mm. in diam., subglobose, with 2 pyrenes; pyrenes similar to the seeds. Seeds 4–4.5 × 3.5–4 × 2–2.5 mm., hemispherical, with U-shaped median fissure on flattened ventral surface but rest of albumen not ruminate.

Zambia. W: 96 km. S. of Mwinilunga on the Kapombo road, fr. 1.vi.1963, *Loveridge* 722 (K; SRGH).

Also in Zaire and Angola. Understorey of *Cryptosepalum* woodland and thicket types in Kalahari Sand woodland; 1260 m.

Petit speaks of glabrous stipules but even some cited material was found to have indumentum.

20. **Psychotria sp. B**

Shrub to 2 m. with distinctly ridged purplish or greyish bark, glabrous, the youngest internodes with dense surface rhaphides as also have the stipules. Leaves mostly distinctly recurved, up to 4.5 × 1.8 cm., narrowly elliptic, acute at apex, cuneate at base; petiole c. 3 mm. long; nodules restricted to junction of petiole and lamina and midrib of basal part of lamina and in some leaves obscure traces of scattered ones; stipules with triangular to linear lobes c. 2 mm. long from a base c. 2.5 mm. long. Inflorescence small, 1–1.5 cm. long and wide, smaller and narrower in bud, branched; peduncle attaining 1.5 cm. in fruit. Calyx tube c. 0.5 mm. long, the limb about 0.7 mm. long with short triangular teeth. Corolla greenish-yellow, 3 mm. long, funnel-shaped; lobes 2 × 0.7 mm. triangular-lanceolate. Stigma exserted about 1 mm. in long-styled flowers. Fruits red, c. 6 × 5 mm., ellipsoid, grooved when dry; pyrenes c. 4 × 4 × 2 mm., ± hemispherical, densely silvery with rhaphides, ventral face plane with a shallow W-shaped groove, the outer convex with 6 ribs.

Mozambique. GI: Inhambane, Tôfo Beach, fl. & fr. 13.x.1968, *Balsinhas* 1376 (LISC).

Only one specimen seen; littoral dunes.

Related to *P. pumila*, *P. heterosticta* and *P. kirkii* but differs from former in its non-umbellate inflorescence and the other two in the nodules and general habit. Further material is required.

2. GEOPHILA D. Don

Geophila D. Don, Prodr. Fl. Nepal.: 136 (1825). —Hepper in Taxon **9**: 88 (1960) nom. conserv.
Carinta W.F. Wight in Contr. U.S. Nat. Herb. **9**: 216 (1905). —G. Taylor in Exell, Cat. Vasc. Pl. S. Tomé, Suppl.: 25 (1956). —L.B. Smith & Downs in Sellowia **7**: 65 (1956).
Geocardia Standley in Contr. U.S. Nat. Herb. **17**: 444 (1914).

Perennial forest floor herbs, mostly with slender creeping stems which root at the nodes and have fibrous roots. Leaves opposite, with mostly long petioles; blades ovate-cordate to rounded-reniform; stipules interpetiolar, ovate, entire or bilobed at the apex. Flowers hermaphrodite, sometimes heterostylous, mostly in terminal umbels or sometimes solitary, often on long peduncles held erect from the main stems, occasionally with an involucre of quite conspicuous bracts. Calyx tube obovoid, the limb short, 5–7-lobed, the lobes subulate or linear, spreading or reflexed, persistent. Corolla cylindrical or funnel-shaped; lobes 4–7, spreading or recurved; throat pilose inside. Stamens 4–7, inserted in the corolla tube; filaments filiform; anthers dorsifixed, included or exserted. Disk swollen. Ovary 2-locular; ovules solitary in each locule, erect from the base, anatropous; style slender, included or exserted; stigma lobes 2, linear, densely papillate or stigma subcapitate-bifid. Drupe fleshy, containing 2 pyrenes; pyrenes plano-convex, dorsally compressed, obtusely ribbed, rugulose and often with an annular area at junction of ventral and dorsal surfaces, 1-seeded. Seeds the same shape as the pyrenes, the ventral surface plane: testa membranous; endosperm corneous.

A genus of about 10–15 species in the tropics of both the Old and New Worlds; 3 species occur in the Flora Zambesiaca area. Other species previously referred to *Geophila* have been accepted as belonging in the genus *Hymenocoleus* recently described by Robbrecht.

1. Inflorescence 1(rarely 2–3)-flowered, the bracts small and never forming an involucre; leaf blades rounded-reniform - - - - - - - - - - - - - - 1. *repens*
– Inflorescences always several-flowered with a distinct involucre made up of separate bracts; leaf blades ovate-cordate or ovate-reniform - - - - - - - - - - - 2
2. Fruits black, purple or blue; involucral bracts mostly glabrous; stipules not persistent; pyrenes with one dorsal keel - - - - - - - - - - - - - - 2. *obvallata*
– Fruits crimson; involucral bracts pubescent; stipules more or less persistent; pyrenes with 2–3 dorsal keels - - - - - - - - - - - - - - - - - 3. sp. *A*

1. **Geophila repens** (L.) I.M. Johnston in Sargentia **8**: 281, 282 (1949). —Brenan in Mem. N.Y. Bot. Gard. **8**: 453 (1954). —Hepper in F.W.T.A. ed. 2, **2**: 205 (1963). —Steyermark in Mem. N.Y. Bot. Gard. **23**: 395 (1972). —Lind & Tallantire, Fl. Pl. Uganda ed. **2**: 164, fig. 106 (1972). —Verdc. in Kew Bull. **28**: 321, fig. 1/2 A-C (1977). —Agnew, Upland Kenya Wild Fl.: 407 (1974). —Verdc. in F.T.E.A., Rubiaceae 1: 110, fig. 7/1,2 (1976). TAB. **4** fig. B. Types from India and Jamaica.
Rondeletia repens L., Syst. Nat. ed. 10, **2**: 928 (1759).
Psychotria herbacea Jacq., Enum. Pl. Carib.: 16 (1760). Type from W. Indies.
Psychotria herbacea L., Sp. Pl. ed. 2: 245 (1762) non Jacq. nom. illegit. Type from Jamaica.
Geophila reniformis D. Don, Prodr. Fl. Nepal: 136 (1825). —Hiern in F.T.A. **3**: 220 (1877). Type from Bangladesh.
Geophila uniflora Hiern in F.T.A. **3**: 221 (1877). —F. Hallé, Ic. Pl. Afr. 7, No. 156 (1965). Types from Nigeria and Sudan.
Geophila herbacea (L.) K. Schum. in Engl. & Prantl, Pflanzenfam. **4**: 119 (1891). —S. Moore in Fl. Jam. **7**: 111, fig. 32 (1936) (as (Jacq.) K. Schum.). —Bremek. in Pulle, Fl. Suriname **4**: 234 (1934).
Carinta herbacea (Jacq.) W.F. Wight in Contrib. U.S. Nat. Herb. **9**: 216 (1905).
Geocardia herbacea (L.) Standley in Contrib. U.S. Nat. Herb. **17**: 444 (1914).
Carinta uniflora (Hiern) G. Taylor in Exell, Cat. Vasc. Pl. S. Tomé Suppl.: 25 (1956).
Carinta repens (L.) L.B. Smith & Downs in Sellowia **7**: 88 (1956).
Geocardia repens (L.) Bakh.f. in Backer, Beknopte Fl. Java **15**, fam. 173: 144 (1956).

Creeping herb with stems 20–30 cm. long, densely appressed pubescent, rooting at the nodes; leafy and flowering shoots reaching a height of 3–7.5 cm. above ground-level. Leaf blades 1.2–3.6(5) × 1.3–4.3(5.5) cm., rounded-reniform, very rounded at the apex, emarginate-cordate at the base, glabrous above, glabrous or pubescent beneath; petiole 0.4–11.5 cm. long, appressed or spreading pubescent; stipules transversely elliptic, almost truncate, 1.5 × 2-3.5 mm. Flowers not truly heterostylous, solitary, or more rarely up to 2 or even 3–5 in American material; peduncle 0.5–4.3 cm. long, pubescent; bracts 1–2, lanceolate, 2.5–4 mm. long. Calyx tube 1.2–2 mm. long, obconic, pubescent; limb tube 0.8–1.2 mm. long; lobes 1.5–3 × 0.5–0.7 mm. at base, lanceolate. Corolla white; tube 0.5–1.3

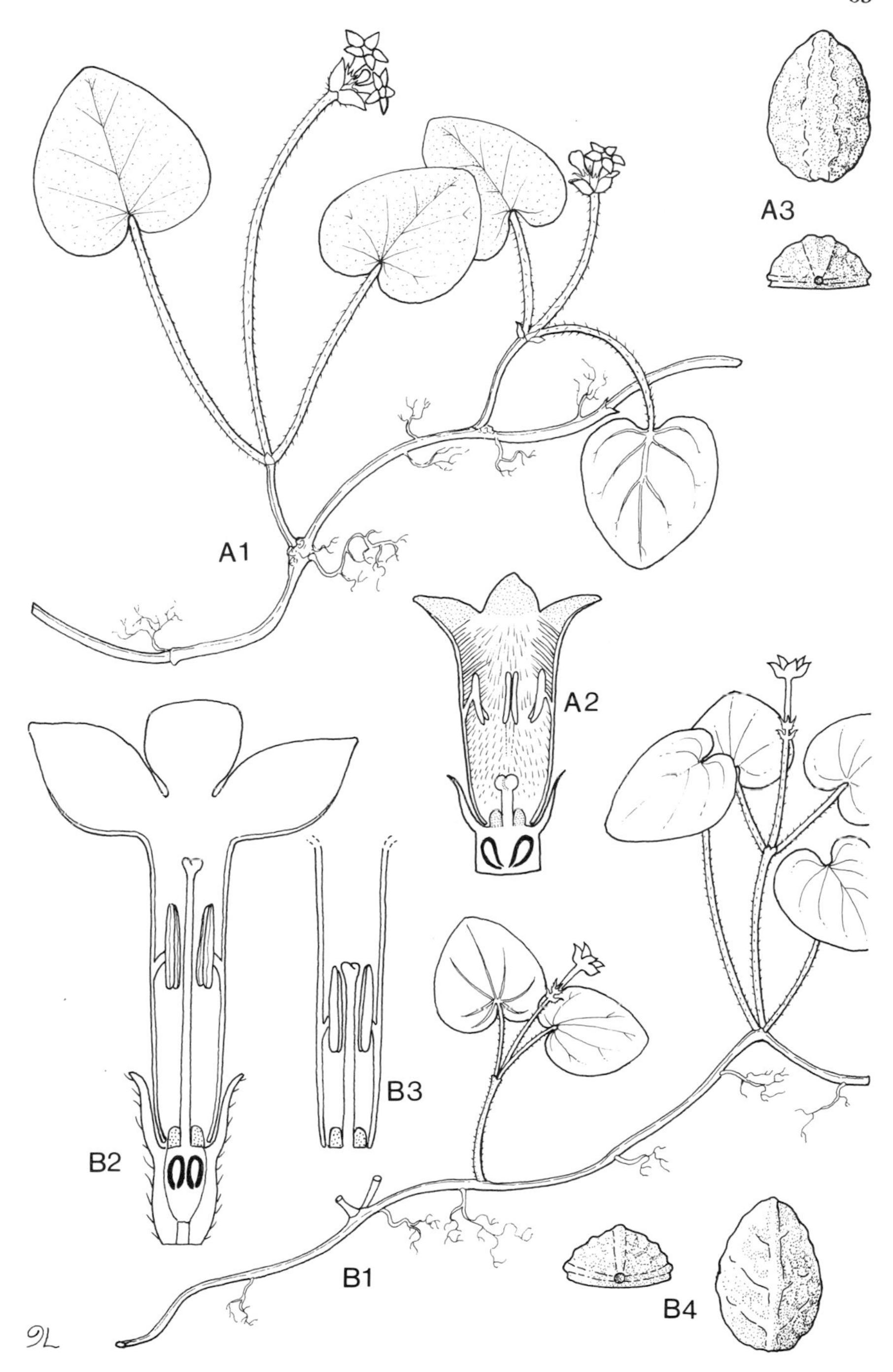

Tab. 4. A. — GEOPHILA OBVALLATA subsp. IOIDES. A1, habit (× $\frac{2}{3}$); A2, flower, longitudinal section (× 8), A1–2 from *Pawek* 1686; A3, pyrene, dorsal and lateral views (× 5), from *Richards* 1513. B. — GEOPHILA REPENS. B1, habit (× $\frac{2}{3}$); B2, flower, longitudinal section (× 4); B3, flower, longitudinal section showing limited heterostyly (× 4), B1–3 from *Biegel* 2822; B4, pyrene, dorsal and lateral views (× 5), from *Chapman* 6092.

cm. long, cylindrical, finely pubescent; lobes 4–9 × 2–5 mm., oblong-elliptic; basal part of limb funnel-shaped. Anthers and style included, the latter 3.5–7 mm. long; stigma 0.25–0.5 mm. long, either level with or reaching beyond the anthers. Berries globose, bright red or orange, glossy, glabrous, 0.5–1.2 cm. in diameter, crowned by the persistent lobes. Pyrenes greyish or straw-coloured, 3.5–4.1 × 3–3.5 mm. wide, semi-ovoid, dorsal side convex, rugose, ventral side flat, rugose, with a narrow impressed smooth annular area where the 2 sides join. Seeds with chestnut-coloured testa easily removable to show the white endosperm, 3 × 2.8 × 1 mm., lenticular, smooth.

Zimbabwe. E: Chimanimani, Haroni-Makurupini forest, fr. 26.v.1969, *Müller* 998 (K; SRGH). **Malawi**. S: Chiradzulu Forest Reserve, Lisau Saddle, fr. 14.i.1982, *Chapman & Patel* 6092 (K; MAL). **Mozambique**. Z: Serra da Morrumbala, fr. 30.iv.1943, *Torre* 5236 (C; COI; LISC; LMU). MS: Dombe, fr. 31.vii.1945, *Simão* 409 (LISC).

Pantropical; in Africa from Guinea Bissau to Angola and Zimbabwe, also in Madagascar. Evergreen forest floors; 300-1140 m.

This species is fairly uniform in the Old World but New World specimens often have up to 5-flowered (and only rarely 1-flowered) inflorescences and hence a rather different appearance. Possibly the Old World material could form a subspecies with a name based on *G. reniformis*. Müller mentions that the fruits turn black but no other reference to this has been found. I am unable to agree with Hepper that *G. lancistipula* Hiern (in F.T.A. **3**: 221 (1877); F.W.T.A. **2**: 128 (1931) type from Gabon?.) is a synonym of *G. repens;* it differs in having dark blue hairy fruits, pyrenes of a very different shape and somewhat differently shaped leaf blades.

2. **Geophila obvallata** (Schumach.) F. Didr. in Vidensk. Meddel. Dansk. Naturhist. Foren. Kjøbenh. **1854**: 186 (1855). —Hiern in F.T.A. **3**: 222 (1877). —Hepper in F.W.T.A. ed. 2, **2**: 206, fig. 243/A (1963). —F. Hallé, Ic. Pl. Afr. 7, No 155 (1965). —Verdc. in Kew Bull. **30**: 265 (1975); in F.T.E.A., Rubiaceae 1: 112 (1976). Type from Ghana.

Psychotria obvallata Schumach., Beskr. Guin. Pl.: 111 (1827).

Carinta obvallata (Schumach.) G. Taylor in Exell, Cat. Vasc. Pl. S. Tomé, Suppl.: 25 (1956).

Creeping herb with prostrate often underground stems (10)30–60 cm. long, glabrous, rooting at the nodes and often forming carpets up to 5 × 5 m. across. Leaf blades 0.8–4(9) × 0.6–4(5.5) cm. triangular-ovate, ovate or ovate-reniform, acute to rounded at the apex, cordate at the base, glabrous above, glabrous or with some pubescence at the sides of the midnerve beneath; petiole 0.5–9.5 cm. long, often with lines of short hairs above or ± densely pubescent above at apex; stipules transversely elliptic, 1–2(3) mm. long, not bifid. Inflorescences 0.5–1.1 cm. across, several-flowered, enclosed in a whorl of bracts; peduncles 1.3–5(9) cm. long, glabrous, finely papillate-pubescent or densely hairy (see note); bracts leafy, 0.35–1.5 × 0.35-1.1 cm., obovate, rounded-elliptic or rhomboid; flowers not heterostylous. Calyx glabrous or rarely hairy; tube and pedicel 1.5–2.5 mm. long; limb tube 0.2–1.5 mm. long; lobes 0.7–6.6 mm. long and up to 1.3 mm. wide, subulate, linear-lanceolate, narrowly triangular or distinctly spathulate; disk 0.3–1.1 mm. tall. Corolla white; tube 3.1–6.5 mm. long, funnel-shaped, glabrous or puberulous outside; lobes 1.2–3 × 0.6–2 mm. ovate-oblong, sometimes shortly joined at the base. Anthers situated near centre of the tube. Style 0.3–1 mm. long, widened above; stigma ± capitate, bifid, 0.3–0.5 mm. long. Berries black, purple or blue, (4)7–8 mm. long and wide, crowned with the persistent calyx lobes; pyrenes dull yellowish brown, 4–4.7 × 3.5–3.7 × 1.7–2.3 mm., semi-ovoid, the ventral surface fairly flat but with 2 depressed areas, bounded by the raised margin and raised median area, dorsal surface rugose, with a longitudinal median keel and a smooth depressed annular area where the dorsal and ventral surfaces meet. Seeds brown, c. 3 mm. × c. 2.6 mm. × c. 1 mm., ± lenticular, dorsally rounded, ventrally flattened, smooth.

Subsp. **ioides** (*K. Schum.*) *Verdc.* in Kew Bull. **30**: 267 (1975); in F.T.E.A., Rubiaceae 1: 113, fig. 7/3.5 (1976). TAB. **4**, fig. A. Type: Mozambique, Quelimane, *Stuhlmann* 711 (B, holotype †).

Geophila ioides K. Schum. in Pflanzenw. Ost-Afr. **C**: 392 (1895).

Geophila cecilae N.E. Br. in Kew Bull. **1906**: 107 (1906). Type: Mozambique, Beira, Dondo, *Cecil* 254 (K, holotype).

Carinta ioides (K. Schum.) Garcia in Mem. Junta Invest. Ultramar, sér. 2, **6**: 44 (1959).

Leaf blades circular-reniform to ovate. Calyx lobes linear or linear-subulate or very slightly spathulate at the apex, 1.4–3.4 mm. long. Corolla tube 3–4 mm. long.

Zambia. N: Mbala Distr., Chilongowelo, fl. 28.i.1955, *Richards* 4264 (K). W: Mwinilunga Distr., just NE. of Dobeka Bridge, fl. 11.xii.1937, *Milne-Redhead* 3606 (K). C: Mkushi Distr., Great North Road, Kashitu R., fl. 3.ii.1973, *Kornaś* 3130 (K: KRA). **Zimbabwe**. E: Chimanimani, between Haroni and Makurupini Forests, fl. 7.i.1969, *Biegel* 2750 (K; LISC; SRGH). **Malawi**. N: Nkhata Bay Distr., 33.6 km. S. on Chinteche road, fl. 2.ii.1969, *Pawek* 1686 (K). **Mozambique**. N: Serra de Ribáuè, Mebalué, fl. 23.i.1964, *Torre & Paiva* 10174 (C; LISC: LMU; SRGH). Z: Lugela, Tacuane, fl. late i, *Faulkner* s.n. (K). MS: Dondo, fl. 23.iii.1960, *Wild & Leach* 5206 (K; SRGH). M: Macocololo, fr. 19.i.1898, *Schlechter* 12056 (BM; K).

Kenya, Tanzania, Burundi, Zaire and Angola. Sandy and grassy places, in bushland, *Brachystegia* and *Cryptosepalum* woodlands, evergreen forest floors; (30)350–1740 m.

The material from all areas save Mozambique having less spreading stems, shorter calyx lobes and mostly very hairy peduncles was suggested by me (F.T.E.A., Rubiaceae 1: 113 (1976)) as possibly a distinct subspecies but I have preferred to take a wide view. The species as a whole occurs from Guinea Bissau to Angola, Zaire, Sudan, Central African Republic, East and South-central Africa with subsp. *involucrata* (Hiern) Verdc. in Sudan, Uganda, Zaire and perhaps Angola and subsp. *obvallata* in W. Africa. The calyx lobe characters are too inconstant to maintain the three reported species formerly always accepted.

3. **Geophila sp. A**

Prostrate herb with underground stems rooting at the nodes, densely pubescent. Leaves thin, 2–6.5 × 2–7.5 cm., rounded-cordate to rounded-reniform, rounded at apex, cordate at the base, glabrescent to pubescent on both surfaces and rhaphides conspicuous; petioles densely pubescent, 3–5.5 cm. long; stipules papery, 3 × 6 mm., triangular, persistent at least at upper nodes. Flowering specimens not known. Fruiting peduncles 1.5–4 cm. long, densely pubescent; bracts 7–10 × 1–8 mm., ovate to rounded or some lanceolate, rounded to acute, usually pubescent. Fruits crimson, up to 9 × 9 mm., 'oval or oblong-globose', fleshy, slightly pubescent. Pyrenes ± yellow, 3.5–4 × 3–3.5 mm., semi-globose to semi-ellipsoid, dorsally (2)3 keeled with ± rugose areas between the keels and margins, the margin smooth; ventral surface with median keel or line and rugose area on each side.

Zambia. W: Ndola, fr. 3.vi.1962, *Fanshawe* 6855 (K; NDO).

Dry evergreen forest, *Erythrophleum suaveolens*; 1370 m.

Although superficially resembling *Geophila obvallata* this differs in a number of ways and appears to be more closely related to *Geophila afzelii* Hiern resembling it in the red fruits, persistent stipules and pyrenes. In fact it might represent a population of this species at the extreme edge of its range. Without adequate flowering material I am unable to make a decision. The flowers are said to be yellow.

3. CHAZALIELLA Petit & Verdc.

Chazaliella Petit & Verdc. in Kew Bull. **30**: 268 (1975). —Verdc. in Kew Bull. **31**: 785 (1977).

Shrubs, often flowering before the leaves are fully developed; stems usually 2-ribbed, all but the youngest shoots usually covered with pale brown soft cork. Leaves opposite or in whorls of 3–4, mostly drying pale, petiolate or subsessile, usually deciduous; nodules absent; domatia small, white-pubescent or absent; stipules mostly short, ovate or triangular, entire, bifid or sometimes with a few teeth. Flowers small, heterostylous, (4)5(6)-merous, hermaphrodite, in mostly small, sessile or pedunculate inflorescences, often in small heads, or sometimes paniculate, ± sessile or pedicellate; bracts and bracteoles very small or absent. Calyx limb usually short, truncate or toothed. Corolla mostly yellow or white; tube shortly cylindrical, hairy at the throat; lobes triangular to elliptic-lanceolate, always valvate. Stamens with long filaments in short-styled flowers and shorter filaments in long-styled flowers, the anthers included or only the tips exserted. Ovary 2-locular, each locule containing a single erect ovule; style filiform, usually papillate-pubescent, divided into 2 thick stigma lobes. Fruit a drupe with 2 pyrenes; pyrenes ± flat ventrally, mostly ± 3-ribbed or 3-lobed dorsally, opening by 2 slits extending along the margins of the ventral face for about half its length. Seeds pale; endosperm not ruminate.

A small genus of 24 species restricted to tropical Africa and previously always included in *Psychotria*, but equally related to *Chassalia*. I had intended to treat the taxon as a subgenus of *Psychotria*, but Petit's detailed revision of the African species of *Psychotria* convinced him that the group of species involved merited generic rank. Only one species occurs in the Flora area.

Chazaliella abrupta (Hiern) Petit & Verdc. in Kew Bull. **30**: 268 (1975). —Verdc. in F.T.E.A., Rubiaceae 1: 117 (1976); in Kew Bull. **31**: 811 (1977). TAB. **5**. Types: Mozambique, Shiramba Dembe & Chigogo [Shigogo], *Kirk* (K, syntypes).

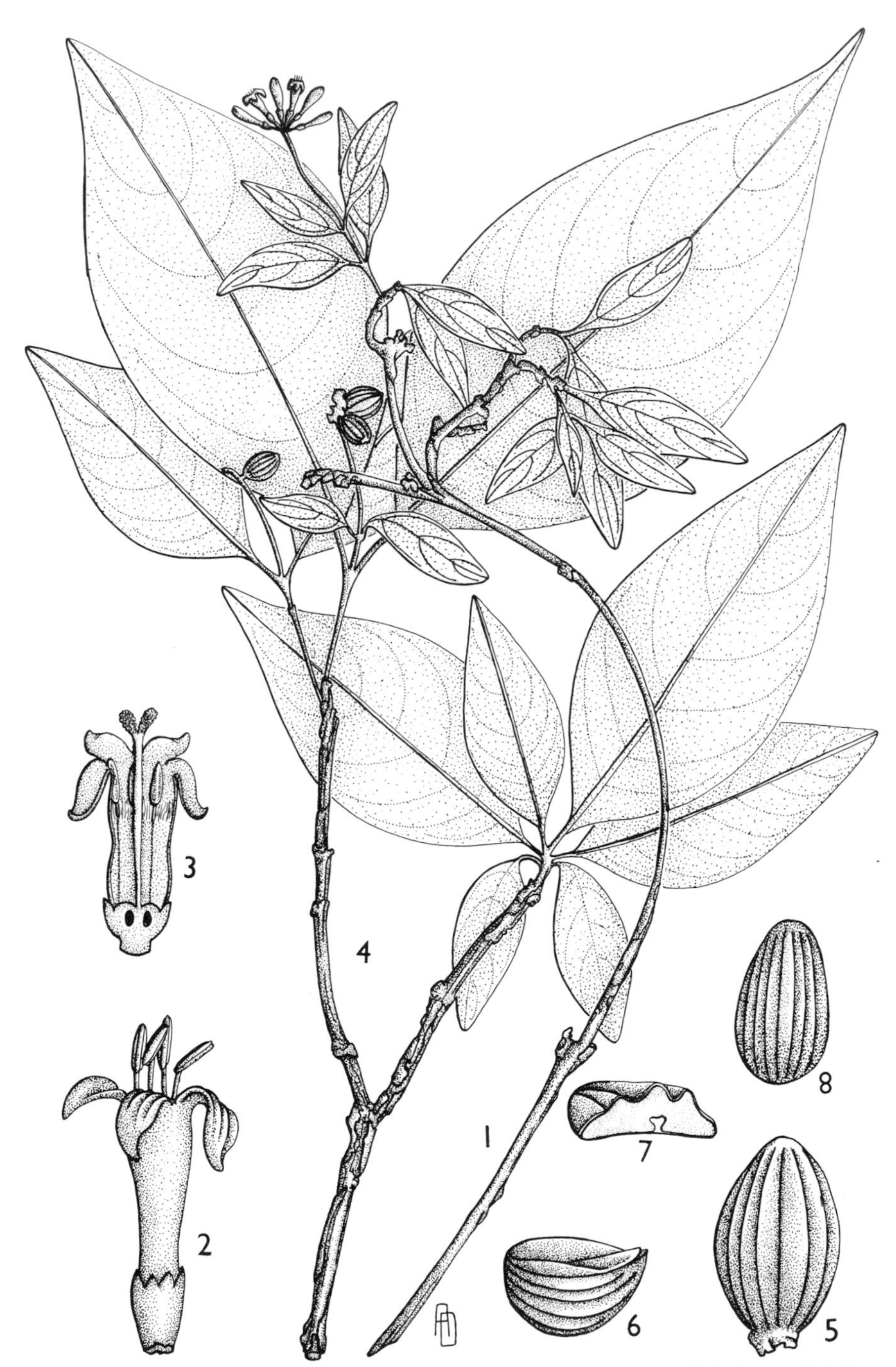

Tab. 5. CHAZALIELLA ABRUPTA var. ABRUPTA. 1, flowering branch (× ⅔); 2, short-styled flower (× 6), 1–2 from *Renvoize & Abdallah* 1486; 3, long-styled flower, longitudinal section (× 6), from *Tanner* 2743; 4, fruiting branch (×⅔); 5, drupe (× 3⅓); 6, pyrene (× 2⅔); 7, section of pyrene (× 4); 8, seed (× 4), 5–8 from *Drummond & Hemsley* 1892. From F.T.E.A.

Psychotria abrupta Hiern in F.T.A. **3**: 205 (1877). —Brenan, T.T.C.L.: 522 (1949). —F. White, F.F.N.R.: 417 (1962).

Psychotria coaetanea K. Schum. in Pflanzenw. Ost-Afr. **C**: 391 (1895). —Brenan, T.T.C.L.: 523 (1949). Type from Tanzania.

Small shrub (rarely a small tree?) 0.6–4.5 m. tall, with pallid (mostly greyish or white) thinly cork-covered stems; youngest parts green, the internodes with 2 longitudinal keels, glabrous, glabrescent or pubescent. Leaf blades 0.8–17.5(22) × 0.4–7.5(10) cm., elliptic to ovate-lanceolate, acute to acuminate at the apex, cuneate at the base, usually rather undeveloped or even not present at the flowering stage and fully expanding in the fruiting stage, glabrous or finely pubescent, pale green but often with a bronze tinge beneath; nodules absent; domatia reduced to small white tufts or quite absent; petiole 1–4(30) mm. long; stipules 2 mm. long, ovate or triangular, obtuse, acute, shortly bifid or even with several teeth. Flowers (4)5(6)-merous, in small 6–20-flowered heads, or sometimes with a few flowers beneath the heads, up to 8 mm. in diameter; peduncle 0.7–3.5 cm. long, ± papillate, finely pubescent or glabrous; pedicels 0.5–1.2 mm. long; bracts minute. Calyx glabrous to pubescent; tube 1 mm. long, oblong-conic; limb very shallow, 0.5 mm. long, truncate or lobes obsolete to distinctly triangular, 1 mm. long. Corolla bright yellow or white, glabrous or very slightly papillate-pubescent outside; tube 2.8 mm. long; lobes 1.5 × 1–1.3 mm., slightly cucullate at the apex. Stamens with filaments 1 mm. long in long-styled flowers, 2–3 mm. long in short-styled flowers. Style 1.8 mm. long in short-styled flowers, 3.5 mm. long in long-styled flowers; stigma lobes thick, 0.8–1.2 mm. long. Drupes 6–9.5 × 4–6.5 mm. ellipsoid, glabrous or slightly pubescent, slightly ribbed in the dry state; pyrenes pale, 6–7 × 4–4.5 × 0.8–2.5 mm., semi-ellipsoidal. Seeds pale brown, 5 × 4–4.5 × 1–1.2 mm., compressed semi-ellipsoid.

Var. **abrupta**. —Verdc. in Kew Bull. **30**: 268 (1971); in F.T.E.A., Rubiaceae 1: 118, fig. 8 (1976); in Kew Bull. **31**: 811, map 1, fig. 1H-K (1977). —Gonçalves in Garcia de Orta, Sér. Bot., **5**: 190 (1982). TAB. **5**.

Psychotria madandensis S. Moore in Journ. Linn. Soc. Bot. **40**: 100 (1911). —De Wild., Pl. Bequaert. **2**: 386 (1924). Type: Mozambique, Manica e Sofala, Madanda Forest, *Swynnerton* 561 (BM, holotype; K).

Mature leaf blades attaining maximum dimensions indicated, very thin, usually narrowly acuminate.

Zambia. N: Musesha, fl. 8.x.1958, *Fanshawe* 4876 (K; NDO). W: Mwinilunga, R. Lunga, fl. 29.xi.1937, *Milne-Redhead* 3432 (K). C: Feira, fl. 5.xii.1968, *Fanshawe* 10476 (K; NDO). E: Petauke, Luangwa R., Beit Bridge, st. 17.iv.1952, *White* 2699 (FHO; K). **Zimbabwe**. N: Gokwe, Copper Queen Native Purchase Area, Muradzi, fl. 20.xii.1963, *Bingham* 976 (K; LISC; SRGH). E: Chipinge, E. escarpment of Sabi R. near Mwangazi Gap, fr. 29.i.1975, *Pope* et al. 1434 (K; SRGH). S: Ndanga, 3.2 km. N. of Chipinda Pools, fl. 17.xii.1959, *Goodier* 28 (K; SRGH). **Malawi**. C: Dedza, Mua, north of Sosola Rest House, fl. 6.i.1965, *Banda* 617 (K; SRGH). S: Chikwawa, Lengwe National Park, fl. 13.xii.1970, *Hall-Martin* 1102 (K; SRGH). **Mozambique**. N: near Malema R., fl. xi.1931, *Gomes e Sousa* 787 (K). Z: Manganja da Costa, forest of Gobene, 43 km from Vila de Manganja, fl. 10.i.1968, *Torre & Correia* 17073 (LISC; LMA; WAG). T: Cabora Bassa, R. Zambezi, fr.ii.1973, *Torre et al.* 19079 (C; LISC; LMU; SRGH; WAG). MS: 5 km. E. of Inhamitanga on road to Lacerdonia, fl. 4.xii.1971, *Müller & Pope* 1869 (K; LISC; SRGH). GI: Manjacaze, environs of Mata de 'Mecrusse', fr. 26.i.1941, *Torre* 2528 (C; COI; J; LISC; LMA; M; SRGH; WAG).

SW. Ethiopia, Kenya, Tanzania (incl. Zanzibar I.), probably also in Angola. Woodland and thicket often on Kalahari Sand, sometimes on termite mounds and base of rocky hills; relict patches of semi-deciduous forest and mixed evergreen forest; 20–1350 m.

Var. **parvifolia** Verdc. in Kew Bull. **30**: 268 (1975); in F.T.E.A., Rubiaceae 1: 117 (1976); in Kew Bull. **31**: 813 (1977). Type from Kenya.

Mature leaf blades mostly very small but up to 4 × 1.7 cm. rather thick, mostly blunt or subacute.

Mozambique. MS: Maringa road, 40 km, from Mt. Espungabera, fl. 22.xi.1960, *Leach & Chase* 10517 (K; SRGH). GI: Gaza, between Chibuto and Alto Changane, fl. 14.xi.1957, *Barbosa & Lemos* 8117 (K; LMA).

Also in Kenya. Sandy hills, open *Brachystegia* woodland; ?–450 m.

Much of the material from Mwinilunga has acuminate small leaves but mostly I think undeveloped. The variation of this very variable species needs field study over a wide area. I have not been able to separate geographical subspecies.

4. CHASSALIA Poir.

Chassalia Poir., Encycl. Méth. Bot., Suppl. **2**: 450, in obs. (1812). —Verdc. in Kew Bull. **30**: 270 (1975).
Chasalia ou *Chassalia* Poir. in Dict. Sci. Nat. **8**: 198 (1817).
Chasallia Juss. in Mém. Mus. Paris **6**: 379 (1820).
Chasalia DC., Prodr. **4**: 431 (1830).
Chasallia A. Rich. in Mém. Soc. Hist. Nat. Paris **5**: 166-7 (1834)
Chazalia DC., Prodr. **9**: 32 (1845). —Petit in Bull. Jard. Bot. Brux. **29**: 378 (1959); **34**: 20 (1964), attributed to A.L. Juss.*

Shrubs or less often small trees or subshrubby herbs, with mostly glabrous or only finely pubescent stems. Leaves opposite or rarely ternate, mostly acuminate, usually quite thin, shortly to distinctly petiolate, usually glabrous; stipules interpetiolar, ovate to triangular or quite short and broad, sometimes united into a small sheath, entire or with 2 short fimbriae, often with colleters and hairs within the base, mostly persistent. Flowers hermaphrodite, 4–5-merous, heterostylous, mostly small, in branched panicles, the ultimate elements usually being small heads but in some few species the flowers are pedicellate; bracts small. Calyx tube mostly ovoid or oblong, ± ribbed, the free limb mostly very short, lobes triangular or linear but mostly very short. Buds often winged. Corolla white, pink or purple, sometimes yellow inside; tube cylindrical, hairy or glabrous inside; lobes often winged; venation of corolla often curiously prominent in dry material. Stamens included or exserted. Disk cylindrical, distinct. Ovary 2-locular; ovules solitary in each locule, erect from the base; style included or exserted; stigma lobes linear. Fruits succulent, with 2 pyrenes; pyrenes pale, semi-globose or semi-ellipsoid, the ventral surface often grooved, often with a median dorsal keel along which dehiscence takes place. Seeds concavo-convex, with a pale testa; endosperm not ruminate.

A genus of about 40–50 species, mostly in tropical Africa and Madagascar, but with a few species in China, India, Burma, Sri Lanka, Malay Peninsular and Malay Is., extending to the Philippine Is. Closely allied to *Psychotria* but with a distinctly different facies, the buds often characteristically winged and the pyrenes with distinct median dorsal dehiscence. Only two species extend into the Flora area.

Corolla tube 1.5–2 cm. long; wings of buds, corolla tube and lobes very marked;
lowland species - - - - - - - - - - - - - - - 1. *umbraticola*
Corolla tube under 1 cm. long; wings of buds, corolla tube and lobes not or scarcely developed save in some populations, perhaps subspecifically distinct; upland species 2. *parvifolia*

1. **Chassalia umbraticola** Vatke in Oest. Bot. Zeitschr. **25**: 230 (1875). —K. Schum. in Pflanzenw. Ost-Afr. **C**: 392 (1895). —Brenan, T.T.C.L.: 490 (1949). —Dale & Greenway, Kenya Trees & Shrubs: 435 (1961). —Verdc. in Kew Bull. **30**: 274 (1975); in F.T.E.A., Rubiaceae 1: 123, fig. 9 (1976). Type from Zanzibar I.
Psychotria zanguebarica Hiern in F.T.A. **3**: 214 (1877). Types from Tanzania (including one from Rovuma Bay, possibly in Mozambique, *Kirk* (K, syntype).
Uragoga zanguebarica (Hiern) O. Kuntze, Rev. Gen. Pl.: 963 (1891) (as "*zangebarica*").
Psychotria umbraticola Williams, Useful Orn. Pl. Zanzibar: 425 (1949) nom. invalid.

Shrub or more rarely a small subshrubby herb, 0.12–4.5 m. tall, rarely stated to be slightly scandent; older stems pale greyish and glabrous, younger glabrous or papillate-pubescent. Leaf blades 3–16 × 1.35–6.3 cm., elliptic, acute or slightly acuminate at the apex but distinctly less so than in other species, cuneate at the base, the blade ± decurrent so that the petiole length is not clear cut, ± discolorous when dry, the margins revolute, glabrous; petioles 0.5–2(2.5) cm. long, usually short and the lower ones distinctly shorter than in related species; stipules broad, 1–1.5 mm. long, undivided, with fine marginal hairs and longer hairs within, sometimes becoming corky. Flowers sweet-scented, in terminal branched inflorescences, the ultimate components 3–several-flowered clusters; inflorescence-components white, tinged purple; peduncles 0.5–2.5(4) cm. long; secondary peduncles 0.3–1.3 cm. long; pedicels actually absent or very short but in reduced inflorescences may appear 1.5–4 mm.; all parts usually finely papillate-puberulous or rarely glabrous; bracts small, but sometimes part of the inflorescence is subtended by a pair of reduced leaves. Calyx cream with purple upper margin; tube 1.2–2

* Petit has accepted this spelling considering that De Candolle had made a valid correction of the spelling, the genus being named after Chazal de Chamarel.

Tab. 6. CHASSALIA PARVIFOLIA. 1, habit (× $\frac{2}{3}$); 2, node showing stipule (× 2); 3, part of inflorescence of long-styled flowers (× 4); 4, calyx (× 8); 5, corolla of long-styled flowers, longitudinal section (× 4); 6, stigma lobes (× 8); 7, ovary, longitudinal section (× 10), 1–7 from *Robson* 561; 8, short-styled flower (× 4); 9, short-styled flower, longitudinal section (× 4), 8–9 from *Pawek* 6177; 10, fruiting branch (× $\frac{2}{3}$); 11, fruit (× 4); 12, two views of pyrene (× 4), 10–12 from *Phillips* 2861.

mm. long, oblong-ovoid, finely papillate-puberulous or rarely glabrous, slightly ribbed; limb tube 0.2–0.5 mm. long; lobes 0.1–0.5 mm long. triangular. Buds distinctly winged in limb portion. Corolla cream or white, often tinged purple sometimes at the base of the tube and tips of the lobes, glabrous or finely puberulous; tube 1.5–2 cm. long; lobes 5–7 × 1–2.5 mm., linear-oblong, conspicuously winged, the wings decurrent on the tube, the venation often curiously raised and prominent in dry material, particularly if flowers were picked in a fading state. Filaments with anthers just completely exserted in short-styled flowers; anthers with tips 3 mm. below the throat in long-styled flowers. Style 10 mm. long in short-styled flowers, with stigma lobes linear, 3 mm. long, flattened, just included; c. 1.6–1.7 cm. in long-styled flowers, with stigma lobes linear, c. 3 mm. long, flattened. Fruits black, subglobose or rounded-ovoid, ± compressed, 4–5(7) mm. long and wide, ribbed, mostly distinctly densely rugulose in the dry state; disk ± persistent; pyrenes pale, 4–5 × 3.8–5 × 1.8–2.5 mm. half-ovoid, vaguely tuberculate to densely covered with pointed rugae. Seeds concavo-convex, basin-shaped.

Mozambique. N: Messalo River Mouth, fl. i.1912, *Allen* 152 (K).
Eastern parts of Kenya and Tanzania (incl. Zanzibar & Pemba Is.) ?Coastal bushland; ? m. (lowland).
I erected a subsp. *geophila* (Kew Bull. **30**: 274 (1975), type from S. Tanzania) for a small subshrubby herb up to 15 cm. tall but K. Vollesen suggests that this is only a response to burning and doubts its validity.

2. **Chassalia parvifolia** K. Schum. in Engl., Bot. Jahrb. **28**: 103 (1899). —Brenan, T.T.C.L.: 490 (1949). —Verdc. in Kew Bull. **30**: 278 (1975); in F.T.E.A., Rubiaceae 1: 130 (1976). TAB. **6**.Type from Tanzania.
Psychotria engleri K. Krause in Engl., Bot. Jahrb. **43**: 153 (1909). —Brenan, T.T.C.L.: 524 (1949). Types from Tanzania.
Psychotria parvifolia (K. Schum.) De Wild., Pl. Bequaert. **2**: 400 (1924) non Oerst. (1852).
Psychotria sp. ? —Brenan in Mem. N.Y. Bot. Gard. **8**: 453 (1954).

Much-branched bushy shrub to small tree 2–4.5(7.5) m. tall, with 2-ribbed or ridged stems, glabrous, soon becoming grey and corky. Leaf blades 1–13 × 0.5–5.2 cm., elliptic to obovate-oblanceolate or elliptic-oblong, abruptly acuminate at the apex, cuneate at the base, glabrous, thin, the margins often ± crinkly in the dry state; petiole 0.2–1.1(1.5) (exceptionally to 4) cm. long; stipules 1–2 mm. long, triangular or short and very rounded, either without cusps or with 2 rather obscure separated cusps. Flowers scented, in trichotomous or much branched ± small inflorescences, the ultimate elements being subcapitate; primary peduncles 0.6–5 cm. long; secondary peduncles 0.3–3 cm. long; pedicels obsolete. Calyx tube 0.8 mm. long, squarish; limb tube c. 0.25 mm. long; lobes 0.25 mm. long, ovate-triangular. Buds not or scarcely winged. Corolla white or greenish-white, often tinged or tipped with pink; tube 4–6(7) mm. long, widened above; lobes 2 × 1.2 mm., ovate. Filaments distinctly exserted in short-styled flowers, the bases of the anthers 0.5–1 mm. above the throat; style 1.4 mm. long in short-styled flowers, stigma lobes 0.6 mm. long. Anthers included in long-styled flowers, the tips 1 mm. below the throat; style exserted 2 mm. in long-styled flowers, the stigma lobes 1.5–2 mm. long. Fruit translucent greenish yellow, pink or shiny black according to field notes, 4.5–5 × 3–3.5 mm. ovoid or ellipsoid, grooved between the pyrenes; pyrenes pale, 4–5 × 3.8 × 2 mm., semi-ellipsoid, with short median dorsal keel at base and shallow ventral grooves.

Zambia. N: Isoka, Mafingi Mts., fl. 21.xi.1952, *White* 3739 (FHO; K). W: Mwinilunga, fl. 6.ix.1955, *Holmes* 1173 (K). E. Lundazi, Nyika Plateau, upper slopes of Kangampande Mt., fr. 6.v.1952, *White* 2729 (FHO; K). **Malawi**. N: Mafinga Mts., above Chisenga, fl. 11.xi.1958, *Robson* 561 (BM; K; LISC). S: Mulanje, Boma Cottage, fr. 8.v.1963, *Wild* 6176 (LISC; SRGH) (cf.). **Mozambique**. Z: Gurue Mts., Estação Pecuária, fr. 27.vii.1979, *de Koning* 7392 (K; LMU).
Also in Kenya & Tanzania. Montane forest including *Podocarpus–Macaranga–Neoboutonia*, sometimes riparian; 1280–2220 m.
I have taken a wide view of this species but material from the Flora Area has longer peduncles and mostly larger leaves than East African populations. In S. Malawi and E. Zimbabwe similar plants are found but the buds are more winged and the inflorescence branches white; moreover the localities are at much lower altitudes. Material so far available is inadequate for any decision. Specimens are cited below.

Var.?

Zimbabwe. E: Chimanimani, Haroni–Makurupini Forest, bud 27.v.1969, *Müller* 1052 (K; SRGH). **Malawi**. S: Mt. Mulanje, The Crater, Mulenza R., fr. 8.v.1980, *Blackmore* et al. 1528 (K; MAL).

Tab. 7. GAERTNERA PANICULATA. 1, habit (× ½); 2, stipule (× ¾); 3, flower (× 3); 4, short-styled flower, longitudinal section (× 6); 5, long-styled flower, longitudinal section (× 6); 6, fruits (× 3); 7, seed (× 6), all from *Loveridge* 882.

Mozambique. MS: S. foothills of Chimanimani Mts., Makurupini Falls, young fr. 25.xi.1967, *Simon & Ngoni* 1307 (K; LISC; SRGH).
Evergreen forest; 390–900 m.

Tribe 2. **GAERTNEREAE**

5. **GAERTNERA** Lam.

Gaertnera Lam., Illustr. Gen. **2**: 273 (1792). —Verdc., Fl. des Mascareignes, Rubiaceae: in press (1989) nom. conserv.

Usually glabrous evergreen trees and shrubs. Leaves opposite, mostly distinctly petiolate, often rather coriaceous. Stipules intrapetiolar, joined, the sheath distinctly leafy or truncate or with 4-many setae. Inflorescences cymose, terminal, usually many-flowered, ± capitate to laxly paniculate; bracts and bracteoles present, the latter often adnate to the calyx. Flowers often (but apparently not always) heterostylous but not always with anthers well-exserted. Calyx tube globose, obovoid or ovoid, the limb campanulate, ± truncate or 5-lobed. Corolla mostly sweet-scented, funnel-shaped or ± salver-shaped, the tube usually elongate and cylindrical but sometimes shorter than the limb; lobes 5, valvate. Stamens inserted beneath the throat; anthers linear, included. Ovary superior, bilocular with one erect ovule in each locule; style filiform, sometimes thickened and puberulous towards the apex, with 2 linear stigmatic lobes. Fruit globose, obovoid, oblong-ellipsoid or fusiform, smooth or ribbed, indehiscent, ± fleshy. Pyrenes and seeds similarly shaped, globose to fusiform; albumen copious; embryo small, erect.

A genus of 40–50 species occurring in tropical Africa, Madagascar, Mascarene Is., Asia, E. Indies and one in Australia (Queensland). Only one occurs in the Flora Area.

Gaertnera paniculata Benth. in Hook., Niger Fl.: 459 (1849). —Bak. in F.T.A. **4**: 543 (1903). —Hutch. & Dalz., F.W.T.A. ed. 1, **2**: 21 (1931). —Petit in Bull. Jard. Bot. Brux. **29**: 42 (1959) (references). —F. White, F.F.N.R.: 407 (1962). —Hepper, F.W.T.A. ed. 2, **2**: 190 (1963). TAB. **7**. Syntypes from Liberia.

Gaertnera occidentalis Baill. in Bull. Soc. Linn. Paris **1**: 235 (1880). —Bak. in F.T.A. **4**: 544 (1903). Types from Guinea and Gabon (not of Hutch. & Dalz. (1931)).

Gaertnera eketensis Wernham in Journ. Bot., Lond. **52**: 30 (1914). Type from Nigeria.

Shrub or small tree (1.2)3–4.5(9?) m. tall with mostly low branching; the stems slightly pubescent or glabrous when young. Leaves 8–18 × 3–7.5 cm., elliptic, oblong-elliptic or oblanceolate-elliptic, abruptly and acutely acuminate at the apex, ± cuneate at the base, thinly coriaceous, ± discolorous, ± glabrous or midrib and nerves sparsely pilose-pubescent beneath; petiole up to 1.5 cm. long; stipules membranous, connate, the sheath c. 1 cm. long, the apices free and ending in long aristate points, later breaking up. Flowers scented (or not sometimes?), numerous in mostly large lax panicles 7–30 cm. long, the main branches supported by narrow bracts 0.5–5 cm. × 1–7 mm. or larger becoming more leafy and lanceolate at lower axes; pedicels obsolete or very short; bracteoles small. Calyx up to 1.7 mm. long, the teeth triangular, c. 1 mm. long. Corolla white or greenish, ± tomentose outside; tube funnel-shaped, 3–4 mm. long; lobes about 2 × 1 mm., triangular. Style and stigmatic lobes exserted c. 2 mm. in long-styled flowers; anthers exserted c. 3 mm. in short-styled flowers. Fruits juicy, violet, purple or blue, c. 9 × 6 mm., ellipsoid, or if with two pyrenes then broader c. 7 × 8 mm. Pyrene wall pale, thin; seed dark, 4 × 3 mm., broadly ovoid, with obscure irregular sulcation.

Zambia. W: Mwinilunga, Lisombo R., fl. 9.vi.1963, *Loveridge* 882 (K).
From Guinea to Cameroon, Gabon and Angola, also Zaire. Gallery forest; 1200–1450 m.

Tribe 3. **MORINDEAE**

Flowers not in compact heads, the calyx tubes not confluent; enlarged lamina-like bracts never present; fruits usually blue, not forming a compound fleshy mass - - **Lasianthus**

Flowers in compact heads, the calyx tubes mostly confluent; outer flowers with coloured stipitate lamina-like bracts in one of the species; fruits fused together to form a fleshy mass - - - - - - - - - - - - - - - - - **Morinda**

6. LASIANTHUS Jack

Lasianthus Jack in Trans. Linn. Soc. **14**: 125 (1823). —Verdc. in Kew Bull. **11**: 450 (1957) nom. conserv.

Shrubs or rarely small trees, sometimes foetid (usually not in Africa), glabrous to hairy or strigose. Leaves opposite, mostly acuminate, thin to coriaceous, petiolate, usually with numerous arching lateral nerves and close venation; stipules interpetiolar, usually broadly triangular or lanceolate, not divided, persistent or deciduous. Flowers hermaphrodite or sometimes unisexual, sometimes heterostylous, mostly small, mostly in sessile axillary fascicles or glomerules or less often in pedunculate, simple or branched inflorescences; pedicels mostly absent; bracts present, usually small. Calyx tube subglobose, ovoid, oblong or urceolate; limb 3–6-toothed or lobed, persistent. Corolla often white or pink, salver-shaped or somewhat funnel-shaped; tube densely hairy at the throat; lobes 4–6, spreading or ± erect. Stamens 4–6, inserted in the throat of the corolla; filaments very short; anthers ± dorsifixed near their base, included or shortly exserted. Disk swollen and fleshy. Ovary 4–12-locular; ovules solitary in each locule, erect from the base, bent, anatropous; style short or elongate, glabrous or hairy, shortly 4–10-lobed at the apex, the lobes linear or obtuse. Fruits ± succulent, very often blue but sometimes pink, purple, white or black, with 4–12 pyrenes; pyrenes cartilaginous or bony, segment-shaped or pyriform, ± 3-angled with flat sides, the dorsal curved face often grooved, keeled or winged, 1-seeded. Seeds narrowly oblong, curved, with membranous testa and fleshy albumen.

A large genus, estimated at 150 species, predominantly in eastern tropical Asia (where the number has been probably much exaggerated) but about 20 species in tropical Africa and 1 in the W. Indies. Only 1 occurs in the Flora Zambesiaca area.

Lasianthus kilimandscharicus K. Schum. in Engl., Pflanzenw. Ost-Afr. **C**: 396 (1895). —Robyns, Fl. Parc Nat. Alb. **2**: 370 (1947). —Brenan, T.T.C.L.: 504 (1949). —Dale & Greenway, Kenya Trees & Shrubs: 450 (1961). —F. White, F.F.N.R.: 410 (1962). —Verdc. in F.T.E.A., Rubiaceae 1: 142, fig. 11 (1971). —Palgrave, Trees S. Afr.: 901 (1977). —Bridson in Troupin, Fl. Pl. Lign. Rwanda: 564, fig. 192/2 (1982). Types from Tanzania.

Shrub or small tree 1.2–7.5 m. tall, with smooth grey bark; fresh wood reported in Uganda to smell unpleasant; shoots glabrous or finely pubescent, drying black above but with pale yellowish bark, the nodes often with some persistent indumentum below the stipules but older stems quite glabrous. Leaf blades (4)9–17(22) × (1.2)2–6(7) cm., oblong, narrowly oblong-elliptic or oblong-lanceolate, narrowly acuminate at the apex, cuneate at the base, firmly papery but not coriaceous, glabrous above and usually beneath save for fine sparse appressed pubescence on the nerves or rarely hairy; lateral nerves (5)8–10 on each side; petiole 0.5–1(1.8) cm. long; stipules 1.5–6 mm. long, narrowly to broadly triangular, hairy, particularly along the margins. Flowers few, sessile in the axils of the leaves, heterostylous; bracts 2–6 × 1–2 mm. ovate to lanceolate, (? sometimes stipules of reduced shoots), with distinctly hairy margins. Calyx pinkish white or tinged purple, particularly on the lobes, glabrous or puberulous; tube 2 mm. long, turbinate; lobes very convex in living state, 0.5–3 × 2 mm., sometimes with pubescent traces of intermediate accessory lobes. Corolla glistening white or violet outside, white inside; tube 2.5–4(5) mm. long, cylindrical, glabrous to slightly pubescent outside, with densely hairy throat but tube glabrous inside or hairy only in upper half; lobes 4–5, 2.2–3.7 × 1.5–2 mm., ovate-oblong, densely hairy inside with white hairs, finely hairy or glabrous outside. Stamens with anther tips just exserted in long-styled flowers but with anthers and c. 1 mm. of filaments exserted in short-styled flowers. Ovary 4–6-locular; style 4.5–5.5 mm. long in long-styled flowers, 2.5 mm. long in short-styled flowers; stigma lobes 4–5, oblong or subcapitate, 0.5–0.6 mm. long. Fruit intense cobalt blue, c. 4.5 (dry) to c. 10 (fresh) mm. in diam., subglobose, prominently 4–6-lobed in dry state but not grooved when fresh, finely puberulous, the persistent calyx lobes oblate, very obtuse, constricted at the base, 2 × 2 mm.; pyrenes chestnut-brown or straw-coloured, basically pyriform but with a marked ventral notch extending from the pointed end for about half the length of the pyrene, 3.2–4 × 2.3–2.5 × 2 mm.

Subsp. **kilimandscharicus**. TAB. 8.

Tertiary venation of leaves very close and with a conspicuous transversely parallel

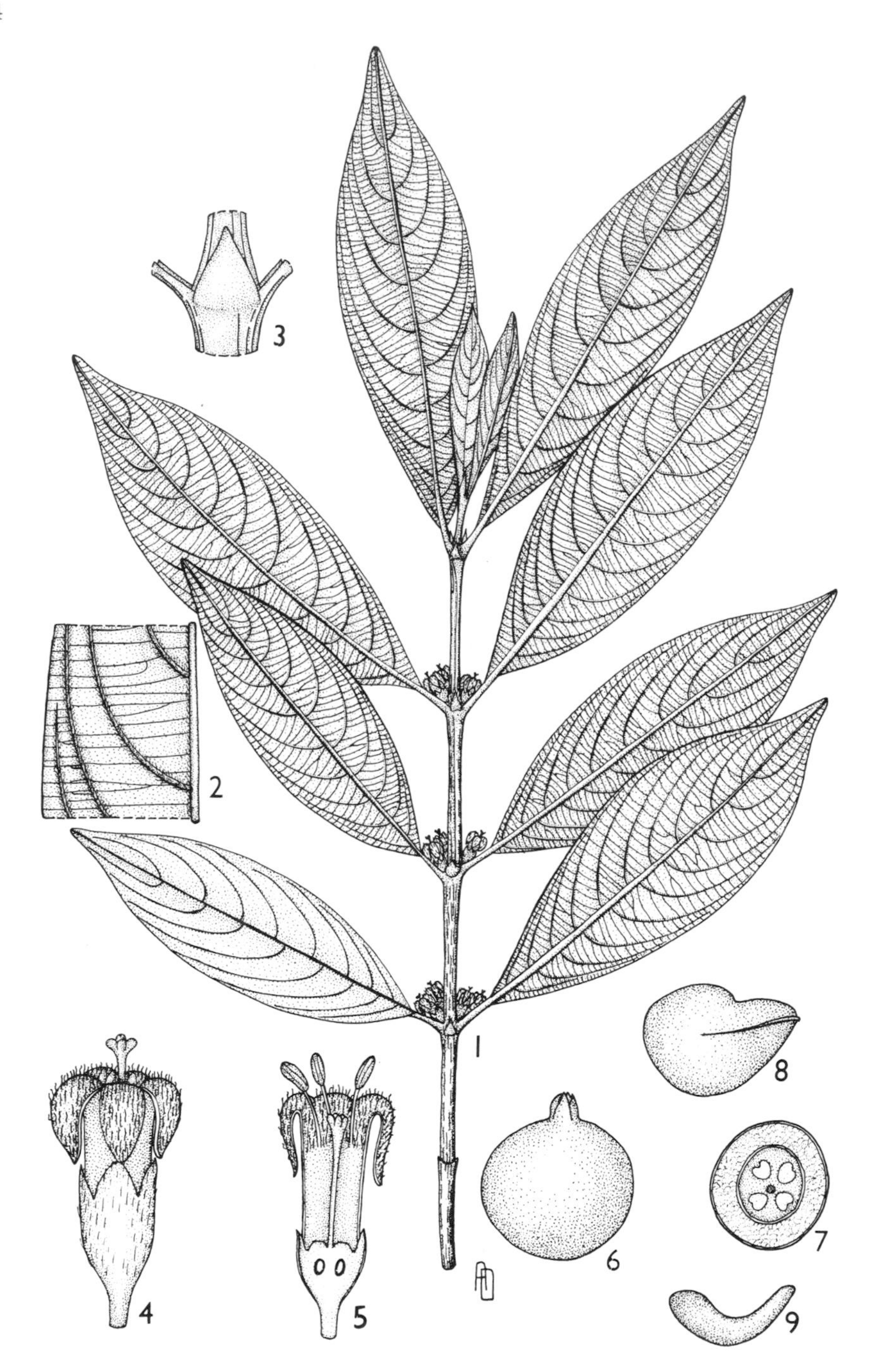

Tab. 8. LASIANTHUS KILIMANDSCHARICUS subsp. KILIMANDSCHARICUS. 1, flowering branch (× ½); 2, detail of inferior leaf surface (×1½), 1–2 from *Osmaston* 2537; 3, node showing stipule (× 4), from *Drummond & Hemsley* 2256; 4, long-styled flower (× 3⅓), from *Drummond & Hemsley* 885; 5, short-styled flower, longitudinal section (× 3⅓); 6, fruit (× 2), from *Verdcourt* 3987a; 7, fruit, transverse section (× 5⅔), from *Drummond & Hemsley* 885; 8, pyrene (× 6), from *Verdcourt* 3987a; 9, seed (× 9), from *Dowson* 76. From F.T.E.A.

element; nerves glabrous or pubescent beneath; leaves attaining maximum dimensions noted in description.

Zambia. N: Kasama Distr., Mungwi, fl. 19.xi.1960, *Robinson* 4083 (K). **Zimbabwe**. E: Inyanga, S. slopes of Inyangani, fl. 7.xi.1967, *Müller* 688 (K; SRGH). **Malawi**. N: Nyika Plateau, Mwenembwe Forest, calyces 17.xii.1981, *Dowsett-Lemaire* 264 (K). S: Mulanje Mt., Litchenya Plateau, fl. 16.x.1941, *Greenway* 6301 (EA; K). **Mozambique**. Z: Serra du Gúruè, Chá Moçambique near R. Malema, fr. 3.i.1968, *Torre & Correia* 16838 (FHO; LISC; LMU; MO). MS: Chimanimani Mts., near St. George's Cave between The Saddle and Poacher's Cave, bud 12.iv.1967, *Grosvenor* 386 (K; LISC; SRGH).

Also in Zaire, Burundi, Rwanda, Uganda, Kenya, Tanzania. Swamp and riverine forest also *Widdringtonia* forest; 900–2300 m.

7. MORINDA L.

Morinda L., Sp. Pl.: 176 (1753); Gen. Pl. ed. 5: 81 (1754).
Appunettia R. Good in Journ. Bot., Lond. **64** Suppl. 2: 30 (1926).

Trees, shrubs or less often lianes, with mostly glabrous, less often hairy or tomentose stems. Stipules leafy, undivided, free or forming a sheath with the petioles. Leaves opposite or rarely in whorls of 3, sometimes only 1 at flowering nodes. Flowers heterostylous (? always), hermaphrodite or rarely unisexual, in tight capitula, the flowers usually joined, at least by the bases of the calyces, the capitula sometimes bearing single large coloured bracts or occasionally many smaller bracts; capitula 1–several at the nodes, frequently arranged in umbels, pedunculate or rarely sessile. Calyx tube urceolate or hemispherical, the limb short, truncate or obscurely to distinctly toothed, persistent. Corolla ± coriaceous, funnel-shaped or salver-shaped; lobes (4)5(7), valvate; throat glabrous or pilose. Stamens (4)5(7), inserted in the throat; filaments short; anthers and style included or exserted. Disk swollen or annular. Ovary 2–4-locular, sometimes imperfectly so; style with 2 short to long linear branches; ovules solitary in the locules, attached to the septum below the middle or near the base, ascending, anatropous or amphitropous. Fruit syncarpous (very rarely scarcely so), succulent, containing several pyrenes; pyrenes cartilaginous or bony, 1-seeded or joined into a 2–4-locular woody structure. Seeds obovoid or reniform, with a membranous testa and fleshy endosperm.

A genus of about 80 species throughout the tropics; 3 species occur in the Flora Zambesiaca area, all very distinct.

1. At least some capitula bearing large whitish or greenish yellow leafy petiolate bracts; corolla green and yellow - - - - - - - - - - - - - - 3. *asteroscepa*
- Capitula not bearing large whitish bracts - - - - - - - - - - - 2
2. Tree or shrub 2.5–18 m. tall - - - - - - - - - - - - - 1. *lucida*
- Small suffrutex to 20 cm. - - - - - - - - - - - - - 2. *angolensis*

1. **Morinda lucida** Benth. in Hook., Niger Fl.: 406 (1849). —Hutch. in Kew Bull. **1916**: 9, fig. on p. 13 (1916). —Brenan, T.T.C.L.: 506 (1949). —Dale & Eggeling, Indig. Trees Uganda, ed. 2: 351 (1952). —Aubrév., Fl. For. Côte d'Ivoire, ed. 2, **3**: 270, t. 348 (1959). —Hepper in F.W.T.A., ed. 2, **2**: 189, fig. 241 (1963). —Keay, Onochie & Stanfield, Nigerian Trees, **2**: 396, fig. 169 (1964). —Verdc. in F.T.E.A., Rubiaceae 1: 146 (1976). Type from Nigeria.
Morinda citrifolia sensu Hiern in F.T.A. **3**: 192 (1877) pro parte non L.

Tree or rarely a shrub 2.4–18 m. tall, with smooth or rough scaly grey or brown bark and crooked or gnarled bole and branches; stems pubescent when young, later glabrous; cork grey but often with some distinct purple layers; slash greenish or yellow. Leaf blades 5.8–18 × 2.2–8.9 cm., elliptic, acute to acuminate at the apex, rounded to broadly cuneate at the base, shining above, glabrous save for tufts of hairs in the axils beneath and some hairs on the midrib, or slightly finely pubescent all over when young; petiole 0.5–1.6 cm. long; stipules short, 1–2.5 mm. long, mucronate at the apex, sometimes splitting into 2 parts, hairy inside, or sometimes ovate or triangular and up to 7 mm. long. Flowers heterostylous; peduncles (1)3 per node opposite a single leaf, alternating at successive nodes, 2.5–7.5 cm. long, often densely pubescent when young; at the base of these 3 peduncles there is a usually stalked cup-shaped gland 2–6 mm. long; capitula (8)10–13(14)-flowered, 4–7 mm. in diam. (excluding corollas). Calyx tube pubescent when young, cupular, c. 2 mm. long, the truncate limb c. 0.5 mm. Corolla heavily scented, white or greenish yellow outside, white inside; tube 1.2–1.6 cm. long, widened at the apex; lobes

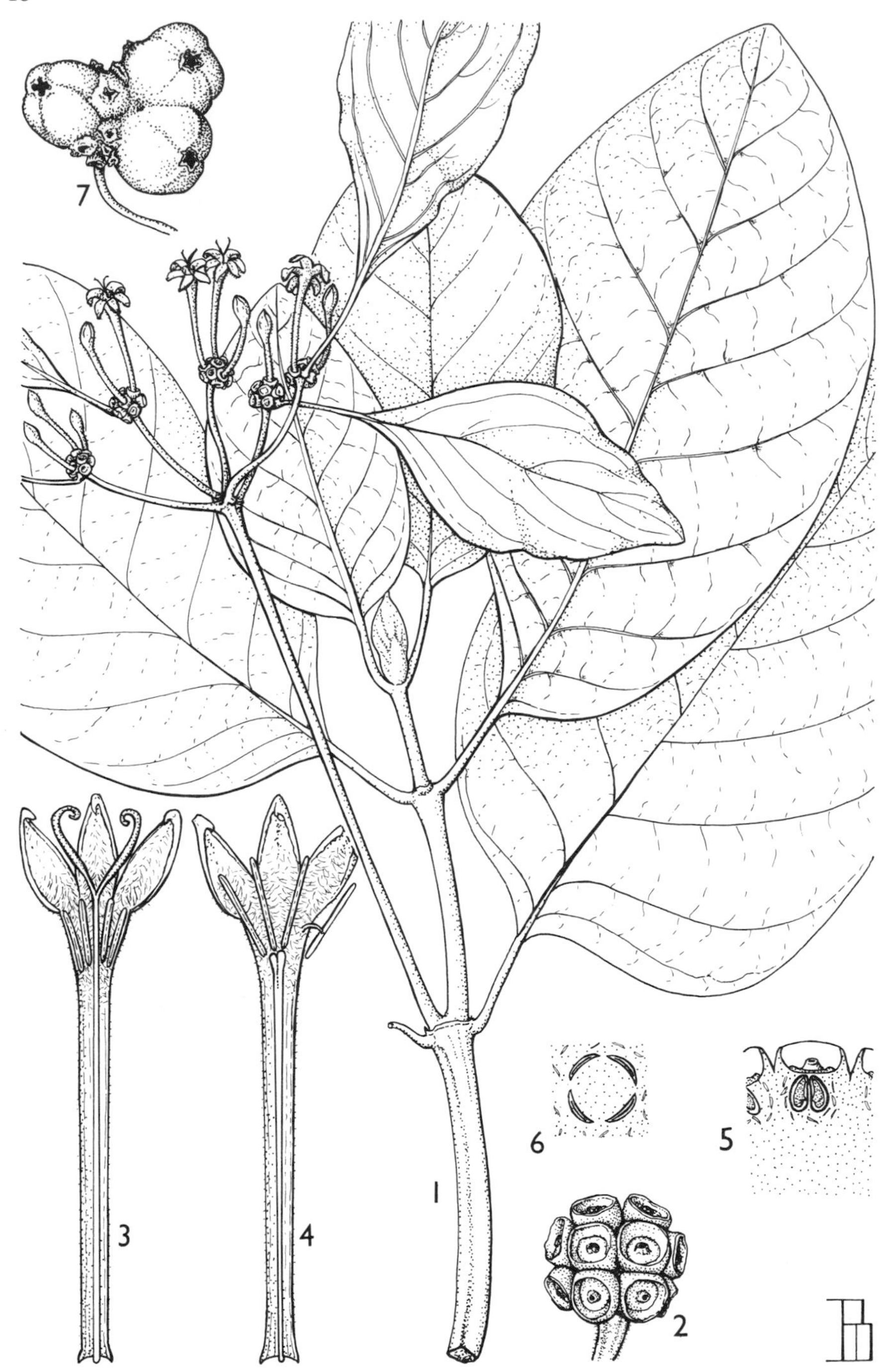

Tab. 9. MORINDA ASTEROSCEPA. 1, flowering branch (× ⅔), from *Hughes* 28; 2, flower-heads with corollas removed (× 3), from *Verdcourt* 1730; 3, long-styled flower, longitudinal section (× 3), from *Peter* K.705; 4, short-styled flower, longitudinal section (× 3); 5, ovaries, longitudinal section (× 4); 6, one ovary, transverse section (× 8), 4–6 from *Verdcourt* 1730; 7, syncarp (× 2), from *Peter* K. 684. From F.T.E.A.

3.5–5 × 1.5–2.5 mm., ovate-lanceolate; throat glabrous. Ovary 2-locular; style 8 or 11 mm. long; stigma lobes c. 4 mm. or 7 mm. long. Syncarps green, hard for a long time but ultimately becoming soft and black, 0.8–2.2(2.5) cm. in diam.; pyrenes dark reddish brown, 5.5–6.5 × 4 mm., compressed ovoid, very hard. Seeds yellowish, soft, about 3.5 × 2 × 0.4 mm., elliptic.

Zambia. N: Lake Mweru, Chelenje, young fr. 4.x.1961, *Lawton* 772 (K; NDO).
W. Africa from Senegal to Angola, Zaire, Uganda and Tanzania.

2. **Morinda angolensis** (R. Good) F. White, F.F.N.R.: 455; 412 (1962). Types from Angola.
Appunettia angolensis R. Good in Journ. Bot., Lond. **64**, Suppl. 2: 30, fig. (1926).

Rhizomatous suffrutex 10–20 cm. tall forming mats up to c. 1 m. diam. drying blackish; shoots annual, glabrous or pubescent; older stems pale brown, ridged, ± corky, glabrous. Leaves 1.5–12 × 0.4–2.3 cm., narrowly oblanceolate to oblong-oblanceolate, rounded to subacute at the apex, tapering cuneate at the base, pubescent to glabrous; petioles up to c. 1 cm. long; stipules 2–3 mm. long and wide, triangular. Inflorescences 3–6-flowered c. 1 cm. diam., the calyx tubes confluent; peduncles (0)2–3 mm. long. Calyx tube c. 2 mm. long, the limb truncate, under 1 mm. tall. Corolla fleshy, fragrant, white or tube greenish inside, glabrous outside; tube slender 2–3 cm. long, pubescent inside; lobes spreading, up to 2.2 cm. × 2.5 mm., linear-oblong, sulcate inside. Anthers 3–4 mm. long, inserted just below throat; style just exserted, the lobes linear, 4 mm. long. Ovary 4-locular. Fruits globose 1.5 cm. diam.

Zambia. W: Mwinilunga, fl. 26.x.1955, *Holmes* 1296 (K; NDO).
Angola. Kalahari sand; open *Brachystegia–Erythrophleum–Pterocarpus–Burkea* woodland or sometimes on bare sand often with other similar suffrutices, *Ochna*, *Parinari capense*, *Combretum platypetalum* etc.; ? ± 1500 m.
The corolla lobes seem extraordinarily variable in length; *Good* gives 8 mm. and *White* 7 mm. but those on the sheet cited exceed 2 cm. In Angola it is said to form patches up to 9 square metres.

3. **Morinda asteroscepa** K. Schum. in Engl., Bot. Jahrb. **34**: 340 (1904). —Brenan, T.T.C.L.: 506 (1949). —Verdc. in F.T.E.A., Rubiaceae 1: 146 (1976). TAB. **9**. Type from Tanzania.

Small to large evergreen tree 6–25 m. tall, much branched, with a dense rounded crown and often branched low down; stems glabrous. Leaf blades 7.8–22.6(25) × 3.5–15.2 cm., elliptic to oblong-ovate, rounded to subacute at the apex, rounded, truncate or broadly cuneate and often unequal-sided at the base, glabrous save for small tufts of hair in the nerve axils beneath; petiole (1)1.7–3 cm. long; stipules 3.6–6 × 0.5–2 cm., lanceolate, connate at the base and enclosing the terminal bud, soon circumscissile and deciduous. Flowers heterostylous, in umbels of (1)2–7(9) capitula, the capitula 4–9 mm. in diam., 5–20-flowered; common peduncles 7–17.5 cm. long; secondary peduncles 0.8–4.5 cm. long, finely puberulous; many of the capitula bear a large coloured petiolate bract, greenish below, cream above with greenish yellow veins, produced on one of the lateral flowers, the lamina (3)6–11.2 × 1.5–5.8 cm., the petiole 0.6–2 cm. long. Calyx cupular, c. 2 mm. tall, of which 1 mm. is a free truncate rim. Corolla cream to yellow, the tube mostly greenish and the lobes bright yellow inside; tube 1.6–2.4 cm. long, widened at the apex, puberulous; lobes 3–7 × 1.2–4 mm., ovate-lanceolate, margined and with thickened apices, hairy inside particularly near throat with yellow hairs; throat densely hairy. Ovary 3–4-locular; style 1.2 or 2.4 cm. long according to form; stigma lobes 3.5–4.5 mm. long, at least exserted ones green. Syncarps c. 1.5 cm. in diam.; pyrenes 5 × 4.5 mm., broadly elliptic, much compressed. Immature seeds 3.5 × 2 mm., probably winged.

Malawi. C: Ntchisi Forest, fl. 19.xi.1963, *Chapman* 1738 (K; LISC; SRGH).
Tanzania. Evergreen forest; 1500–1600 m.
The isolated station from which the only specimen from our area was collected is a considerable distance from the southernmost locality known in Tanzania (Uluguru Mts.).

Tribe 4. TRIAINOLEPIDEAE

8. TRIAINOLEPIS Hook. f.

Triainolepis Hook. f. in Benth. & Hook. f., Gen. Pl. **2**: 126 (1873). —Bremek. in Proc. K. Nederl. Akad. Wetensch., ser. C, **59**: 4 (1956).
Princea Dubard & Dop in Journ. de Bot., Sér. 2, **3**: 2 (1925).

Shrubs or small trees. Leaves opposite, petiolate; blades lanceolate to ovate or oblong-elliptic, often curved; stipules ovate or ovate-triangular, divided into 3–5 lobes. Flowers hermaphrodite, heterostylous, (4)5-merous in terminal corymbs; bracts small or minute. Calyx tube campanulate, the limb cupular, unequally 5–7-toothed. Corolla white, yellowish or tinged red, salver-shaped, tomentose to woolly outside; tube with throat and sometimes upper half inside densely barbate with white hairs; lobes lanceolate, hairy outside, glabrous inside. Stamens inserted at the top of the corolla tube, exserted or included. Disk mostly glabrous. Ovary 4–10-locular, each locule with 2(3) collateral erect anatropous ovules; style filiform, glabrous, exserted or included; stigma lobes 4–10, filiform, straight or twisted, somtimes cohering. Fruits usually red, globose or depressed globose, drupaceous, containing a woody or bony 4–10-celled putamen, which is entire or slightly to deeply incised between the locules; locules 1-seeded. Seeds ellipsoid, compressed, with membranous testa and fleshy albumen.

A small genus occurring in Madagascar, Comoro Islands, Aldabra Group and also on the East African mainland (almost entirely on the coast): Bremekamp has recognised 12 species but this is I feel too high an estimate; many of the species seem very close to *T. africana* but since that is the oldest name the identity of his species has not concerned this account. It has not been found possible to keep up *T. hildebrandtii* at specific level and unfortunately it is the better known name. A distinct species has been found on islands off Mozambique and is described here.

Leaves distinctly tapering acuminate at the apex; petioles developed - - 1. *africana*
Leaves subacute to almost rounded at apex; petioles ± obsolete - - - 2. *sancta*

1. **Triainolepis africana** Hook. f. in Benth. & Hook. f., Gen. Pl. **2**: 126 (1873). —Hiern in F.T.A. **3**: 219 (1877). —Sim, For. Fl. Port. E. Afr.: 76 (1909). —Brenan, T.T.C.L.: 533 (1949). —Bremek. in Proc. K. Nederl. Akad. Wetensch., ser. C, **59**: 13 (1956). —Verdc. in Kew Bull. **30**: 282 (1975); in F.T.E.A., Rubiaceae 1: 149 (1976). Type: Mozambique/Tanzania boundary, Rovuma Bay, *Kirk* (K, holotype (on 2 sheets)).

Weak shrub or rarely a small tree, 1.8–6 m. tall; branches pubescent or covered with dense white hairs, later glabrescent. Leaf blades 3–11.5(16) × 1.2–5.7 cm., elliptic, narrowly ovate or elliptic-lanceolate, narrowly acuminate at the apex, cuneate at the base, almost or completely glabrous to densely appressed pubescent above, more or less velvety beneath with ± white hairs, particularly along the main nerves so that they stand out against the rest of the surface; petioles 0.5–1.7 cm. long; stipules with ovate or triangular bases 1–4 mm. long, with 0–7 (usually 3) unequal subulate fimbriae 0.5–4(6) mm. long, mostly with distinct colleters. Inflorescences hairy, up to 4 cm. across; peduncle 0.35–2.2 cm. long; pedicels 0–2.5 mm. long. Calyx tube 1–1.5 mm. long; limb tube 1–2.4 mm. long; lobes 0.3–2.9 × 0.2–1.3 mm. Corolla with pale green to white tube and white limb; tube 0.7–1.05 cm. long; lobes sometimes tinged pink, 3.5–5.5 × 1.2 mm., the inner and outer layers usually so demarcated that the outer appears as a distinct horn up to 0.6 mm long. Anthers half-exserted in long-styled flowers, well exserted in short-styled flowers. Style 1–1.25 cm. long in long-styled flowers, 4.3–5.5 mm. long in short-styled flowers; stigma lobes 6–8, 0.9–2 mm. long. Drupes white at first, turning dark red, subglobose, 4–6 mm. in diam. when dry, c. 8.5 mm. when fresh, ribbed (at least in dry state); putamen depressed-subglobose, 4–5 mm. in diam., grooved, ± hairy. Seeds pale brown, 2–2.4 × 0.9–1 × 0.5–0.6 mm. narrowly oblong-ellipsoid, often with the undeveloped collateral ovule stuck to the testa.

Subsp. **africana**

Leaf blades distinctly pubescent above, velvety pubescent to ± glabrous beneath.

Malawi. N: Karonga, Lwasyo Stream, fr. 22.iv.1963, *Salubeni* 25 (K; SRGH). **Mozambique**. N: Nampula, Monapo, Sr. Wolf's Forest, fl. & fr. 13.ii.1984, *de Koning et al.* 9638 (LMU).

Also in Tanzania. Probably evergreen coastal thicket; also riverine *Heeria*–Cordyla association; 310–700 m.

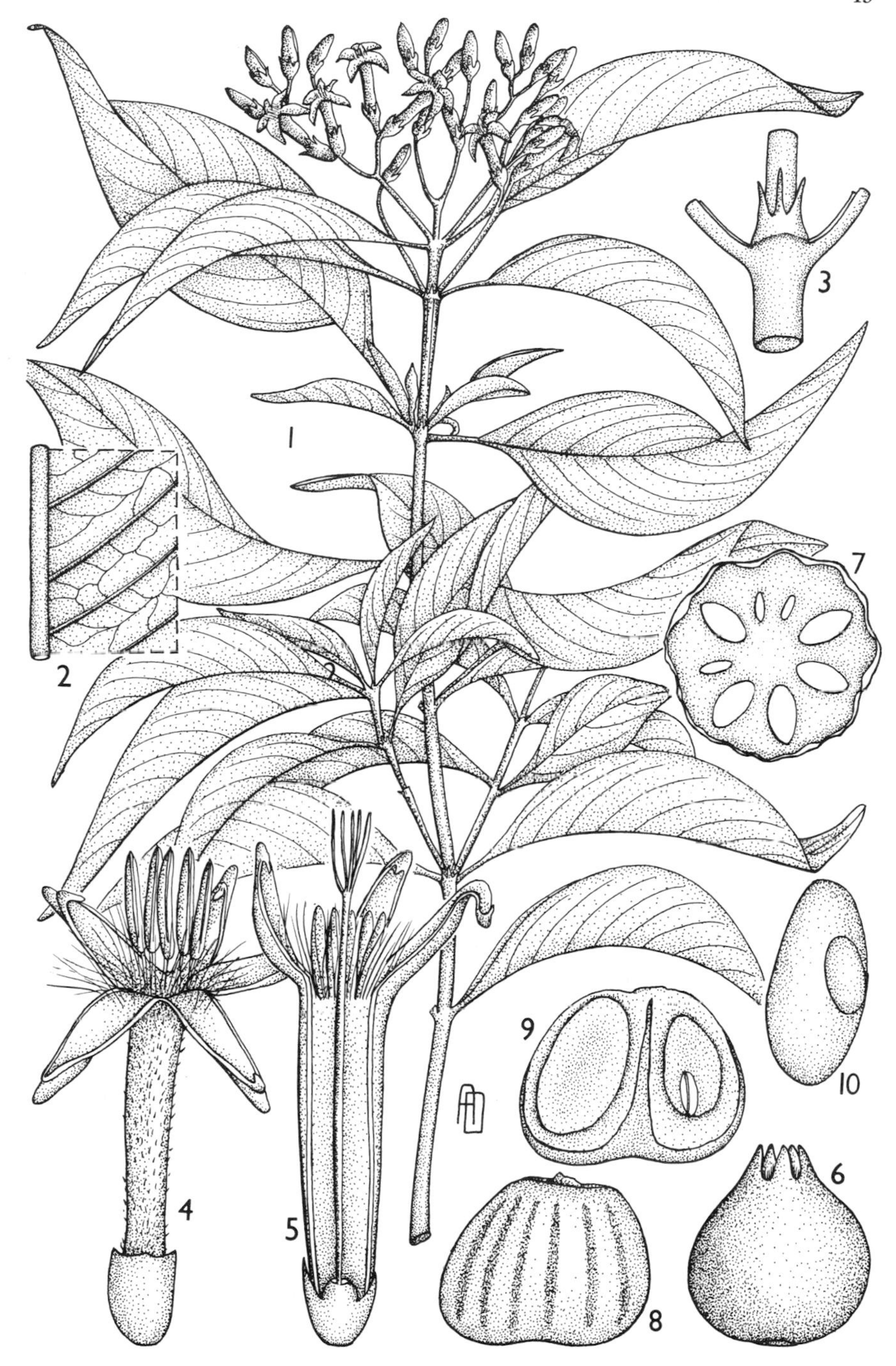

Tab. 10. TRIAINOLEPIS AFRICANA subsp. HILDEBRANDTII. 1, flowering branch ($\times\frac{2}{3}$); 2, detail of inferior leaf surface (× 2); 3, node showing stipule (× 2), 1–3 from *Verdcourt* 3961; 4, short-styled flower (× 4), from *Tweedie* 2377; 5, long-styled flower, longitudinal section (× 4), from *Verdcourt* 3961; 6, fruit (× 6); 7, fruit, transverse section (× 8); 8, putamen (× 6); 9, putamen, vertical section (× 8); 10, seed, with an aborted ovule attached (× 10), 6–10 from *Faulkner* 2316. From F.T.E.A.

It is not possible to localise the *Kirk* type exactly. *Kirk* visited both banks of the Rovuma in Mar. 1861 and the river forms the boundary between Tanzania and Mozambique. See general note.

Subsp. **hildebrandtii** (Vatke) Verdc. in Kew Bull. **30**: 282 (1975); in F.T.E.A., Rubiaceae 1: 150, fig. 13 (1976). —Fosberg & Renvoize, Fl. Aldabra: 163 (1980). TAB. **10**. Type from Zanzibar.

Triainolepis hildebrandtii Vatke in Oest. Bot. Zeitschr. **25**: 230 (1875). —Hiern in F.T.A. **3**: 219 (1877). —Brenan, T.T.C.L.: 533 (1949). —Bremek. in Proc. K. Nederl. Akad. Wetensch., ser. C, **59**: 11 (1956). —Dale & Greenway, Kenya Trees & Shrubs: 475 (1961).

Dirichletia leucophlebia Bak. in Journ. Linn. Soc., Bot. **25**: 321 (1890). Type from NW. Madagascar*

Psathura fryeri Hemsl. in Journ. Bot., Lond. **54**, Suppl. 2: 20 (1916). Type from Aldabra.

Triainolepis fryeri (Hemsl.) Bremek. in Proc. K. Nederl. Akad. Wetensch. ser. C, **59**: 12 (1956).

Triainolepis fryeri var. *latifolia* Bremek. in Proc. K. Nederl. Akad. Wetensch. ser C, **59**: 13 (1956). Type from Grande Comore.

Leaf blades entirely glabrous or with a few obscure hairs above and nerves hairy beneath.

Zambia. N: Mpika, fl. 9.ii.1955, *Fanshawe* 2030 (K; NDO). **Malawi**. N: Nkata Bay Distr., Chinteche, fl. & fr. 21.ii.1961, *Richards* 14433 (K). S: Mangonchi, Monkey Bay, fl. & fr. *Wild* 7692 (K; LISC; SRGH). **Mozambique**. N: Tecomaze I., bud 29.iii.1961, *Gomes e Sousa* 4675 (K).

Coastal Kenya and Tanzania (including Zanzibar and Pemba) also Madagascar amd Comoro Is. Coastal sand, coral rock and also lakeside sand dunes and grass interfaces; open woodland, riparian woodland along rocky steam beds, rocky hillsides; 0–570 m.

This is throughout most of its range a characteristic littoral shrub but in the Flora Zambesiaca area it extends far inland and, moreover, is often a tree to 6 m.; also in this area distinction between the hairy and glabrous variants is more confused and they are scarcely of varietal significance, but elsewhere the distinctions are more valid e.g. nothing but typical subsp. *hildebrandtii* has been found in Madagascar, Comoro Is. or Aldabra.

2. **Triainolepis sancta** Verdc. sp. nov.** Type: Mozambique, Santa Carolina I., near Farolim, fl. & fr. 3.xi.1958, *Mogg* 28806 (K; LISC; PRE, holotype).

Shrub 1–2 m. tall with pale brown stems strongly longitudinally striate when dry, slightly pubescent when young, later glabrous; older stems longitudinally fissured, lenticellate. Leaves rather closely placed, 1.3–5.4 × 0.3–3.2 cm., ovate-elliptic to narrowly elliptic or narrowly oblong, rounded to subacute at the apex, cuneate to rounded at the base, ± wrinkled, probably coriaceous or slightly succulent in life, glabrous; petioles obsolete; stipules with about 3 subulate lobes 1–2 mm. long from a short base c. 1 mm. long. Inflorescences short, c. 3 cm. long, densely pubescent. Calyx tube c. 1 mm. long, glabrous; lobes c. 0.7 mm. long, triangular. Corolla colour not stated, probably white or pink; tube of long-styled flower cylindric, shortly funnel-shaped at the apex, 6.5 mm. long, densely hairy at the throat; lobes 5 × 2 mm., oblong-lanceolate. Ovary 4–6-locular; ovules 1–2 per locule; style 6.5 mm. long with 5–8 stigmatic lobes 3.5 mm. long, 2 arising below the others in flower examined. Anthers about half-exserted in long-styled flower. Fruits purple, c. 6.5 mm. diam.

Mozambique. GI: Santa Carolina I., fl. & fr. 3-5.xi.1958, *Mogg* 28806 (K; LISC; PRE).
Endemic. Open woodland and littoral scrub on sandy soil; 15 m.

* Bremekamp should have taken up this name for *T. fryeri* but did not do so on account of a double error. This much is made clear from a label he has attached to the type of *Dirichletia leucophlebia* which states "this is not *Baron* 5777 which is the type of *Dirichletia leucophlebia* Baker (according to A.M. Homolle in Bull. Soc. Bot. France **83**: 620 (1936) = *D. pervilleana* H. Bn.) It is *Triainolepis fryeri* (Hemsl.) Brem. var. *latifolia* Brem". Firstly Mme. Homolle did not state = *D. pervilleana* but "non *Dirichletia* ni *Carphalea*"; moreover the sheet which bears in pencil "next 5778" is undoubtedly the type of *Dirichetia leucophlebia* being written up in Baker's own hand.

** *T. africanae* affinis sed foliis apice subacutis usque ± rotundatis, internodiis brevioribus ± 1 cm. longis, petiolis obsoletis, ovario 4–6-loculare differt.

Frutex 1–2 m. altus; ramuli pallide brunnei in siccitate longitudinaliter striati, internodiis brevibus. Folia ± dense conferta, ovato-elliptica usque anguste elliptica vel oblonga, 2.5–5.4 × 0.3–3.2 cm.; stipulae lobos 3 subulatos 1–2 mm. longos divisae. Inflorescentiae breves, 3 cm. longae, dense pubescentes. Calycis tubus 1 mm. longus, lobis triangularibus 0.7 mm. longis. Corollae tubus floris brevistyli cylindraceus, superne abrupte breviter ampliatus, 6.5 mm. longus, fauce dense hirsutus, lobis oblongo-lanceolatis 5 × 2 mm. Ovarium 4–6-loculare; stylus 6.5 mm. longus, lobis stigmaticis 5–8, 3.5 mm. longis. Antherae semi-exertae. Fructus purpurei, 6.5 mm. diametro. Also *Mogg* 28500, Bazaruto I., 21.x.1958 (J; LISC) & *Mogg* 28576, Santa Carolina I., 20.x.1958 (LISC).

Tribe 5. **UROPHYLLEAE**

9. **PAURIDIANTHA** Hook. f.

Pauridiantha Hook. f. in Benth. & Hook. f., Gen. Pl. **2**: 69 (1873). —Bremek. in Engl., Bot. Jahrb. **71**: 217 (1940).
Pamplethantha Bremek. in Engl., Bot., Jahrb. **71**: 217 (1940).

Shrubs, small trees or subscandent woody plants. Leaves opposite or ternate, shortly petiolate, the petioles compressed, channelled; blades acuminate or caudate, usually with acarodomatia; midnerve channelled; stipules interpetiolar, triangular or ovate, entire, acute. Flowers hermaphrodite, mostly heterostylous, usually 5-merous, in axillary or terminal, sessile or pedunculate, trichotomously corymbose or subumbellate inflorescences, sometimes reduced to a few or even single flowers; peduncle with 1–2 4-parted involucels situated at the apex or middle; other bracts small or absent. Calyx tube short, denticulate, dentate or lobed. Corolla salver-shaped, white, greenish, violet or lavender; tube short, funnel-shaped or cylindrical, the upper half densely hairy inside; lobes glabrous inside. Stamens with glabrous filaments and dorsifixed anthers, exserted in short-styled flowers. Disk cushion-shaped, papillate or shortly hairy. Ovary 2–3-locular at the base, 4–6-locular at the apex, with 2–3, sometimes lobed placentas affixed at the the middle of the true septum, the false septa incised and broadly cordate; ovules numerous; style glabrous, puberulous or hairy, included or exserted; stigmas 2, globose, mitriform or subcapitate, the apex shortly 2-lobed, the lobes cohering or rarely free and linear or lanceolate. Fruit a globose yellow or red berry, bilocular at the base, 4-locular at the top. Seeds numerous, yellow or yellow-brown, rarely red, ovoid; testa alveolate and sometimes irregularly ribbed; endosperm oily.

A small genus of about 20–25 species confined to tropical Africa and Madagascar; formerly included in *Urophyllum* Wall, but differing from that genus in not being dioecious, in the structure of the ovary and its placentation, and the differently shaped stigma. In his most recent classification Bremekamp even maintains two separate tribes, *Urophylleae* and *Pauridiantheae*.

Leaf blades oblong-elliptic to oblong-lanceolate or oblanceolate; inflorescences usually many-flowered; stigma capitate, lobes not distinctly free - - - - - - 1. *paucinervis*
Leaf blades narrowly lanceolate to lanceolate; flowers 1–3 in each axil with no or only a very short common peduncle; stigma lobes linear, quite free from each other 2. *symplocoides*

1. **Pauridiantha paucinervis** (Hiern) Bremek. in Engl., Bot. Jahrb. **71**: 212 (1940). —Hepper, F.W.T.A. ed 2, **2**: 168 (1963). —Verdc. in F.T.E.A., Rubiaceae 1: 153 (1976). Type from Fernando Po.
Urophyllum paucinerve Hiern in F.T.A. **3**: 74 (1877). —Hutch. & Dalz., F.W.T.A. **2**: 104 (1931).

Forest shrub or small tree, apparently sometimes somewhat scandent, 1.5–9(12) m. tall, with finely appressed pubescent to distinctly hairy twigs; older stems glabrescent. Leaf blades, 3.5–15.5 × 0.9–5.2 cm., oblong-elliptic to oblong-lanceolate or-oblanceolate, acuminate at the apex, cuneate at the base, glabrous above save for short hairs on the narrowly impressed midnerve, entirely glabrous or pilose on the main nerves beneath; lateral nerves ± 12 on each side, together with other venation prominent to very prominent beneath; petioles short, 2.5–10 mm. long, usually pubescent; stipules 0.3–1.35 cm. × 0.8–2.5 mm., lanceolate, appressed pubescent. Inflorescences axillary, short, the cymes c. 1 cm. long; peduncles 1–5 mm. long; pedicels 0–3.5(4) mm. long; bracts 1.5–2.5 mm. long, lanceolate; all parts pubescent. Calyx glabrous, pubescent or hairy; tube 0.6–1.1 mm. long; limb tube 0.3–0.9 mm. long; lobes yellow, red or purple, 0.4–2.8 mm. long, lanceolate-subulate. Corolla glabrous (or rarely puberulous) outside, greenish, yellow, cream or white but sometimes reported to be orange or yellow turning red and certainly often drying red; tube 2–5 mm. long, longest in short-styled flowers; lobes 1.7–3.1 × 0.8–1 mm., oblong-lanceolate. Stamens as long as the corolla lobes in short-styled flowers, just included in long-styled flowers. Disk depressed, glabrous. Style glabrous or sparsely papillate, 3–4.6(5.5) mm. long in long-styled flowers, 1.4–2.8 mm. long in short-styled flowers; stigma 0.45–0.8 mm. long. Berries orange, red or tinged purple, subglobose, 3–5 × 2.5–4 mm. Seeds orange-brown, 0.8 × 0.5 mm. wide, ellipsoid, very strongly pitted.

Subsp. **holstii** (K. Schum.) Verdc. in Kew Bull. **30**: 283 (1975). —Verdc. in F.T.E.A. Rubiaceae 1: 153, fig. 14 (1976). TAB. **11**. Type from Tanzania.

Tab. 11. PAURIDIANTHA PAUCINERVIS subsp. HOLSTII. 1, flowering branch (× ½); 2, node showing stipule (× 2); 3, long-styled flower (× 4½); 4, ovary, longitudinal section (× 8), 1–4 from *Verdcourt* 2303; 5, short-styled flower, longitudinal section (× 4½), from *Brasnett* in F.D. 1506; 6, berry (× 4); 7, seed (× 20), 6–7 from *Kerfoot* 3874. From F.T.E.A.

Urophyllum holstii K. Schum. in Engl., Pflanzenw. Ost-Afr. **C**: 379 (1895).
Pauridiantha holstii (K. Schum.) Bremek. in Engl., Bot. Jahrb. **71**: 212 (1940). —Brenan, T.T.C.L.: 508 (1949). —Dale & Eggeling, Ind. Trees Uganda, ed. 2: 354 (1952) pro parte. —Brenan in Mem. N.Y. Bot. Gard. **8**: 449 (1954). —Dale & Greenway, Kenya Trees & Shrubs: 453 (1961). —F. White, F.F.N.R.: 414 (1962).

Leaf blades with venation not quite so prominent beneath as in subsp. *lyallii;* inflorescences typically lax (but not in some populations): calyx lobes 1–2.8 mm. long; seeds orange-brown.

Zambia. N: 32.6 km. from Mporokoso towards Kasama, fl. 17.x.1947, *Brenan & Greenway* 8138 (EA; FHO; K). W: Mwinilunga, fl. 4.x.1955, *Holmes* 1222 (K; NDO). **Malawi**. N: Misuku Hills, Wilindi Forest, st. 4.iii.1983, *Dowsett-Lemaire* 682 (K). **Mozambique**. Z: Serra do Guruè via fábrica Junqueiro, a Oeste dos Picos Namuli, near R. Malema, fr. 6.xi.1967, *Torre & Correia* 15965 (L; LISC; LMU) (probably).
Also in Burundi, Zaire (intermediates), Kenya and Tanzania. Riverine evergreen forest including swamp forest with *Syzygium* etc and also upland evergreen forest; 1500–2350 m. (fide Dowsett-Lemaire).

Subsp. *paucinervis* occurs in W. Africa and subsp. *lyallii* (Bak.) Verdc. in Madagascar but there is great variability and a tendency to produce distinctive local variants and many intermediates occur. Most Zambian material had been named *Pauridiantha pyramidata* (K. Krause) Bremek. and F. White (F.F.N.R.: 414 (1962)) sinks this into *holstii* but true *pyramidata* (Type from Cameroon) is in general rather different in foliage and I have not included it in synonomy although one specimen *Robson* 600 (Malawi N: Mafinga Mts., N. of Chisenga (BM; K; LISC)) is certainly very similar indeed save for the glabrous flowers.

2. **Pauridiantha symplocoides** (S. Moore) Bremek. in Engl., Bot. Jahrb. **71**: 212 (1940). —Verdc. in F.T.E.A., Rubiaceae 1: 156 (1976). —Palgrave, Trees S. Afr.: 845 (1977). Type: Zimbabwe, Chimanimani, Mt. Pene, *Swynnerton* 1278 (BM, holotype; K).
Urophyllum symplocoides S. Moore in Journ. Linn. Soc., Bot. **40**: 79 (1911).

Shrub or small tree, usually much-branched, 1.8–9 m. tall; stems glabrous or nearly so. Leaf blades 4.5–13.3 × 0.9–2.8 cm., lanceolate, narrowly acuminate at the apex, cuneate at the base, glabrous above save for a few short hairs on the narrowly impressed midnerve, entirely glabrous or with a few hairs on the venation beneath; lateral nerves ± 10–12 on each side, prominent beneath; petioles 2–8 mm. long, pubescent; stipules 1.5–5 × 0.5–2 mm., triangular-lanceolate, pubescent. Flowers heterostylous, 1–3 in each axil with no or only a short common peduncle 0–2.5 mm. long; pedicels 1.5–2.5 mm. long, pubescent. Calyx pubescent; tube 1.6–1.7 mm. long; limb tube 0–0.2 mm.; lobes 1–4.1 mm. long, lanceolate-subulate. Corolla white, yellowish white or cream, glabrous outside; tube 4–4.3 mm. long; lobes 2–4.9 × 1.1–2 mm., triangular-lanceolate. Anthers about three-quarters exserted in short-styled flowers, just included in long-styled flowers. Disk depressed, glabrous. Style glabrous, 5.8 mm. long in long-styled flowers, 1.5 mm. long in short-styled flowers; stigma 0.9–1.2 mm. long, the lobes quite distinct, linear. Berries 3–3.7 mm long, 4–5 mm. across, subglobose, glabrous. Seeds typically orange-brown, 1.2 × 0.95 mm. ovoid.

Zimbabwe. E: Mutare Distr., Nyamkwarara Valley, SE. Boundary, Stapleford Forest Reserve, Rupesi Peak, fl. 25.ix.1952, *Chase* 4638 (BM; K; LISC; SRGH). **Malawi**. S: Mt. Mulanje, Sombani Hut to Fort Lister Depot path, fl. 13.xii.1979, *Blackmore & Kupicha* 1040 (K: MAL). **Mozambique**. MS: Mountain Home Farm near bauxite mine, fl. 13.iv.1969, *Chase* 8552 (K; LISC; SRGH).
What may be a variety occurs in Tanzania. Evergreen forest; 1000–2000 m.

Tribe 6. **CRATERISPERMEAE**

10. **CRATERISPERMUM** Benth.

Craterispermum Benth. in Hook., Niger Fl.: 411 (1849). —Hook. f. in Benth. & Hook. f., Gen. Pl. **2**: 112 (1873). —Verdc. in Kew Bull. **28**: 433 (1973).

Glabrous trees or shrubs mostly with yellow-green foliage. Leaves opposite, petiolate, the blades mostly oblong or elliptic, often coriaceous; venation mostly closely reticulate; stipules intrapetiolar, broad, connate to form a tube, made up of 2 triangular parts joined by thinner tissue, undivided, persistent or deciduous. Flowers hermaphrodite, heterostylous, 5-merous, in small subcapitate or somewhat elongated occasionally 2-branched cymes, the peduncles short or less often long and slender, strongly compressed, axillary or more usually supra-axillary; bracteoles present. Calyx tube obconic or turbinate; limb cupular, truncate, sinuate or shortly 5-dentate, persistent. Corolla salver-shaped or somewhat funnel-shaped with short or elongated tube and densely hairy or less often glabrous throat. Stamens either included in the throat or exserted; anthers linear-oblong, dorsifixed. Disk annular, thick. Ovary 2-locular; ovules solitary in each locule, pendulous from the apex; style filiform; stigma divided into 2 linear papillate branches or fusiform, bifid. Fruit subglobose, pea-like, sessile or pedicellate, 1(2)-locular, 1-seeded; endocarp chartaceous. Seeds pendulous, hemispherical or almost bowl-shaped, dorsally convex, ventrally deeply excavated; albumen fleshy; embryo small with a superior radicle.

A small genus with 15–20 species widespread in tropical Africa and also in the Seychelles and Madagascar. Only a single species occurs in the Flora Zambesiaca area.

Craterispermum schweinfurthii Hiern in F.T.A. **3**: 162 (1877). —K. Schum. in Engl., Pflanzenw. Ost-Afr. **C**: 386 (1895). —F.W. Andr., Fl. Pl. Sudan **2**: 433 (1952). —Verdc. in Kew Bull. **28**: 434 (1973); in F.T.E.A., Rubiaceae 1: 162, fig. 16 (1976). —Palgrave, Trees S. Afr.: 859 (1977). TAB. **12**. Type from Sudan.

Craterispermum reticulatum De Wild. in Ann. Mus. Congo IV [1]: 158 (1903). Type from Zaire.

Craterispermum laurinum sensu auctt. mult., Dale & Eggeling, Indig. Trees Uganda ed. 2: 341 (1952). —Brenan in Mem. N.Y. Bot. Gard. **8**: 453 (1954). —Dale & Greenway, Kenya Trees & Shrubs: 437 (1961). —F. White, F.F.N.R.: 405 (1962). —Gomes e Sousa, Dendrol. Moçamb.: 678, t. 226 (1967) non (Poir.) Benth.

Shrub or small to medium tree 1.8–15 m. tall, glabrous; bark greyish white, rough warty with old swollen nodes. Leaf blades usually yellow-green when dry, (5)7–18 × 2–7.3 cm., elliptic, oblong, obovate or oblanceolate, ± obtuse to distinctly shortly acuminate at the apex, cuneate at the base, often ± coriaceous; venation closely reticulate; petiole 1–1.7 cm. long; stipules 2.5–5 mm. long, the thicker deltoid part often with a few stiff hairs at its apex. Inflorescences supra-axillary (sometimes only slightly so), compact and subcapitate, several-flowered; peduncles mostly stout, compressed, 2–10 mm. long, thickened apically; bracts and bracteoles triangular, keeled, c. 1.5 mm. long, acuminate, very congested. Calyx tube 0.7–1.5 mm. long; limb 0.9–1.4 mm. long, slightly toothed, the teeth 0.3–0.5 mm. long. Corolla white, sometimes tinged pink in bud, sweetly scented; tube 3.5–5.6 mm. long, densely hairy inside at the throat; lobes 3–5.8 mm. long, oblong-lanceolate to oblong-ovate, hairy inside at least at the base. Anthers with tips just exserted in long-styled flowers, completely exserted and reaching to or nearly to the tips of the corolla lobes in short-styled flowers. Style 7–7.5 mm. long in long-styled flowers, well exserted, 2.6–4 mm. long in short-styled flowers, included; stigma lobes linear-clavate, 1.3–2.5 mm. long. Fruit brown when dry but described as green or black, 5–6(7 in spirit material) mm. long and wide, subglobose or ellipsoid, sessile. Seeds dark brown, shining, bowl-shaped, longest dimension 3.4 mm., shorter dimension 2 mm., the depression deep and rounded.

Zambia. N: Mbala, Ningi Pans, fl. 3.ix.1960, *Richards* 13190 (K). W: Mwinilunga, Matonchi Farm, fl. 3.ix.1930, *Milne-Redhead* 1043 (K). C: Serenje Distr., Kundalila Falls, fl. 14.x.1963, *Robinson* 5727 (K). **Zimbabwe**. E: Chimanimani Distr., Nyambamba Falls, Lower Ngorima Road, fl. 14.x.1963, *Masterson* 238 (K; SRGH). **Malawi**. N: Mzimba Distr., Mzuzu, Marymount, fl. 11.xi.1969, *Pawek* 2898 (K). C: Nkhota-Kota Distr., Chia area, fl. 5.ix.1946, *Brass* 17542 (K; NY). S: Mulanje, Ruo Gorge, fl. 2.x.1983, *Johnston-Stewart* 149 (K). **Mozambique**. N: between Lichinga and Maniamba, fl. 10.x.1942, *Mendonça* 721 (BR; EA; LISC; LMU; PRE; SRGH). Z: Quelimane, Munguluni Mission, fl. ix. *Faulkner* Kew 60 (K). MS: Mafuci Forest, fl. x–xi.1911, *Dawe* 473 (K).

Also in Nigeria, Cameroon, Central African Republic, Zaire, Burundi, Sudan, Ethiopia, East Africa and Angola. Riverine, streamside and lake-shore fringing dry evergreen forest also relict mist-forest, often on laterite escarpments and sometimes on termite mounds; 20 (in Beira area) 540–1770 m.

Throughout its range *Craterispermum schweinfurthii* is distinguished from *Craterispermum laurinum* (with which it was confused) partly by its sessile fruits but one specimen from Zimbabwe E, Lusitu River, fr. 8.i.1969, *Mavi* 835 (K; LISC; SRGH) is anomalous in having ± pedicellate fruits — the species is here at the extreme SE limit of its range.

Tribe 7. **KNOXIEAE**

Slender annual herb; flowers 4-merous, with corolla tube 3 mm. long; ovary 2-locular; style arms 2; fruit breaking off to leave a small cup formed of the persistent woody flanged pedicel - - - - - - - - - - - - - - - - **11. Paraknoxia**

Perennial herb; flowers mainly 5-numerous, with corolla tube exceeding 3 mm. in length; ovary 2–5-locular; style arms 2–5; fruit not leaving a cup-like remnant after falling **12. Pentanisia**

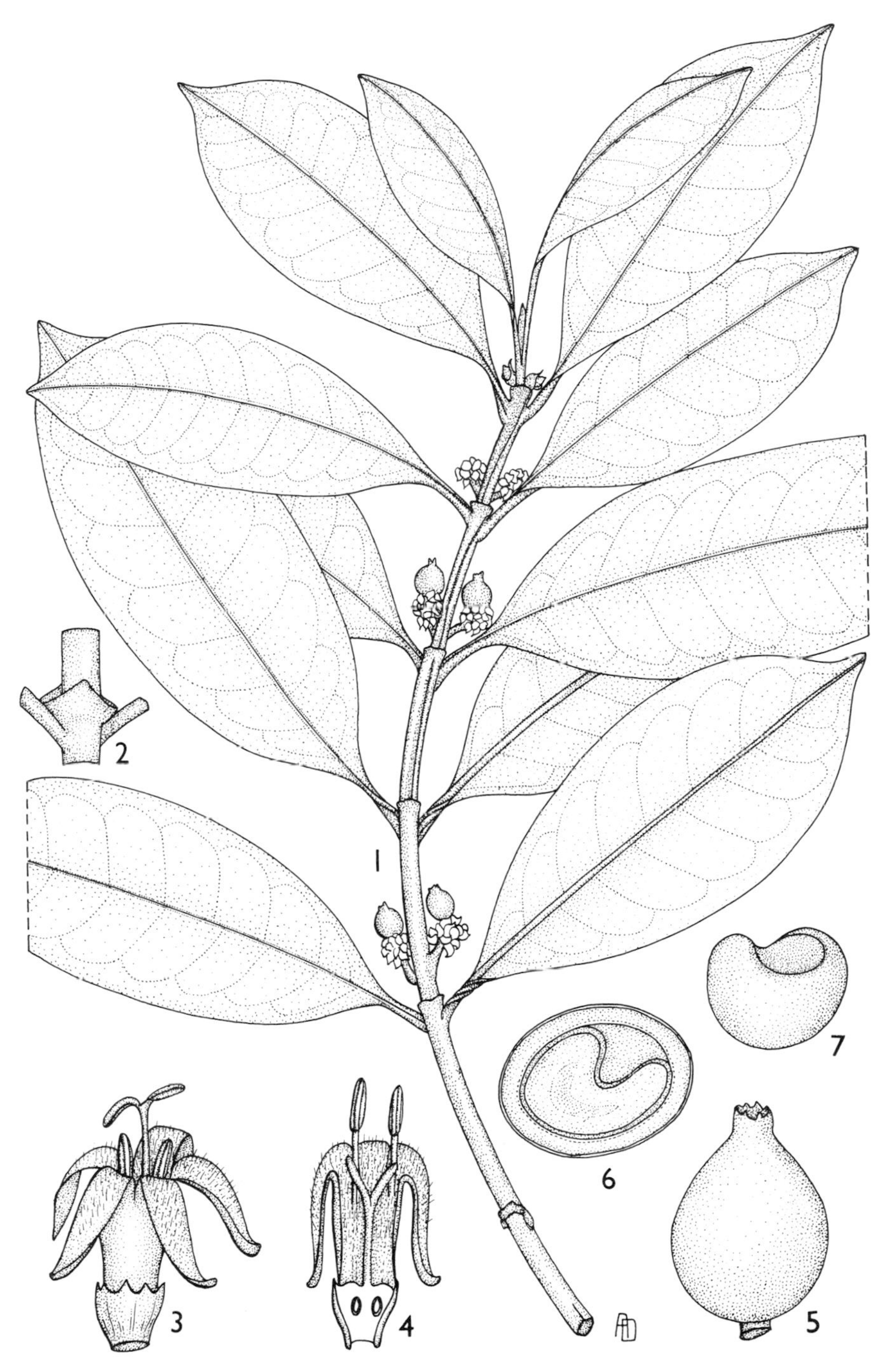

Tab. 12. CRATERISPERMUM SCHWEINFURTHII. 1, habit (×$\frac{1}{2}$); 2, node showing stipule (× 2), 1–2 from *Ritchie* 1; 3, long-styled flower (× 4), from *Haarer* 2170; 4, short-styled flower, longitudinal section (× 4), from *Lye* 3452; 5, fruit (× 4); 6, fruit, transverse section (× 4); 7, seed (× 4), 5–7 from *Purseglove* 1265. From F.T.E.A.

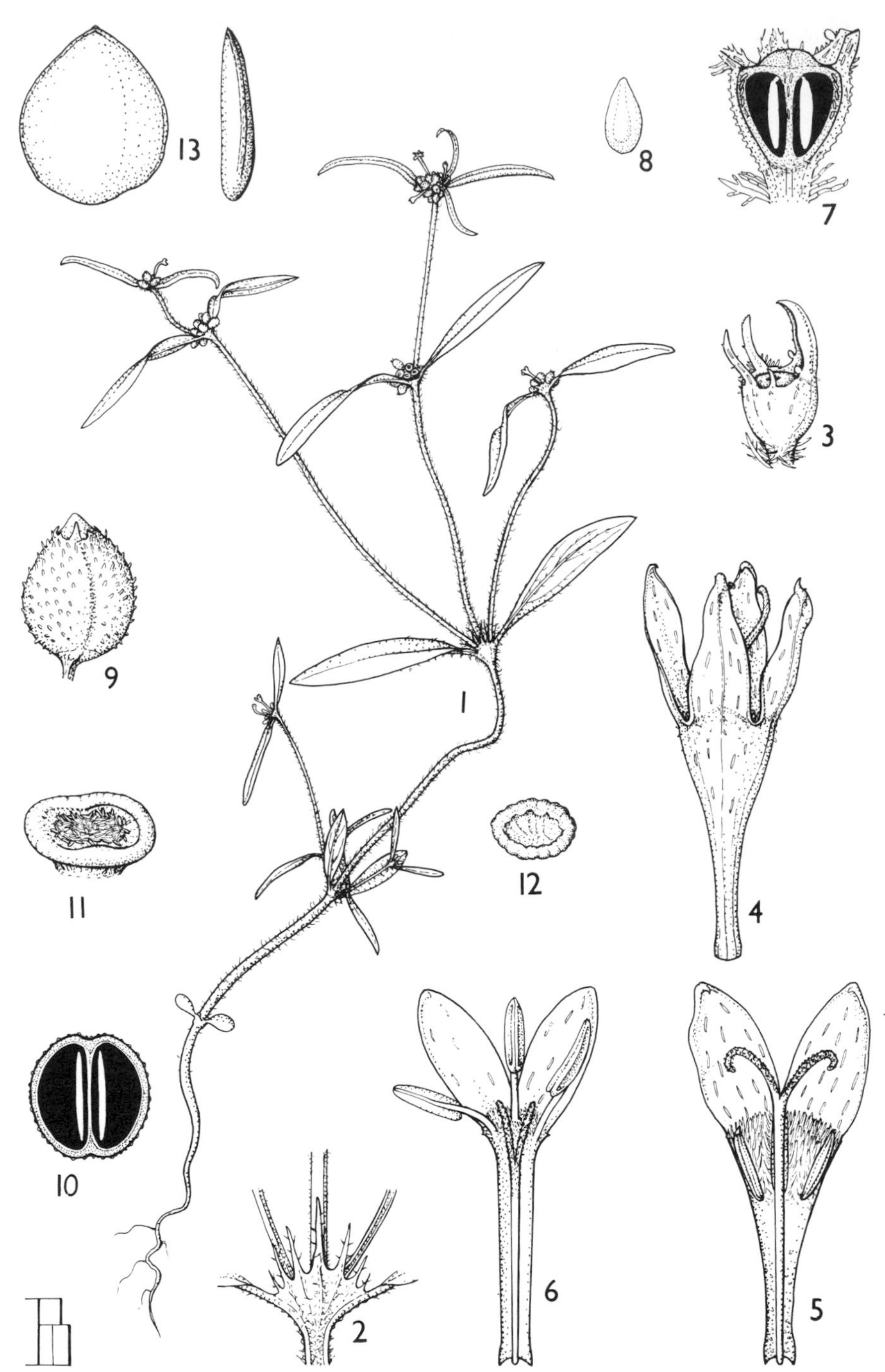

Tab. 13. PARAKNOXIA PARVIFLORA. 1, habit (× 1), from *Polhill & Paulo* 1986; 2, node showing stipule (× 4), from *Maitland* 1310; 3, calyx (× 14); 4, corolla of long-styled flower (× 10); 5, long-styled flower, longitudinal section (× 10); 6, short-styled flower, longitudinal section (× 10); 7, ovary, longitudinal section (× 20); 8, ovule (× 20), 3–8 from *Lewis* 5967; 9, fruit (× 14), from *Whyte*; 10, fruit, transverse section (× 14); 11, pedicel cup after fall of fruit with rhaphides still present (× 14), from *Polhill & Paulo* 1986; 12, pedicel cup after fall of fruit (× 14); 13, seed, two views (× 20), 12–13 from *Lugard* 99. From F.T.E.A.

11. **PARAKNOXIA** Bremek.

Paraknoxia Bremek. in Bull. Jard. Bot. Brux. **22**: 77 (1952).
Pentanisia Harv. subgen. *Micropentanisia* Verdc. in Bull. Jard. Bot. Brux. **22**: 262 (1952).

Annual herbs. Leaves paired, shortly petiolate; stipules small, with several deltoid segments from a short base. Flowers small, hermaphrodite, dimorphic, some flowers with anthers included and style exserted and others completely vice versa, in small sessile terminal heads. Calyx tube short; lobes small, 3–4, sometimes 1 enlarged and rest minute, or all minute. Corolla tube narrowly funnel-shaped; lobes 3–4, oblong; throat hairy. Ovary 2-locular; ovules solitary in each locule, pendulous; style filiform; stigma bifid, the lobes filiform. Fruit ovoid, indehiscent, but longitudinally grooved and where the fruit breaks off a little cup is left being the persistent woody flanged pedicel; between this flange and the base of the fruit are masses of rhaphides. Seeds narrowly ellipsoid or ovate in outline, strongly compressed.

A monotypic genus occurring in eastern and central Africa. I formerly considered it best retained in *Pentanisia*, but agree that there are sufficient characters to separate it from that genus.

Paraknoxia parviflora (Stapf ex Verdc.) Verdc. ex Bremek. in Kew Bull. **8**: 439 (1953). —Agnew, Upland Kenya Wild Fl.: 405, fig. (1974). —Verdc. in F.T.E.A., Rubiaceae 1: 166 (1976). TAB **13**. Type from Uganda.

Pentanisia parviflora Stapf ex Verdc. in Kew Bull. **6**: 383 (Jan. 1952); in Bull. Jard. Bot. Brux. **22**: 263, fig. 31/H, I (1952).

Paraknoxia ruziziensis Bremek. in Bull. Jard. Bot. Brux. **22**: 77 (June 1952). Type from Zaire.

Paraknoxia parviflora Verdc. in Bull. Jard. Bot. Brux. **22**: 265 (1952) (nomen eventuale).

Erect herb 4–38 cm. tall; stems slender, often branched, hairy. Leaf blades 1.3–4.2 × 0.2–1(1.7) cm., narrowly elliptic, lanceolate, linear or linear-oblanceolate or ovate-elliptic, ± acute at the apex, cuneate at the base, covered with short hairs; petioles up to 3 mm. long; stipules with c. 3–7 deltoid segments 1–2 mm. long from a pale chestnut-coloured base 1–1.5 mm. long. Inflorescences c. 2–4 mm. wide, supported by the apical leaf pair; there are usually 2 axillary flowering branchlets which overtop the true terminal one. Calyx tube 1 mm. long, covered with small scaly hairs; foliaceous lobe 0.8 × 0.5 mm. or absent, the others very minute, crowned with hairs. Corolla white or tinged bluish mauve; tube 1.75–4 mm. long, dilated in long-styled flowers to 1–1.5 mm. wide for the apical third, glabrous outside: lobes 1.3–1.5 × 0.4–0.9 mm., hairy outside. Style 3.3 mm. long in long-styled flowers, tomentose; stigma lobes 0.6–0.8 mm. long, papillate. Fruit 1.3–1.75 × 1.25 mm., somewhat acute at the apex, densely covered with scaly hairs. Seeds chestnut-coloured, 0.8–0.9 mm. long.

Zimbabwe. N: Mazoe, Mbebe Beacon, fl. & fr. 26.ii.1952, *Wild* 3771 (K; SRGH). **Malawi**. N: Karonga Distr., fl. & Fr. iii.1954, *Jackson* 1255 (K).

Zaire, Central African Republic, Uganda, Kenya, Tanzania and Angola. Ironstone Crags, CuO ore with dolomite; 1500 m.

12. **PENTANISIA** Harv.

Pentanisia Harv. in Hook., Lond. Journ. Bot. **1**: 21 (1842). —Verdc. in Bull. Jard. Bot. Brux. **22**: 233 (1952).

Perennial herbs or subshrubs, with glabrous or hairy, erect, procumbent or decumbent stems. Leaves variable, small to moderate, linear to round, paired or rarely in whorls of 3, sessile or shortly petiolate; stipules with base connate with the petiole, apically fimbriate. Flowers mostly blue, small to medium-sized, hermaphrodite, dimorphic, some flowers with anthers included and style exserted and others completely vice versa, usually in few- to many-flowered terminal capitate inflorescences, frequently becoming more elongate in fruit, often spicate. Calyx tube ovoid or ± rectangular; lobes mostly 5, 1–3 enlarged and often foliaceous, the rest small or even obsolete. Corolla tube narrowly cylindrical, with an apical cylindrical dilation in long-styled flowers but narrowly funnel-shaped in short-styled flowers; lobes mostly 5, ovate to oblong; throat densely hairy. Ovary 2–5-locular, each locule with 1 pendulous ovule attached close to the apex; style filiform, the stigmas divided into 2–5 filiform lobes corresponding with the number of locules in the ovary. Fruit dry, globular, slightly lobed, ovoid or compressed obcordate, indehiscent or tardily dehiscent into mericarps, or subglobose and rather succulent. Seeds small, compressed.

A genus of 15* species confined to tropical Africa and Madagascar; 7 are known from the Flora Zambesiaca area. Species 1–3 belong to subgen. *Pentanisia* section *Pentanisia*, 4 to subgen. *Pentanisia* section *Axillares* Verdc., and species 5–7 to subgen. *Holocarpa* (Bak.) Verdc. Robbrecht & Puff have recently raised section *Axillares* to generic rank as *Chlorochorion* on account of the fruit structure and axillary inflorescences.

1. Ovary 2-locular, very unusually abnormally 3-locular - - - - - - - - - 2
– Ovary 3–5-locular - - - - - - - - - - - - - - - - - 5
2. Stems long, weak and straggling; rootstocks not distinctly woody; flowers in 2 axillary pedunculate inflorescences at the terminal node and often at the lower nodes also - - - - - - - - - - - - - - - - - 4. *monticola*
– Stems erect or procumbent, caespitose from a woody rootstock; main inflorescences terminal, although axillary ones are often present as well - - - - - - - - - 3
3. Foliaceous calyx lobes 1–2.5(3.5) mm. long; fruit ovoid, with thin walls 3. *schweinfurthii*
– Foliaceous calyx lobes 0.3–1.2 cm. long; fruit black, compressed obcordate, with woody walls - - - - - - - - - - - - - - - - - - - 4
4. Leaves orbicular to oblong, often hirsute, more rarely glabrous; inflorescence often capitate; leaf ratio (length/breadth) 1–5.35(8.4) - - - - - - - - 1. *prunelloides*
– Leaves linear to lanceolate or oblong, glabrous or margins ciliate; inflorescence often spicate in fruit; leaf ratio (5)6.35–54 - - - - - - - - - - - - 2. *angustifolia*
5. Corolla tube 3.5–5.5 cm. long; leaves lanceolate to oblong-lanceolate 7. *confertifolia*
– Corolla tube 0.7–1.9 cm. long - - - - - - - - - - - - - 6
6. Leaves lanceolate, narrowly elliptic, elliptic-oblong or ± oblanceolate 5. *sykesii*
– Leaves broadly elliptic or ovate to reniform or oblate - - - - - - 6. *renifolia*

1. **Pentanisia prunelloides** (Eckl. & Zeyh.) Walp., Repert. **2**: 941 (1843). —O. Kuntze, Rev. Gen. Pl. **3**: 122 (1898). —Schinz, Viert. Nat. Ges. Zürich **68**: 437 (1923). —Verdc. in Bull. Jard. Bot. Brux. **22**: 248 (1952); in F.T.E.A., Rubiaceae 1: 168 (1976). Type from South Africa.
Declieuxia prunelloides Eckl. & Zeyh., Enum. Pl. Afr. Austr. Extratrop.: 363 (1837).
Diotocarpus prunelloides (Eckl. & Zeyh.) Hochst. in Flora **26**: 71 (1843).

Erect or semi-prostrate to completely prostrate herb 7–60 cm. tall, with mostly numerous hairy or glabrous stems from a thick woody rootstock up to 7 cm. wide. Leaf blades 1.3–8.5 × 0.2–3.5 cm., oblong-elliptic or ovate to almost round, acute or subobtuse at the apex, rounded to subcordate at the base, mostly pubescent to densely villous, less often glabrous; petiole obsolete; stipules with ± 5 narrow segments, 3–7 × 0.3–2.5 mm., from a short base to 3 mm. long. Inflorescence capitate, ± villous, branched, or spicate below, up to 7 × 4 cm.; peduncle 3.5–34 cm. long, often densely hairy. Calyx tube c. 1 × 0.8–1 mm., pubescent; lobes unequal, 1–2 larger and foliaceous, 0.3–1.2 cm. × 0.5–1.5(2) mm. the rest small. Corolla bright blue; tube 1.1–1.8 cm. long, 1–1.5(2) mm. wide at the apex, pubescent outside; lobes 2.5–5 × 1–2.5 mm.; throat densely hairy. Style in long-styled flowers exserted 2–3 mm.; stigma 2-fid, the lobes filiform, 1–1.5 mm. long. Fruit black, 2.5–4.5 × 2–3.3 mm., obcordate, laterally compressed, composed of 2 indehiscent cocci which eventually separate, pubescent. Seeds brown, almost round in outline, 2.5 × 2 mm., thin.

Subsp. **prunelloides**

Leaves oblong, 1.7–7.5 × 0.3–2.8 cm. usually hairy; leaf blade length to breadth ratio 2.35–5.35.

Mozambique. M: Maputo, Goba Fronteira, fl.9 fr. 12.xii.1947, *Barbosa* 694 (COI; LISC; LMU; SRGH; WAG).
Also in South Africa. Rocky orange-brown soil; 650 m.

Subsp. **latifolia** (Hochst.) Verdc. in Bull. Jard. Bot. Brux. **22**: 250 (1952). Type from S. Africa.
Declieuxia latifolia Hochst. in Flora **26**: 70 (1843).
Pentanisia prunelloides var. *latifolia* (Hochst.) Walp., Repert. **2**: 941 (1843).
Pentanisia variabilis var. *latifolia* (Hochst.) Sond. in Harv. & Sond., F.C. **3**: 24 (1865).
Pentanisia longisepala Krause in Engl., Bot. Jahrb. **39**: 532 (1907). Type from S. Africa.

* The transference of *P. parviflora* Verdc. to a separate genus *Paraknoxia* Bremek. is upheld in this work — see genus No. 11.

Leaves 1.6–8.5 × 1.1–3.5 cm., broadly ovate to round, or less often oblong, usually densely hairy; leaf blade length to breadth ratio 0.8–2.8.

Zambia. N: Mpika, fl. 11.ii.1955, *Fanshawe* 2055 (K; NDO). **Malawi**. N: Rumphi Distr., N. Viphya, 29 km. from Mzuzu on Ususya road, fl. 23.iii.1969, *Pawek* 1870 (K).
Also in S. Tanzania and S. Africa. Plateau woodland, grassy places in *Brachystegia–Uapaca* woodland;

See the general note after the next species.

2. **Pentanisia angustifolia** (Hochst.) Hochst. in Flora **27**: 555 (1844). —Verdc. in Bull. Jard. Bot. Brux. **22**: 252 (1952). —Letty, Wild Fl. Transvaal: 324, t. 162/2 (1962). Type from S. Africa.
Diotocarpus angustifolius Hochst. in Flora **26**: 71 (1843).
Pentanisia variabilis var. *glaucescens* Cruse ex Sond. in Harv. & Sond., F.C. **3**: 24 (1894). Types from S. Africa.

Erect straggling or prostrate plant, 20–60 cm. tall with several hairy or glabrous stems from a thick woody rootstock usually 2 cm. wide. Leaf blades variable, 2–15 cm. × 2–10 mm., linear to oblong or lanceolate sometimes even up to 50 times as long as wide, acute at the apex, cuneate at the base, glabrous or rarely hairy or margins above ciliate; petiole 1.5–2 mm. long; stipules with 3–5 linear segments 2.5–6 mm. long from a base 3–4 mm. long. Inflorescence a head or short spike, usually becoming a branched or unbranched spike up to 9 cm. long; peduncle 9–17 cm. long, elongating in fruit. Calyx tube c. 1 mm. long, the foliaceous lobe 4–10 × 0.3–1.5 mm., glabrous or ciliate. Corolla blue or violet-blue; tube 1.2–1.6 cm. long, pubescent outside with short appressed hairs; lobes 3.5–5 × 1–2 mm. Style in long-styled flowers as in previous species. Fruit as in last species.

Mozambique. M: Goba, near the R. Maiuaua, fl. & fr. 3.xi.1960, *Balsinhas* 174 (BM; K; LISC; LMA; LMU).
Also in NE. S. Africa. Herb layer of *Androstachys johnsonii* woodland; ? –800 m.
The *Pentanisia prunelloides* complex is one of the most variable in existence. A perusal of several hundred specimens shows such extremes that it is difficult to believe at first, that the material does not represent at least five species. There is, however, one constant factor — the fruit is identical in all forms and much weight has been attached to this fact. All the variation may be reduced to variation in four characters namely — (a) ratio of length to breadth of leaves (b) nature of the inflorescence (c) hairyness, and (d) habit. The latter character does not seem to be correlated to the others and both narrow-and wide-leaved forms can be prostrate. The leaves vary from orbicular to linear and glabrous to hirsute. Hairyness shows a definite tendency to be associated with broad leaves. The inflorescence varies from a compact hemispherical head on a relatively stout hairy peduncle which does not elongate in fruit, to a cylindrical head or spicate inflorescence, which often becomes an elongated spike when fruiting. Elongated inflorescences are associated with narrow leaves but there are indefinite intermediates. Extremes are so distinct that the complex has been divided into two species, one with narrow usually glabrous or ciliate leaves and an elongated inflorescence, and another with broader often hairy leaves and a capitate inflorescence. Specimens with almost orbicular leaves are so distinctive that a subspecific name has been used for them. Between *P. prunelloides* and its subspecies *latifolia* every possible intermediate occurs and there are numerous intermediates between *P. prunelloides* and *P. angustifolia*. The two extreme forms often occur together in the same locality, e.g. at Sabie in S. Africa (*Rogers* 23741) where specimens with leaves 29 mm. long by 21 mm. broad, and others with leaves 77 by 2.5 mm. grow within a foot of each other.
The correlation between length and breadth of leaves and hairyness is shown in fig. 29 of my original revision and at the ends of the range the correlation is quite good.

3. **Pentanisia schweinfurthii** Hiern in F.T.A. **3**: 131 (1877). —Verdc. in Bull. Jard. Bot. Brux. **22**: 254 (1952). —Brenan in Mem. N.Y. Bot. Gard. **8**: 451 (1954). —Agnew, Upland Kenya Wild. Fl.: 405, fig. (1974). —Plowes & Drummond, Wildfl. Rhod., t.175 (1976). —Verdc. in F.T.E.A., Rubiaceae 1: 168 (1976). —Gonçalves in Garcia de Orta, Sér. Bot. **5** (2): 201 (1982). TAB. **14**. Type from Sudan.
Pentanisia crassifolia K. Krause in Engl., Bot. Jahrb. **39**: 531 (1907). Type: Zimbabwe, Harare, Norton, *Engler* 3022 (B, holotype †).
Pentanisia rhodesiana S. Moore in Journ. Bot., Lond. **40**: 252 (1902). Type: Zimbabwe, Harare, *Rand* 575 (BM, holotype).
Pentanisia sericocarpa S. Moore in Journ. Bot., Lond. **40**: 251 (1902). Type: Zimbabwe, Harare, *Rand* 619 (BM, holotype).
Pentanisia schweinfurthii var. *pubescens* Verdc. in Bull. Jard. Bot. Brux. **22**: 258 (1952). Type: Malawi, Gt. North Road, 26 km. from Chitipa towards Katumbi Camp, *Burtt* 6095 (BM; EA; K, holotype).
Pentanisia variabilis sensu auctt. non Harv.

Perennial pyrophytic herb, 3.8–24 cm. tall, from a woody rootstock; stems up to 25 from

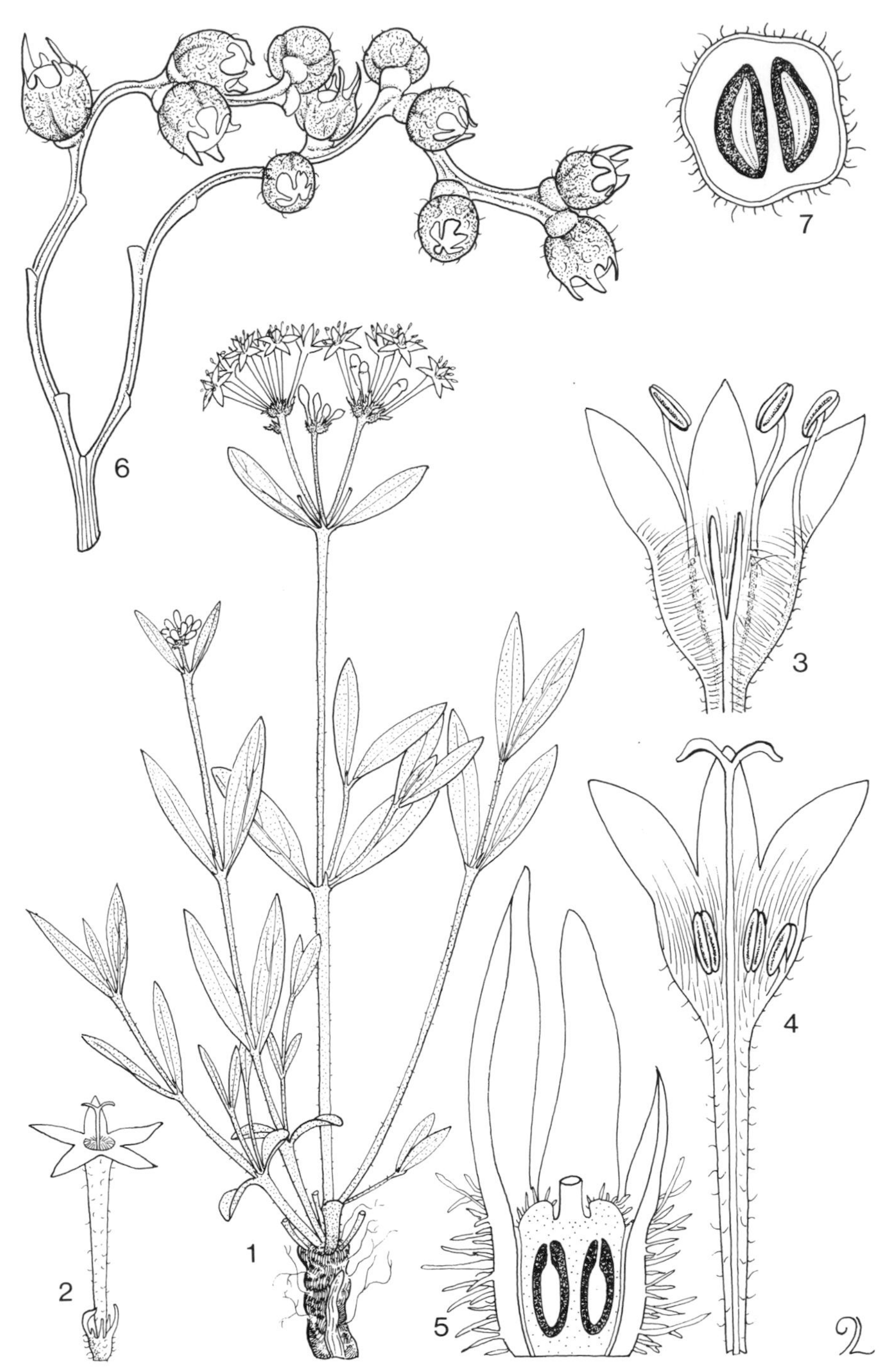

Tab. 14. PENTANISIA SCHWEINFURTHII. 1, habit (× 1); 2, flower (× 3); 3, part of short-styled flower, longitudinal section (× 9); 4, long-styled flower, longitudinal section (× 9); 5, ovary (× 30); 6, infructescence (× 3); 7, fruit, transverse section (× 12), all from *Leach & Brunton* 10032.

each root (about 3–5 from each apical branch of the rootstock), glabrous or bifariously hairy. Leaf blades very variable, 0.3–6 × 0.2–1.8(2) cm., the lower round, elliptic or elliptic-obovate, the upper linear to round, acute to obtuse at the apex, cuneate at the base, mostly drying yellowish green, glabrous or rarely shortly hairy; petiole up to 2 mm. long; stipules with 2–4 deltoid lobes or 1 trifid lobe, 1–4.5(7) mm. long from a base 1–3 mm. long. Inflorescences capitate, 1–2.5 × 0.3–1.3 cm. or branched and spike-like, 2–3.5 cm. long; peduncle up to 7.4 cm. long. Calyx tube squarish 0.5–1(1.5) mm. long and wide, glabrous or sometimes covered with bristly white hairs; lobes unequal, the longest 1–3.5 × 0.3–0.8 mm., the rest minute. Corolla bright blue, white, pale lilac or purple; tube 0.6–1.3 cm. × 1.2–2(3) mm. at the apex, glabrous or pubescent with short white hairs; throat densely hairy; lobes 2–6 × 1–2 mm., ovate-oblong to ovate-lanceolate. Style exserted 2–4 mm. in long-styled flowers; stigma lobes linear, 0.5–1.5(2) mm. long. Fruiting inflorescence spicate, up to 4 cm. long; fruits 1.5–2.5 × 1.5–2 mm., ovoid, broadest at the middle, glabrous or covered with white pubescence, 2-locular, the locules often unequal, thin-walled, not ribbed, borne on ledges on the rhachis. Seeds yellow-brown, 2 × 1.5 × 0.6 mm., broadly elliptic in outline, concavo-convex, finely marked with brown.

Zambia. B: 14.4 km. ESE. of Mankoya, near Luene R., fl. 21.xi.1959, *Drummond & Cookson* 6709 (K; SRGH). N: 48 km. S. of Mbala, fl. 18.vii.1930, *Hutchinson & Gillett* 3853a (K). W: Mwinilunga Distr., slope E. of R. Lunga, fl. & fr. 23.xi.1937, *Milne-Redhead* 3368, 3369 (K). C: 16 km. SW. of Serenje, fl. 14.vi.1960, *Leach & Brunton* 10032 (K; LISC; SRGH). E: Nyika Plateau, 3.2 km. SW. of Rest House, fl. & fr. 21.x.1958, *Robson & Angus* 184 (BM; K). **Zimbabwe**. C: Makoni District, 13 km. W. of Rusape, Lawrencedale Estate, fl. 15.x.1961, *Chase* 7543 (K; LISC; SRGH). E: Chimanimani, Pork Pie Mt., fl. 21.ix.1960, *Rutherford-Smith* 135 (K; LISC; SRGH). **Malawi**. N: Nyika Plateau, top of ridge off the Chelinda Road, fl. 15.xi.1967, *Richards* 22554 (K). C: Dedza Distr., Chongoni Forest Reserve, foot of Chencherere Hill, fl. 25.iv.1970, *Brummitt* 10143 (K; MAL). S: Zomba Plateau, fl. 28.v.1946, *Brass* 16042 (K; NY). **Mozambique**. N: Niassa, N. of Mandimba, fl. 25.xi.1941, *Hornby* 3493 (K). T: Angonia, Ulongue, Cauzuzu, fl. & fr. 17.xii.1980, *Macúacua* (K; LMA; SRGH). MS: Nyamakwarara Valley, Gorongo Mt., fl. 2.xi.1967, *Mavi* 443 (K; LISC; SRGH).

Widespread in tropical Africa from Nigeria and Sudan to Zaire, Angola and Zimbabwe. Submontane and lower grassland, clearings in *Brachystegia–Julbernardia–Uapaca* woodland particularly on bare soil and other exposed areas and poorish grassland subject to burning; 780–2270 m.

Pubescent forms are now known to occur in many populations.

4. **Pentanisia monticola** (K. Krause) Verdc. in Kew Bull. **7**: 362 (1952); in Bull. Jard. Bot. Brux. **22**: 260, fig. 31/J, K (1952); in F.T.E.A., Rubiaceae 1: 170 (1976). Type from Tanzania.
Otomeria monticola K. Krause in Engl., Bot. Jahrb. **57**: 26 (1920).

Perennial erect weak herb 0.9–1.2 m. tall, with striate stems dichotomously branched at many of the nodes, glabrescent save for nodes and young shoots. Leaf blades 4.6–10 × 1.1–3.3 cm., lanceolate to ovate-lanceolate, acute at the apex, rounded to cuneate at the base, with long appressed pubescence above and on the nerves beneath; petiole 1–4(7) mm. long; stipules with 5–6 setae, often reflexed when older, 4–9 mm. long from a base 12—2 mm. long. Flowers white, pinkish white or very pale purple, dimorphic, in slender axillary spikes, 1–5 cm. long, 2 from most nodes even the lowest, usually those where dichotomous branching occurs; peduncle 1–9.5 cm. long. Calyx tube 1 × 0.8–1 mm., ovoid; largest lobes 1.5–2(3) × 0.5(1) mm., rest minute. Corolla tube 6–8(10) mm. long, c. 0.5 mm. wide at the base, 1.2–2 mm. wide at the apex, hairy inside; lobes 2 × 0.9 mm. Style in long-styled flowers exserted 2–2.2 mm.; stigma bifid, the lobes filiform c. 1 mm. long. Fruit 1.5–2 × 1.8–2 mm., globose-reniform, cordate at the base, bilocular, each coccus oblique when separated, the interface densely covered with rhaphides, thin-walled. Seeds yellow-brown, ellipsoid, 1.4 × 0.8 mm.

Zambia. N: Mbala Distr., Kalambo Road, fl. & fr. 10.iv.1970, *Sanane* 1166 (K). **Malawi**. N: Chitipa Distr., Misuku Hills; Mughesse rain forest, fl. & fr. 25.iv.1972, *Pawek* 5217 (K).

Also in Burundi and S. Tanzania. Grassland, bushland, *Brachystegia* woodland, also edges of evergreen forest, roadsides etc.; 1350–1800m.

5. **Pentanisia sykesii** Hutch. in Kew Bull. **1906**: 248 (1906). —Verdc. in Bull. Jard. Bot. Brux. **22**: 266, fig. 32/H (1952); in F.T.E.A., Rubiaceae 1: 170 (1976). Type: Zambia, Batoka Plateau, *Sykes* in *Herb. Allen* 225 (K, holotype; SRGH).

Perennial herb with 5–20 erect, spreading or prostrate stems 8–30(40) cm. tall from a woody rootstock c. 3 cm. in diam., or sometimes forming mats 1.2 m. in diam.; stems narrowly bifariously hairy or scaly puberulous. Leaves paired or sometimes appearing verticillate due to short axillary branchlets, 0.65–6.0 × 0.2–1.5 cm., the lowest ovate, the rest lanceolate, narrowly elliptic, elliptic-oblong or almost oblanceolate, obscurely to distinctly acute at the apex, cuneate at the base, glabrous or at most puberulous on the lateral nerves beneath, often drying yellow-green; petiole up to 2 mm. long, adnate to the stipular sheath; stipules with 3 flat linear or rarely spathulate lobes, 0.2–1.2 cm. long from a base 1.5–4 mm. long. Inflorescence dense, capitate or spicate, sometimes branched, 1.5–3 cm. long and wide; peduncle 2.5–10.5 cm. long. Flowers blue or rarely white, dimorphic. Calyx tube blackish, quadrate, 1–1.2 mm. long and wide, glabrous; lobes unequal, the 1–2 foliaceous ones 0.3–1.2 cm. × 0.9–3 mm., lanceolate, ovate-lanceolate or oblong-lanceolate, the rest small, c. 1 mm. long, all glabrous. Corolla tube (0.7)1.05–1.7 cm. long, 0.3–1.2 mm. wide at the base, 1–2.5 mm. wide at the throat for a distance of 2–5 mm. in long-styled flowers, the dilation sometimes constricted at the throat so its orifice is narrow, glabrous to scaly puberulous outside; lobes (3)4–7 × 1–3.5 mm., broadly elliptic, ovate-oblong or oblong, hairy outside; throat densely hairy. Ovary and fruit 3–5-locular; style in long-styled flowers exserted 2.5–4 mm.; stigma lobes 2–5, filiform 0.5–1 mm. long. Fruit black, globose, ± fleshy and succulent, 4.5–8 × 3.1–7.5 mm., glabrous, narrowed at the apex, very shallowly lobed, crowned by the persistent calyx lobes; walls thickish and woody, with regularly spaced channels just within the calyx tube layer. Seeds yellow, 2.5 × 1.5 mm., ovoid, laterally much compressed, with a reddish brown pattern.

Subsp. **sykesii**. —Verdc. in Bull. Jard. Bot. Brux. **22**: 266, fig. 32/H (1952); in F.T.E.A., Rubiaceae 1: 171 (1976).

Lateral nerves 4–6 pairs prominent on both surfaces.

Zambia. N: Mbala Distr., escarpment above Chilongowelo, fl. 15.xii.1951, *Richards* 42 (K). C: Chikakata R., 24 km. NW. of Kabwe, fl. 21.i.1973, *Kornaś* 3064 (K; KRA). S: between Choma and Mochipapa Research Farm, fl. 14.i.1963, *van Rensburg* 1204 (K). **Malawi**. N: Chitipa Distr., 0.8 km. W. of Misuku–Karonga–Chendo crossroads, fr. 25.iv.1977, *Pawek* 12672 (K; MAL; MO; SRGH; UC).

Also in Tanzania. Dambos in grassland and *Brachystegia* woodland; 1150–1500 m.

Subsp. **otomerioides** Verdc. in Bull. Jard. Bot. Brux. **22**: 268 (1952). Type: Zimbabwe, Rusape, *Eyles* 7575 (K; SRGH, holotype).

Lateral nerves obsolete or visible only with difficulty, and then perfectly plane.

Zimbabwe. C: Rusape, fl. v.1931, *Doyle* 16 in *Eyles* 7575 (K; SRGH). E: Chimanimani, plateau S. of Mt. Peni Beacon, fl. xii. 1971, *Goldsmith* 31/71 (K; SRGH). **Mozambique**. MS: Báruè, Serra de Choa, 29 km. from Catandica (Vila Gouveia), fl. 9.xii.1965, *Torre & Correia* 13450 (COI; LISC; LMU).

Also in S. Africa (Transvaal). Grassland, roadsides etc.; 1050–1810 m.

Very similar to *Pentanisia schweinfurthii* and mixed gatherings have been made indicating that the two taxa grow together but they are absolutely distinct and the present one is easily distinguished by its numerous styles, broader corolla lobes, larger and more foliaceous calyx lobes and much larger fruits. Further work is needed on the fruits of this species, those of *sykesii* are undoubtedly brown and corky sometimes and Dr. Wild confirmed that the fruits of presumably subsp. *otomerioides* were black and succulent. There are, however, large numbers of herbarium specimens in flower and few in fruit and the possibilities that specific rank might be more appropriate needs investigation.

6. **Pentanisia renifolia** Verdc. in Bull. Jard. Bot. Brux. **22**: 271 (1952); in Kew Bull. **12**: 354 (1957). Type from Zaire.

Perennial herb with hairy ± prostrate stems up to 1.5 m. long, from a woody rootstock which is orange inside. Leaves 1.5–4 × 1–5 cm., broadly elliptic or ovate to reniform or oblate, acute or subacute to ± rounded at the apex with often a short acumen, rounded to emarginate at the base, pubescent to hairy, particularly on the c. 10 arcuate subbasal or pinnate nerves; petiole obsolete or up to c. 2(5) mm. long; stipules with c. 4 linear fimbriae 1.5–4 mm. long. Inflorescences congested, ± capitate, mostly at least ± bifid and somewhat spicate, mostly 3–4 cm. long and wide; peduncle 2–17 cm. long. Calyx tube oblong, 1.5–2 mm. long; lobes 5, the foliaceous one 2.5–5 × 1–2.5 mm., oblong-elliptic, the rest 0.5–1 mm. long. Corolla bright blue, the tube more violet-blue; tube 1.4–1.9 cm. long, pubescent outside; lobes oblong, 4–8 × 1–3 mm., pubescent outside, mostly acuminate. Filaments exserted c. 2 mm. in short-styled plants and style about 11 mm. long with lobes c. 2 mm. long. In long-styled plants style exserted c. 1.5 mm. but lobes shorter, about 0.5 mm. long. Fruit drying dark, ellipsoid, 8 × 6.5 mm., ribbed in dry state, with 4–5-locular straw-coloured putamen.

Zambia. W: Solwezi Distr., 12 km. W. of Mutanda, fl. 19.i.1975, *Brummitt et al.* 13849 (K).
Also in S. Zaire. *Brachystegia* woodland, bare ground at edge of limestone dambo; 1300 m.

7. **Pentanisia confertifolia** (Bak.) Verdc. in Kew Bull. **7**: 361 (1952); in Bull. Jard. Bot. Brux. **22**: 272, fig. 31/F–G (1952). Type: Zambia, Lake Tanganyika, Fwambo, *Carson* 23 (K, holotype).
Pentas confertifolia Bak. in Kew Bull. **1895**: 87 (1895).

Erect herb with 4–several stems 40–75 cm. tall from a woody rootstock, pubescent to hairy or glabrous save for lines of short white hairs on some internodes. Leaves paired but appearing pseudoverticillate due to arrested leafy branchlets, 2–8 × 0.3–1.5 cm., lanceolate to oblong-lanceolate, drying a distinct yellow-green, acute at the apex, cuneate at the base, hairy or glabrous; petioles obsolete; stipules with 3–9 linear-lanceolate flattened capitellate green segments 0.9–2 cm. long from a base 2 mm. long. Flowers apparently trimorphic in dense heads up to c. 10 cm. wide; peduncle 5.5–11 cm. long. Calyx tube 1.5 mm. long, densely hairy, the limb tube up to 2.5 mm. long, the foliaceous lobe 0.9–1.4 cm. × 1.2–2.3 mm., lanceolate, remainder minute or linear, up to about 3 × 0.5 mm. Corolla white; tube 3.5–5.5 cm. long, densely hairy outside; lobes 6–9 × 1.7–3 mm., narrowly oblong, glabrous to pubescent outside, the acuminate tips inflexed. Filaments exserted 1 mm. in short-styled flowers, anthers livid blue. Style exserted 2–6(9) mm. in long-styled flowers the 4–5 lobes 1.5–2 mm. long; there also seem to be flowers with both style and stamens included. Fruiting inflorescence becoming spicate c. 4 cm. long; fruits drying dark, subglobose, 7–8 mm. long and wide.

Zambia. N: Mbala, fl. 15.vi.1951, *Bullock* 3966 (K).
Apparently endemic. Coarse grassland, bushland and *Brachystegia* woodland, sometimes amongst boulders on sandy ground, also old cultivations; 1500–1650 m.
Forma *glabrifolia* Verdc. (Bull. Jard. Bot. Brux. **22**: 273 (1952)). Type: Zambia, 8 km. Mbala to Lungwa, *Greenway & Brenan* 8238 (EA, holotype; K; PRE) occurs in normal populations and is not worth retaining.

Tribe 8. **PAEDERIEAE**

Shrubs, dwarf shrubs or perennial herbs, often with a foetid odour. Flowers ⚥, occasionally also ♀; corolla cylindrical to campanulate; filaments inserted above middle of corolla tube; disk present; ovary 2–5-locular, each locule with a single basally attached anatropous ovule. Fruit fleshy to dry, indehiscent or variously dehiscent. Seed with membranous testa and copious endosperm, remaining enclosed in endocarp.

A tribe centred in the N. hemisphere; only 2 genera occur in Africa south of the Sahara and only 1 (sub)species in the Flora Zambesiaca area.

13. **PAEDERIA** L.

Paederia L., Mant. Pl. **1**: 7, 52 (1767) nom. conserv.
Lygodisodea Ruiz & Pav., Prodr. 32, t.5 (1794).
Lecontea A. Rich. ex DC., Prodr. **4**: 470 (1830). —A. Rich. in Mém. Fam. Rubiac.: 115 (1830); in Mém. Soc. Hist. Nat. Paris **5**: 195 (1834).
Siphomeris Boj., Hort. Maurit.: 170 (1837) nom. superfl.

Shrubs, mostly climbing, foetid-smelling. Leaves usually petiolate; stipules entire, ± triangular, deciduous. Flowers in axillary or terminal thyrsoid inflorescences, or solitary. Calyx lobes 4–5, persistent. Corolla tube cylindrical to narrowly campanulate, hairy inside, base often splitting into 4–5 parts; lobes 4–5, shorter than tube, valvate in aestivation. Stamens 4–5, inserted near or at mouth of corolla tube, anthers included or exserted and usually at different heights; filaments short. Ovary 2(3)-locular, with a solitary erect ovule in each chamber, crowned by a small disk; styles filiform, free or joined; stigma lobes 2(3), filiform. Fruit subglobose to compressed-ovoid or -ellipsoid; exocarp thin, papery and brittle, at length breaking open and falling off, exposing 2(3) round to elliptic, laterally compressed and often distinctly winged 1-seeded pyrenes which are pendulous from a detached persistent median vein. Seed compressed, similar in shape to unwinged part of pyrene; testa thin; endosperm copious, embryo large, with conspicuous cordate cotyledons.

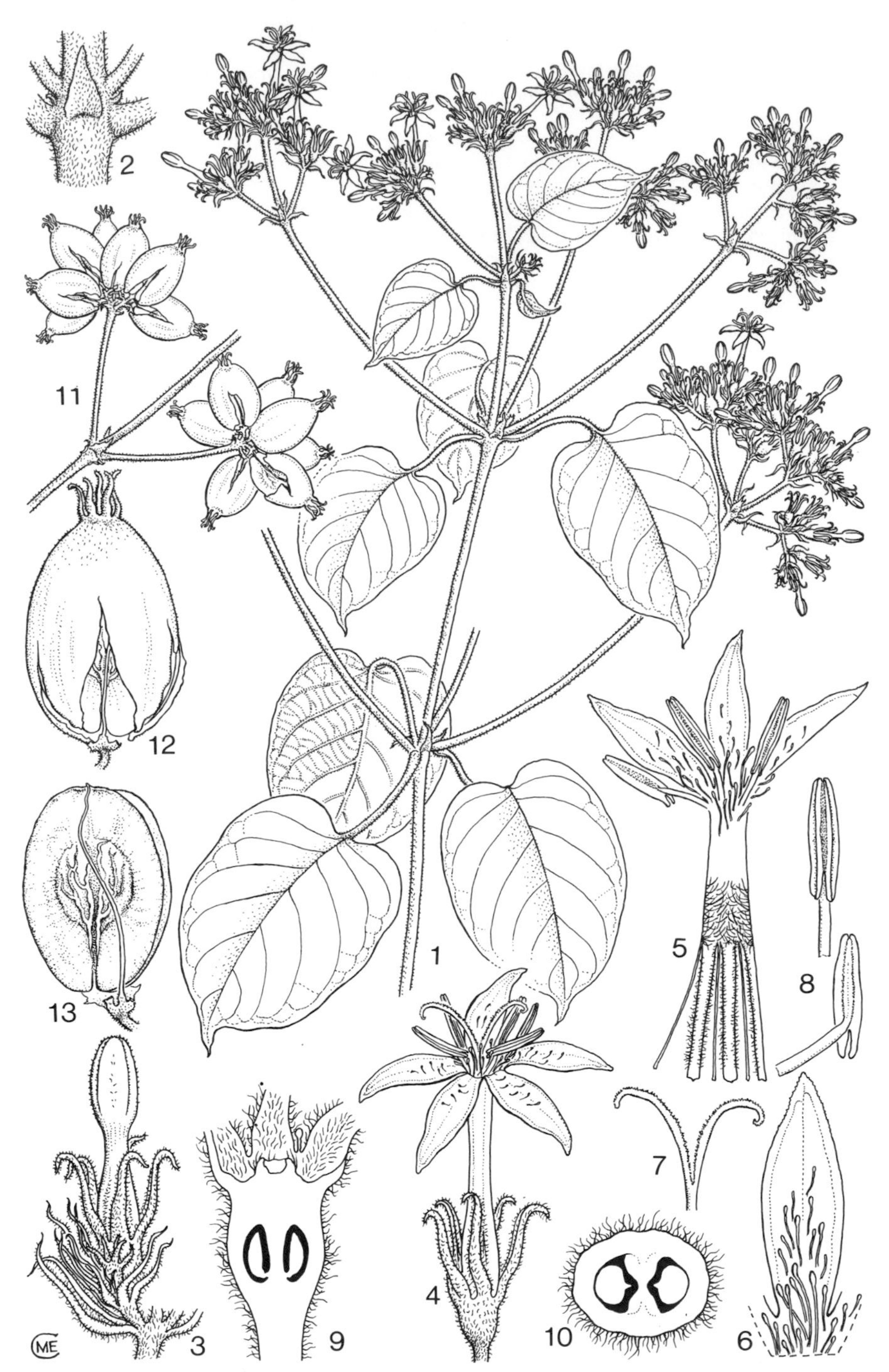

Tab. 15. PAEDERIA BOJERIANA subsp. FOETENS. 1, habit ($\times \frac{2}{3}$), from *Leach & Rutherford-Smith* 10951; 2, stipule (× 2), from *Allen* 439; 3, part of inflorescence showing bracts and flower buds (× 4); 4, flower (× 3); 5, corolla, longitudinal section (× 4); 6, corolla lobe (× 6); 7, stigma lobes (× 4); 8, two views of stamen (× 6); 9, ovary, longitudinal section (× 10); 10, ovary, transverse section (× 12), 3–10 from *Leach & Rutherford-Smith* 10951; 11, fruits ($\times \frac{2}{3}$); 12, fruit (× 2); 13, pyrenes (× 2), 11–13 from *Faulkner* 247.

A tropical to subtropical genus of c. 50 species; only 2 taxa are native to mainland Africa.

Paederia bojeriana (A. Rich.) Drake in Grandidier, Hist. Phys. Madagascar 36 [Hist. Nat. Pl. 6]: t. 412A (1900). —Verdc. in F.T.E.A., Rubiaceae 1: 176 (1976), as *P. bojerana*. Type from Madagascar.
Paederia lingun Sweet, Hort. Brit., App.: 487 (1827) nom. nud.
Lecontea bojeriana A. Rich. ex DC., Prodr. **4**: 471 (1830). —A. Rich. in Mém. Fam. Rubiac.: 115, t. 10/1 bis (1830); in Mém. Soc. Hist. Nat. Paris **5**: 195, t. 20/1 bis (1834).
Siphomeris lingun (Sweet) Boj., Hort. Maurit.: 170 (1837) nom. superfl.

Subsp. **foetens** (Hiern) Verdc. in Kew Bull. **30**: 285 (1975). —Gonçalves in Garcia de Orta, Sér. Bot. **5**: 199 (1982). —Puff in Fl. Southern Afr. **31**, 1(2): 3(1986). TAB. **15**. Syntypes: Mozambique, N. of Sena, N'Keza and by the Shire R. about the cataracts, *Kirk* (K) & 'Zambesiland', *Stewart* (BM).
Siphomeris foetens Hiern in F.T.A. **3**: 229 (1877). Type as above.
Paederia foetens (Hiern) K. Schum. in Engl. & Prantl, Pflanzenfam. **4**, 4: 125 (1891); in Engl., Pflanzenw. Ost-Afr. **C**: 393 (1895). Type as above.

Shrub, climbing, foetid-smelling. Stems to several m. long, slender and flexuous, pubescent to velvety hairy. Leaves decussate or in whorls of 3, petiolate; blades (2)5–11(18) × (2)3.5–8(13) cm., elliptic to ovate, acuminate at apex, cordate or rounded at base, somewhat pubescent above, densely velvety-tomentose beneath; petioles (1)2.5–15 cm. long, (velvety) pubescent; stipules to 1 cm. long. Inflorescences mostly axillary, on stalks c. 4–15(20) cm. long, often ± condensed and many-flowered, sometimes rather extensive; pedicels to 1 mm. long. Flowers 5-merous. Calyx lobes c. 3–8 mm. long, subulate, recurved above. Corolla whitish to greenish-yellow, somewhat pubescent outside; tube 6–10(12) mm. long, ± cylindrical, splitting into 5 segments near base, hairy inside at least near throat; lobes 2.5–6 × 1–2 mm., oblong-lanceolate, hairy inside at least near base. Anthers exserted. Ovary c. (1.5)2–3 mm. long, ovoid, pubescent; style to c. 1 cm. long; stigma lobes to c. 5(6.5) mm. long. Fruit 10–12(14) × 8–12 mm., elliptic in outline, strongly compressed; exocarp shiny, yellowish-brown to brownish, bearing persistent calyx lobes; pyrenes slightly smaller than exocarp, conspicuously winged. Seed black, c. 3–6 × 3–5 mm.

Zambia. B: Kataba, fl. 3.vii.1963, *Fanshawe* 7900 (K; NDO). C: Mpika Distr., Luangwa Game Reserve, Mfuwe, fl. 6.v.1965, *Mitchell* 2859 (K; SRGH). E: Luangwa Valley, Mashanga, Lupande R., fl. & fr. 19.iii.1968, *Phiri* 103 (K). S: Mapanza West, fl. & fr. 22.v.1953, *Robinson* 258 (K). **Zimbabwe**. N: Chipuriro, Hunyani R., 9.6 km. S. of Dande Mission, fl. 15.v.1962, *Wild* 5750 (K; SRGH; WAG). W: Hwange, fr. vii.1920, *Rogers* 7364 (K). C: Charter Distr., E. end of Devuli R. bridge, fl. 15.iv.1963, *Chase* 7986 (BR; K; SRGH). E: Chipinge, Tanganda R. Valley, gateway to 'New Year's' Gift Tea Plantations, fl. 17.v.1962, *Chase* 7720 (K; SRGH). S: Mwenezi, SW. Mateke Hills, Malangwa R., fr. 2.v.1958, *Drummond* 5533 (K; SRGH). **Malawi**. C: Nkotakota Distr., Ntchisi, fl. & fr. 3.viii.1946, *Brass* 17120 (K; NY; SRGH). S: Kasupe Distr., Liwonde National Park, between Chiungune Hill and Shire R., fl. 20.iii.1977, *Brummitt et al.* 14883 (K; MAL). **Mozambique**. N: Mutuali, fl. & fr. 22.vi.1947, *Hornby* 2792 (K; SRGH). Z: Massingire, Vila Bocage, fr. 3.x.1944, *Mendonça* 2337 (LISC). T: 11 km. from Estima towards Marueira, fl. 8.iv.1972, *Macêdo* 5165 (K; LISC; LMA; SRGH). MS: Chimoio, Bandula, fl. & fr. 9.iii.1948, *Barbosa* 1144 (LISC). GI: Vilanculos, Mapinhane, fr. 25.v.1941, *Torre* 2729 (LISC).

Subsp. *foetens* is also known from Tanzania and S. Africa (Transvaal). In scrub, woodland or forest, often riverine; 50–1100 m.

Subsp. *bojeriana* is confined to Madagascar, Mauritius and the Comoro Islands.

Tribe 9. **HEDYOTIDEAE**

1. Calyx limb spreading, eccentric, entire or shallowly lobed, venose, accrescent, up to 2.8 cm. wide in fruit - - - - - - - - - - - - - - - - - **17. Carphalea**
– Calyx limb not as above, usually with 4–5 distinct lobes or teeth - - - - - - 2
2. Flowers mostly 4-merous; leaf blades often narrow and uninerved or with lateral nerves obscure - - - - - - - - - - - - - - - - - - - 3
– Flowers mostly 5-merous; leaf blades often broad and with very obvious lateral and tertiary venation save in *Pentodon* - - - - - - - - - - - - - - - 11
3. Anthers and stigma included, the latter always overtopped by the former; corolla tube narrowly cylindrical - - - - - - - - - - - - - - - - - **18. Kohautia**
– Anthers and/or stigmas exserted or if both included then anthers almost always overtopped by the stigmas; corolla tube cylindrical or funnel-shaped - - - - - - - - 4
4. Corolla tube cylindrical, at least usually 2–3 cm. long; anthers included and style exserted; flowers not heterostylous - - - - - - - - - - - - **19. Conostomium**

– Corolla tube cylindrical or more often funnel-shaped, always less than 2 cm. long, sometimes with included anthers and exserted style but in that case plant mostly heterostylous or if not corolla small - - - - - - - - - - - - - - - - - 5
5. Flowers solitary or fascicled at numerous nodes forming long interrupted spike-like inflorescences; leaves subulate or filiform, in one species in whorls of 3–4 but appearing verticillate with a greater number due to abbreviated shoots - - **24. Manostachya**
– Not as above - - - - - - - - - - - - - - - - - - 6
6. Capsule opening both septicidally and loculicidally - - - - - - - - 7
– Capsule opening only loculicidally - - - - - - - - - - - 9
7. Beak of capsule as long as or longer than the rest of the capsule - - - - - 8
– Beak of the capsule shorter than the rest of the capsule - - **22. Agathisanthemum**
8. Fragile annual herb, with capsule of a very characteristic shape (TAB. 23, fig. 7), emarginate at the base, the beak much exceeding the rest of the capsule; stipules triangular with 2 lobes - - - - - - - - - - - - - - - **20. Mitrasacmopsis**
– Robust subshrubby herb, mostly drying blackish and with rather coriaceous leaves; beak as long as the rest of the capsule; stipules 3–7-fimbriate - - - - - - **21. Hedythyrsus**
9. Rush-like plants with linear or filiform leaves; stipule sheath tubular, truncate or with 2 minute teeth; seeds dorsiventrally flattened - - - - - - - - - **23. Amphiasma**
– Plants not rush-like even if leaves linear; if seeds dorsiventrally flattened then leaves not linear - - - - - - - - - - - - - - - - - - 10
10. Capsule with thick woody wall and a solid beak, tardily dehiscent - - **26. Lelya**
– Capsule with a horny wall, with or without a beak but never solid, early dehiscent - - - - - - - - - - - - - - - - - **27. Oldenlandia**
11. Leaves uninerved, mostly fairly small; decumbent plant of wet places with small white or blue flowers in very lax elongated axillary cymes - - - - - - - **25. Pentodon**
– Leaves larger, pinnately nerved; flowers mostly terminal; mostly erect herbs or subshrubs 12
12. Flowers solitary or paired, axillary or at apices of branchlets; small subshrubs with small leaves sometimes ± pseudoverticillate - - - - - - - - - - **16. Batopedina**
– Flowers in more complicated inflorescences - - - - - - - - - - 13
13. Flowering inflorescences capitate or lax much-branched complicated cymes; individual branches sometimes becoming spicate in fruit; plant never climbing; fruits subglobose or obtriangular; corolla lobes narrower, not or scarely connate at the base **14. Pentas**
– Flowering inflorescences capitate, later elongating into a long simple "spike", rarely with axillary spikes from the upper axils, and frequently with solitary flowers at the lower nodes, or if not elongating into a spike, then plant climbing; fruits oblong; corolla lobes more rounded, joined shortly at the base to form an annulus bearing hairs around the orifice of the tube (in subgenus in the Flora Zambesiaca area) - - - - - - - - - - - - **15. Otomeria**

14. PENTAS

Pentas Benth. in Curtis, Bot. Mag. **70**, t. 4086 (1844). —Verdc. in Bull. Jard. Bot. Brux. **23**: 237–371 (1953).

Mostly perennial (rarely biennial) herbs or shrubs, with erect or straggling stems from a fibrous or woody rootstock. Leaves paired or in whorls of 3–5; stipules divided into 2-many filiform colleter-tipped segments. Flowers small to very large, hermaphrodite, mono-, di- or tri-morphic, mostly in much-branched terminal complicated cymose inflorescences, the individual branches often becoming spicate in fruit. Calyx tube ovoid or globose, sometimes with a free annular part at the top; lobes usually 5, either equal or unequal, 1–3 being larger than the others or sometimes foliaceous. Corolla tube shortly cylindrical to narrowly tubular, 2–40 times as long as wide, hairy in the throat; lobes ovate or oblong. In monomorphic flowers stamens enclosed in an abrupt apical dilation of the tube and style exserted; in dimorphic flowers the tube is gradually dilated at the apex in short-styled flowers and abruptly dilated in long-styled flowers; in rare cases, trimorphism is shown and the third form has both stamens and style included in the tube. Ovary bilocular, with numerous ovules in each attached to placentas affixed to the septum. Capsule obtriangular or ovoid, ribbed, beaked, opening at the apex, the beak splitting into 4 valves; capsule sometimes separating into 2 cocci. Seeds minute, brownish, irregularly globose or tetrahedral, with reticulate testa. Rhaphides are plentiful in most of the tissues.

A genus of about 40 species, widely distributed throughout tropical Africa from W. Africa and Somali Republic to Angola and S. Afr. (Natal), also in tropical Arabia, Madagascar and the Comoro Is.

Key to subgenera of Pentas

1. Calyx with 1–2 lobes enlarged into a white stipitate membranous lamina in the majority of flowers - - - - - - - - - - - - - - Subgen. 1. *Phyllopentas* (p. 67)
– Calyx lobes often foliaceous but then green and never enlarged into a stipitate lamina 2
2. Calyx lobes flat, foliaceous or deltoid, nearly always 1–3 enlarged and the rest much smaller - - - - - - - - - - - - - - - - - - - Subgen. 5. *Pentas* (p. 72)
– Calyx lobes subulate, narrowly spathulate or linear, subequal - - - - - - - 3
3. Leaves in whorls of 3–5 or if paired then plant a short pyrophyte c. 16 cm. tall - - - - - - - - - - - - - - Subgen. 4. *Longiflora* (p. 70)
– Leaves paired save in one subsp. of 11; plant never a pyrophyte - - - - - 4
4. Corolla tube 0.4–1.4 cm. long - - - - - - Subgen. 2. *Vignaldiopsis* (p. 67)
– Corolla tube 2.5–16 cm. long - - - - - - - - - - - - - - 5
5. Calyx lobes mostly over 1 cm. long or if shorter then corolla lobes 1.5–2 × 0.5–1 cm. - - - - - - - - - - - - - Subgen. 3. *Megapentas* (p. 68)
– Calyx lobes under 1 cm. long (mostly under 0.7 cm. long); corolla lobes up to 1.2 cm. × 2.5 mm. - - - - - - - - - - - - - - Subgen. 4. *Longiflora* (p. 70)

Subgen. 1. PHYLLOPENTAS

Verdc. in Bull. Jard. Bot. Brux. **23**: 254 (1953).

Shrubs or subshrubs. Calyx lobes subequal or 1 often enlarged into an ovate petaloid stipitate lamina. Flowers dimorphic. (sp. 1).

1. **Pentas schumanniana** K. Krause in Engl., Bot. Jahrb. **39**: 521 (1907). —Brenan, T.T.C.L.: 518 (1949). —Verdc. in Kew Bull. **30**: 287 (1975); in F.T.E.A., Rubiaceae 1: 186 (1976). Type from Tanzania.
Pentas ionolaena subsp. *schumanniana* (K. Krause) Verdc. in Bull. Jard. Bot. Brux. **23**: 256, fig. 31/F (1953).

Shrubby herb 1.2–3 m. tall; stems with rusty indumentum on the young parts. Leaf blades 3.5–17 × 1.5–8.4 cm., ovate or ovate-oblong, acuminate at the apex, narrowly and often unequally cuneate at the base, pubescent or with scattered hairs above and on the nerves beneath; petiole 1–4.7 cm. long; stipules with c. 5 setae up to 0.7–1.2 cm. long. Inflorescences terminal together with axillary ones, small and compact when flowering, 4–11 cm. wide, enlarging in fruit to 20 cm. long, 16 cm. wide; peduncles 0.7–7 cm. long. Calyx lobes very unequal, the foliaceous one white tinged green and green-veined 2–6 × 1–3 mm., elliptic to rounded-ovate, mucronate or acuminate, membranous and markedly venose, pubescent on the nerves, with the stipe 1.5–2 mm. long (rarely none or very few of the calyces with a lobe developed into a lamina); other lobes filiform, 1.5–2.5 mm. long. Corolla scented, lilac or white; tube (3)5–6 mm. long, dilated at the apex for 2 mm. in long-styled flowers; lobes 2–3 mm. long, pubescent outside, especially at the apex. Style exserted for c. 2–3 mm. in long-styled flowers. Fruit 2 × 3 mm., the base of the persistent calyx lobes with a net-veined pattern. Seeds c. 0.5 mm. long.

Malawi N: Nkhata Bay Distr., 8 km. E. of Mzuzu, Roseveare's, fl. 1.iii.1969, *Pawek* 1778 (K). C: Ntchisi Forest Reserve, fl. 25.iii.1970, *Brummitt* 9376 (K).
Also in S. Tanzania. Evergreen forest including riverine associations; 1200–1710 m.
A *Whyte* specimen from Masuku Plateau states 6500–7000 ft. but nothing else has been seen from so high an altitude. Some Tanzanian specimens occasionally have none of the calyx lobes enlarged. *Pawek* 5212 (Misuku Hills) has congested (but immature) inflorescences and less acuminate leaves but further material is needed. The 1710 m. record is from this specimen.

Subgen. 2. VIGNALDIOPSIS

Verdc. in Bull. Jard. Bot. Brux. **23**: 261 (1953).

Shrubs or subshrubs covered with ferruginous hairs (at least when dry). Leaves with up to 22 lateral nerves on each side and venation reticulate beneath. Calyx lobes subulate, subequal, never foliaceous. Flowers dimorphic. (sp. 2).

2. **Pentas schimperiana** (A. Rich.) Vatke in Linnaea **40**: 192 (1876). —Hiern in F.T.A. **3**: 45 (1877). —K. Krause in Notizbl. Bot. Gart. Berl. **10**: 602 (1929). —Robyns, Fl. Parc Nat. Alb. **2**: 327 (1947).

—Brenan, T.T.C.L.: 518 (1949). —Verdc. in Bull. Jard. Bot. Brux. **23**: 263, fig. 31/G, H & I (1953). —Hepper in F.W.T.A. ed. 2, **2**: 216 (1963). —Agnew, Upland Kenya Wild Fl.: 404 (1974). —Verdc. in F.T.E.A., Rubiaceae 1: 187 (1976). Type from Ethiopia.

Vignaldia schimperiana A. Rich., Tent. Fl. Abyss. **1**: 359 (1847).

Pentas schimperi Engl., Pflanzenw. Ost-Afr. **A**: 92 (1895) nom. superfl.

Shrub or woody herb (0.3)1.3–2.7(5) m. tall, with wrinkled blackish or purplish-black woody stems, at first densely rusty pubescent (at least when dry*), later glabrescent. Leaf blades 6–21 × 1.8–8 cm., ovate, ovate-lanceolate or ovate-oblong, acute to acuminate at the apex, cuneate to round at the base, appressed rusty hairy above and on the characteristically reticulate (but not prominent) venation beneath; petiole 0–1.5 cm. long, velvety rusty pubescent; stipules large, with c. 6–10 linear-lanceolate brown hairy setae 0.2–1.5 cm. long from a variable base 3–8 × 4–8 mm.; these bases connate and forming persistent cups at the nodes on old stems. Inflorescences branched, 2.5–15 cm. wide; peduncles up to 1.5 cm. long, velvety pubescent. Calyx tube hairy, 1.5–3 mm. long; lobes subequal, 0.3–1.2 cm. × 1 mm., linear, hairy outside, either one third–half the length of the corolla tube or approximately equalling it. Corolla musky scented, white, often tinged pink, or entirely pinkish, glabrous or pubescent; in long-styled flowers, tube funnel-shaped 0.5–1.3 cm. long, dilated at the apex for 2 mm.; in short-styled flowers the tube is cylindrical, 0.5–1.4 cm. long; lobes 3–7 × 0.75–2.5 mm., the tips thickened and inflexed. Style exserted for 0.5–5 mm. long in long-styled flowers. Fruit 4–6 mm. long and wide, the persistent calyx lobes often reflexed, costate, valvular, opening c. 1–2 mm. wide, or rarely separating loculicidally into 2 cocci.

Subsp. **schimperiana**. —Verdc. in Bull. Jard. Bot. Brux. **23**: 262 (1953).

Pentas thomsonii Scott Elliot in Journ. Linn. Soc. Bot. **32**: 435 (1896). Type from Kenya.

Neurocarpaea thomsonii (Scott Elliot) S. Moore in Journ. Linn. Soc. Bot. **37**: 157 (1905).

Calyx lobes approximately equalling the corolla tube.

Zambia. E: Nyika Plateau, below rest house on path to N. Rukuru Waterfall, fr. 27.x.1958, *Robson & Angus* 397 (BM; LISC; K). **Malawi**. N: Nkhata Bay Distr., N. Viphya, Camp Gordon road, km. 41.6, fl. & fr. 21.ix.1969, *Pawek* 2728 (K).

Ethiopia, Zaire, Burundi, Uganda, Kenya & Tanzania, Montane forest, *Hagenia–Podocarpus* etc., particularly at forest/grassland boundaries, scrub, *Pteridium–Protea–Kotschya* thicket, thicket at streamside marshy edges, ± open ground; 1710–2350 m.

Subsp. *occidentalis* (Hook.f.) Verdc. occurs in Fernando Po, S. Tomé, Cameroon and also in the Ituri Forest of Zaire; records from S. Arabia (Robyns, Fl. Parc Nat. Alb. **2**: 327 (1947)) and South West Africa (Verdc. in Bull. Jard. Bot. Brux. **23**: 268 (1953)) are almost certainly based on wrongly localised material.

Subgen. 3. MEGAPENTAS

Verdc. in Bull. Jard. Bot. Brux. **23**: 269 (1953).

Subsucculent herbs or subshrubs. Leaves opposite, often large, petiolate or sessile. Calyx lobes subequal, linear-deltoid or narrowly spathulate, never leafy. Flowers mostly large, with corolla tube up to 16 cm. long, often dimorphic. Style tomentose. (sp. 3).

3. **Pentas nobilis** S. Moore in Journ. Bot. Lond. **46**: 37 (1908). —Brenan, T.T.C.L.: 516 (1949). —Verdc. in Bull. Jard. Bot Brux. **23**: 274 (1953); in Fl. Pl. Afr., t. 1690 (1974); in F.T.E.A., Rubiaceae 1: 192 (1976). —Gonçalves in Garcia de Orta, Sér. Bot. **5**(2): 202 (1982). TAB. **16**. Type: Zimbabwe, Mazoe, Iron Mask Hill, *Eyles* 248 (BM, lectotype; SRGH).

Pentas longituba sensu De Wild. & Th. Dur. in Bull. Soc. Bot. Belge **37**: 117 (1898). —Engl. in Engl., Bot. Jahrb. **30**: 412 (1901) non K. Schum.

Pentas nobilis var. *grandifolia* Verdc. in Bull. Jard. Bot. Brux. **23**: 275 (1953). Type from Tanzania.

Shrub or herb 0.6–1.5(2.4) m. tall, the stems woody, ridged, grey-black, the epidermis ridged or rucked up, pubescent all over, often drying orange-ferruginous. Leaf blades 5–20 × 1.6–8.8 cm., ovate, ovate-elliptic or elliptic-oblong, acute at the apex, rounded or abruptly and then gradually cuneate, subscabrid above with very minute bristly hairs and on the costa and nerves beneath; petiole 0.2–1 cm. long; stipules with 3–5 subulate

* Often pallid in living material.

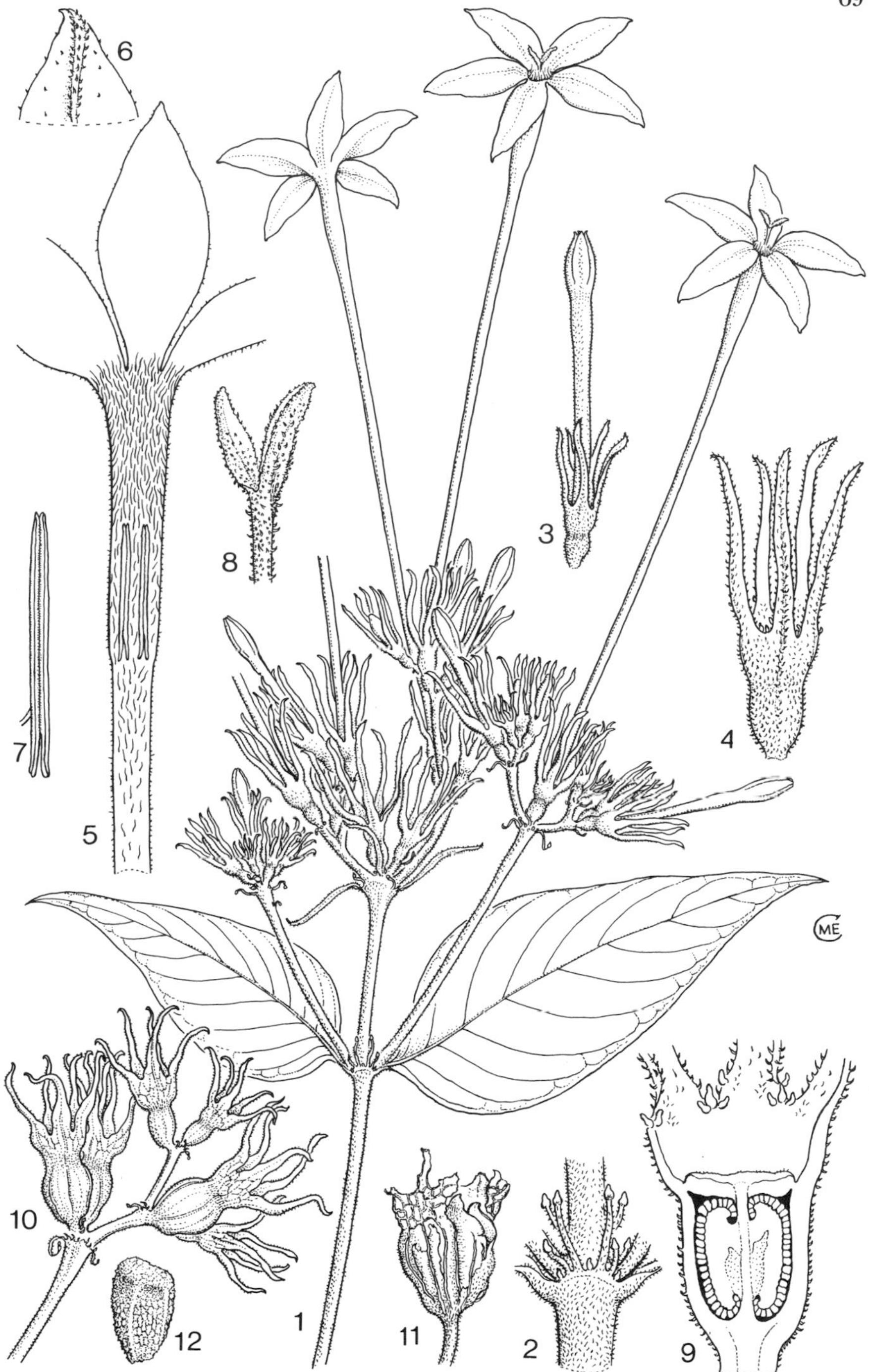

Tab. 16. PENTAS NOBILIS. 1, habit (× ⅔); 2, node showing stipule (× 2), 1–2 from *Richards* 19043; 3, flower bud (× 1); 4, calyx (× 2); 5, upper part of corolla, longitudinal section (× 2); 6, outer surface upper part of corolla lobe (× 4); 7, stamen (× 4); 8, stigma lobes (× 4); 9, ovary, longitudinal section (× 6), 3–9 from *Richards* 4492a; 10, fruits (× ⅔), from *Fanshawe* 2777; 11, dehisced capsule (× 1); 12, seed (× 8), 11–12 from *Fanshawe* 1814.

subequal setae 0.6–1.4(2) cm. long from a short base, with conspicuous colleters. Inflorescence small and c. 6-flowered or larger, lax, up to 25 × 15 cm. Flowers whitish, sometimes tinged reddish, sweetly scented, showing limited dimorphism. Calyx shortly hairy; tube 5–7 mm. long; lobes 0.6–2.4 cm × 0.5–2 mm., subulate or minutely spathulate, equal, if a trifle spathulate then apical portion acute, 1.7 × 1.5 mm. Corolla tube pubescent, 7.5–14.2 cm. long, 1–3.5 mm. wide at base, dilated to 3–7 mm. for apical 1.2–2 cm.; lobes ovate-oblong, 1.2–2.1 cm. × 3–7 mm., acute; throat densely hirsute. Flowers with style exserted 0–1.7 cm. and anthers included or with both included, but none with exserted anthers. Fruiting inflorescence large, lax, with peduncle 10–12 cm. long. Fruits straw-coloured 1–1.7 cm. × 0.9–1.4 cm. × 8–9 mm., oblong, subglobose or dorsally compressed, with a very nervose frill formed by the limb tube of the calyx extending above the disk for 2.5–8 mm., minutely pubescent, the bases of the calyx lobes persistent.

Zambia. N: Shiwa-Ngandu, Machipara Hill, fl. 16.i.1959, *Richards* 10690 (K). W: Chingola, fl. & fr. 16.ii.1956, *Fanshawe* 2777 (K). C: Serenje Distr. 53 km. ENE. of Serenje, Kundalila Falls, fl. 4.ii.1973, *Strid* 2899 (K; O). **Zimbabwe**. N: Mtoko, fl. 28.i.1941, *Hopkins* in SRGH 7903 (K; SRGH). C: Harare, fl. 17.ii.1927, *Eyles* 4693 (K; SRGH). **Malawi**. C: Lilongwe, Dzalanyama Forest Reserve, 6 km. SE. of Chaulongwe Falls, fr. 22.iii.1970, *Brummitt* 9267 (K). **Mozambique**. T: Mt. Furancungo, fr. 15.iii.1966, *Pereira et al.* 1690 (LMU).

Also in Zaire and S. Tanzania. Mostly in rocky places especially vast granite outcrops in *Brachystegia* woodland but also in grassy places; 870–1740 m.

Subgen. 4. LONGIFLORA

Verdc. in Bull. Jard. Bot. Brux. **23**: 281 (1953).

Erect, often woody herbs. Leaves opposite or often in whorls of 3–5. Calyx lobes subequal, linear-deltoid. Flowers not dimorphic, the style always exserted. (sp. 4–6).

1. Plant a short pyrophyte 13–30 cm. tall; corolla tube 6.5–13 cm. long; lobes elliptic, 1.2–2 cm. × 5–10 mm. - - - - - - - - - - - - - - - - 6. *lindenioides*
– Plant a tall herb, 0.3–2 m. tall; corolla lobes 0.3–1.3 cm. × 1.5–4 mm. - - - - - 2
2. Leaves usually paired; corolla tube 2–4.2 cm. long, 0.5–1.5 mm. wide at the base 4. *longiflora*
– Leaves in whorls of 3–5 (in Flora Zambesiaca area); corolla tube 3–13 cm. long, 1.5–3.3 mm. wide at the base - - - - - - - - - - - - - - - - 5. *decora*

4. **Pentas longiflora** Oliv. in Trans. Linn. Soc. Ser. 2, **2**: 335 (1887). —Robyns, Fl. Parc Nat. Albert **2**: 328 (1947). —Brenan, T.T.C.L.: 517 (1949). —Verdc. in Bull. Jard. Bot. Brux. **23**: 282, fig. 32/D & I (1953). —Lind & Tallantire, Fl. Pl. Uganda ed. 2: 160 (1972). —Agnew, Upland Kenya Wild Fl.: 404, fig. (1974). —Verdc. in F.T.E.A., Rubiaceae 1: 195 (1976). Type from Tanzania.
   Pentas longiflora var. *nyassana* Scott Elliot in Journ. Linn. Soc. Bot. **32**: 433 (1896). Type: Malawi, *Buchanan* 475 (BM; E; K, holotype; US).
   Neurocarpaea longiflora (Oliv.) S. Moore in Journ. Linn. Soc. Bot. **37**: 157 (1905).
   Pentas longiflora forma *glabrescens* Verdc. in Bull. Jard. Bot. Brux. **23**: 286 (1953). Type from Kenya.

Shrubby herb to 2 m. tall, with 2–3 main stems from a woody rootstock; stems glabrous to densely covered with whitish or rusty-orange hairs. Leaves paired or in whorls of 3; leaf blades 5–15(16.5) × 0.7–3(5) cm., lanceolate or rarely ovate-lanceolate, acute at the apex, narrowed at the base, glabrous to velvety or pubescent, often ferruginous, sessile or petiole 0.5–1(2.5) cm. long; stipules with 3–7 linear setae 0.1–1.3 cm. long from a base 0.5–2.5 mm. long. Flowers white or bluish white or tinged purplish. Inflorescences 20–100-flowered, up to 14 cm. wide; primary peduncles up to 15 cm. long; bracteoles filamentous, up to 3 mm. long. Calyx tube glabrescent to velvety, often ferruginous, c. 1.5–2 mm. long and wide; lobes subequal, 3–7.5 mm. long, linear. Corolla tube 2–4.2 cm. long, 0.5–1.5 mm. wide at the base, dilated at the apex to 1.5–2.5 mm. wide for a distance of 3–5.5 mm., pubescent outside; lobes 3–6 × 1.5–2.9 mm., oblong or elliptic, acute or rather blunt; throat hairy. Stamens entirely included. Style exserted 1–6 mm., tomentose with white scaly papillae. Fruit 3–6.5 × 4–7.5 mm., oblong, depressed, ribbed, shortly pubescent with white or orange-brown hairs.

Malawi. S: Zomba Plateau, near Chingwe's Hole, fl. 18.iv.1970, *Brummitt* 9941 (K). **Mozambique**. Z: Mulanje, fl. & fr. 10.ix.1941, *Torre* 3396 (LISC; LMU; MO).

Also in Zaire, Burundi, Rwanda, Uganda, Kenya & Tanzania. Forest clearings, plantation edges,

Pteridietum, sometimes in rocky places; 1300–1930 m.
Purves 67 from Somba (Zomba) has the altitude given as 2900 ft. but this seems very low for the species.

5. **Pentas decora** S. Moore in Journ. Bot., Lond. **48**: 219 (1910). —Verdc. in Kew Bull. **7**: 362 (1952); in Bull. Jard. Bot. Brux. **23**: 287 (1953). —Agnew, Upland Kenya Wild Fl.: 404, fig. (1974). —Verdc. in F.T.E.A., Rubiaceae 1: 195 (1976). Type from Zaire.

Herb 0.3–1.5 m. tall; stems usually single, mostly somewhat woody, glabrous to sparsely pubescent or densely hairy with short hairs. Leaves in whorls of 3–5 or rarely opposite; leaf blades 3.9–13 × 0.9–5.6 cm., narrowly to broadly elliptic, elliptic-lanceolate, lanceolate or narrowly ovate, subacute or rounded at the apex, rounded to cuneate at the base, glabrous to densely velvety; petiole obsolete; stipules with 1–5 ciliate setae, 0.5–8(10) mm. long from a short base; often one deltoid lobe with a colleter on either side of it between the leaves in plants with 4–5 leaves in a whorl. Inflorescences consisting of 1 terminal and 2–4 axillary ones amalgamated, up to 12 cm. wide, each cluster with c. 30 flowers; peduncles 0–7.5 cm. long. Calyx glabrous to hairy; tube 2–5 × 2–4 mm.; lobes 3–11 × 1–1.5 mm. at base, deltoid to filiform, subulate at the apex, ciliate. Flowers white, sweet-scented. Corolla tube 3–13 cm. long, 1.5–3.3 mm. wide at the base, scarcely dilated at the apex to 2.5–3.5 mm. wide for a distance of 1.2– 1.7 cm., shortly hairy outside; throat densely hairy with long hairs; lobes sometimes yellowish, (0.5)0.8–1(1.8) cm. × 1.5–4.5 mm., linear-oblong to elliptic-lanceolate, acute, hairy outside. Stamens entirely included. Style exserted 0.3–1.8 cm., densely tomentose with white papillae; stigma lobes blackish, elliptic, 1–3 mm. long, thickened. Fruit 1–1.5 × 0.7–1.1 cm., obovoid-obtriangular, prominently ribbed, pubescent; calyx lobes reflexed, the limb tube nervose, 3.5–4 mm. tall, exceeding the 2 mm. long beak. Seeds brown, ovoid, 1–1.5 mm. long, conspicuously reticulate.

Var. **decora**. —Verdc. in Bull. Jard. Bot. Brux. **23**: 287 (1953); in F.T.E.A., Rubiaceae 1: 197 (1976).
Pentas verticillata Scott Elliot in Journ. Linn. Soc. **32**: 431 (1896) nomen nudum.
Pentas homblei De Wild. in Fedde Repert. **13**: 109 (1914). Type from Zaire.
Pentas longituba sensu K. Krause in Notizbl. Bot. Gart. Berl. **10**: 603 (1929) non K. Schum.

Leaves mostly glabrous, less often sparsely pubescent or rarely hairy. Inflorescence of 1–5 clusters, each c. 30-flowered. Corolla tube 3–12 cm. long. Capsule obovoid-obtriangular, strongly ribbed.

Zambia. N: Mbala, Pans, fl. 20.i.1955, *Richards* 4186 (K).. W: Mwinilunga Distr., Kalenda Plain, fl. 24.xii.1937, *Milne-Redhead* 3804 (BM; K). **Malawi**. N: Nyika Plateau, Makanga Hill E., fl. 25.i.1976, *Phillips* 809A (K; MO).
Also in Zaire, Sudan, Uganda, Kenya, Tanzania & Angola. Grassland often with *Protea* and sometimes boggy, bushland with *Uapaca*, occasionally in woodland; 900–1740 (?2250) m.
Other varieties extend the distribution to Nigeria, Central African Republic and Ethiopia. It is possible that this species should be called *P. liebrechtsiana* De Wild., the type of which (from Zaire) has very slender corolla tubes and is intermediate between *P. decora* and *P. longiflora* (see Verdc. in Bull. Jard. Bot. Brux. **23**: 287 (1953)). Two specimens from W. Tanzania are virtually indistinguishable from *P. liebrechstiana*, e.g. *Pirozynski* 236 (Buha District, Gombe Stream Reserve, Kakombe Valley, 10 Jan. 1964). They may also be equivalent to *P. decora* var. *lasiocarpa*. Further material is needed from the Katanga to show variation before the name *decora* is given up.

6. **Pentas lindenioides** (S. Moore) Verdc. in Kew Bull. **30**: 344 (1975); in F.T.E.A., Rubiaceae 1: 198 (1976). Type: Malawi, Nyika Plateau, *Henderson* (BM, holotype).
Heinsia lindenioides S. Moore in Journ. Linn. Soc. Bot. **37**: 301 (1906).
Pentas geophila Verdc. in Bull. Jard. Bot. Brux. **23**: 293 (1953). Type from S. Tanzania.

Pyrophytic herb 13–20(30) cm. tall, but otherwise very similar to *P. decora*; stems simple, erect, hairy. Leaves opposite or in whorls of 3; leaf blades 3.2–11 × 1.4–3 cm., elliptic, ± acute at the apex, rounded or cuneate at the base, hairy; petioles 0–3 mm. long; stipules small, with 1–5 deltoid segments 4 mm. long. Inflorescences terminal, dense, 1–8-flowered; peduncles 1–1.5 cm. long. Flowers white, greenish or blue. Calyx tube 2–4 mm. long, hairy; lobes deltoid to narrowly triangular, subequal, 4–6.5 mm. long. Corolla tube 6.5–13 cm. × 1–3 mm., hairy outside; throat densely hairy, the hairs spreading over the inside of the lobes; lobes 1.2–2.3 cm. × 5–10 mm., elliptic or oblong-obovate, hairy outside. Stamens entirely included. Style exserted for 4–8(15) mm., minutely tomentose; stigma lobes 2 mm. long. Capsule not seen.

Zambia. E: Nyika Plateau, fl. 13.i.1974, *Fanshawe* 12176 (K; NDO). **Malawi**. N: Nyika Plateau, between Kasaramba and Chelinda, fl. 10.i.1967, *Hilliard & Burtt* 4407 (E; K; NU).

Also in S. Tanzania. Montane grassland; 2280–2440 (?) m.

This could be treated as a local subspecies of *P. decora* S. Moore since intermediate plants have been recorded, e.g. Tanzania, Mbeya Mt., 13 May 1956, *Milne-Redhead & Taylor* 10235!; in fact plants very like *P. decora*, but with longer corolla tubes and larger corolla lobes, i.e. differing from *P. lindenioides* only in stature, are common in S. Tanzania. The correct answer may be to recognise a larger-flowered variant of *P. decora* which may have a small montane pyrophytic ecotype. This must be decided by studying populations in the field. Nevertheless it is a very distinctive taxon.

Subgen. 5. PENTAS

Verdc. in Bull. Jard. Bot. Brux. **23**: 294 (1953).

Herbs or subshrubs. Leaves usually paired, in whorls of 3–4 only in three species. Calyx lobes ± unequal, not subulate, 1–3 foliaceous, enlarged, elliptic, lanceolate, oblong or deltoid. Flowers small to large, the corolla tube 0.4–9 cm. long, usually dimorphic, white or brightly coloured. (sp. 7–14).

1. Flowers bright vermilion-scarlet; leaves with a very fine characteristic indumentum beneath; stem indumentum mostly ferruginous when dry; capsule ovoid-oblong, a little contracted above - - - - - - - - - - - - - - - - - 7. *bussei*
– Flowers white, mauve, blue or pink, only rarely red and then of a deeper crimson shade; leaf indumentum coarser and capsule obtriangular - - - - - - - - - 2
2. Flowers white, not dimorphic*, style always exserted; leaves thin in texture: - - 3
– Flowers white or coloured, dimorphic - - - - - - - - - - - - 5
3. Corolla tube mostly 4.5 mm. long but occasionally up to 7 mm.; leaves large and thin; inflorescence branches spicate in fruit - - - - - - - - - 9. *herbacea*
– Corolla tube usually longer; inflorescence more condensed - - - - - - 4
4. Corolla tube (5)7–9.5(11) mm. long; leaves tenuous, elliptic-ovate - - 8. *micrantha*
– Corolla tube (10)14–18 mm. long; leaves narrowly lanceolate - - - 10. *angustifolia*
5. Corolla tube 1–4 cm. long - - - - - - - - - - - - 14. *lanceolata*
– Corolla tube usually under 10 mm. long, rarely up to 11 mm. - - - - - - 6
6. Corolla white or whitish (3)4–5 mm. long (Mozambique, Gúruè Mts.) 13. *pubiflora*
– Corolla white or coloured, nearly always longer, up to 1 cm. long - - - - - 7
7. Flowers and fruits in laxer cymose inflorescences; corolla lobes devoid of hairs on the inner face - - - - - - - - - - - - - - - - - - 11. *zanzibarica*
– Flowers and fruits in very dense globular heads, or if rather laxer then throat hairs extending up over the inner surface of the corolla lobes which are often erect; flowers usually indigo or deep mauve, sometimes paler - - - - - - - - - - - - - 12. *purpurea*

Section COCCINEAE

Verdc. in Bull. Jard. Bot. Brux. **23**: 296 (1953).

Shrubs or subshrubs with woody erect or scrambling stems. Corolla distinctly vermilion-scarlet. Flowers dimorphic. Capsule eventually dividing into 2 cocci and showing little apical dehiscence. (sp. 7).

This section has distinct affinities with the genus *Otomeria* (see Verdc. in Bull. Jard. Bot. Brux. **23**: 246 and fig. 30) and is well-characterised.

7. **Pentas bussei** K. Krause in Engl., Bot. Jahrb. **43**: 134 (1909). —Brenan, T.T.C.L.: 517 (1949). —Verdc. in Bull. Jard. Bot. Brux. **23**: 297, fig. 33/C, E-H (1953); in F.T.E.A., Rubiaceae 1: 200 (1976). Type from Tanzania.

Pentas klotzschii sensu Vatke in Oest. Bot. Zeitschr. **25**: 231 (1875) quoad *Hildebrandt* 1124.

Pentas coccinea Stapf in Curtis, Bot. Mag. **149**, t. 9005 (1924). Type a specimen cultivated at Kew from Tanzanian material.

Pentas flammea Chiov., Fl. Somala **2**: 231 (1932). Type from S. Somalia.

* This is not an easy character; from single specimens it may be impossible to tell unless the anthers are exserted, but it is a valuable character often easily observed in populations in the field.

Erect or somewhat scrambling shrub or herb 0.6–4 m. tall; stems with sparse to dense white or brown hairs above, usually drying ferruginous or yellowish, sometimes densely velvety. Leaf blades 3.5–15 × 1.7–7 cm., ovate-lanceolate or ovate-oblong, acute or acuminate at the apex, cuneate at the base, mostly very discolorous, sparsely shortly hairy above, pubescent to velvety beneath with very fine short white hairs; petiole up to 2 cm. long; stipules with 3–9 linear setae 0.4–1.5 cm. long from a triangular base 1–6 mm. long. Inflorescences terminal and axillary, dense or lax, up to 8 cm. wide, many-flowered; peduncles up to 4 cm. long. Calyx tube c. 1.5 mm. long and wide, glabrous or velvety; lobes foliaceous, very unequal; 1–3 much enlarged, 0.5–1.8 cm. × 1–4 mm., narrowly lanceolate, acute, 3-nerved, pubescent, the rest small, 1.5–7 × 0.3–1.3 mm., linear, deltoid or lanceolate. Flowers vermilion-scarlet. Long-styled flowers with corolla tube 0.7–2 cm. long, 0.5–1.5 mm. wide at the base, dilated at the apex to 1.5–3 mm. wide for an apical cylindrical portion of 2.5–5(7) mm., glabrous to hairy outside, with white hairs inside save at the throat where they are scarlet; lobes (2.5)3–10(12) × 1–4.5 mm., narrowly oblong, elliptic or oblong-lanceolate; stamens entirely included; style usually bluish purple, exserted 1–9 mm., the stigmas scarlet or purple, 1.5–4.5 mm. long. Short-styled flowers very similar; corolla tube 0.7–1.7 cm. long, 0.5 mm. wide at the base, gradually expanding to the throat which is 1–2.5 mm. wide; exserted anthers usually blue-purple. Fruiting inflorescences with individual branches ± spicate. Capsules 3–6 × 2.25–3.5 mm., oblong to obovoid, 10-ribbed, crowned with the persistent calyx lobes; beak only slightly raised.

Zambia. N: Mbala Distr., Lunzua Valley, Kafakula, fl. 5.iii.1955, *Richards* 4807 (K). **Malawi**. N: Chitipa Distr. 14.4 km. E. of crossroads towards Karonga, Songa Stream, fl. 19.iv.1969, *Pawek* 2255 (K). S: Namwera Escarpment, near Mangochi, fl. & fr. 15.iii.1955, *Exell et al.* 891 (BM; LISC). **Mozambique**. N: Cabo Delgado, Palma monte a ca. 2 km. S. of R. Rovuma, 16 km. from Nangade, fl. 18.iv.1964, *Torre & Paiva* 12145 (COI; K; LISC; LMU; SRGH; WAG) (extreme form).

S. Somalia, Zaire, Burundi, Kenya, Tanzania, also cultivated but much confused with red forms of *Pentas zanzibarica*. *Brachystegia* and other woodland, bushland, often riverine or by streams but sometimes in rocky areas; 750–1500 m. (see note).

A record by *A. Whyte* from 7000 ft. (2133 m.) is probably inaccurate. All the Zambian material is of the form with long internodes and fistulose stems; although distinctive and occurring in several areas of Tanzania I have not been able to satisfy myself that it is distinct enough for a name, *Pawek* 13480 and *Jackson* 1170, although clearly this species have small leaves approaching those of *Pentas parvifolia* Hiern. Description of the flowers of many Malawi specimens as crimson is I think an error for scarlet.

There are some nomenclatural difficulties concerned with this species. The possibility that it should be called *P. zanzibarica* (Klotzsch) Vatke has been discussed in detail by me (Bull. Jard. Bot. Brux. **23**: 300 (1953)). The evidence for this is a fragment of type preserved at Kew, but I think some confusion may have occurred and refuse to upset the nomenclature on inadequate evidence. The name *zanzibarica* has been used consistently for another species for a very long while and the utmost confusion would result from any changes. In Bull. Jard. Bot. Brux. **23**: 302 (1953) I described three forms of *P. bussei:* forma *brevituba* Verdc., forma *minor* Verdc. and forma *glabra* Verdc. but these are not worth retaining. Some specimens said to have been collected in Angola by *Gossweiler* in 1907 must I think be wrongly labelled.

Section MONOMORPHI

Verdc. in Bull. Jard. Bot. Brux. **23**: 307 (1953).

Biennial or perennial herbs with often thin leaves. Flowers small, mostly white, not dimorphic; style always exserted. (sp. 8–10).

8. **Pentas micrantha** Bak. in Journ. Linn. Soc. Bot. **21**: 408 (1885). —Verdc. in Bull. Jard. Bot. Brux. **23**: 307 (1953); in F.T.E.A., Rubiaceae 1: 202 (1976). Type: from Madagascar.

Herb 45–90 cm. tall, ? annual, biennial or a short-lived perennial (but habit of typical race not certain), stems hairy when young. Leaves 5.5–14.5 × 1.5–6.3 cm., ovate to ovate-oblong or elliptic, acute at the apex, cuneate at the base, thin and membranous, sparsely or moderately pubescent above and on the venation beneath; petiole 0.3–3.5(7) cm. long; stipules with 4–7 setae up to 6 mm. long from a base 0.5–3.5 mm. long, hairy. Inflorescence a small lax cluster c. 1.5 cm. wide or trichotomous with individual cymes lax, few-flowered, 2–3 cm. wide; flowers white or lavender-blue; peduncles 0–6.5 cm.; pedicels c. 1.5 cm. long. Calyx tube 1–1.5 mm. long and wide, hairy; lobes unequal, the longest lanceolate, (3)4–8(10) × 1–3.5 mm., rest 1–2.3 × 0.25–0.5 mm. Buds characteristically clavate. Corolla tube (5)6.5–10(11) mm. long, 0.8–1.5 mm. wide at the base, abruptly expanded at the apex to 1.5–2.5 mm. for 1.5–2 mm., the resulting dilated portion being urceolate, all hairy outside; lobes 1.8–3 × 1–1.75 mm., ovate, acute, hairy on the main nerve outside. Stamens completely included. Style glabrous, exserted 0–2.5 mm.; lobes of stigma 0.75–1.5(2) mm., mostly practically immersed in the dense throat hairs, sometimes the lower parts immersed in the corolla tube. Capsules 2.5–5 mm. long and wide, obtriangular or oblong-ovoid, prominently ribbed, glabrous or pubescent. Seeds minute, 0.5 mm. long, bluntly angular.

Subsp. **wyliei** (N.E. Br.) Verdc. in Bull. Jard. Bot. Brux. **23**: 308, fig. 33/I (1953); in Kew Bull. **12**: 355 (1957); in Kew Bull. **30**: 287 (1975); in F.T.E.A., Rubiaceae 1: 203 (1976). Type from S. Africa (Natal).

Pentas wyliei N.E. Br. in Kew Bull. **1901**: 123 (1901). —Wood, Natal Plants **4**, t. 344 (? 1904).

Pentas zanzibarica var. *membranacea* Verdc. in Bull. Jard. Bot. Brux. **23**: 323 (1953) *quoad typum solum*. Type from Tanzania.

Pentas zanzibarica var. *pembensis* Verdc. in Bull. Jard. Bot. Brux. **23**: 324 (1953). Type from Tanzania.

Lower petioles mostly c. 1–4(7) cm. long. Corolla tube (5)7–9.5(11) mm. long.

Zimbabwe. E: Chimanimani, Haroni–Lusitu river confluence, Ngorima Reserve, fl. & fr. 23.xi.1967, *Ngoni* 50 (K; LISC; SRGH). **Mozambique**. N: Malema, Serra Merripa, fl. & fr. 5.ii.1964, *Torre & Paiva* 10470 (COI; LISC; LMU; SRGH; WAG). Z: Macuze, fl. & fr. 28.viii.1949, *Barbosa & Carvalho* 3860 (LISC; LMA). MS: near Beira, Dondo swamps, fl. & fr. xii.1899, *Cecil* 252 (K). GI: 'Machisugu' [Machingas = Massinga fide Jessop, Journ. S. Afr. Bot. **30**: 129-142 (1964)], fl. & fr. 10.ii.1898, *Schlechter* 12118 (BM; K).

Also in Tanzania and S. Africa (Natal). Probably bushland; *Cecil* 252 reports "in the swamps"; ± 0–450 m.

9. **Pentas herbacea** (Hiern) K. Schum. in Just Jahresb. **26**: 392 (1900). —F.W. Andr., Fl. Pl. Angl. Egypt. Sudan **2**: 459 (1952).—Verdc. in Bull. Jard. Bot. Brux. **23**: 311 (1953); in Kew Bull. **12**: 355 (1957). Type from Angola.

Pentas sp. nov. Hiern MS in Scott Elliot in Journ. Linn. Soc. Bot. **32**: 437 (1896).

Neurocarpaea herbacea Hiern, Cat. Afr. Pl. Welw. **1**(2): 439 (1898).

Otomeria herbacea (Hiern) Hiern ex Good in Journ. Bot. Lond. **64** suppl. 2: 3 (1926).

Branched herb, 0.45–1.2 m. tall; root-system of small extent, 9 × 6 cm., annual or biennial. Stem 3 mm. in diam. at the base. Leaves very thin, 3.5–14 × 1–6.5 cm., oblong-elliptic to ± lanceolate, acuminate at the apex, cuneate at the base, sparsely appressed pubescent above, minutely puberulous on venation beneath; petiole 0.3–2 cm. long; stipule setae up to 5 mm. long. Inflorescence a small head. Calyx tube 2 mm. long and wide; lobes deltoid, unequal, largest 3–4 × 1.5–2 mm. Corolla white or yellowish, glabrous or minutely puberulous on tube angles; tube 4–4.5 mm. long, 1.5 mm. wide at the base, 0.8 mm. wide in the middle, expanded apically to 1.8 mm.; lobes 0.9 × 0.6 mm., tips inflexed. roughened, 3.8 mm. long overall; stigma lobes 0.9 mm. long the tips just exserted. Anthers 1 mm. long, just included; throat densely hairy and tube hairy below, papillate at the base. Fruiting inflorescence branched consisting of an umbel of 4 spikes, together with several truly terminal flowers in the axil; spikes 9–26 cm. long. Fruits 3.4–4 mm. long and broad, oblong-obtriangular, not elongate, calyx lobes persistent, reticulated and venose at their bases; beak 1.5 mm. tall.

Zambia. W: Chingola, fl. & fr. 17.iv.1954, *Fanshawe* 1091 (K; NDO).

Also in Guinea, Central African Republic, Zaire, Sudan and Angola. Relict evergreen thicket, bushland, termite mounds in *Brachystegia* woodland; 1200–1500 m.

10. **Pentas angustifolia** (A. Rich. ex DC.) Verdc. in Bull. Jard. Bot. Brux. **23**: 312, fig. 34/A & B (1953). —Gonçalves in Garcia de Orta, Sér. Bot. **5**(2): 202 (1982). Type from Angola.

Sipanea angustifolia A. Rich. ex DC., Prodr. **4**: 414 (1830). —A. Rich., Mém. Fam. Rubiac.: 196 (1830); in Mém. Soc. Hist. Nat. Paris **5**: 276 (1834).

Pentas woodii Scott Elliot in Journ. Linn. Soc., Bot. **32**: 434 (1896). Type from S. Africa (Transvaal).

Pentas transvaalensis Bär in Viert. Nat. Ges. Zürich **68**: 433 (1923). Type from S. Africa (Transvaal).

Pentas carnea var. *welwitschii* Scott Elliot in Journ. Linn. Soc. Bot. **32**: 434 (1896). Type from Angola.

Neurocarpaea lanceolata sensu Hiern, Cat. Afr. Pl. Welw. **1**(2): 438 (1898) non (Forssk.) Britten.

Perennial herb, 30–100 cm. tall; rootstock little branched, with about three main roots, 6–12 cm. long. Stems woody, about 4 mm. in diam., brown or green above, terete, striate, branched, tomentose with white or brownish hairs; internodes 1.5–17 cm. long. Leaves 5.5–16 × 1–3 cm., lanceolate, frequently narrowly so, acute, cuneate basally, thickish or membranous, pubescent above and on the venation below, lateral nerves 6–14, strongly ascending at the margins, prominent below, venation a little reticulate; stipules with 5–8 setae, 4–10 mm. long from a short base; petiole 0.3–1.6 cm. long. Inflorescence capitate in flower, becoming spicate and laxer in fruit, up to 6 × 4 cm. in extent. Flowers geminate, not dimorphic. Calyx tube hairy, 2 mm. long; lobes unequal, 2.5–16.5 × 1–3 mm., lanceolate to elliptic, pubescent on the face, ciliate at the margins, venation prominent on the back. Corolla tube 1.2–1.9 cm. long, glabrescent or pubescent, 0.5–0.9 mm. in diam. at the middle, expanded above to 2–3 mm. wide forming a dilation 2–4 mm. long, lobes 2–6 × 0.8–2.5 mm., oblong-ovate, acute or rounded, glabrous or pubescent; throat and tube densely hairy. Style glabrous, exserted 1.25–3.5 mm. or rarely included; stigma lobes often purple, 1.5 mm. long, always exserted. Anthers often purple, 2 mm. long. Fruits pubescent 4–5.5 × 3.5–4.5 mm., somewhat constricted above, beak 1–1.5 mm. tall; finally dehiscent into two cocci; the truly terminal solitary one largest, 6 mm. tall and 4.5 mm. wide.

Zambia. W: Chibuluma, fr. 15.iii.1955, *Fanshawe* 2140 (K; NDO). C: 36 km. N. of Lusaka, by Kamaila Forest Rest House, fl. & fr. 29.i.1975, *Brummitt, Chisumpa & Polhill* 14161 (K; NDO). S: Nega Nega Hills, top of Munali Pass, fl. & fr. 26.ii.1964, *Angus* 3872 (FHO; K). **Zimbabwe**. N: Mazoe, about 20 km. N. of Umvukwes, fl & fr. 16.iv.1972, *Pope* 610 (K; LISC; SRGH). W: Matopos, *Hopkins* SRGH 7936 (K; SRGH). C: Gweru Kopje, fl. 5.iii.1967, *Biegel* 1978 (K; SRGH). E: Mutare, Commonage, fl. 2.ii.1955, *Chase* 5458 (BM; K; LISC; SRGH). **Malawi**. S: Machinga, Chikala Hills, fl. 26.i.1979, *Blackmore & Masiye* 189 (K; MAL) (identification needs confirming from further material). **Mozambique**. T: about 6 km. from Marueira towards Songo, fl. 5.ii.1972, *Macêdo* 4790 (K; LISC; LMA).

Also in Angola and S. Africa (Transvaal). *Brachystegia, Isoberlinia–Diplorhynchus, Terminalia–Combretum–Piliostigma* and similar woodland particularly in rocky places, granite kopjes etc., also in wooded grassland and streamside woodland; 730–1440 m.

Section PENTAS

Verdc. in Bull. Jard. Bot. Brux. **23**: 319 (1953).

Herbs or subshrubs. Flowers white or brightly coloured, mostly lilac to indigo, markedly heterostylous, always with forms with either style or anthers strongly exserted and occasionally with a third form with both included. (sp. 11–14).

11. **Pentas zanzibarica** (Klotzsch) Vatke in Oest. Bot. Zeitschr. **25**: 232 (1875). —Brenan, T.T.C.L.: 518 (1949). —Verdc. in Bull. Jard. Bot. Brux. **23**: 319, fig. 33/A, B (1953). —Agnew, Upland Kenya Wild Fl.: 405 (1974). —Verdc. in F.T.E.A., Rubiaceae 1: 203 (1976). Type from Zanzibar.
Pentanisia zanzibarica Klotzsch in Peters, Reise Mossamb., Bot. **1**: 286 (1861).

Herb, shrubby herb or rarely a shrub 0.3–2.6 m. tall, with 1–2(6) stems from a somewhat woody rootstock; stems greenish or purple-tinged, mostly strict and unbranched, often densely hairy above. Leaf blades (2.8)4–14.5 × (0.5)1.4–6 cm., lanceolate to ovate or elliptic, acute at the apex, cuneate at the base, hairy on both surfaces; petiole 0–1 cm. long; stipules with c. 7 setae 4–9(14) mm. long from a short base. Inflorescences lax or somewhat globose, terminal and axillary, 2–6.5 cm. across; peduncles 0–15 cm. long; bracts 6 cm. long. Flowers white, pink or mostly lilac, bluish mauve or in one variety bright crimson-red. Calyx tube hairy, 1–1.3 × 1–1.5 mm.; lobes unequal, 1–9 × 0.5–1.5(2.5) mm. Long-styled flowers; corolla tube (4)5–9(11) mm. long, hairy; lobes 1.5–5.5(6.5) × 0.8–2.7(3) mm., elliptic-oblong; style exserted 0.5–5.5 mm.; stigma lobes 1.5–2.5 mm. long; stamens completely included; throat densely hairy. Short-styled flowers very similar, the anthers exserted 2–3 mm. and style and stigma completely enclosed. Capsule pubescent, (2)3–4(5.5) mm. long and wide, with beak 1–2 mm. tall.

Subsp. **zanzibarica**

Stems less leafy with longer internodes; leaves opposite; plant less hairy with leaf venation not so evident beneath.

Var. **zanzibarica**. —Verdc. in F.T.E.A., Rubiaceae 1: 204 (1976).

Inflorescences more condensed. Corolla usually coloured, the tube often longer, up to 10 mm. Foliage usually less tenuous and more pubescent.

Malawi. S: Manganja Country, Mbame village, fl. 8.iii.1862, *Kirk* (K).
Also in Uganda, Kenya, Tanzania, E. Zaire; 900 m.

Var. **haroniensis** Verdc. var. nov.* Type: Zimbabwe, Haroni/Makurupini Forest, *Müller & Gordon* 1844 (K, holotype & isotype; SRGH).

Inflorescences very lax. Corolla white, tube 5–6 mm. long. Foliage thin and glabrescent.

Zimbabwe. E: Haroni R., fl. & fr. iv.1969, *Goldsmith* 37/69 (K; SRGH).
Riverine evergreen forest; 270 m.

Subsp. **milangiana** (Verdc.) Verdc. comb. nov. Type: Malawi, Mulanje, Luchenza Plateau, *Brass* 16447 (K, isotype; NY, holotype).
Pentas zanzibarica var. *milangiana* Verdc. in Bull. Jard. Bot. Brux. **23**: 327 (1953). —Brenan in Mem. N.Y. Bot. Gard. **9**: 115 (1954).

Stems leafy with shorter internodes; leaves sometimes in whorls of 3; plant usually densely hairy, the leaf venation mostly distinctly raised and hairy beneath. Corolla tube always short, about 5 mm. long.

Malawi. S: Mulanje Mt., Luchenza Plateau–Chembe Basin Path, fl. 13.vi.1962, *Richards* 16654 (K). **Mozambique**. N: Ribáuè, Serra Mepáluè, fl. 9.xii.1967, *Torre & Correia* 16421 (LISC; LMU; M; SRGH). Z: Milange, Serra Tumbine, 19.i.1966, *Correia* 520 (LISC).
Not occurring elsewhere. Forest edges, in scrub, grassland with bracken etc. and persisting in pine plantations; 1500–2100 m.
Considered within the context of the Flora Zambesiaca area alone these infraspecific variants might easily be considered separate species but var. *intermedia* Verdc. shows rather a strong resemblance to var. *haroniensis* but certainly has intermediates with var. *zanzibarica*. *Pentas zanzibarica*, when considered throughout its range, is extremely variable.
Southern populations of *Pentas zanzibarica* intergrade with *Pentas purpurea* Oliv. A deep crimson-flowered variant from East Africa, var. *rubra* Verdc. with corolla tube 8–13 mm. long has been grown in Harare, *Müller* 369 and 371. The former is typical *rubra* but the latter with 13 mm. corolla tube and deep pink flowers is technically nearer *P. lanceolata* but I think only a robust form of var. *rubra*.

12. **Pentas purpurea** Oliv. in Trans. Linn. Soc. **29**: 83 (1873). —Verdc. in Kew Bull. **7**: 363 (1952); in Bull. Jard. Bot. Brux. **23**: 330 (1953). —Hepper in F.W.T.A. ed. 2, **2**: 215 (1963). —Verdc. in F.T.E.A., Rubiaceae 1: 207 (1976). Type from Tanzania.
Pentas stolzii Schum. & Krause in Engl., Bot. Jahrb. **39**: 522 (1907). Type from Tanzania.
Pentas zanzibarica sensu auctt. mult. non (Klotzsch) Vatke.

Herb 0.4–1.3 m. tall, mostly c. 30–45 cm., with 1–4 hairy or glabrescent mostly unbranched stems from a woody rootstock. Leaves paired; blades (3)4.5–13.5(16.5) × 1–4.5(5.5) cm., elliptic-, ovate- or oblong-lanceolate or elliptic, ± acute at the apex, rounded or cuneate at the base, pubescent above and on the venation beneath where the hairs have a bristly appearance; petiole obsolete or up to 7 mm. long; stipules with 1–10 setae, 1–9(12) mm. long from a short base. Flowers mostly deep purple-violet or indigo but sometimes paler, or even white particularly inside lobes. Inflorescences a small dense head 1.3–3(4) cm. in diam., scarcely enlarging in fruit, or sometimes laxer up to 6.5 cm. wide, sessile or peduncle 2–20 cm. long. Calyx tube 1–2 mm. long and wide; lobes 1.5–8.5 × 0.5–2 mm., oblong, deltoid or minute, strigosely ciliate. Long-styled flowers; corolla tube 4–8(9) mm. long, often split longitudinally at the base along the filament sutures, cylindrical and scarcely dilated, glabrescent to densely pubescent outside; lobes 1.25–4 × 0.5–1.25 mm. linear-oblong, tips inflexed and hairy, and with the long matted throat hairs extending over the lower third of the interior surface; anthers either with tips 0.5 mm. below the throat orifice or situated nearly in the middle of the tube. Style exserted 1.5–2.5(4) mm., the stigma lobes 1–1.5 mm. long. Short-styled flowers very similar but anthers exserted 1.5 mm. (often hidden by the throat hairs). Capsule 3–4 × 3 mm., with a rounded triangular beak 1.5–2 mm. tall.

* *P. zanzibarica* (Klotzsch) Vatke var. *haroniensis* Verdc. var. nov., var. *intermedia* Verdc. similis sed foliis tenuioribus majoribus lanceolatis usque ellipticis glabrescentibus, usque 14 × 6 cm. differt; a *P. zanzibarica* var. *zanzibarica* characteribus isdem, inflorescentiis valde laxis, corolla alba distinguenda.

Subsp. **purpurea**. —Verdc. in Bull. Jard. Bot. Brux. **23**: 330 (1953).

Anthers in long-styled flowers with tips c. 0.5 mm. below the throat.

Zambia. N: Mbala Distr., Ndundu, fl. 10.v.1968, *Richards* 22895 (K). **Zimbabwe**. C: Wedza Mt., fr. 27.ii.1964, *Wild* 6339 (K; LISC; SRGH). E: 3.2 km. from Chimanimani on new Mutare Road, fl. 26.xi.1955, *Drummond* 5032 (K; SRGH). S: Masvingo, Great Zimbabwe National Park, fl. 28.iii.1973, *Chiparawasha* 640 (K; SRGH). **Malawi**. N: Nkhata Bay Distr., Viphya, 120 km. S. of Mzuzu, Luwawa Dam, fl. 8.ii.1971, *Pawek* 4393 (K; MAL). C: Dedza Mt., fl. 24.iv.1970, *Brummitt* 10108 (K). S: Blantyre Distr., Matenje Road, 1 km. N. of Limbe, fl. 5.ii.1970, *Brummitt & Banda* 8414 (K). **Mozambique**. N: Maniamba Valley, fl. 3.i.1942, *Hornby* 3535 (K). Z: Quelimane Distr., *Faulkner* Pretoria 354A (K; PRE). MS: between Beira and Massi Kessi (= Macequece), *Cecil* 13 (K).

Also in Guinea, Nigeria, Cameroon, Zaire and Tanzania. *Brachystegia* and other mixed open woodland, grassland with scattered trees and open grassland; (600?) 780–2100 m.

Subsp. **mechowiana** (K. Schum.) Verdc. in Bull. Jard. Bot. Brux. **23**: 334, fig. 35A (1953). Type from Angola.

Pentas mechowiana K. Schum. in Engl., Bot. Jahrb. **23**: 420 (1897).

Neurocarpaea purpurea sensu Hiern, Cat. Afr. Pl. Welw. **1**: 438 (1898) non (Oliv.) Hiern sensu stricto.

Anthers in long-styled flowers ± in middle of corolla tube.

Zambia. W: 24 km. N. of Mwinilunga, fl. 11.xii.1963, *Robinson* 5893 (K). **Zimbabwe**. E: 24 km. S. Chimanimani, Skyline, fl. 11.i.1947, *Fisher* 1255 (K; SRGH).

Also in Angola. *Cryptosepalum*, *Brachystegia* and 'Chipya' woodland; 1140–1500 m.

Variation in this species is complex. The difference in anther position is genuine but is variable in Zimbabwe E. whereas in Zambia W. it seems constant. The status of subsp. *mechowiana* needs examination in the field. I have included var. *buchananii* Scott Elliot in Journ. Linn. Soc. Bot. **32**: 436 (1896) (Type: Malawi, S., without locality, *Buchanan* 156 (K, lectotype)) in subsp. *purpurea* — it differs from typical *purpurea* in its taller more leafy habit, more densely shortly pubescent leaves and often paler flowers — in fact some specimens approach *Pentas zanzibarica* in flower structure. There are, however, numerous intermediates with typical *Pentas purpurea*. Again a reappraisal of status should be made in the field. Occasional specimens of *Pentas purpurea* have the open inflorescences of *Pentas zanzibarica* but mostly *Pentas purpurea* is a well-defined species.

13. **Pentas pubiflora** S. Moore in Journ. Linn. Soc. Bot. **38**: 254 (1908). —Robyns, Fl. Parc Nat. Alb. **2**: 331 (1947). —Verdc. in Bull. Jard. Bot. Brux. **23**: 327, fig. 35/B, C (1953). —Hepper, F.W.T.A., ed. 2, **2**: 216 (1963). —Verdc. in F.T.E.A., Rubiaceae 1: 206 (1976). Type from Uganda.

Herb or subshrub 0.6–1.5(3) m. tall; stems erect or somewhat decumbent, woody below, hairy or glabrescent. Leaf blades (3)9.5–18 × (1.2)3–6.5 cm., ovate to lanceolate, acute at the apex, cuneate at the base, pubescent; petiole 0–0.6(3) cm. long; stipules with c. (5)7–9 filiform setae 0.5–1.2 cm. long from a short base. Inflorescences of terminal and axillary components, usually wide and corymbose, up to 12 cm. wide (smaller, 2–5.5 cm. wide, in some forms); peduncles c. 2–4.5 cm. long. Flowers white, rarely tinged pale blue or pinkish, or in W. African race blue. Calyx tube 1.5 m. long; lobes unequal, 1 foliaceous, 2–6.5 × 0.75–2 mm., ovate, the rest deltoid or triangular, 1–2 mm. long. Long-styled flowers: corolla tube funnel-shaped, (3)4–5 mm. long, 1–1.25 mm. wide below, dilated above to 1.75 mm.; lobes 2–3 × 1–1.25 mm., oblong-lanceolate; stamens entirely included; style exserted (1)2–3.5 mm. Short-styled flowers: similar, tube up to 5.5 mm. long, 2 mm. wide at the throat; lobes 1.5–2 mm. long; anthers exserted 0.5–1.25 mm. Capsule 2–4.5 × 2–3 mm., hairy, ribbed; beak 1–2 mm. tall.

Mozambique. Z: Gúruè Mt., fl. & fr. 8.iv.1943, *Torre* 5117 (C; LD; LISC; LMU; MO; SRGH).

Also in Zaire, Uganda and Kenya, also some variants in Tanzania, Burundi, Nigeria and Cameroon. Mountain slope grassland with scattered trees, often by streams; 1100–1500 m.

Typical material occurs in W. Kenya to E. Zaire and, despite the wide disjunction, the Mozambique material appears correctly placed, although further material may show a subspecies can be delimited by the narrower more acuminate capsule beak. Atypical glabrescent variants occur in W. Tanzania and also in Burundi, Nigeria and Cameroon. Formerly I confused them with another taxon under the name *P. pubiflora* subsp. *bamendensis* but the type of this proves to be the same as *P. ledermanii* K. Krause which I mistakenly synonymised with *P. schimperiana* subsp. *occidentalis* (Hook. f.) Verdc. More study is needed before a name can be given to this glabrescent variant as *P. pubiflora*.

14. **Pentas lanceolata** (Forssk.) Deflers, Voy. Yemen: 142 (1889). —Verdc. in Kew Bull. **6**: 377 (1951); in Bull. Jard. Bot. Brux. **23**: 339, fig. 35/D, G (1953); in F.T.E.A., Rubiaceae 1: 208 (1976). —Lind & Tallantire, Fl. Pl. Uganda ed. 2: 160, fig. 103 (1972). —Agnew, Upland Kenya Wild. Fl.: 404, fig. (1974). Type from Arabia.

Ophiorrhiza lanceolata Forssk., Fl. Aegypt. Arab.: 42 (1775).
Manettia lanceolata (Forssk.) Vahl, Symb. Bot. **1**: 12 (1970).
Neurocarpaea lanceolata (Forssk.) R. Br. in Salt, Voy. Abyss. App. **4**: 64 (1814). —Britten in Journ. Bot. Lond. **35**: 129 (1897) excl. syn.

Herb or subshrub with erect or straggling mostly woody stems 0.5–1.3 m. tall, hairy. Leaf blades 3–13 × 1–6 cm., ovate, lanceolate, ovate-lanceolate or elliptic, acute at the apex, cuneate at the base, pubescent to densely velvety on both surfaces; petioles 0.5 cm. long; stipules with 3–9(14) setae, 2–9 mm. long, bearing small colleters, from a short base. Inflorescence with terminal and axillary components combined into a single cluster. Calyx tube hairy, 1–3 × 1.5–2.5 mm. lobes very unequal, the largest 0.5–1.3 cm. × 0.5–3 mm., lanceolate, the smallest 1–3 mm. long. Flowers often trimorphic, either with style exserted and anthers included, anthers exserted and style included or both included. Long-styled flowers: corolla tube (1)2.3–4 cm. long, dilated at the apex to 3(6) mm. wide for a distance of 4–8 mm., hairy or glabrous outside; lobes 0.3–1 cm. × (1)1.5–4.5 mm., oblong-ovate to elliptic; throat hairy within; anthers completely included; style exserted 1.5–5.5 mm.; stigma 2–5 mm. long. Short-styled flowers very similar; corolla lobes 4.5–8 mm. long; anthers exserted 2.5–4 mm.; style and stigma usually completely enclosed or rarely tips of stigma lobes exserted 2.5 mm. Flowers with both style and stigma and anthers included; tube 2–4 cm. long; lobes 0.5–1.1 cm. × 2.5–3.3 mm., ovate-oblong; anthers sometimes with tips exserted 0.25 mm. but usually included; style and stigma always included. Fruit 4–6 mm. tall and wide, obtriangular; beak 1–2 mm. tall.

Subsp. **quartiniana** (A. Rich.) Verdc. in Bull. Jard. Bot. Brux. **23**: 34 (1953); in F.T.E.A., Rubiaceae 1: 210 (1976). —Gonçalves in Garcia de Orta, Sér. Bot. **5**:(2): 202 (1982). Type from Ethiopia.
Vignaldia quartiniana A. Rich., Tent. Fl. Abyss. **1**: 357 (1847).
Pentas quartiniana (A. Rich.) Oliv. in Trans. Linn. Soc. **29**: 82, t. 46 (1873).
Pentas verruculosa Chiov. in Atti R. Accad. Ital., Mem. Cl. Sc. Fis. Mat. Nat. **11**: 35 (1940). Type from Ethiopia.
Pentas concinna sensu Chiov., R. Accad. Ital. Col. Stud. Afr. Or. Ital. 4, IV: 226 (1939) non K. Schum.
Pentas carnea sensu auctt. non Benth.

Corolla usually not white, tube 1.2–1.8 cm. long.

Var. **oncostipula** (K. Schum.) Verdc. in Bull. Jard. Bot. Brux. **23**: 351 (1953); in F.T.E.A., Rubiaceae 1: 211 (1976). Type from NE. Tanzania.
Pentas oncostipula K. Schum. in Engl., Bot. Jahrb. **34**: 329 (1904).

Inflorescences with components remaining congested even when fruiting, in no way spicate.

Malawi. C: Dedza, Chencherere Hill, fl. 18.i.1959, *Robson & Jackson* 1243 (BM; K; LISC). **Mozambique**. T: Angonia, Mt. Dómuè, fl. 9.iii.1964, *Torre & Paiva* 11097 (COI; LISC; LMU; SRGH; WAG).

Also in Tanzania. *Albizia* and *Brachystegia* woodlands, *Pinus* plantations, rocky hillside grassland; 1550–2230 m.

Curiously restricted to the Dedza area. *Exell, Mendonça & Wild* 1082 has very short corollas.

15. OTOMERIA Benth.

Otomeria Benth. in Hook., Niger Fl.: 405 (1849). —Verdc. in Bull. Jard. Bot. Brux. **23**: 5–34, 249 (1953).

Annual or perennial erect, subprostrate or twining herbs with mostly hairy stems. Leaves paired; stipules with base divided into several narrow segments. Flowers small and white or rather large and coloured, hermaphrodite, monomorphic or dimorphic, in cymose heads, which in fruit develop into a long simple spike with the fruits geminately arranged and a solitary remotely placed flower in the axil at the base of the spike (save in subgen. *Volubiles* where the cluster remains dense). Calyx tube ovoid or elongate-oblong; lobes unequal, 5, 1–3 foliaceous and larger than the rest, alternating with small colleters. Corolla tube long and narrow, with a markedly ovoid-oblong apical dilation in long-styled forms; throat densely hairy; lobes in small-flowered species elliptic, but in large-flowered species broader, ovate to orbicular, narrowing to the base where the lobes are often

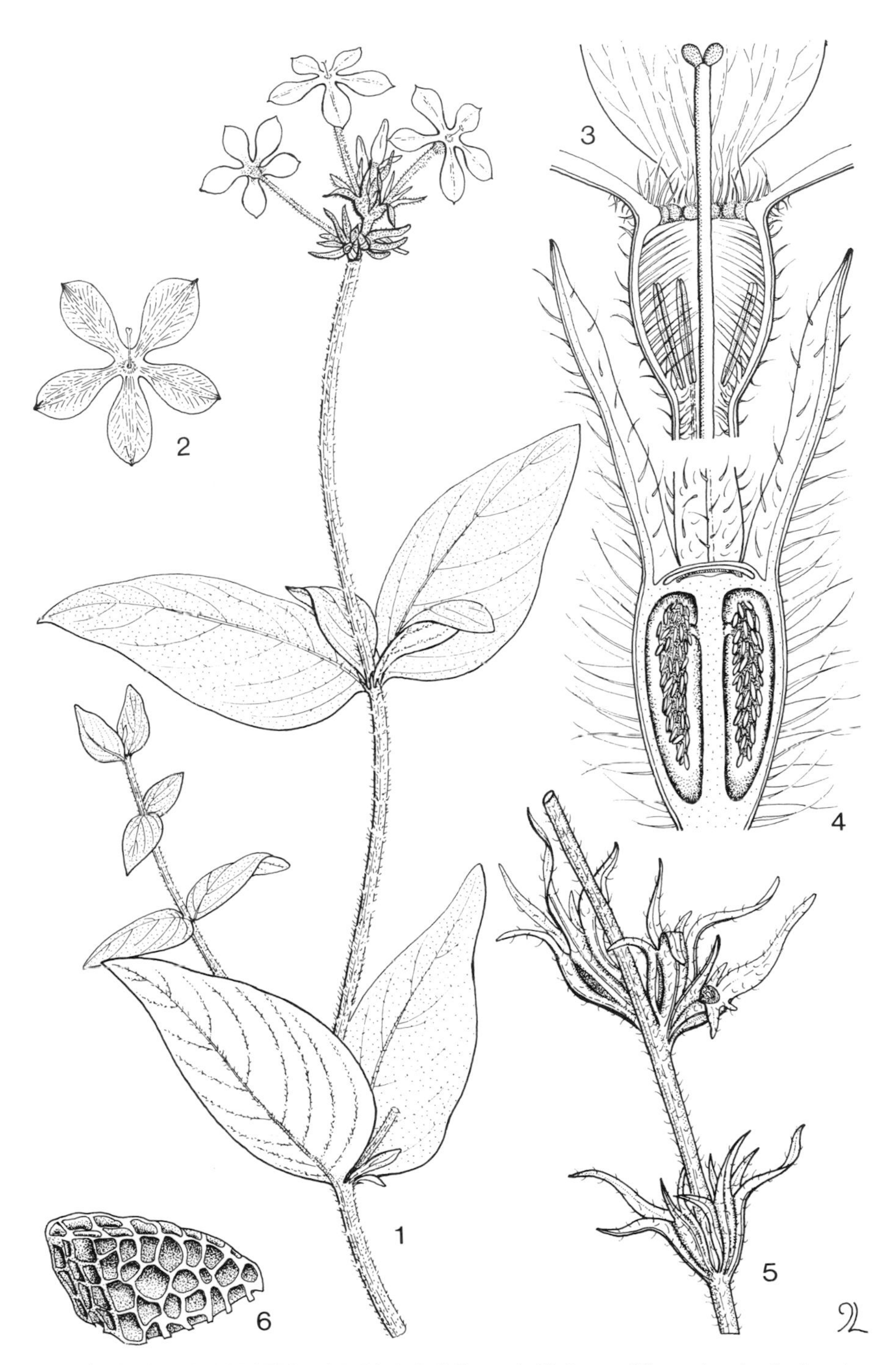

Tab. 17. OTOMERIA ELATIOR. 1, habit (× 1); 2, flower (× 1½); 3, part of flower, longitudinal section (× 9); 4, ovary, longitudinal section (× 9), 1–4 from *Robson* 1443; 5, fruits (× ½); 6, seed (× 45), 5–6 from Richards 1466.

connate for a short distance. Stamens completely exserted in short-styled forms. Style exserted in long-styled forms, the stigma bifid with filiform lobes; anthers completely included in the apical dilation of the corolla tube. Capsule oblong, compressed, ribbed, opening by apical valves and also frequently splitting longitudinally. Seeds small, reticulate.

A small genus of 8 species widely distributed in tropical Africa, 1 of which occurs in the Flora Zambesiaca area.

As has been mentioned in my revision the circumscription of this genus is very unsatisfactory and forms a reticulate pattern with *Pentas;* it could be combined with that genus but the result would be no more satisfactory. Subgen. *Otomeria*, with small white flowers, is very closely related to *Pentas* subgen. *Pentas* sect. *Monomorphi* Verdc., whereas subgen. *Neotomeria* Verdc. is closely related to *Pentas* subgen. *Pentas* sect. *Coccineae* Verdc. Nevertheless these two subgenera of *Otomeria* are united by their inflorescence and capsule structure. It seems practical to retain the classification adopted in my revision.

Otomeria elatior (A. Rich. ex DC.) Verdc. in Bull. Jard. Bot. Brux. **23**: 18, fig. 3/A-D (1953). —Hepper in F.W.T.A. ed. 2, **2**: 214 (1963). —Hallé, Fl. Gabon 12, Rubiacées: 117 (1966). —Lind & Tallantire, Fl. Pl. Uganda ed. 2: 160 (1972). —Agnew, Upland Kenya Wild Fl.; 405 (1974). —Verdc. in F.T.E.A., Rubiaceae 1: 214 (1976). TAB. **17**. Type from Angola.

Sipanea elatior A. Rich. ex DC., Prodr. **4**: 415 (1830). —A. Rich., Mem. Fam. Rubiacées: 196 (1830); in Mem. Soc. Hist. Nat. Paris **5**: 276 (1834).

Pentas elatior (A. Rich. ex DC.) Walp., Repert. **6**: 57 (1846).

Otomeria dilatata Hiern in F.T.A. **3**: 50 (1877). Type from Nigeria.

Erect or rarely straggling herb 0.35–3 m. tall, with single unbranched or sparsely branched glabrescent pubescent or hairy stem. Leaf blades 1.5–9.5 × 0.7–3.2 cm. wide, ovate-elliptic or rarely linear, ± acute at the apex, rounded or cuneate at the base, pubescent to densely hairy on both surfaces, or glabrous above and pubescent beneath on the nerves, or altogether glabrous save for a very minute pubescence near the edges above and on the venation beneath; petiole obsolete or 1–6 mm. long; stipules with 1–3 flat linear setae 1–5 × 0.2–1 mm., and also 2–6 short setae or sessile colleters from a short base. Flowering inflorescence 1–6.5 cm. long, becoming 4–37 cm. long in fruit, with a peduncle 0–30 cm. long; solitary nodal flower with pedicel up to 2 cm. long. Calyx tube c. 2 mm. long, glabrous to hirsute; lobes very unequal, 1–3 foliaceous, lanceolate, 0.5–2.4 cm. × 1.2–4.8 mm., the rest 1–4 × 0.5–1.5 mm. Corolla rose-scarlet or -crimson; tube 1.7–2.7 cm. long, sparsely hairy or glabrous below, 0.25–2.3 mm. wide below, dilated above to 1.25–3 mm. wide for a distance of 3–5 mm., the dilation urceolate and usually rounded at the base, externally hairy with long multicellular crimson or purple hairs; lobes 0.5–1.8 × 0.25–1 cm, ovate, orbicular or elliptic-spathulate, mucronate from the emarginate tip, connate at the base for 0.5–1.5 mm.; throat orifice 0.5–1.5 mm. in diam.; throat densely hairy, the apical hairs crimson and spreading over the orifice and connate parts of the corolla lobes. Anthers completely included. Styles exserted 0–5 mm.; stigmas elliptic, 0.5–1 mm. long. Fruit chestnut or purple-brown, 0.6–1.2 cm. × 4–6 mm. × 2.5–5 mm. (rarely smaller, 5 × 3.5 mm.), oblong, compressed, strongly ribbed, pubescent or hairy; dehiscence apical and accompanied by some longitudinal splitting. Seeds angular, c. 0.7 mm. long.

Zambia. N: Mbala, Chilongowelo, fl. & fr. 18.xii.1951, *Richards* 83 (K). W: 72 km. from Mwinilunga towards Solwezi, fl. 22.xi.1972, *Strid* 2618 (K; S). C: 100–129 km. E. of Lusaka, Chakwenga Headwaters, fl. 27.x.1963, *Robinson* 5782 (K). E: Petauke, fl. & fr. 15.xii.1958, *Robson* 952 (BM; K; LISC). **Zimbabwe**. N: Goromonzi, Umfuleni, Arcturus, fl. 2.i.1947, *Jack* in SRGH 15631 (K; SRGH). C: Marondera, fl. 25.x.1948, *Corby* 105 (K; SRGH). E: Mutare, fl. xii.1899, *Cecil* 233 A (K). S: Masvingo Road, fl. xii.1920, *Mainwaring* in *Eyles* 2807 (K; SRGH). **Malawi**. N: Mzimba Distr., Mzuzu Govt. School, Lunyangwa River, fl. 14.iii.1969, *Pawek* 1823 (K). C: Dedza Distr., Chongoni Forestry School, base of Chiwao Hill, fl. 4.ii.1959, *Robson* 1443 (BM; K; LISC). S: Mt. Mulanje, fl. & fr. iii.1897, *Adamson* 448 (E; K). **Mozambique**. N: Maniamba, about 45 m. from Vila Cabral towards Unango, fl. & fr. 2.iii.1964, *Torre & Paiva* 10938 (BR; LISC; LMU; SRGH; WAG). Z: Gúruè, near pico Namuli, fl. & fr. 9.iv.1943, *Torre* 5168 (COI; LISC; LMU; PRE). MS: Mossurize, Mafusi, fl. & fr. 22.ii.1907, *W.H. Johnson* 142 (K).

Widespread; W. Africa from Mali to Cameroon and Angola, Central African Republic, Sudan to Zimbabwe and Mozambique. Nearly always in wet places, dambos, stream-banks and riversides, swamps, vleis, often near evergreen fringes in *Brachystegia* woodland; occasionally in drier areas but still near water; 510–1900 m.

In SW. Tanzania and adjoining part of Zambia a variant occurs with a more showy corolla, the lobes being 1.6–1.8 × 1 cm., narrowed at the base to 2.5 mm. In my revision this was formally

Tab. 18. BATOPEDINA LINEARIFOLIA. 1, habit, long-styled (× $\frac{2}{3}$), from *Fanshawe* 3116; 2, habit, short-styled (× $\frac{2}{3}$), from *Fanshawe* 4144; 3, stipule (× 4), from *Fanshawe* 3116; 4, flower bud (× 3), from *Fanshawe* 4144; 5, long-styled flower (× 2); 6, calyx (× 6); 7, long-styled flower, longitudinal section (× 3); 8, stigma (× 12); 9, ovary, longitudinal section (× 16); 10, ovules on one placenta (× 2), 5–10 from *Fanshawe* 3116; 11, short-styled flower (× 3); 12, short-styled flower, longitudinal section (× 3), 11–12 from *Fanshawe* 1738; 13, fruit (× 6); 14, seed (× 20), 13–14 from *Robinson* 3395.

recognized as forma *speciosa* (Bak.) Verdc. (in Bull. Jard. Bot. Brux. **23**: 23 (1953); *Pentas speciosa* Bak. in Kew Bull. **1895**: 67 (1895); *Otomeria speciosa* (Bak.) Scott Elliot in Journ Linn. Soc. Bot. **32**: 437 (1896). Type: Zambia, Fwambo, *Carson* (K, holotype)). It seems scarcely worth maintaining despite the distinctiveness of extreme forms, e.g. *Bullock* 2151 (Mbala to Kambole).

16. BATOPEDINA Verdc.

Batopedina Verdc. in Bull. Jard. Bot. Brux. **23**: 29 (1953). —Robbrecht in Bull. Jard. Bot. Nat. Belg. **51**: 174 (1981).

Subshrubs with many branches from a thick woody rootstock, often tufted. Leaves opposite but sometimes appearing verticillate, linear to ovate; stipules with 3–5 short lobes from a short base. Flowers heterostylous, solitary or paired, axillary or at apices of branchlets; calyx tube ovoid; lobes unequal, one foliaceous and rest minute or of varying sizes. Corolla tube filiform, enlarged at apex; throat hairy inside; lobes 5, ovate. Stamens 5, included or exserted. Ovary 2-locular with many ovules. Style included or exserted; stigma lobes filiform. Capsule ovoid, splitting into 2 cocci with numerous seeds. Testa finely striate.

A genus of 3 species from Upper Volta, Ghana, Zambia and Zaire which I am now very doubtful should be separated from *Otomeria*. Robbrecht figures the testa cells of all 3 showing wavy walls, lumina not punctate and fine striae more or less at right angles to the radial walls. This, however, is not helpful since testa cell sculpture and shape vary considerably in *Otomeria*, the walls being straight or wavy, the lumina smooth or punctate and in 2 species there are similar striae. *Batopedina linearifolia* var. *glabra* Petit is I think a distinct species and has the wavy walls, striations and unpunctured lumina of *Batopedina* but distinctly spicate inflorescences of *Otomeria*.

Batopedina linearifolia (Brem.) Verdc. in Bull. Jard. Bot. Brux. **23**: 29 (1953). —Robbrecht in Bull. Jard. Bot. Nat. Belg. **51**: 186, fig. 7B (1981). TAB. **18**. Type: Zambia, Mumbwa, Nambala Hill, *Macaulay* 775 (K, holotype).
Otomeria linearifolia Brem. in Kew Bull. **3**: 461 (1949).

Subshrub 10–25 cm. tall often with an ericoid appearance with few to many branched or ± unbranched erect or decumbent stems from a woody rootstock, blackish below where epidermis eventually peels, reddish brown above, spreading white pubescent when young; internodes short. Leaves paired but appearing verticillate due to very abbreviated leafy axillary shoots, 0.6–1.7 cm. × 0.8–4 mm., linear, linear-lanceolate or narrowly elliptic, acute at the apex, cuneate at the base, revolute, uninerved, densely white pubescent especially beneath; petioles ± obsolete; stipules chestnut with 3–5 small lobes 1.5–2 mm. long from a short base. Flowers solitary or mostly paired at the apices of the branchlets and also in the second and third and even lower axils. Calyx tube ovoid, about 2 mm. × 1.5 mm., pubescent, the largest lobes 5.5 × 0.25–1.3 mm. the smallest about 1 mm. Corolla white or pale blue, the tube often pinkish, pubescent outside; tube 1.1–2.4 cm. long, 0.6 mm. wide at base, dilated in long-styled flowers to 1.5 mm. for 3.5–4 mm.; lobes 5–9 × 2.5–4.2 mm. Style exserted in long-styled flowers 1.6–7 mm. and stamens completely included; stigmatic lobes 1.5–3 mm. long, filiform. Filaments exserted 4–5 mm. in short-styled flowers, the style included. Fruit 2.5 mm. long and wide, ovoid, splitting longitudinally septicidally and loculicidally, without a beak.

Zambia. W: Luanshya, Muva Hill, fl. 21.xii.1957, *Fanshawe* 4143 (K; NDO). S: near Mumbwa, Nambala Hill, *Macaulay* 775 (K).
Not known elsewhere. Crevices in granite and quartzite rocks on kopjes; 1250 m.
I have not retained Petit's var. *glabra* described from Zaire within the above circumscription since I consider it to be a distinct glabrous species with a distinct spike-like inflorescence and is better placed in *Otomeria* but my original erection of the genus may have been an error. The two may have to be combined.

17. CARPHALEA Juss.

Carphalea Juss., Gen.: 198 (1789). —Homolle in Bull. Soc. Bot. Fr. **83**: 613 (1937). —Verdc. in Kew Bull. **28**: 423 (1974).
Dirichletia Klotzsch in Monatsber. Königl. Preuss. Akad. Wiss. Berl. **1853**: 494 (1853); in Peters, Reise Mossamb., Bot. **1**: 292, t. 47, 48 (1862).

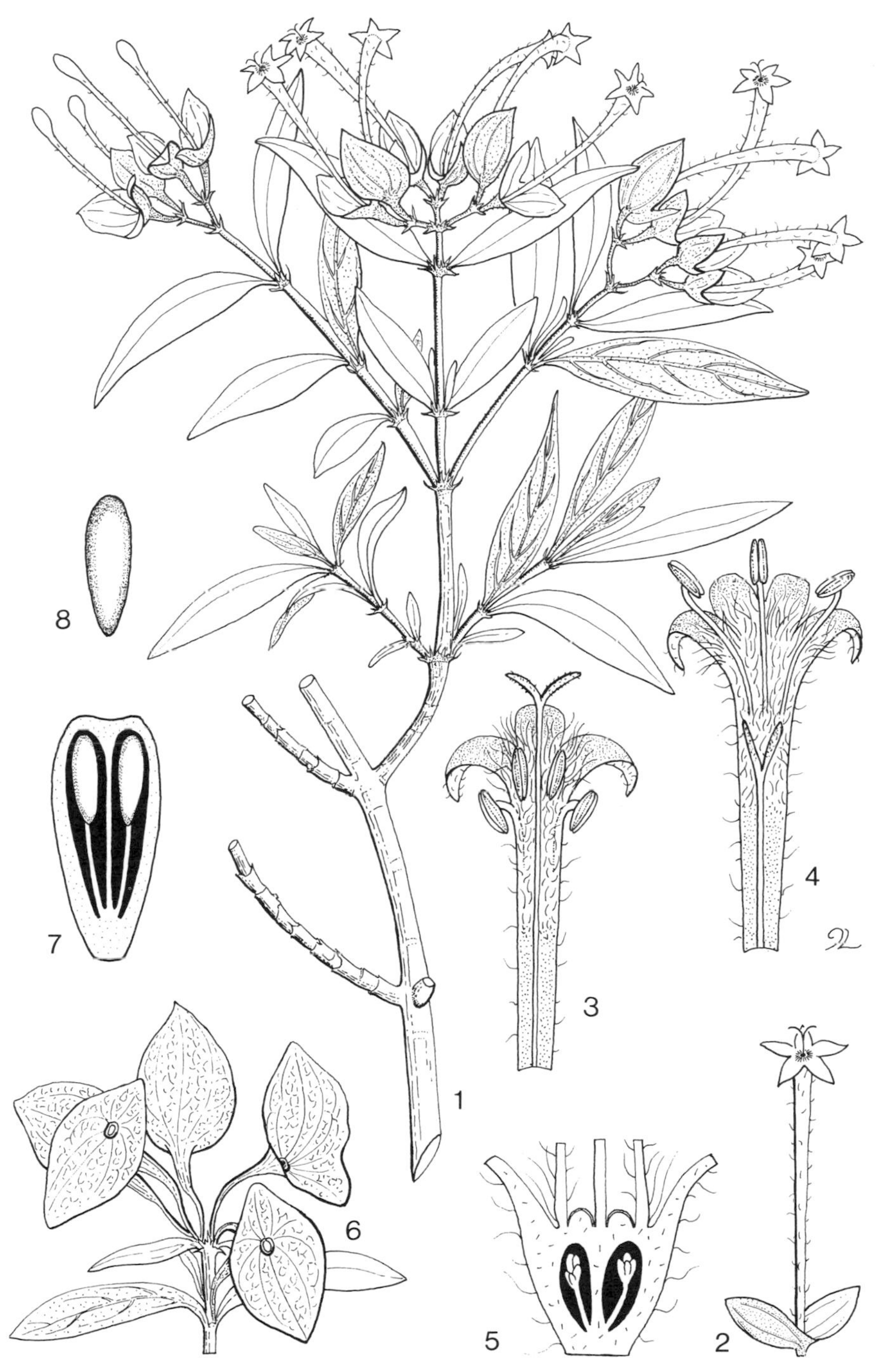

Tab. 19. CARPHALEA PUBESCENS. 1, habit (× $\frac{2}{3}$); 2, flower (× 1); 3, long-styled flower, longitudinal section (× 3); 4, short-styled flower, longitudinal section (× 3); 5, ovary, longitudinal section (× 9); 6, infructescence (× $\frac{2}{3}$); 7, fruit, longitudinal section (× 4); 8, seed (× 6), all from *Rogers* 5414.

Small shrubs with erect branched stems. Leaves paired or in whorls of 3, petiolate. Stipules with (1)3–5 linear or filiform colleter-tipped setae from a short base adnate to petioles. Flowers medium-sized, hermaphrodite, dimorphic, some flowers with anthers included and style exserted and others completely vice versa, in rather dense few–many-flowered terminal corymbose inflorescences. Calyx tube narrowly obconic or turbinate, ribbed; limb variously deeply 4–5-lobed or, in sect. *Dirichletia*, nearly always eccentrically elliptic, the tube ± placed at one of the foci, sometimes shallowly 3-lobed. Corolla tube very narrowly cylindrical; lobes 4–5, oblong or ovate; throat densely hairy. Ovary 2–3-locular, each locule with a slender basal placenta bearing c. (3)4–6 ovules; style filiform, the stigma divided into 2(4) filiform lobes. Fruit obconic, sometimes curved, of bony texture, strongly ribbed, the ribs running out into the strongly nervose accrescent calyx limb or lobes, apparently always indehiscent, 1–2-seeded. Seeds narrowly oblong-obconic.

A small genus of about 15 species confined to eastern and central tropical Africa, Socotra and Madagascar. The 3 tropical African species belong to the section *Dirichletia* (Klotzsch) Verdc. and are very poorly defined.

Carphalea pubescens (Klotzsch) Verdc. in Kew Bull. **28**: 426 (1974). —Palgrave, Trees S. Afr.: 842 (1977). —Gonçalves in Garcia de Orta Sér. Bot. **5**(2): 188 (1982). TAB. **19**. Type: Mozambique, Tete, Upper Zambezi, near Nhampazaza, *Peters* (B, holotype†; K).

Dirichletia pubescens Klotzsch in Monatsb. Königl. Preuss. Akad. Berl. **1853**: 495 (1853); in Peters, Reise Mossamb., Bot. **1**, 292, t. 48 (1861). —Hiern in F.T.A. **3**: 51 (1877). —F. White, F.F.N.R.: 406 (1962).

Dirichletia glabra Klotzsch in Monatsb. Königl. Preuss. Akad. Berl. **1853**: 495 (1853); in Peters, Reise Mossamb., Bot. **1**: 292, t. 47 (1861). Type: Mozambique, Querimba I. & the mainland, *Peters* (B, ? syntypes †).

Dirichletia rogersii Wernham in Journ. Bot. Lond. **54**: 231 (1916). Type: Zimbabwe, Victoria Falls, *Rogers* 5533 (BM, holotype).

Dirichletia duemmeri Wernham in Journ. Bot. Lond. **55**: 78 (1917). Type: Mozambique, Pemba, *Dummer** 68 (BM, holotype).

Shrub (0.3)0.9–4.5 (rarely described as a woody herb or even a small tree); young stems pale brown or straw-coloured, glabrous or pubescent at first and becoming glabrous, eventually developing corky longitudinally furrowed bark. Leaves 3.5–11 × 0.8–4.5 cm., elliptic or elliptic-lanceolate to lanceolate, narrowly tapering to acuminate at the apex, cuneate at the base, glabrous or with hairs on midrib beneath only or quite densely hairy particularly on the nerves beneath; petiole 0–6 mm. long; stipules with 3–5 setae up to 3 mm. long from a base up to 5 mm. long. Inflorescences about 4–8 cm. wide; peduncles and secondary peduncles c. 1 cm. long; pedicels 6–10 mm. long. Calyx tube obconic, 2 mm. long, glabrous to densely hairy; limb white or whitish green, 2–2.7 × 1.5–2 cm., usually eccentrically elliptic, the larger part occasionally 3-lobed, glabrous or hairy particularly on the venation. Corolla slightly fragrant, white or tube greenish; tube (2.3)3–5 cm. long, glabrous or pubescent outside; lobes up to 7 × 3 3mm., triangular. Fruit 6–10 × 2.5–5 mm., the accrescent calyx limb 2.5–3.5 × 2–5 cm.

Caprivi Strip. Mpilila I., fl. 13.i.1959, *Killick & Leistner* 3358 (K; PRE). **Botswana**. N: Pandamatenga, fl. ii.1876, *Holub* 958.9.60 (K). **Zambia**. C: Mpika Distr., 4.8 km. S. of Mfuwe, fl. 15.i.1969, *Astle* 5413 (K). S: Livingstone Distr., Katambora, fl. 9.i.1956, *Gilges* 533 (K; SRGH). **Zimbabwe**. N: Mtoko Distr., Mkota Reserve, fl. 29.xii.1950, *Whellan* 485 (K; SRGH). W: 8 km. W. of Matetsi, fl. 6.iii.1964, *Leach* 12120 (K; SRGH). C: Kwekwe Distr., Umniati, fl. 27.xii.1959, *Leach* 9734 (K; LISC; SRGH). E: Inyanga Distr., St. Swithin's Tribal Trust Land, fl. 17.i.1967, *Biegel* 1776 (K; SRGH). **Malawi**. S: Blantyre Distr., Native Authority Simon, fl. 10.i.1956, *Jackson* 1799 (FHO; K). **Mozambique**. N: 3 km. S. of Pemba, fl. 11.iii.1960, *Gomes e Sousa* 4528A (K). T: between Lupata & Tete, ii.1859, *Kirk* 'Rub(3)' (K).

Not known elsewhere. In various types of woodland including *Colophospermum*, *Bauhinia petersiana*, *Commiphora–Terminalia–Kirkia* etc., bushland and even grassland on coastal and Kalahari Sand, also in rocky places on Karroo basalt etc.; 0–1200 m.

Mostly well defined from the eastern African *C. glaucescens* (Hiern) Verdc. but some populations from Tanzania, Mpwapwa approach *C. pubescens* closely.

* At that date Dummer used an umlaut hence specific name.

18. **KOHAUTIA** Cham. & Schlecht.

Kohautia Cham. & Schlecht. in Linnaea **4**: 156 (1829), nom conserv. —Bremek. in Verh. K. Nederl. Akad. Wet., Afd. Natuurk., ser. 2, **48**(2): 56 (1952). —Verdc. in F.T.E.A., Rubiaceae 1: 228 (1976).
Hedyotis L. sect. *Kohautia* Wight & Arn., Prodr. **1**: 417 (1834).
Oldenlandia L. subg. *Kohautia* Hook. f. in Benth. & Hook. f., Gen. Pl. 2: 59 (1877).
Duvaucellia S. Bowd. in T. Bowd., Exc. Madeira (ed. S. Bowd.): 259 (1825).

Annual or perennial mostly erect herbs, occasionally subshrubs, more rarely dwarf shrubs, sometimes with short woody subterranean stems. Leaves sessile, opposite, mostly linear to narrowly elliptic-lanceolate, apex ± acute, midrib prominent below, rarely 3-nerved from the base; stipular sheath ± membranaceous, produced either into 1 or 2 ± fimbriated triangular lobes becoming narrower and often displaced towards leaves in apical parts or into 2–8 fringing ± subulate rigid fimbriae. Flowers ⚥, 4- (rarely 3- or 5-) merous, medium to small, stigmas and anthers always included in the corolla tube, never heterostylous, in terminal thrysic, sometimes corymbose, extensive panicle-like, subcapitate or capitate inflorescences, occasionally rather few-flowered; subtending bracts at the base ± leaf-like reduced gradually to small triangular rudiments towards the top of the partial inflorescence branches. Calyx tube mostly ovoid, ovoid-elliptic to globose, occasionally hemispherical; calyx lobes 4 (5), small, (±) equal, narrowly triangular, subulate or ovate-lanceolate, sometimes keeled. Corolla tube narrowly cylindrical, with dilated barrel-shaped or more rarely narrowly funnel-shaped apical part; entrance to and throat inside glabrous or bearded with short flattened papillate hairs; corolla lobes broadly elliptic to narrowly linear, completely parted or shortly fused at the base, acute, sometimes mucronate or apiculate. Stamens contained within the upper dilated portion of the tube, completely included or only sterile anther connectives emergent; anthers ± sessile. Ovary bilocular with numerous ovules embedded in a fleshy peltate placenta attached by a short stalk to the middle of the septum. Style glabrous either bearing 2 filiform stigmatic lobes or a single fused cylindrical or ovoid stigma; stigma usually held well below the anthers or occasionally just touching anthers. Capsules hemispherical, globose, subglobose or ellipsoid, crowned by the permanent calyx lobes, splitting loculidally at the top, beak not conspicuous. Seeds numerous, roundish, angular or conical to subconical, light brown to blackish.

A genus of 31 species (34 taxa) occurring in India, Pakistan, Iran, the Arabian peninsula and throughout Africa; also in Madagascar, the Cape Verde Islands and Socotra.

A specimen from Zimbabwe (*Philcox* et al. 8782 (K, SRGH), Urungwe National Park, 306 km. Harare to Chirundu, 20.ii.1981) has been excluded from *Kohautia* by Mantell on the grounds that the style is exserted. In one flower the style and long lobes are exserted well beyond the corolla lobes but in others the stigmatic lobes reach only to the centre of the stamens. This variation is disturbing and needs further examination in the field. Bridson and Verdcourt are convinced the specimen is a *Kohautia* — it has exactly the facies of that genus — and probably a variant of *K. caespitosa* with a short corolla closely allied to subsp. *brachyloba*. (Note added by B. Verdcourt).

Key to the subgenera of Kohautia

Stigma comprised of 2 free ± filiform lobes; stipular sheaths in the mid-stem region mostly produced into 1 or 2 ± fimbriated triangular lobes - - - - - - - Subgen. *Kohautia*
Stigma cylindrical or ovoid, not divided into 2 lobes; stipular sheath in the mid-stem region mostly produced into 2–8 fringing, ± rigid, subulate fimbriae - - Subgen. *Pachystigma*

Subgen. KOHAUTIA

Verdc. in F.T.E.A., Rubiaceae, 1: 230 (1976).
Kohautia subgen. *Eu-kohautia* Bremek. in Verh. K. Nederl. Akad. Wet., Afd. Natuurk., ser. 2, **48**(2): 81 (1952) nomen non valide publ. TAB. **20**, fig. A.

Stipular sheath in the mid-stem region mostly with 1 or 2 soft ± fimbriated triangular lobes; corolla lobes above and corolla throat always glabrous; stigma lobes 2, filiform.

The subgen. *Kohautia* comprises two series, *Kohautia* and *Diurnae*. These, however, are not used in the following key as the diagnostic characters are sometimes difficult to spot (especially on herbarium material) and some characters (i.e. dimensions and colour of the corolla lobes) may overlap especially in smaller flowered species.

Key to the species of the Subgen. Kohautia

1. Partial inflorescences distinctly capitate; roots fibrous, often with bead-like nodules - - - - - - - - - - - - - - - - 4. *amatymbica*
- Partial inflorescences not capitate; 1 or 2 flowers at a node (sessile, subsessile or pedicellate) - - - - - - - - - - - - - - - - - 2
2. Flowers mostly small; corolla tubes 2–6 mm. long in the Flora Zambesiaca area; stigma lobes held just below or touching the anthers; annuals - - - - - - - - - 3
- Flowers larger; corolla tubes (6)6.5–11(16) mm. long; stigma lobes held well below the anthers; perennials sometimes flowering in first season - - - - - - - - - - 7
3. Corolla lobes brightly coloured above (scarlet, red, pink, orange, mauve, very rarely white), broadly elliptic, 1.4–4.3(5.6) × (0.6)0.8–2.7 mm. - - - - - - - - - - 4
- Corolla lobes usually white, cream or blueish-, pinkish-, brownish- or greenish-white above, narrowly ovate-lanceolate (0.3)0.6–1.9(2.3) × 0.2–1.1 mm. - - - - - - - 5
4. 1 or usually 2 distinctly pedicellate flowers at a node; pedicels slender (4.6)8.4–16.6(30) mm. long, lengthening slightly after anthesis - - - - - - - - - - 9. *confusa*
- 1 sessile and 1 (pseudo-) pedicellate flower at a node; (pseudo-) pedicels ± robust, (0.6)1–3.6(6.7) mm. long - - - - - - - - - - - - - - - 7. *coccinea*
5. Flowers mostly in subsessile or sessile pairs at the nodes - - - - - - - 6
- Flowers 1 or 2 at a node, the terminal one ± sessile, the lateral one on a distinct pseudo-pedicel*; flowers small, greenish and inconspicuous - - - - - - - - - 3. *microflora*
6. Corolla tubes (2.7)3.3–5.8(6) mm. long, the dilated part ± one quarter to one third the length of narrow part; narrow part glabrous; dilated part and back of corolla lobes scabrid; calyx lobes ± one quarter the length of the narrow part; calyx tube densely beset with long papillae or hirsute - - - - - - - - -2. *subverticillata* subsp. *subverticillata*
- Corolla tubes (2.1)2.3–3.1(3.4) mm. long, the dilated part ± equal in length to the narrow part; narrow part with sparse minute papillae; dilated part ± glabrous; calyx lobes nearly as long as narrow part; calyx tube densely beset with coarse rounded papillae - - - - 8. *aspera*
7. Calyx tubes and capsules ± distinctly hemispherical-obconic, glabrous or minutely verrucose; leaves mostly filiform to very narrowly linear, to 1.5(2) mm. wide; plants often turning black(ish) on drying; pollen grains 3- (rarely 4) colporate - - - - - - - - - - 8
- Calyx tubes ovoid-ellipsoid; capsules subglobose, sparsely to densely beset with coarse rounded, conical or flattened triangular white papillae or hairs; leaves mostly narrowly linear-lanceolate to lanceolate-elliptic, to 8(9.5) mm. wide, rarely linear; plants mostly a whitish green on drying; pollen grains (3), 4- and 5-colporate - - - - - - 1. *caespitosa* subsp. *brachyloba*
8. 1–2 shortly to distinctly pedicellate flowers at a node; pedicels to 17 mm. long; corolla lobe margins in mature flowers permanently revolute; leaves much shorter than internodes, at most to 18(23) mm. long, caducous or reduced to 2 mm. long vestiges on lower parts of plants - - - - - - - - - - - - - - - - 6. *ramosissima*
- 1 or mostly 2 flowers at a node, one subsessile, the other pseudo-pedicellate; pedicels to 12(20) mm. long; corolla lobe margins never revolute; leaves half to as long or slightly longer than internodes, to 40(65) mm. long - - - - - - - - - - - - 5. *cynanchica*

Series 1. KOHAUTIA

Verdc. in F.T.E.A., Rubiaceae 1: 230 (1976).
Noctiflorae Bremek. in Verh. K. Nederl. Akad. Wet., Afd. Natuurk., ser. 2, **48**(2): 91 (1952) nomen non valide publ.

Corolla lobes narrow, white or whitish above, tube outside and lobes beneath darker; the majority of species phalaenophilous, rarely micro-melittophilous or ? cleistogamous; corolla lobes mostly completely parted to the base, rarely slightly fused, in all cases the swollen intrusions, common in the series *Diurnae,* (i.e. at the base between the corolla lobes) absent.

* Frequently terminal flowers may be subtended by 1–2 sterile, minute bracts below which there is a discrete axis. These to the naked eye appear to be ordinary pedicellate flowers and are, for convenience, described as being 'pseudo-pedicellate'.

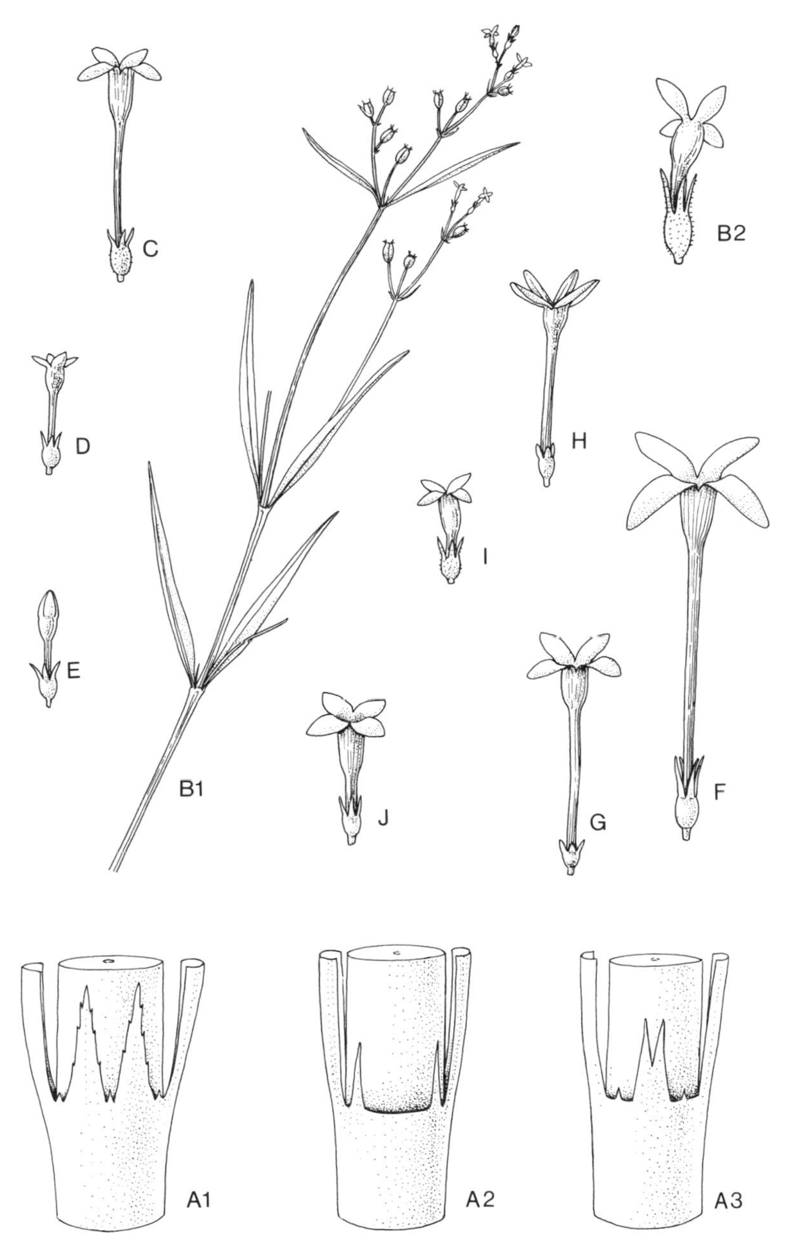

Tab. 20. A. —KOHAUTIA subgen. KOHAUTIA. Schematic representation of stipular sheath. A1, two narrowly triangular lobes borne near the middle of the sheath; A2, two narrowly triangular fimbriae borne near the leaves; A3, stipular sheath with a single, bifurcate lobe borne on the middle of the sheath. B. —K. COCCINEA. B1, habit (× $\frac{1}{2}$); B2, flower, B1–2 from *Richards* 14981. C. —K. CAESPITOSA subsp. BRACHYLOBA. Flower, from *Balsinhas & Macuacua* 566. D. —K. SUBVERTICILLATA subsp. SUBVERTICILLATA. Flower, from *Phipps* 2382. E. —K. MICROFLORA. Flower, *Brown* 7946. F. —K. AMATYMBICA. Flower, from *Goodier* 631. G. —K. CYNANCHICA. Flower, *Hansen* 3516. H. —K. RAMOSISSIMA. Flower, *Timberlake* 2076. I. —K. ASPERA. Flower, from *Plowes* 1659. J. —K. CONFUSA. Flower, *Robinson* 1292. Flowers all (× 2).

1. **Kohautia caespitosa** Schnizl. in Flora **25**, Beibl. 1: 145 (1842). —Bremek. in Verh. K. Nederl. Akad. Wet., Afd. Natuurk., ser. 2, **48**(2): 104 (1952); in Kew Bull. **8**: 439 (1953). —Cufod., Enum.: 986 (1965). —Launert & Roessler in Merxm. Prodr. Fl. SW. Afr. **115**: 16 (1966). —Wickens in For. Bull. **14** (n.s.): 27 (1969). —Migahid, Fl. Saudi Arabia, ed. 2, **2**: 425 (1973). —Agnew, Upland Kenya Wild Flowers: 398 (1974). —Verdc. in Kew Bull. **30**: 290 (1975); in F.T.E.A., Rubiaceae 1: 238 (1976). —Wickens, Fl. Jebel Marra: 130 (1976). Type from the Sudan.

According to my recent investigations and revision of the genus, this somewhat variable species can be divided into 3 subspecies of which only 1 (subsp. *brachyloba)* occurs in the Flora Zambesiaca area. Subsp. *caespitosa* occurs in N.E. Africa (Sudan, N. Ethiopia, N. Somalia, Egypt), Sinai and the Arabian peninsula, whereas subsp. *amaniensis* is found in S. Somalia, S. Ethiopia, E. Uganda, Kenya and N. Tanzania.

Subsp. **brachyloba** (Sonder) D. Mantell comb. et stat. nov. Type from S. Africa.

Hedyotis thymifolia C. Presl in Bot. Bemerk.: 85 (1844) nom. illeg. non Ruiz & Pav. (1794). TAB. **20**, fig. C.

Hedyotis brachyloba Sonder in Linnaea **23**: 50 (1850). —Walp., Ann. Bot. Syst. **2**: 771 (1851). —Sonder in Harv. & Sond., F. C. **3**: 10 (1865). Type as for subsp. *brachyloba.*

Kohautia lasiocarpa Klotzsch in Peters, Reise Mossamb. Bot. **1**: 296 (1862). —Bremek. in Verh. K. Nederl. Akad. Wet., Afd. Natuurk., ser. 2, **48**(2): 110 (1952) pro parte. —Launert & Roessler in Merxm. Prodr. Fl. SW. Afr. **115**: 18 (1966). —Verdc. in F.T.E.A., Rubiaceae 1: 240 (1976) pro parte. Type: Mozambique, Sena, *Peters* s.n. (B†); Type: Zambia, Mazabuka, *Rogers* 8726 (K, neotype).

Oldenlandia lasiocarpa (Klotzsch) Hiern, in F.T.A. **3**: 53 (1877). Type as for *K. lasiocarpa.*

Oldenlandia brachyloba (Sonder) Kuntze, Rev. Gen. Pl. **1**: 292 (1891). Type as for *H. brachyloba.*

Oldenlandia thymifolia (C. Presl) Kuntze, Rev. Gen. Pl. **1**: 298 (1891) nom. illegit. Type from S. Africa.

Oldenlandia papillosa K. Schum. in Engl., Bot. Jahrb. **23**: 416 (1897). —Hiern, Cat. Afr. Pl. Welw. **2**: 443 (1898). Type from Angola.

Oldenlandia welwitschii Hiern, Cat. Afr. Pl. Welw. **1**: 442 (1898). —Broun & Massey, Fl. Sudan: 258 (1929). —F.W. Andr., Fl. Pl. Anglo-Egypt. Sudan **2**: 451 (1952). Type from Angola.

Oldenlandia delagoensis Schinz in Mém. Herb. Boiss. **10**: 64 (1900). Type: Mozambique, Delagoa Bay, Rikalta, *Junod* 203 (Z, holotype; BR).

Oldenlandia stenosiphon K. Schum. ex S. Moore in Journ. Linn. Soc., Bot. **37**: 300 (1906). Type from Angola.

?*Oldenlandia schaeferi* K. Krause in Bot. Jahrb. **48**: 405 (1912). Type from Namibia (B†).

Kohautia brachyloba (Sonder) Bremek. in Verh. K. Nederl. Akad. Wet., Afd., Natuurk., ser. 2, **48**(2): 98 (1952). —Launert & Roessler in Merxm. Prodr. Fl. SW. Afr. **115**: 18 (1966). Type as for *H. brachyloba.*

Kohautia caespitosa var. *dolichostyla* Bremek. in Verh. K. Nederl. Akad. Wet., Afd. Natuurk., ser. 2, **48**(2): 108 (1952). Type: Zambia, along the Zambezi R., *Borle* 223 (PRE).

Kohautia caespitosa var. *thymifolia* (Kuntze) Bremek. in Verh. K. Nederl. Akad. Wet., Afd. Natuurk., ser. 2, **48**(2): 111 (1952). —Launert & Roessler in Merxm. Prodr. Fl. SW. Afr. **115**: 18 (1966). Type as for *O. thymifolia.*

Kohautia lasiocarpa var. *breviloba* Bremek. in Verh. K. Nederl. Akad. Wet., Afd., Natuurk., ser. 2, **48**(2): 112 (1952). —Verdc. in F.T.E.A., Rubiaceae 1: 240 (1976). Type: Zimbabwe, Trelawney, *Jack* 184 (K; holotype; SRGH).

Kohautia latibrachiata Bremek. in Verh. K. Nederl. Akad. Wet., Afd. Natuurk., ser. 2, **48**(2): 96 (1952). Type: Zambia, Upper Zambezi, *Kiener* s.n. (P, holotype).

Kohautia longiscapa Bremek. in Verh. K. Nederl. Akad. Wet., Afd. Natuurk., ser. 2, **48**(2): 99 (1952). Type from Angola.

Kohautia longiscapa var. *scabridula* Bremek. in Verh. K. Nederl. Akad. Wet., Afd. Natuurk., ser. 2, **48**(2): 99 (1952). Type from Angola.

Kohautia stenosiphon (K. Schum. ex S. Moore) Bremek. in Verh. K. Nederl. Akad. Wet., Afd. Natuurk., ser. 2, **48**(2): 97 (1952). Type as for *O. stenosiphon.*

Kohautia densifolia Bremek. in Kew Bull. **8**: 438 (1953). Type: Zimbabwe, Sebungwe Distr., Dongamusa R., (trib. of the Lutope R.), 16 km. N. of Nkoka's Kraal, *Wild* 3847 (K, holotype; SRGH; UPS).

Annual or perennial erect or suberect herbs (7)10–80 cm. tall, unbranched or few to several slender to robust stems ascending from often contracted basal, often ± woody nodes to 7 mm. in diam. (occasionally subterranean); stems sparsely to densely scabrid, papillose, ± hirtellous or more rarely ± glabrous. Leaves (10)14–51(70) × (0.6)1.3–8(9.5) mm., linear to narrowly linear-lanceolate or narrowly lanceolate-elliptic, rarely both surfaces or mostly upper surface and/or only margins and midrib beneath sparsely scabrid-papillose or sparsely hirtellous, apex acute, narrowed to the base, rarely ± fleshy, margins mostly revolute; stipular sheath (0.5)0.9–1.8(2.8) mm. long, scabrid-papillose, sparsely hirtellous or ± glabrous, stipular lobes (0.7)1–4.5(8) mm. long, occasionally with additional fimbriae on either side. Inflorescences extensive, usually 2 flowers at a node, either both ± sessile or one sessile and the other shortly to distinctly pseudo-pedicellate, rarely single at a node; peduncles (0)5–95 mm. long; (pseudo-) pedicels (0.5)1–10(17) mm. long; all parts scabrid-papillose, pubescent or ± glabrous like the stem. Calyx tube (0.5)0.7–1.8(2.2) × (0.7)0.9–1.7(2) mm., ovoid-ellipsoid, densely beset with coarse round, conical or flattened triangular papillae, or ± densely hirtellous, very rarely glabrous; lobes narrowly triangular-lanceolate (0.4)0.6–1.9(2.1) mm. long and 0.2–0.6(0.7) mm. wide at the base, scabrid or only margins scabrid. Flowers scented from the evening to early morning; corolla lobes above white, cream, greyish or yellowish, darker beneath, the tube slightly lighter – brownish, reddish or greenish, glabrous or back of corolla lobes ± sparsely papillose especially along midveins and tips; tube altogether (6)6.5–14(16.1) mm. long, the apical part abruptly widened, slightly constricted above, (1)1.5–2.6(2.8) × (0.6)0.7–1.3(1.6) mm., the narrow part 0.3–0.7(0.9) mm. wide; lobes narrowly oblong-lanceolate to very narrowly oblong-elliptic (2)2.8–6(7.8) × 0.5–1.5(2.3) mm., not quite parted to the base, ± acute. Anthers ovoid-oblong, c. one half to two thirds the length of the widened tube, rarely as long but then the tips included, sterile connnectives often darkly discoloured; pollen (3)4–(5) colporate. Style and stigma together one third to two thirds the length of the narrow tube; stigma lobes 1–2 mm. long, very rarely tips touching the anthers. Capsules (1.1)1.8–3.9(4.5) × (2.2)2.7–4.5(6.1) mm., subglobose, sparsely to densely coarsely verrucose, papillose, hirtellous or ± glabrous. Seeds light to dark brown, angular-subconic, c. 0.4–0.7 mm. long.

Caprivi Strip. Mpilila I., near banks of Chobe R., 923 m., fl. & fr. 15.i.1959, *Killick & Leistner* 3390 (K; PRE; SRGH; WIND). **Botswana**. N: Okavango Delta, Mboma I., fl. & fr. 14.ii.1974, *Smith* 831 (K; PRE; SRGH). SW: Kgala-gadi Distr., W. border of Molopo ranches, along road from Lobatse to Werda, fl. & fr. 31.i.1978, *Skarpe* 225 (K; SRGH). **Zambia**. B: Kaunga near Mashi R., Kazila Plant, fl. & fr. 27.viii.1962, *Mubita & Reynolds* 169 (SRGH). N: Chinsali Distr., Shiwa Ngandu, 1538 m., fl. & fr. 2.vi.1956, *Robinson* 1548 (K; SRGH). C: Serenje Distr., Kundalila Falls, c. 2 km. N. of Falls on road from Kanona, 1450 m., fl. & fr. 12.iii.1975, *Hooper & Townsend* 681 (K; SRGH). S: Namwala Distr., Nkala R. at Ngoma, Kafue Nat. Park, fl. & fr. 7.xii.1963, *Mitchell* 2420 (SRGH). **Zimbabwe**. N: Gokwe Distr., 2.4 km. N. of Gokwe, fl. & fr. 1.v.1962, *Bingham* 249 (K; SRGH). W: Hwange Distr., Denda farm, Victoria Falls, 969 m., fl. & fr. 20.iii.1974, *Gonde* 7674 (SRGH). C: Makoni Distr., between Rusape and Maidstone, fl. & fr. 30.xii.1930, *Fries, Norlindh & Weimarck* 4030 (BR; K; SRGH). E: Mutare Distr., 'Clogheen' farm, W. of Old Umtali, 954 m., fl. & fr. 6.i.1956, *Pole-Evans* s.n. (BM; BR; K; LMA; SRGH). **Malawi**. S: Chikwawa Distr., Lengwe Game Reserve, 123 m., fl. & fr. 14.xii.1970, *Hall-Martin* 1132 (K; SRGH). **Mozambique**. MS: Chemba Distr., Chiou, CICA research station, fl. & fr. 9.iv.1962, *Balsinhas & Macuacua* 566 (BM; K; LMA). M: Bela Vista, Inhaca I., 36.8 km. E. of Maputo, near Saco, 30 m., fl. & fr. 28.xii.1956, *Mogg* 27015 (K; PRE).

Also in S. Angola, E. Zaire, S. Tanzania, Namibia/SW. Africa and S. Africa (N.W. Cape Prov., Transvaal, Orange Free State, N. Natal, Swaziland). Open flat sandy areas, sand veld, thin bushland in shady places, in rocky or stony areas, near or in ephemeral water-courses, floodplains, also in secondary grass-land, open thickets and wooded grassland, in disturbed areas along paths and roads; mainly in well-drained sandy soils or occasionally in seasonally water-logged areas, e.g. clay soils near swamps, etc.; 0–1800(2450) m.

2. **Kohautia subverticillata** (K. Schum.) D. Mantell comb. nov. Type from Angola.

Oldenlandia subverticillata K. Schum. in Engl., Bot. Jahrb. **23**: 419 (1897). —Hiern, Cat. Afr. Pl. Welw. **2**: 444 (1898). Type as for *K. subverticillata*.

Kohautia lasiocarpa var. subverticillata (K. Schum.) Bremek. in Verh. K. Nederl. Akad. Wet., Afd. Natuurk., ser. 2, **48**(2): 112 (1952). —Verdc. in F.T.E.A., Rubiaceae 1: 421 (1976). Type as for *K. subverticillata*.

K. subverticillata is divided into 2 subspecies. Subsp. *eritreensis* (Bremek.) D. Mantell comb. et stat. nov. is concentrated in N.E. Africa and on the Indian subcontinent, while subsp. *subverticillata* is widespread in E., S. central and southern Africa.

Subsp. **subverticillata**. TAB. **20**, fig. D.

Oldenlandia sordida K. Krause in Engl., Bot. Jahrb. **43**: 133 (1909). Type from Namibia.

?*Oldenlandia trothae* K. Krause in Engl., Bot. Jahrb. **43**: 133 (1909). Type from Namibia* (not traced).

Oldenlandia xerophylla Schinz in Viert. Nat. Ges. Zürich **68**: 431 (1923). Type: Mozambique, Region of the Zambezi, on the Komadzi, *Menyharth* 611 (Z, holotype; WU).

* Bremekamp (1952) included *O. trothae* provisionally under *K. aspera*, but the length of the corolla tube in Krause's description (7–8 mm.) makes it more likely that *O. trothae* belongs to either *K. subverticillata* or to *K. caespitosa* subsp. *brachyloba*.

Oldenlandia setulosa Wilson in Kew Bull. **1924**: 256 (1924). Type from S. Africa.
Kohautia lasiocarpa var. *xerophila* (Schinz) Bremek. in Verh. K. Nederl. Akad. Wet., Afd. Natuurk., ser. 2, **48**(2): 113 (1952). Type as for *O. xerophylla.*

Annual, erect, slender to ± robust, unbranched to much branched herbs 11–60 cm. tall; stems ± densely scabrid with long triangular papillae or hirtellous. Leaves (20)22–54(70) × (0.7)2–8(9) mm., linear-lanceolate to narrowly lanceolate-elliptic, apex acute, narrowed to the base, rarely ± rounded, upper surfaces and/or only margins and midrib beneath ± papillose, margins often revolute; stipular sheath (0.5)0.6–2(2.8) mm. long, sparsely scabrid like the stem, stipular lobes (0.4)0.6–4.5(5) mm. long. Inflorescences ± compact, the flowers in pairs at the nodes, both sessile or subsessile or occasionally one on a distinct pseudo-pedicel; peduncles (0)2–100 mm. long; (pseudo-) pedicels (0.5)2–19 mm. long; all parts papillose. Calyx tube (0.6)0.8–1.3(1.5) × 0.6–1.2(1.7) mm., ovoid-globose, ± densely beset with long papillae or hirsute; lobes 0.5–1.7(1.9) mm. long and 0.15–0.4 mm. wide at the base, narrowly linear-lanceolate, slightly keeled, papillose at least along margins. Corolla lobes above white, off-white, grey, yellowish-green, blueish-white or brownish, tube and lobes beneath darker, glabrous or tube outside sparsely (especially along ribs) and/or only lobes beneath (especially along mid-vein) sparsely to densely papillose; tube altogether (2.7)3.6–5.8(6) mm. long, the widened part (0.6)0.8–1.3 × (0.3)0.4–10.17(0.9) mm., the narrow part 0.2–0.45 mm. wide; lobes (0.6)0.9–1.9(2.3) × (0.3)0.4–0.8(1.1) mm., narrowly triangular-lanceolate, not quite parted to the base, ± acute. Anthers narrowly ovate, c. 1 mm. long; pollen 4–5-colporate. Style and stigma together as long as or slightly longer than the narrow part of the tube; stigma lobes c. 1 mm. long. Capsules (1.2)1.6–3.4(3.8) × (1.5)2.7–4.3(5.3) mm., subglobose, sometimes slightly vertically constricted along the septum, ± sparsely beset with longish papillae or hairs. Seeds beige-brown angular-subconic, c. 0.5 mm. long.

Caprivi Strip. Grootfontein North Distr., 16 km. W. of Andara, fl. & fr. 11.iii.1958, *Merxmüller & Giess* 2070 (PRE: WIND). **Botswana**. N: 69 km. W. of Nokaneng, fl. & fr. 12.iii.1965, *Wild & Drummond* 6894 (SRGH). SW: Ghanzi commonage, Mamuno Gate, 953 m., fl. & fr. 18.i.1970, *Brown* 7925 (K; SRGH). SE: Central Kalahari Game Reserve, 12 km. W. of E. boundary, Rakops–Kala traverse, fl. & fr. 24.iii.1979, *Kreulen* 647 (PRE; SRGH). **Zambia**. S: Sinazongwe, 600 m., fl. & fr. 28.xii.1958, *Robson & Angus* 985 (BR; K). **Zimbabwe**. N: Darwin Distr., S.E. corner of Chiswiti Reserve, 695 m., fr. 22.i.1960, *Phipps* 2382 (BR; K; LMA; SRGH). W: Shangani, 1415 m., fl. & fr. iii.1918, *Eyles* 950 (K; SRGH). C: Harare Distr., Groombridge, Mt. Pleasant, 8 km. N. of Harare, fl. & fr. 6.iv.1959, *Drummond* 6045 (BR; K; LMA: SRGH). S: Beitbridge Distr., Tuli Breeding Station, fl. & fr. 5.i.1961, *Wild* 5299 (K; SRGH). **Malawi**. N: Mzimba, Rumphi Gorge, 1077 m., fl. & fr. 29.v.1976, *Pawek* 11332 (K). S: Blantyre Distr., near Matope Mission N. of Shire R., 480 m., fl. & fr. 2.ii.1970, *Brummitt & Banda* 8530 (K). **Mozambique**. T: Cahobra Bassa, Mucangadzi R., 330 m., fl. & fr. 17.v.1972, *Pereira & Correia* 2661 (WAG).

Also in Angola, Tanzania, Burundi, Namibia and S. Africa (Transvaal and N.W. Cape Prov.). Open areas in grassland, open mixed woodland, in sandveld, in rocky areas, near ephemeral watercourses, also in disturbed or cultivated land, along foot-paths and roads; mainly in sandy, stony soils overlying granite; 480–1750 m.

3. **Kohautia microflora** D. Mantell sp. nov.* TAB. **20**, fig. E. Type: Botswana, Khalakhati (Kgalagadi), Ditatso Pan, 958 m., fl. & fr. 25.ii.1963, *Leistner* 3073 (PRE, holotype).

Annual, erect, unbranched to very sparsely branched fragile herbs 5–23 cm. tall; stems at base 0.4–1.7 mm. in diam., densely scabrid with triangular white papillae. Leaves (14)20–27(52) × (0.8)1.5–2.6(3.9) mm., linear-lanceolate, apex acute, narrowed to the base, margins and midrib beneath sparsely papillose, margins revolute; stipular sheath (0.7)1.2–1.8(3) mm. long, stipular lobes (0.7)0.9–1.5(1.9) mm. long. Inflorescences ± lax, 1 or 2 flowers at a node, if 2 then one subsessile the other pseudo-pedicellate; peduncles 10–45 mm. long; pseudo pedicels (3)8–12(20) mm. long; all parts densely papillose like the stem. Calyx tube (0.7)0.8–1.2 × (0.4)0.5–1.3(1.5) mm., ovoid-elliptic to ± globose, densely covered with rounded white papillae; lobes (0.4)0.5–0.9(1) mm. long and 0.2–0.45(0.6) mm. wide at the base, narrowly triangular, scabrid with papillae along margins and midrib. Corolla greenish, glabrous or sparsely papillose along ribs; tube altogether (2.1)2.2–3.1(3.2) mm. long, the widened part 0.5–0.7(0.9) × 0.3–0.5(0.7) mm., the narrow

* *K. subverticillatae* subsp. *subverticillatae* habitu annuo et floribus plus minusve isostylosis similis sed differt floribus minoribus viridulis (probabiliter cleistogamis), inflorescentia dissimili et ovariis rotundato-papillosis (nec ovariis hispido-papillosis).

part 0.2–0.3(0.5) mm. wide; lobes (0.3)0.4–0.65(0.85) × (0.2)0.3–0.5 mm., oblong-lanceolate. Anthers ovate, c. 0.6 mm. long; pollen 4-colporate. Style and stigma as long as narrow part; stigma lobes c. 0.5 mm. long. Capsules (1.6)2–2.8(3) ×(2.5)2.6–3.4(3.8) mm., subglobose, ± distinctly constricted along the septum, papillose. Seeds light brown, angular-subconic, c. 0.5 mm. long.

Botswana. N: Aha Hills, fr. 28.iv.1980, *Smith* 3468 (SRGH). SW: Ghanzi, Ghanzi Hide Store, 958 m., fl. & fr. 18.i.1970, *Brown* 7946 (K; SRGH). SE: Botletle delta, NE. of Mopipi, 850 m., fl. & fr. 19.iv.1973, *Tyers* 19 (SRGH).

Also in Namibia and S. Africa (N.W. Cape Prov. and N. Transvaal). In arid areas on margins of dried-up seasonal watercourses, on flat sandy areas, in grazed grassland around water pans and in blackthornveld; in clay-sand liable to seasonal water-logging and in calcareous white silty sand over limestone, occasionally in shallow soil in rocky places; c. 600–900 m.

The presence of unopened corolla remnants on fully developed, seed-bearing capsules, as well as the minute greenish, ± isostylous flowers, suggests that *K. microflora* may well be cleistogamous. This suspicion, however, still needs to be confirmed.

4. **Kohautia amatymbica** Eckl. & Zeyh., Enum. Pl. Afr. Austr. Extratrop.: 360 (1836). —Bremek. in Verh. K. Nederl. Akad. Wet., Afd. Natuurk., ser. 2, **48** (2): 116 (1952). TAB. **20**, fig. F. Type from S. Africa.

Hedyotis amatymbica (Eckl. & Zeyh.) Steudel, Nom. Bot. ed. 2: 726 (1840). —Hochst. in Flora **27**: 552 (1844). —Sonder in Harv. & Sond., F. C. **3**: 11 (1865). Type as above.

Oldenlandia amatymbica (Eckl. & Zeyh.) Kuntze, Rev. Gen. Pl. **1**: 292 (1891). Type as above.

Annual or perennial strict erect herbs 15–60 cm. tall, with one to several unbranched stems from a short subterranean woody base; roots long, fibrous, occasionally with bead-like nodules; stems mostly glabrous to shortly pubescent at the base. Leaves (3)11–45(61) × 0.4–1.8(2) mm., narrowly linear to narrowly linear-lanceolate, rarely filiform, apex acute, slightly tapered to the base, ± scabrid above otherwise glabrous; stipular sheath (0.5)0.6–1.9(2.2) mm. long, glabrous, stipular lobes (0.3)0.9–3.5(4) mm. long. Inflorescences many-flowered with 5–12 flowered partial inflorescences capitate; peduncles (18)32–75(170) mm. long, glabrous. Calyx tube (1.4)1.6–2.8(3.2) × (1.1)1.3–2(2.4) mm., ovoid-elliptic, scabrid or ± glabrous; lobes (1.8)2–2.8(4) mm. long and 0.3–1.3(1.8) mm. wide at the base, narrowly triangular, glabrous. Flowers strongly scented in the afternoon and evening; corolla lobes above white, pinky white, cream, pale yellow or khaki yellow, tube and lobes beneath darker – usually olive green to brownish, glabrous or back of lobes scabridulous; tube often slightly bent, altogether (10.3)11–19.2(23.1) mm. long, the widened part (2.1)2.3–3.1(3.5) × 0.8–1.9(2.2) mm., the narrow part (0.3)0.5–1(1.2) mm. wide; lobes (3.9)4.4–8.5(8.8) × 0.9–2.7(3) mm., narrowly linear-elliptic to linear-oblong, parted to the base, ± acute. Anthers narrowly linear, c. 2–4 mm. long; pollen (3)4(5)-colporate. Style c. 3–5 mm. long; stigma lobes 1.5–3 mm. long. Capsules (2.1)2.4–3.2(3.9) × (3.1)3.4–4.2(4.8) mm., subglobose, glabrous. Seeds dark brown c. 0.8–1 mm. long, slightly flattened, ellipsoid.

Zimbabwe. E: Bundi Valley, Chimanimani Mts., 1630 m., fl. 15.xi.1959, *Goodier* 631 (BM; LMA; SRGH). **Mozambique**. MS: Chimanimani Mts., fl. 6.vi.1949, *Munch* 175 (SRGH). M: Maputo, 4 km. from Namaacha, fl. & fr. 1.xi.1982, *Marime & Mauhica* 93 (LMA).

Also in S. Africa (Transvaal Highveld, Natal, Orange Free State, Transkei, Lesotho and Swaziland, S.E. Cape Prov.). Mostly in recently burned grassland, rocky slopes and "koppies", moist depressions, disturbed or cultivated ground on roadsides.

5. **Kohautia cynanchica** DC., Prodr. **4**: 430 (1830). —Bremek. in Verh. K. Nederl. Akad. Wet., Afd. Natuurk., ser. 2, **48**(2): 121 (1952). —Launert & Roessler in Merxm. Prodr. Fl. SW. Afr. **115**: 17 (1966). TAB. **20**, fig. G. Type from S. Africa.

Kohautia longiflora DC., Prodr. **4**: 430 (1830). Type from S. Africa.

Kohautia rigida Benth. in Hook. f., Niger Fl.: 402 (1839). —Bremek. in Verh. K. Nederl. Akad. Wet., Afd. Natuurk., ser. 2, **48**(2): 118 (1952). Type from Angola.

Hedyotis cynanchica (DC.) Steudel, Nom. Bot., ed. 2, **1**: 727 (1840). Type as for *K. cynanchica.*

Hedyotis longiflora (DC.) Steudel, Nom. Bot., ed. 2, **1**: 727 (1840) non Schumach. & Thonn. (1827). Type as for *K. longiflora.*

Hedyotis rigida (Benth.) Walp., Ann. Bot. Syst. **2**: 772 (1851). Type as for *K. rigida.*

Hedyotis stricta Smith *sensu* Sonder in Harv. & Sond., F. C **3**: 11 (1865) quoad spec. citata.

Oldenlandia rigida (Benth.) Hiern in F.T.A. **3**: 55 (1877); Cat. Afr. Pl. Welw. **1**: 442 (1894) pro parte. Type as for *K. rigida.*

Oldenlandia cynanchica (DC.) K. Schum. ex Kuntze, Rev. Gen. Pl. **3**: 121 (1893). Type from S. Africa.

?*Oldenlandia seineri* K. Krause in Engl., Bot. Jahrb. **43**: 131 (1909). Type: Botswana, Kalahari, N. of Matschabing, *Seiner* 257 (not traced)*.

Oldenlandia omahekensis K. Krause in Engl., Bot. Jahrb. **48**: 404 (1912). Holotype and neotype from Namibia.

Oldenlandia neglecta Schinz in Viert. Nat. Ges. Zürich **68**: 430 (1923). Type from Namibia.

Oldenlandia stricta K. Schum. *sensu* Wordsworth, Hutch., F. Bolus & L. Bolus in Ann. Bol. Herb. **3**: 23 (1923).

Oldenlandia breviflora Chiov. in Bull. Soc. Bot. Ital. **2**: 39 (1924). Type from Angola.

Oldenlandia graminifolia Chiov. in Bull. Soc. Bot. Ital. **2**: 39 (1924) nom. illegit. non (L.) DC. (1830). Type from Angola.

Oldenlandia calcitrapifolia Pearson ex Bremek. in Verh. K. Nederl. Akad. Wet., Afd. Natuurk., ser. 2, **48**(2): 121 (1952) nom. non valide publ., pro synon.

Kohautia gracilifolia Bremek. in Verh. K. Nederl. Akad. Wet., Afd. Natuurk., ser. 2, **48**(2): 119 (1952). —Launert & Roessler in Merxm. Prodr. Fl. SW. Afr. **115**: 18 (1966). Type: Zimbabwe, Lower Sabi, *Wild* 2429 (K, holotype; BR; SRGH).

Kohautia omahekensis (K. Krause) Bremek. in Verh. K. Nederl. Akad. Wet., Afd. Natuurk., ser. 2, **48**(2): 120 (1952). —Launert & Roessler in Merxm. Prodr. Fl. SW. Afr. **115**: 17 (1966) pro synon. Types as for *O. omahekensis*.

Kohautia raphidophylla Bremek. in Verh. K. Nederl. Akad. Wet., Afd. Natuurk., ser. 2, **48**(2): 123 (1952). —Launert & Roessler in Merxm. Prodr. Fl. SW. Afr. **115**: 17 (1966). Type from Namibia.

Kohautia thymifolia C. Presl ex Bremek. in Verh. K. Nederl. Akad. Wet., Afd. Natuurk., ser. 2, **48**(2): 121 (1952) nom. non valide publ., pro synon.

Kohautia desertorum Welw. in sched. nom. non valide publ., pro synon.

Annual or perennial occasionally suffrutescent erect to suberect herbs or rarely small dwarf shrubs (6)10–60(100) cm. tall, with woody base to 5(10) mm. in diam.; stems ± quadrangular, papillose, scabrid or glabrous, becoming glabrous towards the top, often with sparse rounded papillae along ridges. Leaves (10)16–40(65) × (0.4)0.6–1.5(2) mm., filiform to narrowly linear, rarely narrowly linear-lanceolate, apex acute, very slightly narrowed to the base, margins mostly revolute, ± glabrous or lower surfaces especially midrib and margins with rounded loosely arranged papillae, numerous short shoot leaves often giving nodes a fasciculate appearance; stipular sheath (0.5)0.6–1.5(1.7) mm. long, glabrous, stipular lobes 0.4–1.2(2) mm. long. Inflorescences ± spreading, 1 or mostly 2 flowers at a node, one subsessile, the other pseudo-pedicellate; peduncles (2)10–45 mm. long; pseudo-pedicels 3–12(20) mm. long; glabrous, scabrid or ± verrucosely papillose. Calyx tube (0.8)1–1.6(1.8) × (0.7)1–1.5(1.7) mm., hemispherical to ± ovoid, glabrous or minutely verrucose; lobes (0.5)1–2(2.5) mm. long and 0.2–0.6 mm. wide at the base, broadly to narrowly triangular, rarely ± ovate-lanceolate, glabrous or ± scabrid with minute papillae along margins and midribs or minutely verrucose occasionally with short filiform elements on either side. Flowers scented during late afternoon and evening; corolla lobes above white, cream, more rarely pinkish mauve, tube and lobes darker below – brown, reddish brown, olive green, olive brown or dirty cream, glabrous or outside of dilated tube and midvein on back of lobes with sparse small round papillae; tube altogether (8.5)9.6–14.2(16) mm. long, the widened part (1.4)1.5–2.6(3.4) × 1–1.8(2) mm., the narrow part (0.4)0.5–1 mm. wide; lobes 2.8–5.8(7) × (0.8)1–2(2.5) mm., narrowly oblong to ovate, not quite parted to the base, ± acute. Anthers narrowly ovate, c. 1–2.2 mm. long, often sterile, connectives darkly discoloured; pollen 3- (very rarely 4-)colporate. Style 1.5–2.6 mm. long; stigma lobes 1.4–2.6 mm. long. Capsules (1)2.2–3 × (2.2)2.8–3.8(4.3) mm., obconic-hemispherical rarely subglobose, glabrous or slightly scabrid. Seeds brown c. 0.6 mm. long, angular-subconic.

Botswana. N: Orapa, Baobab Drive, fl. & fr. 20.iii.1974, *Allen* 41 (J; SRGH). SW: Kghoti (Kgalagadi), Mabua Sefhubi Pan, 1015 m., fl. & fr. 28.ii.1963, *Leistner* 3098 (K; LISU; PRE). SE: 6 km. SSE. of Mahalapye, fl. & fr. 4.xi.1978, *Hansen* 3522 (K; PRE; SRGH; WAG). **Zimbabwe**. N: Near Darwendale, vicinity of the Umvukwe Mts., fl. & fr. 20.iv.1948, *Rodin* 4338 (WAG). W: Bulawayo Distr., Hillside Farm, 1385 m., fl. & fr. v. 1958, *Miller* 5284 (K; SRGH). S: 16 km. N. of Beitbridge, fl. & fr. 22.iii.1967, *Mavi* 273 (K; LMA: SRGH). **Mozambique**. GI: Gaza Distr., Massingir, 49 km. NW. of Lagoa Nova along the Elefantes R., fl. & fr. 8.iii.1973, *Lousa & Rosa* 338 (LMA).

Also in SW. Angola, Namibia and S. Africa (mainly NW. Cape Prov., Transvaal and Orange Free State). In arid areas common along river banks, near or in ephemeral watercourses, etc., on coastal

* Bremekamp (1952) placed *O. seineri* in *K. virgata* but from Krause's description, i.e. corolla tube 9–11 mm. long, corolla lobes white above and plant turning black when dried, indicates that it is better placed in *K. cynanchica*.

or stabilized sand dunes and sandy plains; in less arid areas mostly in open (disturbed) or wooded grassland, in bushveld or in thornveld, also in disturbed ground in farmland, waste places and along paths and roads; mostly in well drained rocky or sandy ground, very rarely on clay soils; 0–1600(1800) m.

6. **Kohautia ramosissima** Bremek. in Verh. K. Nederl. Akad. Wet., Afd. Natuurk., ser. 2, **48**(2): 124 (1952). —Launert & Roessler in Merxm. Prodr. Fl. SW. Afr. **115**: 17 (1966). TAB. **20**, fig. H. Type from Namibia.

?*Oldenlandia dinteri* K. Krause in Engl., Bot. Jahrb. **39**: 518 (1907). Type from Namibia (not traced).

Oldenlandia filifolia K. Krause in Engl., Bot. Jahrb. **43**: 130 (1909) nom. illegit. non Elmer (1906). Types from Namibia.

Oldenlandia heynii sensu Dinter in Fedde, Repert. **19**: 319 (1924) nom. illegit. non G. Don (1834).

Kohautia aphylla Dinter ex Bremek. in Verh. K. Nederl. Akad. Wet., Afd. Natuurk., ser. 2, **48**(2): 125 (1952). —Launert & Roessler in Merxm. Prodr. Fl. SW. Afr. **115**: 18 (1966) pro synon. Type from Namibia.

Oldenlandia ramosissima Dinter ex Bremek. in Verh. K. Nederl. Akad. Wet., Afd. Natuurk., ser. 2, **48**(2): 124 (1952) nom. non valide publ.: pro synon., non (Spreng.) DC. (1830). Type as for *K. ramosissima*.

Annual or perennial ± erect suffrutescent herbs or occasionally small dwarf shrubs 20–60 cm. tall with 1 to several ascending ± rigid, divaricate or pseudo-dichotomously branched stems from a woody base to 6 mm. in diam.; stems ± quadrangular, glabrous and often ± polished, basal woody parts often with a yellowish or brownish papery cortex (bark). Leaves much shorter than the internodes,(1)3–18(23) × (0.2)0.4–1.3(1.7) mm., filiform or rarely narrowly linear, apex acute, not noticeably narrowed to the base, margins revolute, glabrous or with sparse minute papillae (especially younger leaves), caducous or reduced to small triangular vestiges to 2 mm. long on lower parts especially; stipular sheath 0.3–0.8(1) mm. long, absent on lower parts, otherwise produced into 2 filiform fimbriae nearer the leaves (0.2)0.4–0.8(1.3) mm. long. Inflorescences spreading, 1–2 shortly to distinctly pedicellate flowers at a node; peduncles 2–35 mm. long; pedicels (1)3–12(17) mm. long, becoming stiff and brittle in the fruiting stage, held at an angle of 25°–45°(90°) to each other and to the axis, glabrous or immature inflorescence branches scabrid. Calyx tube (0.7)1–1.3(1.5) × (0.5)1–1.2(1.5) mm., hemispherical, glabrous or rarely minutely verrucose; lobes (0.3)0.5–1(1.3) mm. long and 0.2–0.6(1) mm. wide at the base, broadly triangular to ovate-triangular, glabrous, occasionally with short filiform elements on either side. Corolla lobes above white, cream, brownish white or yellow, or rarely purplish, tube and lobes darker below – olive green, brown or purplish brown, only back of lobes (especially margins and tips) slightly scabrid with minute papillae otherwise glabrous; tube altogether (6)6.5–9.2(10.8) mm. long, the widened part (1)1.2–2(2.2) × (0.8)1–1.4(1.6) mm., the narrow part 0.3–0.7(1) mm. wide; lobes (1.3)1.5–3.1(3.8) × 0.4–1.2(1.6) mm., linear-oblong, rarely shortly and broadly triangular, parted to the base, ± acute, mostly margins permanently revolute giving lobes a narrow triangular appearance from above. Anthers c. 1.3–2 mm. long, narrowly linear, often sterile connectives emergent; pollen 3-colporate. Style 1–2 mm. long; stigma lobes c. 1–1.5 mm. long. Capsules (1.4)1.7–3(3.3) × (1.9)2.4–4.2(4.8) mm., hemispherical-obconic, rarely subglobose, glabrous. Seeds brownish, c. 0.4 mm. long, angular-subconic.

Botswana. SW: Bokspits, fl. & fr. 3.xii.1979, *Timberlake* 2076 (SRGH).
Otherwise mainly in southern Namibia and S. Africa (NW. Cape Prov.). In desert and arid sandy areas, sand dunes, near or in ephemeral watercourses, etc., also in dry sandveld or sparse grassland; 0–960 mm.

Series 2. DIURNAE

Bremek. in Verh. K. Nederl. Akad. Wet., Afd. Natuurk., ser. 2, **48**(2): 81 (1952). —Verdc. in F.T.E.A., Rubiaceae 1: 230 (1976).

Corolla lobes broadly elliptic or ovate, mostly brightly coloured above (red, lilac, pink, blue or rarely white) tube outside and lobes below paler, or if lobes above white or whitish-blue, then flowers usually very small with stigma lobes touching the anthers; psychophilous or micro-melittophilous; corolla lobes fused at the base, the fused basal portions 'swollen' and intruding into the throat.

7. **Kohautia coccinea** Royle, Ill. Bot. Himal.: 241, t. 53/1 (1839). —Bremek. in Verh. K. Nederl. Akad. Wet., Afd. Natuurk., ser. 2, **48**(2): 82 (1952). —Hepper, F.W.T.A., ed. 2, **2**: 210 (1963). —Cufod., Enum.: 988 (1965). —Wickens in For. Bull. **14** (n.s.): 26 (1969). —Agnew, Upland Kenya Wild Flowers: 398, fig. (1974). —Verdc. in F.T.E.A., Rubiaceae 1: 253 (1976). —Wickens, Fl. Jebel Marra: 130 (1976). TAB. **20**, fig. B. Type from India.

Hedyotis abyssinica Hochst. ex A. Rich., Tent. Fl. Abyss. **1**: 363 (1847). —Walp., Ann. Bot. Syst. 771 (1851). Type from Ethiopia.

Hedyotis senegalensis sensu A. Rich., Tent. Fl. Abyss. **1**: 362 (1847) non Cham. & Schlechtend. (1829).

Oldenlandia abyssinica (Hochst. ex A. Rich.) Hiern, F.T.A. **3**: 57 (1877). —K. Schum. in Engl., Pflanzenw. Ost-Afr. **C**: 376 (1905). —Hutch. & Dalz., F.W.T.A., ed. 1, **2**: 131 (1931). Type as for *H. abyssinica*.

Oldenlandia coccinea (Royle) Hook. f., Fl. Brit. Ind. **3**: 69 (1882). —Woodrow in Journ. Bomb. Nat. Hist. Soc. **11**: 644 (1898). —T. Cooke, Fl. Bombay **1**: 591 (1903). —Duthie, Fl. Gangetic Plain: 415 (1905). Type as for *K. coccinea*.

Oldenlandia macrodonta Baker in Kew Bull. **1895**: 67 (1895). Type: Zambia, Lake Tanganyika, Fwambo, fl. & fr. 1894, *Carson* 107 (K, holotype, BR, photogr.).

Pentas modesta Baker in Kew Bull. **1895**: 290 (1895). Type: Zambia, Mweru, Kalungwizi R., fl. & fr. 1894, *Carson* 33 (K, holotype).

Oldenlandia debeerstii De Wild. & Th. Dur. in Ann. Mus. Cong., sér. 2, **1**: 27 (1900). Type from unknown locality.

?*Oldenlandia ledermannii* K. Krause in Engl., Bot. Jahrb. **48**: 403 (1912). Type from the Cameroon Republic.

Annual erect unbranched to sparsely branched herbs 5–45(70) cm. tall; stems ± papillate. Leaves (9.2)12–55(65) × (0.5)1.1–4(4.9) mm., linear to linear-lanceolate, margins and lower midrib papillate, lower surface scabrid or glabrous, apex acute, narrowed to the base; stipular sheath (0.2)0.7–2.8(3.8) mm. long, stipular lobes (0.5)0.8–3.7(6.7) mm. long. Inflorescences ± many-flowered, 2 flowers at a node, one ± sessile, the other pedicellate; peduncles (3.2)6.8–11.9(16.5) mm. long, papillate; (pseudo-) pedicels (0.6)1–3.6(6.7) mm. long, densely papillate. Calyx tube (1.2)1.4–2.2(2.5) × (0.9)1.1–1.6(1.8) mm., ellipsoid, densely papillate; lobes (1.4)1.8–4.2(5.1) mm. long and (0.2)0.3–0.5(0.75) mm. wide at the base, narrowly triangular to linear, margins scabrid. Corolla lobes above scarlet, red, orange, pink, lilac or rarely light blue and very infrequently white, paler beneath with dark venation, intrusions between lobes paler or white; tube greenish red to purplish white, tube and lobes beneath sparsely papillate, tube often slightly bent, altogether (2.4)3–5(5.5) mm. long, the dilated part (1.3)1.6–2.6(3)× (0.4)0.5–0.9(1.1) mm., the narrow part (0.2)0.3–0.7(0.8) mm. wide; lobes (1.4)1.5–4.3(5.6) × (0.8)0.9–2.3(2.5) mm., oblong-elliptic, subacute. Anthers narrowly ovoid, c. 1–2 mm. long; pollen (3)4–5-colporate. Style and stigma as long as narrow part of the tube; stigma lobes 0.8–1.8 mm. long. Capsules (2.1)2.6–5(5.2) × (2.3)2.8–4.3(4.7) mm., oblong-elliptic, scabrid to sparsely papillose, ribs often purplish. Seeds light brown, c. 0.4–0.5 mm. long, angular-subconic.

Zambia. N: Mbala Distr., Mwenzo, 8 km. NE. of Mbala, 1692 m., fl. & fr. 10.vi.1951, *Bullock* 3959 (BR; K). W: Ndola, fl. & fr. 4.v.1954, *Fanshawe* 1157 (K). C: Lusaka Distr., Mt. Makulu Research Station, 17.6 km. S. of Lusaka, fl. & fr. 16.iv.1956, *Angus* 1228 (K; SRGH). S: Mumbwa Distr., on the Mumbwa–Namwala road, banks of the Luabale R., fl. & fr. 22.iii.1963, *Van Rensburg* 1780 (K; SRGH). **Zimbabwe**. N: Lomagundi Distr., Chininga farm, 1107 m., fl. & fr. 18.vii.1921, *Eyles* 3151 (K; SRGH). C: 8 km. N. of Harare, Groombridge, Mt. Pleasant, fl. & fr. 6.iv.1959, *Drummond* 6043 (BR; K; LMA; SRGH). **Malawi**. N: Rumphi waterworks, c. 6.4 km. NE. of Rumphi on Chelinda R., 1076 m., fl. & fr. 2.vi.1976, *Pawek* 11351 (K). C: Dedza Distr., E. of Bembeke Mission, 1692 m., fl. & fr. 25.iv.1971, *Pawek* 4685 (K). S: Chilwa Island, 1060 m., fl. & fr. 11.iii.1955, *Exell, Mendonça & Wild* 799 (BM; SRGH). **Mozambique**. N: Niassa Distr., Massangulo, 1100 m., fl. & fr. iv.1933, *Gomes e Sousa* 1377 (K; LMA).

Also in N. India, Pakistan, Yemen Arab Republic, Ethiopia, Sudan, Uganda, Kenya, E. Zaire, Burundi, Rwanda, Tanzania and in W. Africa (Senegal, Cameroon and Nigeria). In grassland and open woodland as well as in abandoned fields or wasteland and along paths and roadsides; 500–2800(3130) m.

8. **Kohautia aspera** (Heyne ex Roth) Bremek. in Verh. K. Nederl. Akad. Wet., Afd. Natuurk., ser. 2, **48**(2): 113 (1952). —Hepper, F.W.T.A., ed. 2, 2: 210 (1963). —Launert & Roessler in Merxm. Prodr. Fl. SW. Afr. **115**: 16 (1966). —Lebrun in Mitt. Bot. Staatss. München **10**: 444, Fig. 5 (1971). —Agnew, Upland Kenya Wild Flowers: 389 (1974). —Verdc. in F.T.E.A., Rubiaceae 1: 241 (1976). TAB. **20**, fig. I. Type from India.

Hedyotis aspera Heyne ex Roth, Nov. Pi. Sp.: 94 (1821). —Wight & Arn., Prodr. **1**: 417 (1834). —Dalz. & Gibs., Bombay Fl.: 115 (1861). Type as for *K. aspera*.

Oldenlandia aspera (Heyne ex Roth) DC., Prodr. **4**: 428 (1830). —Hook. f., Fl. Brit. Ind. **3**: 68

(1882) pro parte. —E.H.C. Krause in Engl., Bot. Jahrb. **14**: 413 (1892). —Woodrow in Journ. Bomb. Nat. Hist. Soc. **11**: 644 (1898) pro parte. —T. Cooke, Fl. Bombay **1**: 591 (1903) pro parte. —Duthie, Fl. Gangetic Plain: 415 (1905) pro parte. —A. Chev. in Rev. Bot. Appl. **15**: 892 (1935). Type as for *K. aspera.*

Hedyotis strumosa Hochst. ex A. Rich., Tent. Fl. Abyss. **1**: 364 (1847). —Walp., Ann. Bot. Syst. **2**: 77 (1851). Type from the Sudan.

Oldenlandia strumosa (Hochst. ex A. Rich.) Hiern, F.T.A. **3**: 58 (1877). —K. Schum. in Engl., Pflanzenw. Ost-Afr. **C**: 376 (1895). —Broun & Massey, Fl. Sudan: 258 (1929). —F.W. Andr., Fl. Pl. Anglo-Egypt. Sudan **2**: 452, fig. 164 (1952). Type as for *H. strumosa.*

Oldenlandia rhodesiana S. Moore in Journ. Bot. **40**: 250 (1902). Type: Zimbabwe, Harare, fl. & fr. xii. 1897, *Rand* 122 (BM holotype).

Oldenlandia cyanea Dinter in Fedde, Repert. **19**: 318 (1924). Type from Namibia.

Oldenlandia leclercii A. Chev. in Bull. Mus. Hist. Nat. Paris, sér. 2, **5**: 162 (1933). —Lebrun in Mitt. Bot. Staatss. München **10**: 444 (1971) pro synon. Type from Mali.

Erect unbranched to sparsely branched annual herbs 10–30 cm. tall; stems ± densely papillose often reddish at base and below nodes. Leaves (23)28–39(47) × (0.5)0.7–2.5(4) mm., linear, linear-lanceolate to narrowly elliptic-lanceolate, margins and midrib beneath papillose, otherwise surfaces scabrid to glabrous, apex acute, narrowed to the base; stipular sheath (0.7)1–1.6(2) mm. long, stipular lobes (0.5)0.7–1.2(2.5) mm. long. Inflorescences spike-like, many-flowered, mostly 2 subsessile flowers at a node, the flower in a fork of a dichasium distinctly pedicellate; peduncles 1–60 mm. long; pedicels (0)2–10(25) mm. long, all parts densely papillose like the stem. Calyx tube (0.6)1-1.4(1.5) × (0.5)0.7–1.5(1.7) mm. ovoid to elliptic, densely covered with rounded papillae; lobes (0.8)1–1.4(1.7) mm. long and (0.2)0.3–0.5(0.6) mm., narrowly triangular to linear-lanceolate, margins and midrib scabrid with stiff minute papillae. Corolla lobes above mostly white or blueish or pinkish white, bluish or mauve below, tube light green or greeny white, narrow part of tube outside with rounded minute papillae; tube slightly curved, altogether (2.1)2.3–3.1(3.4) mm. long, the dilated part ovoid-elliptic in shape, (0.8)1–1.3(1.5) × (0.4)0.5–0.8(0.9) mm., the narrow part (0.2)0.3–0.6(0.7) mm. wide; lobes (0.7)1–1.5(1.8) × (0.3)0.4–0.7(0.9) mm., ovate-triangular to ovate-oblong, not parted to the base, acute. Anthers ± oval, c. two thirds the length of the dilated tube; pollen 4–5-colporate. Style and stigma as long as the narrow part of the tube, stigma lobes 0.7–1(1.4) mm. long, tips touching the anthers. Capsules (2)2.2–3(3.4) × (2.8)3.4–4.3(4.7) mm., subglobose to globose, ± sparsely papillose. Seeds light brown, c. 0.4–0.5 mm. long, angular-subconic.

Zimbabwe. N: Lake Kariba, Dinosaur Island, 540 m., fl. & fr. i.1978, *Kerfoot, Thomas, Schramm & Dyer* 104 (J). W: Nyamandhlovu, Pasture Res. Station, fr. 10.i.1954, *Plowes* 1659 (SRGH).

Widespread in the drier parts of tropical Africa and S. Africa, also in SW. Arabia, Pakistan and India. In grassland, in open woodland and in disturbed areas, in cultivations and on roadsides; frequently in seasonally water-logged black clay (cotton) soils; (400)700–2150 m.

9. **Kohautia confusa** (Hutch. & Dalz.) Bremek. in Verh. K. Nederl. Akad. Wet., Afd. Natuurk., ser. 2, **48**(2): 89 (1952). —Hepper, F.W.T.A., ed. 2, **2**: 209 (1963). —Verdc. in F.T.E.A., Rubiaceae 1: 236 (1976). TAB. **20**, fig. J. Type from Senegal.

Kohautia stricta sensu DC., Prodr. **4**: 430 (1830) non Smith (1811).

Oldenlandia effusa sensu Hiern in F.T.A. **3**: 59 (1877) non Oliver (1873).

Oldenlandia confusa Hutch. & Dalz., F.W.T.A., ed. 1, **2**: 131 (1931) pro parte. Type as for *K. confusa.*

Kohautia ubangensis Bremek. in Verh. K. Nederl. Akad. Wet., Afd. Natuurk., ser. 2, **48**(2): 90 (1952). —Hepper, F.W.T.A., ed. 2, **2**: 209 (1963). Type from Chad.

Annual, slender laxly branched herbs 20–70 cm. tall; stems ± glabrous to roughly papillose. Leaves 10–45 × (0.9)1.2–2(2.2) mm., linear to filiform, apex acute to apiculate, slightly narrowed to the base, ± glabrous with lower midrib papillose; stipular sheath (0.6)0.9–2(2.7) mm. long, ± sparsely papillose, the lobes (0.5)0.6–1.8(2.6) mm. long. Inflorescences lax, extensive, 2 distinctly pedicellate flowers at a node; peduncles c. 10–60 mm. long; pedicels slender (4.6)8.4–16.6(30) mm. long, lengthening after anthesis, glabrous to slightly scabrid. Calyx tube (0.9)1.2–2.8(2.9) × (0.7)1.1–2.7(3) mm., ellipsoid, verruculose to ± densely papillose, rarely ± glabrous; calyx lobes (0.9)1.3–2(2.3) mm. long and (0.3) 0.4–0.6(0.8) mm. wide at the base, narrowly triangular, margins scabrid. Corolla lobes above scarlet, red or light violet, paler beneath, tube outside and back of lobes ± densely papillose to ± glabrous; tube altogether (3.1)4.3–5.5(9.4) mm. long, the widened part (1.3)2–2.4(2.6) × (0.4)0.7–1.5(1.65) mm., the narrow part (0.2)0.4–0.75(0.9) mm. wide; lobes (1.5)1.7–3.9(4) × (0.6)1–2.5(2.7) mm., triangular-elliptic, not parted to the base, acute.

Anthers c. 1–1.5 mm. long, narrowly ovoid; pollen 4(5)-colporate. Style and stigma as long as the narrow part of the tube; stigma lobes c. 1–1.8 mm. long. Capsules (2.7)2.8–3.6(3.8) × (2.9)3.1–3.8(4.1) mm., subglobose, ± glabrous. Seeds brownish, c. 0.4 mm. long, angular-subconic.

Zambia. C: Chilanga, 12.8 km. S. of Lusaka, 1230 m., fl. & fr. 10.vi.1955, *Robinson* 1292 (K). S: Namwala Distr. Lochinvar Nat. Park, 989 m., fl. & fr. 5.i.1972, *Van Lavieren, Sayer & Rees* 390 (SRGH). **Malawi**. N: between Kondowe and Karonga, 615–1846 m., fl. & fr. vii. 1896, *Whyte* s.n. (K).

Also in Senegal, Gambia, Guinea, Nigeria, Cameroon Republic, Central African Republic and Tanzania. In open disturbed areas in grassland or open woodland; mainly found as a weed of cultivated land and on roadsides; 200–1000(1800) m.

Subgen. PACHYSTIGMA

Bremek. in Verh. K. Nederl. Akad. Wet., Afd. Natuurk., ser. 2, **48**(2): 66 (1952). —Verdc. in F.T.E.A. Rubiaceae, 1: 230 (1976). TAB. **21**, fig. A.
Kohautia series *Barbatae* Bremek. loc. cit. —Verdc. loc. cit.
Kohautia series *Imberbae* Bremek. tom. cit.: 77 (1952). —Verdc. loc. cit.

Stipular sheath in the mid-stem region mostly with 2–8(11) fringing, ± rigid, subulate fimbriae (lobes); base of corolla lobes and corolla throat inside mostly shortly hairy or papillate; stigma ovoid or cylindrical.

Key to the species of the Subgen. Pachystigma

1. Corolla throat distinctly bearded with flattened hairs or papillae - - - - - 2
– Corolla throat not bearded or, in long-tubed forms (6–7 mm.), throats sometimes with a few sparse inconspicuous papillae; stipular lobes fringe-like; leaf apices with a distinct arista to 1 mm. - - - - - - - - - - - - - - - - - - 10. *virgata*
2. Partial inflorescences subcapitate, 5–10 ± sessile flowers per capitulum 13. *cuspidata*
– Partial inflorescences not subcapitate; flowers all distinctly pedicellate, 2(5) per node 3
3. Calyx tube and dilated part of corolla tube glabrous or pubescent (not papillose); usually 1-several stemmed perennials - - - - - - - - - - - 11. *longifolia*
– Calyx tube and dilated part of corolla tube papillose (not pubescent or glabrous); erect sparsely to much branched annuals - - - - - - - - - - - - 12. *microcala*

10. **Kohautia virgata** (Willd.) Bremek. in Verh. K. Nederl. Akad. Wet., Afd. Natuurk., ser. 2, **48**(2): 77 (1952). —Hepper, F.W.T.A., ed. 2, **2**: 209 (1963). —Cufod., Enum.: 989 (1965). —Launert & Roessler in Merxm. Prodr. Fl. SW. Afr. **115**: 13 (1966). —Agnew, Upland Kenya Wild Flowers: 398 (1974). —Verdc. in F.T.E.A., Rubiaceae 1: 234 (1976). —Gonçalves in Garcia de Orta, Sér Bot. **5**: 195 (1982). TAB. **21**, fig. C. Type probably from Ghana.
Hedyotis virgata Willd., Sp. Pl. ed. 4, **1**: 567 (1797). Type as for *K. virgata*.
Oldenlandia virgata (Willd.) DC., Prodr. **4**: 426 (1830). —Hiern, F.T.A. **3**: 59 (1877). —Broun & Massey, Fl. Sudan: 258 (1929). —Hutch. & Dalz., F.W.T.A., ed. 1, **2**: 132 (1931). —F.W. Andr., Fl. Pl. Anglo-Egypt. Sudan **2**: 453 (1952). Type as for *K. virgata*.
Kohautia setifera DC., Prodr. **4**: 430 (1830). Type from S. Africa.
Oldenlandia caffra Eckl. & Zeyh., Enum. Pl. Afr. Austr. Extratrop.: 360 (1836). —Hiern, F.T.A. **3**: 58 (1877) pro parte; Cat. Afr. Pl. Welw. **1**: 44 (1898). Type from S. Africa.
Hedyotis setifera (DC.) Steudel, Nom. Bot., ed. 2: 728 (1840). —Sonder in Harv. & Sond., F.C. **3**: 10 (1865). —Schinz in Mém. Herb. Boiss. **1**: 65 (1900). Type as for *K. setifera*.
Kohautia parviflora Benth. in Hook. f., Niger Fl.: 403 (1849). Type from Ghana.
Hedyotis gerrardii Harvey ex Sonder in Harv. & Sond., F. C. **3**: 11 (1865). Type from S. Africa.
Hedyotis setifera var. *pubescens* Sonder in Harv. & Sond., F. C. **3**: 10 (1865). Type from S. Africa.
Oldenlandia parviflora (Benth.) Oliver in Trans. Linn. Soc. **29**: 84 (1873). —Hiern, F.T.A. **3**: 60 (1877) quoad typum solum. Type as for *K. parviflora*.
Oldenlandia moandensis De Wild. in Ann. Mus. Congo Belge, Bot. Sér. 5, **2**: 180 (1907). Type from Zaire.
Kohautia virgata var. *oblanceolata* Bremek. in Verh. K. Nederl. Akad. Wet., Afd. Natuurk., ser. 2, **48**(2): 80 (1952). Type from S. Africa.

Annual or perennial, erect to suberect, ascending or straggling herbs, (6)15–50(60) cm. tall; one to many branched slender stems from a ± woody base; stems slightly ribbed, glabrous, scabrid or ± densely hirsute, often becoming glabrous towards the apex. Leaves (14)19–29(32) × (0.5)1–5(7) mm., narrowly linear, linear-lanceolate to lanceolate-elliptic, rarely filiform, acute and shortly aristate, arista 0.5–1 m. long, glabrous or hirsute, mostly ±

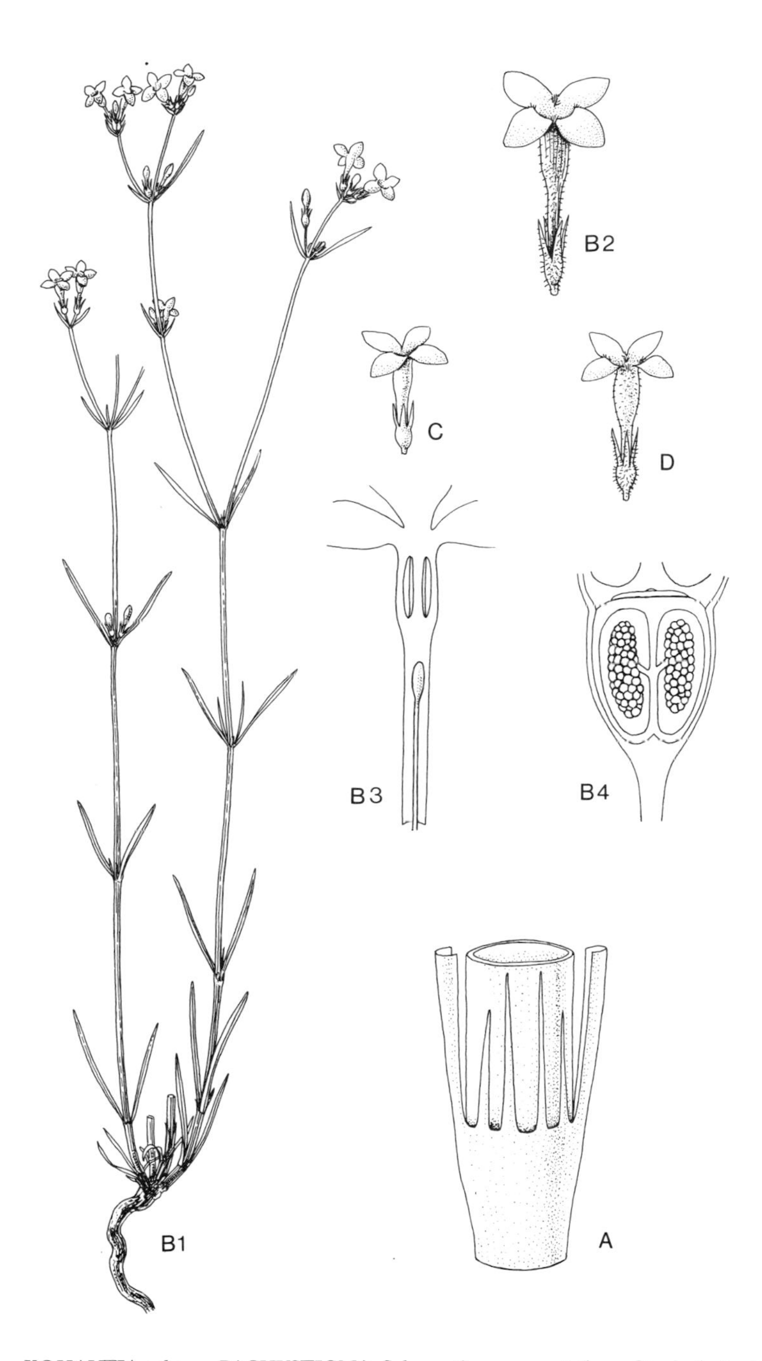

Tab. 21. A. —KOHAUTIA subgen. PACHYSTIGMA. Schematic representation of typical stipular sheath with four rigid subulate fimbriae. B. —K. CUSPIDATA. B1, habit (× ½); B2, flower (× 2); B3, flower, longitudinal section (× 3); B4, ovary, longitudinal section, showing placenta attachment (× 3), B1–4 from *Robinson* 5510. C. —K. VIRGATA. Flower (× 2), from *Chase* 1912. D. —K. LONGIFOLIA. Flower (× 2), from *Pereira, Sarmento & Marques* 1824.

scabrid to hirsute on upper surface along margins and midrib beneath, upper leaves ± glabrous or hirsute at the base, margins revolute; stipular sheath (0.9)1.2–2(2.4) mm. long, hirsute, produced into (2, 3) 4–6 reflexed filiform fimbriae (1.4)1.8–4.5(6) mm. long. Inflorescences usually with 2 or, more rarely, 3 with additional accessory, flowers at a node, pedicellate or rarely subsessile; peduncles (3)24–34(79) mm. long; pedicels (1)3–7(15) mm. long, glabrous. Calyx tube (0.5)0.7–1.3(1.5) × (0.4)0.9–1.4(1.6) mm., obovate-subglobose glabrous; lobes narrowly triangular to narrowly ovate-triangular, ending in a short seta or mucro, margins with short stiff hairs, otherwise glabrous. Corolla lobes variously coloured, white, cream, yellowish, blueish, mauve, purple, pink or red, sometimes bicoloured on the same plant; buds usually darker in colour (red or pink), open flowers becoming paler with age?; back of lobes and tubes paler or brownish to pinkish green or greenish white, interior and exterior glabrous or very rarely in longer tubed forms throat inside very slightly papillose; tube altogether (2.7)3–6.5(7) mm. long, the dilated part (0.7)1–2.1(2.4) × (0.7)0.8–1.2(1.45) mm., the narrow part (0.4)0.5–0.9(1.1) mm. wide; lobes (0.9)1–2.2(2.4) × (0.6)0.8–1.2(1.3) mm., ovate-triangular to ovate-elliptic, subacute to slightly apiculate. Anthers narrowly ovate, 1–1.5 mm. long; pollen (3)4(6) colporate. Style and stigma together as long as the narrow tube, stigma tip just touching anthers or ± isostylous. Capsules (1)1.2–2.1(2.5) mm. long and (1.8)2–2.9(3.3) mm. wide, hemispherical to subglobose, glabrous. Seeds brownish, c. 0.4 mm. long, round-subconic.

Botswana. N: On Linyati road, 5 km. SE. of Hunter's Camp and 17 km. SW. of Hyaena Camp, fl. & fr. 28.vi.1972, *Biegel, Pope & Gibbs Russell* 4089 (K; SRGH). SE: 3.2 km. S. of Lobatsi, E. of railway, Springfield Farm, fl. & fr. 17.i.1960, *Leach & Noel* 161 (K; SRGH). **Zambia**. B: Senanga Distr., Sioma Falls, 76 km. S. of Senanga, W. bank of Zambezi R., 1050 m., fl. & fr. 1.ii.1975, *Brummitt, Chisumpa & Polhill* 14217 (K). N: 48 km. S. of Mbala, 1538 m., fl. & fr. 18.vii.1930, *Hutchinson & Gillett* 3853 (K). S: c. 0.8 km. from Chirundu bridge (Otto Beit bridge) near Lusaka road, fl. & fr. 5.ii.1958, *Drummond* 5483 (BR; IIAM; K; LMA; SRGH). **Zimbabwe**. N: Chipuriro Distr., Mwanzamtanda R. area, W. of Kanyemba airfield, c. 550 m., fl. & fr. 29.i.1966, *Müller* 248 (K; SRGH). W: Bulawayo, 1385 m., fl. & fr. xi.1902, *Eyles* 1220 (BM, SRGH). C: Harare, Cranborne Park, 1507 m., fl. & fr. 18.i.1948, *Wild* 2287 (BR; K; SRGH). E: Mutare Commonage, E. of Darlington, 1169 m., fl. & fr. 17.i.1950, *Chase* 1912 (BM; K; SRGH). S: c. 10 km. SW. of Bukwa, fl. & fr. 4.v.1973, *Pope, Biegel & Simon* 1101 (SRGH). **Malawi**. S: Mt. Zomba, 769 m., fl. & fr. xii.1896, *Whyte* s.n. (K). **Mozambique**. Z: Namagoa Distr., Vila de Mocuba, 61–123 m., fl. & fr. v.1944, *Faulkner* 173 (K; LMA; SRGH). T: between Lupata and Tete, fl. & fr. ii.1859, *Kirk* s.n. (K). MS: near Vila de Sena (Senna), fl. & fr. xi.1859, *Kirk* s.n. (K). GI: Gaza Distr., Bilene do Chipenhe, near Maniquenique, 1 km. from Licile, fl. & fr. 9.vi.1960, *Lemos & Balsinhas* 52 (K; LMA; SRGH). M: Maputo Distr., Moamba, from Chinhanguanine to Manhica, fl. & fr. 14.xii.1979, *De Konig* 7784 (K; SRGH).

Also in S. Africa, Namibia, E. tropical Africa, S. Zaire, Burundi, S. Sudan and coastal areas in W. Africa (Dahomey Gap), as well as in Madagascar and the Comoro Islands. Always in moist localities, in open, secondary, swampy or wooded grasslands, on coastal sandflats and stabilized sand dunes, in rocky or stoney ground, in waste or cultivated land, along paths and roads; c. 0–1700(1900) m.

11. **Kohautia longifolia** Klotzsch in Peters, Reise Mossamb. Bot. **1**: 297 (1862). —Bremek. in Verh. K. Nederl. Akad. Wet., Afd. Natuurk., ser. 2, **48**(2): 68 (1952). —Verdc. in F.T.E.A., Rubiaceae 1: 232 (1976). —Gonçalves in Garcia de Orta, Sér. Bot. **5**: 194 (1982). TAB. **21**, fig. D. Type: Mozambique, Sena, *Peters* s.n. (B, holotype†); neotype: Gonubi Hill, *Schlechter* 12181 (K, neotype; BM; BR; E; W).

?*Kohautia macrophylla* Klotzsch in Peters, Reise Mossamb. Bot. **1**: 297 (1862). Type: Mozambique, Cabaceira, *Peters* s.n. (B, holotype†).

Oldenlandia effusa Oliver, Trans. Linn. Soc. **29**: 84, t. 48 (1873). —Hiern in F.T.A. **3**: 59 (1877) pro parte. Type from Tanzania.

Oldenlandia caffra sensu Hiern in F.T.A. **3**: 58 (1877) non Eckl. & Zeyh. (1836).

Oldenlandia longifolia (Klotzsch) K. Schum. in Engl., Pflanzenw. Ost-Afr. **C**: 376 (1895) nom. illegit. non DC. (1830).

Kohautia effusa (Oliver) Bremek. in Verh. K. Nederl. Akad. Wet., Afd. Natuurk., ser. 2, **48**(2): 72 (1952). Type from Tanzania.

Kohautia effusa var. *hirtella* Bremek. loc. cit. Type: Mozambique, Western Zone, *Hornby* 3564 (PRE).

Kohautia longifolia var. *macrocalyx* Bremek. tom. cit. 70 (1952). —Verdc. in F.T.E.A., Rubiaceae 1: 233 (1976). Type from Tanzania.

Kohautia longifolia var. *psilogyna* Bremek. loc. cit. (1952). —Verdc. loc. cit. —Gonçalves in Garcia de Orta, Sér. Bot. **5**: 195 (1982). Type: Mozambique, Moribane, *Dawe* 490 (K, holotype).

Kohautia longifolia var. *vestita* Bremek. loc. cit. (1952). —Verdc. loc. cit. —Gonçalves loc. cit. Type: Zambia, 54.6 km. NE. of Livingstone, *Hutchinson & Gillett* 3496 (K, holotype; BM).

Kohautia sennii Bremek. tom. cit. 71 (1952). —Cufod., Enum.: 989 (1965). Type from Somalia.

Kohautia longifolia var. *effusa* (Oliver) Verdc. in Kew Bull. **30**: 290 (1975); in F.T.E.A., Rubiaceae 1: 233 (1976). —Gonçalves in Garcia de Orta, Sér. Bot. **5**: 195 (1982). Type as for *O. effusa.*

Annual or perennial erect, straggling or decumbent herbs, (8)12–95(110) cm. tall, 1 to several stems from woody base and basal nodes, basal stems quadrangular, glabrous with ribs sparsely pubescent or papillate becoming ± densely hirsute towards the top. Leaves (9)12–90(95) × (1)2.2–10(13) mm., linear, linear-lanceolate to elliptic-lanceolate, apex acute to apiculate, narrowed to the base, in broader leaves ± rounded at the base; usually midrib beneath, margins and upper surface pubescent or papillate and lower surface glabrous to slightly scabrid, or entire leaf glabrous, scabrid or pubescent, margins often revolute; stipular sheath (0.7)1–4.5(6) mm. long, produced into a short broad lobe bearing 2–8(11) fimbriae (1.5)2.3–7.8(8.5) mm. long. Inflorescences strict to very lax, cymose, mostly 2 distinctly pedicellate flowers at a node, in well developed specimens additional accessory inflorescence branches from upper vegetative nodes and lower inflorescence nodes; peduncles (4)7–10(42) mm. long; all parts ± densely pubescent or occasionally glabrous. Calyx tube (0.6)1–1.8(1.9) × (0.7)0.8–1.6(1.75) mm., subglobose to ± ovoid-elliptic, glabrous to densely pubescent; lobes (1.2)1.4–4.5(7) mm. long and (0.3)0.4–0.8(1) mm. wide at the base, narrowly triangular to lanceolate, often ± distinctly keeled, ± glabrous to sparsely pubescent especially along keel; margins always fringed with short stiff hairs. Corolla lobes above from bright red through to varying shades of purple, lilac, pink, blue or white (becoming paler with age?), often hairs in the throat a contrasting darker or lighter colour, occasionally corolla tips, margins and/or midveins darker; lobes beneath paler; tube outside mostly yellowish green to pale green, altogether (2.3)2.5–6.3(7) mm. long, the dilated part (1)1.2–2.6(3) × (0.6)0.8–1.2(1.4) mm., the narrow part (0.4) 0.5–0.9(1.1) mm. wide, usually the narrow part glabrous and the dilated part ± sparsely pubescent, hairs extending sometimes on to lower parts of lobes beneath or entire tube glabrous or ± pubescent, throat inside with short hairs; lobes (1.3)1.8–6.4(7.3) × (0.9)1.2–3.7(4.1) mm., ovate-elliptic to lanceolate-elliptic, joined at the base for 0.2–1.3 mm., acute to apiculate. Anthers 0.9–1.3 mm. long, narrowly ovate, acute; pollen 4–8-colporate. Style 0.5–1.4 mm. long; stigma 0.6–1.2 mm. long. Capsules (1.4)1.6–2.9(3.8) × (2.2)2.4–3.4(3.9) mm., hemispherical to ± globose, glabrous to ± sparsely pubescent. Seeds darkish brown, c. 0.4–0.5 mm. long, round-subconic.

Zambia. B: 14.4 km. ESE. of Kaoma (Mankoya), near Luene R., fl. & fr. 21.xi.1959, *Drummond & Cookson* 6710 (SRGH). N: Mbala Distr., Lake Chila, 1630 m., fl. & fr. 20.vi.1956, *Robinson* 1699 (BR; K; SRGH). C: Luangwa Valley Game Reserve, 4.8 km. S. of Lube R., 1846 m., fl. & fr. 12.iv.1967, *Prince* 468 (K; SRGH). E: Lundazi, 1107 m., fl. & fr. 1.vi.1954, *Robinson* 803 (K; SRGH). S: Mochipapa, near Choma, 1230 m., fl. & fr. 10.iii.1962, *Astle* 1496 (K; SRGH). **Zimbabwe**. N: Mazoe Distr., Umvukwe Mts. near Umvukwes fl. & fr. 28–29.iv.1948, *Rodin* 4474 (BR; K; MO; WAG). C: Harare Distr., Rumani, 1385 m., fl. & fr. 10.v.1948, *Wild* 2530 (BR; K; SRGH). **Malawi**. N: Mzimba Distr., 16 km. NE. of Mzuzu, 1230 m., fl. & fr. 6.v.1977, *Phillips* 2216 (WAG). C: Nkhota Kota, grounds or resthouse by airfield, 520 m., fl. & fr. 16.vi.1970, *Brummitt* 11480 (K; SRGH). S: Zomba Distr., Namitambo Estate, fl. & fr. 30.v.1963, *Salubeni* 45 (K; SRGH). **Mozambique**. N: Massangulo, 1100 m., fl. & fr. iii.1930, *Gomes e Sousa* 1340 (BR; K; LMA). Z: Quelimane Distr., Namagoa, Mocuba, 62–123 m., fl. & fr. i-iii.1943, *Faulkner* 154 (BR; K; SRGH). T: Macanga (Furancungo) Distr., between the base of Mt. Furancungo and stream, 1265–1140 m., fl. & fr. 17.iii.1966, *Pereira, Sarmento & Marques* 1824 (WAG). M: 32 km. W. of Moamba, 62 m., fl. & fr. 28.v.1949, *Gerstner* 7091 (K; PRE; SRGH).

Also in N. Somalia, SE. Kenya and Tanzania, as well as bordering Zaire and Madagascar. Mainly in moist areas in grassland and woodland (often seasonally waterlogged), also common in disturbed areas; (0)30–2000(2461) m.

12. **Kohautia microcala** Bremek. in Verh. K. Nederl. Akad. Wet., Afd. Natuurk., ser. 2, **48**(2): 73 (1952). —Verdc. in F.T.E.A., Rubiaceae 1: 231 (1976). Type: Zambia, near Kalungwizi R., *Walter* 5 (K, holotype; BR, photogr.).

Kohautia leucostoma Bremek. tom. cit. 74 (1952). Type from Zaire.

Erect annual (occasionally biennial?) herbs 9–35 cm. tall, unbranched to much branched; stems slender, sparsely to densely scabrid-papillate or ± glabrous. Leaves (15)23–40(54) × (0.6)1–3(3.6) mm., filiform or narrowly linear to narrowly linear-lanceolate, apex acute, very slightly narrowed to the base; upper surface, margins and midrib beneath scabrid-papillose, lower surface ± glabrous, margins distinctly revolute; stipular sheath (0.9)1.3–2.4(3.7) mm. long bearing 2–5 fimbriae (1.7)2.1–3.4(4.9) mm. long. Inflorescences ± lax, cymose, flowers all distinctly pedicellate, often additional axillary inflorescence branches from vegetative and inflorescence nodes; peduncles

30–85 mm. long; pedicels slender elongating slightly and becoming ± rigid after anthesis (5)12–40(57) mm. long; all parts scabrid-papillose to ± densely puberulous-papillate. Calyx tube (0.9)1–1.5(1.7) × (0.9)1–1.3(1.5) mm., obovate-elliptic, sparsely to densely papillate; lobes (1.2)1.6–2.8(3.1) mm. long and (0.3)0.4–0.6(0.7) mm. wide at the base, narrowly triangular to triangular-lanceolate, acute, keeled, margins and keel with short stiff fringing hairs, otherwise papillate. Corolla lobes red, orange-red or violet above, paler beneath; tube altogether (3.9)4.5–6.4(7.4) mm. long, the dilated part (1.4)1.9–2.6(2.8) × (0.8)0.9–1.2(1.35) mm., the narrow part (0.4)0.5–0.8(0.9) mm. wide, exterior glabrous or dilated part scabrid, entrance to throat with short hairs; lobes ± round to broadly elliptic, (3.9)4.4–6.4(6.8) × (1.6)1.8–3.6(4.7) mm., acute and very shortly acuminate, joined at the base for c. 1.5 mm. Anthers c. 1.3–1.6 mm. long, narrowly lanceolate; pollen 4–5-colporate. Style c. 1.4 mm. long; stigma 0.8–1 mm. long. Capsules (1.9)2.1–2.7(3.2) × (2.3)2.9–3.6(3.9) mm., subglobose, glabrous to ± scabrid-papillose. Seeds dark brown, c. 0.4 mm. long, roundly angular.

Zambia. N: Mporokoso Distr., Nsama, 1200 m., fl. & fr. 3.iv.1957, *Richards* 8990 (BR; K).

Also in W. Tanzania and SE. Zaire. In grassland and open woodland; along rivers and streams and damp sandy ground; 800–1330 m.

More material is needed to decide whether *K. microcala* and *K. longifolia* are indeed distinct, or whether *K. microcala* is merely an extreme form of the very variable *K. longifolia*. *K. microcala* is only distinguishable from its very near ally by its apparently always annual habit, its ± roundly elliptic corolla lobes and the papillate (not glabrous or pubescent) indumentum of the calyx tube.

13. **Kohautia cuspidata** (K. Schum.) Bremek. in Verh. K. Nederl. Akad. Wet., Afd. Natuurk., ser. 2, **48**(2): 74 (1952). —Launert & Roessler in Merxm. Prodr. Fl. SW. Afr. **115**: 17 (1966). TAB. **21**, fig. B. Type from Angola.

Oldenlandia cuspidata K. Schum. in Engl., Bot. Jahrb. **23**: 413 (1897). —Hiern, Cat. Afr. Pl. Welw. **1**: 443 (1898). Type as for *K. cuspidata*.

Oldenlandia acudentata Wight in Kew Bull. **1898**: 145 (1898). Type: Malawi, Mt. Zomba, *Whyte* s.n. (K. holotype!).

Oldenlandia capituliflora K. Krause in Engl., Bot. Jahrb. **39**: 517 (1907). Type from Angola.

Annual or perennial erect herbs, 15–20 cm. tall with few to many robust stems from a robust, woody, rarely ± fleshy subterranean rhizome-like base; stems either ± sparsely beset with white papillae or ± glabrous. Leaves (22)31–46(55) × (1)1.4–4(8) mm. narrowly elliptic to broadly lanceolate, apex acute, narrowed to the base; lower midrib and upper surfaces ± sparsely covered with long white hairs or papillae, otherwise glabrous; stipular sheath (1.4)2.4–4.3(5) mm. long, ± densely pubescent with long white hairs, bearing 2–6 fimbriae (1.9)3.2–6.7(7.8) mm. long. Inflorescences comprised of dichasially arranged 5–15-flowered subcapitate partial inflorescences, occasionally a pair of ± sessile flowers at nodes preceding 'capitula', pubescent. Calyx tube (1.1)1.3–2.2(3.4) × (1)1.2–1.9(2) mm., subglobose densely pubescent with long white silky hairs; lobes subulate (2.8)2.9–5.7(7.6) mm. long and (0.5)0.6–0.9(1.1) mm. wide at the base, keeled; margins with short stiff hairs. Corolla lobes above red, purplish-red, bright pink or very rarely white, often hairs in throat paler or white; lobes beneath paler; tube outside whitish-red, -violet or -pink, altogether (4.2)4.7–10.9(13.5) mm. long, the dilated part (1.9)2–3.8(5) × (0.8)1–1.7(1.8) mm. the narrow part (0.4)0.6–1(1.9) mm. wide; tube outside sparsely to densely pubescent with long white hairs; throat inside with flattened papillae; lobes (4)4.1–8(8.5) × (2)2.3–5.3(6) mm., obovate-elliptic, acute to acuminate. Anthers 1.9–2.3 mm. long ± ovoid; pollen (3)4–5-colporate. Style 3.2–3.8 mm. long; stigma 1.1–1.4 mm. long. Capsules (1.3)1.6–3.1(3.8) × (1.8)2.1–3.6(4.1) mm., subglobose, ± densely covered in long white hairs. Seeds brownish, c. 0.4–0.5 mm. long, roundly subconic.

Botswana. N: just beyond Tsimanemeha Pan on track to Shishikola, fl. & fr. 27.i.1978, *Smith* 2273 (K). SE: Eastern Bamangwato Territory, Limpopo R., c. 332 m., fl. & fr. 4.x.1876, *Holub* s.n. (K). **Zambia**. B: Mongu, fl. & fr. 1.xii.1962, *Robinson* 5510 (K; SRGH). S: Namwala Distr., Shakalonga Plain, Kafue Nat. Park, fl. & fr. 5.xii.1962, *Mitchell* 1543 (BR). **Zimbabwe**. W: Hwange Game Reserve, 83 km. SE. of Main Camp, Makololo Pan, fl. & fr. 17.iv.1972, *Grosvenor* 704 (BR; K; SRGH). **Malawi**. S: Zomba Mt., fl. & fr. 13.i.1960, *Banda* 374 (BM; SRGH).

Also in N. Namibia and Angola. In dried-up seasonal watercourses, and in seasonally water-logged short grassland, on roadsides, more rarely from open rocky ground; 360–2000 m.

Doubtful Species.

14. **Kohautia obtusiloba** (Hiern) Bremek. in Verh. K. Nederl. Akad. Wet., Afd. Natuurk., ser. 2, **48**(2): 66 1952). —Jex-Blake, Gard. E. Afr., ed. 4, t. 4/4 (1957). —Verdc. in F.T.E.A., Rubiaceae 1: 230 (1976). Lectotype from Tanzania.

?*Oldenlandia zanguebariae* Lour., Fl. Cochinch.: 78 (1790). —DC., Prodr. **4**: 429 (1830). Type not traced.

?*Hedyotis zanguebariae* (Lour.) Roem. & Schult. in Syst. **3**: 192 (1818). Type as for *O. zanguebariae.*

Oldenlandia obtusiloba Hiern in F.T.A. **3**: 56 (1877). —K. Schum. in Engl., Pflanzenw. Ost-Afr. **C**: 376 (1895). Type as for *K. obtusiloba.*

It is extremely doubtful whether *K. obtusiloba* occurs in the Flora Zambesiaca area. The species has a very well defined lowland coastal distribution in SE. Kenya and NE. Tanzania. The locality of the syntype *Forbes* 358 (K) from Mozambique sine loc. (also the only specimen known from Mozambique) appears, therefore, to be somewhat suspect.

19. CONOSTOMIUM (Stapf) Cufod.

Conostomium (Stapf) Cufod. in Nuov. Giorn. Bot. Ital. n.s. **55**: 85 (1948); in Phyton **1**: 134 (1949). —Bremek. in Verh. K. Nederl. Akad. Wet., Afd. Natuurk., ser. 2, **48**(2): 125 (1952).
Oldenlandia L. sect. *Conostomium* Stapf in Journ. Linn. Soc., Bot. **37**: 517 (1906).

Perennial herbs or branched shrublets mostly with erect stems. Leaves sessile, linear to lanceolate, the lateral nerves rather obscure; stipules with the truncate sheath bearing hairs and colleters or 2–several teeth. Flowers small to fairly large, never truly heterostylous, either in terminal corymbs, spike-like inflorescences or solitary or in pairs or fascicles in the axils of the leaves. Calyx lobes 4, linear-subulate. Corolla salver-shaped with a long narrowly cylindrical tube; lobes 4; throat and lobes within glabrous or sparsely pilose. Stamens always included in the tube or only the tips of the anthers exserted. Ovary 2-locular; ovules numerous, immersed in the peltate placentas; style usually exserted or rarely included, in which case the stigmas do not reach the bottoms of the anthers; stigma lobes oblong or linear-oblong, densely papillate. Capsule ovoid or subglobose, crowned by the persistent calyx lobes and with a usually well-developed beak, loculicidally dehiscent at the apex. Seeds numerous, angular, smooth.

A small genus of 9 species confined to the eastern and central parts of tropical Africa and also in S. Africa. Bremekamp divides the genus into 3 subgenera one of which, *Hochstetteria* Bremek. contains all the species described below.

1. Leaves wider, 10–23 mm. wide, ± rounded at the base; pedicels mostly very short; corolla tube not exceeding 2 cm. - - - - - - - - - - - - - - - 1. *natalense*
- Leaves narrower, 1–7 mm. wide, strongly narrowed at the base; pedicels 2–7(15) mm. long; corolla tube c. 3 cm. long - - - - - - - - - - - - - - 2
2. Leaves lanceolate, narrowly attenuate to slender acute apex; corolla lobes lanceolate - - - - - - - - - - - - - - 2. *zoutpansbergense*
- Leaves linear-elliptic, more abruptly acute; corolla lobes ovate - - - - - 3. *gazense*

1. **Conostomium natalense** (Hochst.) Bremek. in Verh. K. Nederl. Akad. Wet. Afd. Natuurk., ser. 2, **48**(2): 136 (1952). TAB. **22**. Type from S. Africa (Natal).

Hedyotis natalensis Hochst. in Flora **27**: 552 (1844). —Sond. in Harv. & Sond., F.C. **3**: 12 (1865).

Oldenlandia natalensis (Hochst.) Kuntze, Rev. Gen. Pl. **1**: 292 (1891). —Phillips in Fl. Pl. S. Afr. 10, t. 364 (1930).

Oldenlandia natalensis var. *hirsuta* Bär in Viert. Nat. Ges. Zürich **68**: 430 (1923). Type from S. Africa (Natal).

Conostomium natalense var. *glabrum* Bremek. in Verh. K. Nederl. Akad. Wet. Afd. Natuurk., ser. 2, **48**(2): 137 (1952). Type from S. Africa (Natal).

Herb 0.2–1.2 m. tall, the stems quadrangular and woody at the base, pubescent or ± glabrous save at nodes. Leaves 3.5–7 × 1–2.3 cm., lanceolate, narrowly tapering acute at the apex, ± rounded at the base, ± pubescent or glabrous save for midrib beneath; petiole up to 3 mm. long; stipule sheath ± 2 mm. long with 2 short fimbriae 0.5–2 mm. long. Flowers in condensed terminal inflorescences on most upper branches, 2–4 cm. wide, closely surrounded by leafy bracts; pedicels up to 2 mm. long in fruit; occasionally an isolated axillary flower at lower node with pedicels to 5 mm., perhaps a reduced axillary branch. Calyx tube subglobose, 1.5–2 mm. wide, glabrous to sparsely pubescent; lobes 2–5.5 × 0.8–1.5 mm., lanceolate, acute, ciliate. Corolla pale blue, lilac or purplish pink; tube 1–1.7 cm. long, slightly widened at the apex for 4 mm., sparsely pilose-pubescent or ±

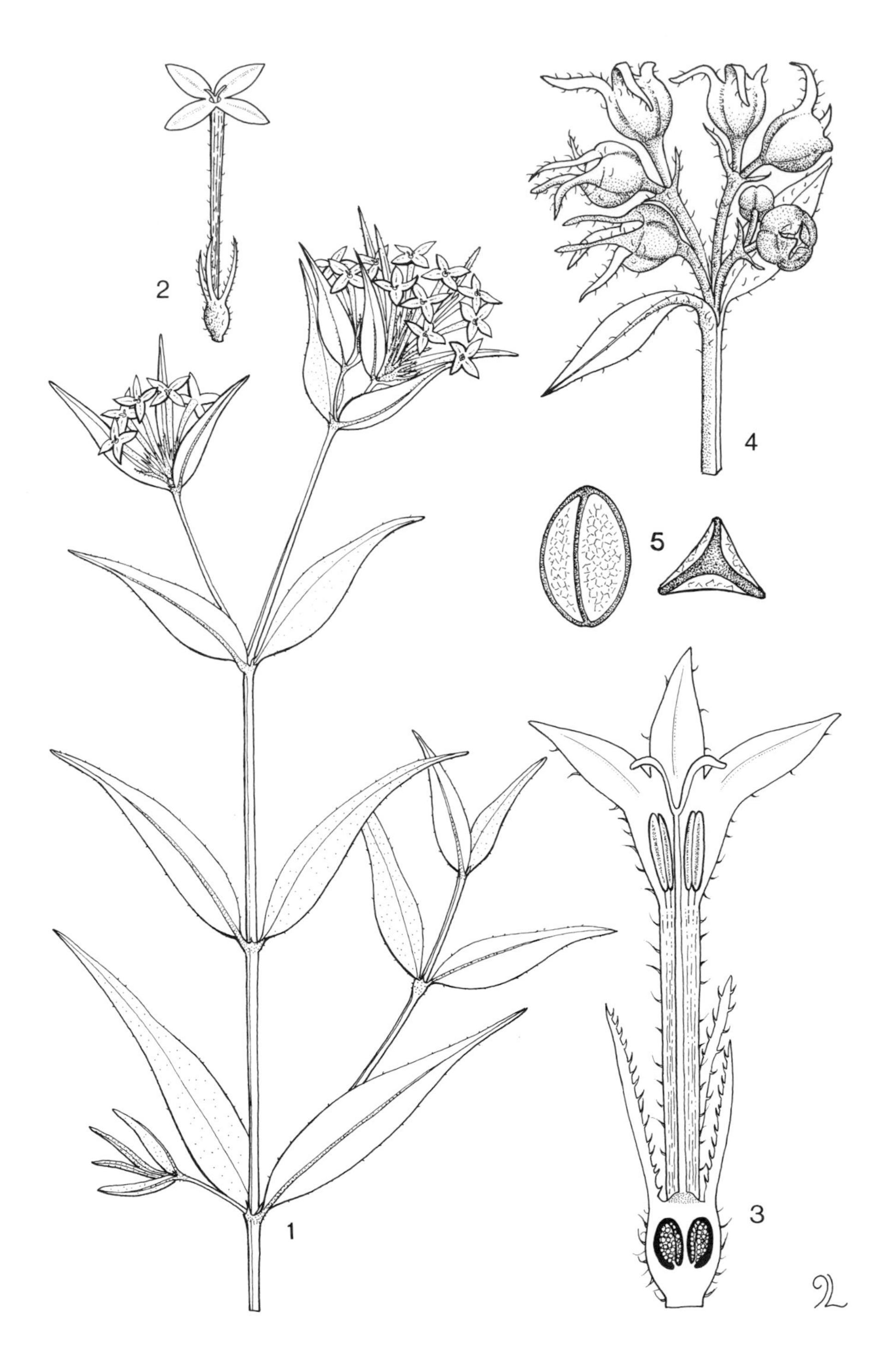

Tab. 22. CONOSTOMIUM NATALENSE. 1, habit (× ½); 2, flower (× 1½); 3, flower, longitudinal section (× 3); 4, infructescence (× 1½); 5, pyrene (× 30), all from *Rutherford-Smith* 523.

glabrous, glabrous inside but with a ring of hairs at throat rim; lobes 2.5–4.5 × 1.5–2 mm., elliptic. Anthers included but stigmatic branches 1–2 mm. long, just included or shortly exserted. Fruit 5.5–7 × 4.5–6 mm. including the hemispherical beak, ovoid-globose; calyx-lobes ± persistent.

Zimbabwe. E: Inyanga, Pungwe View, fl. 13.ii.1961, *Rutherford-Smith* 523 (K; SRGH). **Mozambique**: MS: Mossurizi, Mafusi, fl. 23.ii.1907, *W.H. Johnson* 156 (K). Z: Serra do Gúruè ao km. 3 a seguir à cascata picada para o pico Namúli, fl. 23.ii.1966, *Torre & Correia* 14798 (C; LISC; LMA).

Also in S. Africa and Swaziland. Open grassland to dense scrub, often on granite soils; 960–2100 m.

Bremekamp recognises four varieties, three based on indumentum and one on leaf shape but the variation is such as to suggest they are unnecessary. All the Flora Zambesiaca material comes under var. *glabrum*.

2. **Conostomium zoutpansbergense** (Bremek.) Bremek. in Verh. K. Nederl. Akad. Wet. Afd. Natuur., ser. 2, **48**(2): 138 (1952). Type from S. Africa (Transvaal).
Oldenlandia zoutpansbergensis Bremek. in Ann. Transv. Mus. **15**: 256 (1933).

Probably ± straggling herb at least 50 cm. tall, the upper stems quadrangular, minutely scabrid on the edges but otherwise glabrous; lower stems brown, woody, the bark peeling. Leaves 1–4.5 cm. × 1–5 mm., lanceolate or linear-elliptic, tapering acute at the apex, narrowed at the base, the petiole obsolete, glabrous or scabrid on margins at extreme base; stipule sheath under 1 mm. long, scabrid, with 2 minute fimbriae. Flowers in lax corymbs; pedicels 2–7 mm. long, elongating up to 1.5 cm. in fruit but mostly about 6 mm., glabrous. Calyx tube 2 × 1.5 mm., ellipsoid, glabrous or ± scabrid on the ribs; lobes 3.5 × 1 mm., lanceolate, sparsely scabrid on margins near base. Corolla white, blue or pale mauve-pink; tube 2.5–3.5 cm. long, very slender, slightly widened at apex, glabrous or slightly scabrid outside near throat, pubescent inside throat; lobes 5.5–7 × 1.5–2 mm., lanceolate. Anthers included; style papillate, shortly to distinctly exserted, the stigmatic lobes ± 2 mm. long. Capsule ovoid, 5–2 × 4–5 mm. including the conical beak.

Mozambique. M: Mangulane, fr. 15.iv.1931, *Gomes e Sousa* 546 (K).

Also in S. Africa (Transvaal). Sandy area; 80 m.

3. **Conostomium gazense** Verdc. sp. nov.* Type: Mozambique, Gaza, between Chibuto and Gomes da Costa, *Barbosa & Lemos* 8127 (LISC holotype; LMA).

Much branched shrubby herb, woody at the base, glabrous; older stems with flaking bark. Leaves subsessile, 0.7–4 cm. × 1–7 mm., linear-elliptic to linear, acute at the apex, cuneate at the base, blackish on drying, minutely wrinkled above when dry and lateral venation obscure; stipule sheath c. 0.5 mm. long with 2 filiform setae c. 0.5 mm. long. Flowers in 3's or 5's at the apex of the stem with a solitary terminal flower and solitary axillary flowers; pedicels 3–6 mm. long. Calyx tube c. 1.5 mm. long, ovoid; lobes 2–3 mm. long, lanceolate, the limb tube scarcely developed. Corolla greenish white; tube c. 2.8 cm long; lobes 5–6 × c. 3 mm., ovate. Style just exserted, the stigmatic lobes linear-oblong c. 2 mm. long; stamens included in the throat. Capsule 3–4 × 3 mm., subglobose or broadly ovoid, ribbed, the beak prominent, 2 mm. tall, laterally compressed, the valves ultimately ± reflexed.

Mozambique. GI: Gaza, between Chibuto and Gomes da Costa, fl. & fr. 14.xi.1957, *Barbosa & Lemos* 8127 (LISC; LMA).

Endemic. Ecology unknown. Although obviously close to *C. zoutpansbergense* it has a distinctly different facies.

20. MITRASACMOPSIS Jovet

Mitrasacmopsis Jovet in Archiv. Mus. Nat. Hist. Paris. ser. 6, **12**: 589 (1935).
Diotocranus Bremek. in Verh. K. Nederl. Akad. Wet., Afd. Natuurk., ser. 2, **48**(2): 148 (1952).

Annual herb with short erect branched stems. Leaves paired, narrowly (rarely broadly) elliptic or elliptic-lanceolate; stipules with 2–4 fimbriae from a short base. Flowers small,

* *C. zoutpansbergensis* valde affinis sed foliis apice minus attenuatis haud distincte lanceolatis, corollae lobis ovatis satis differt.

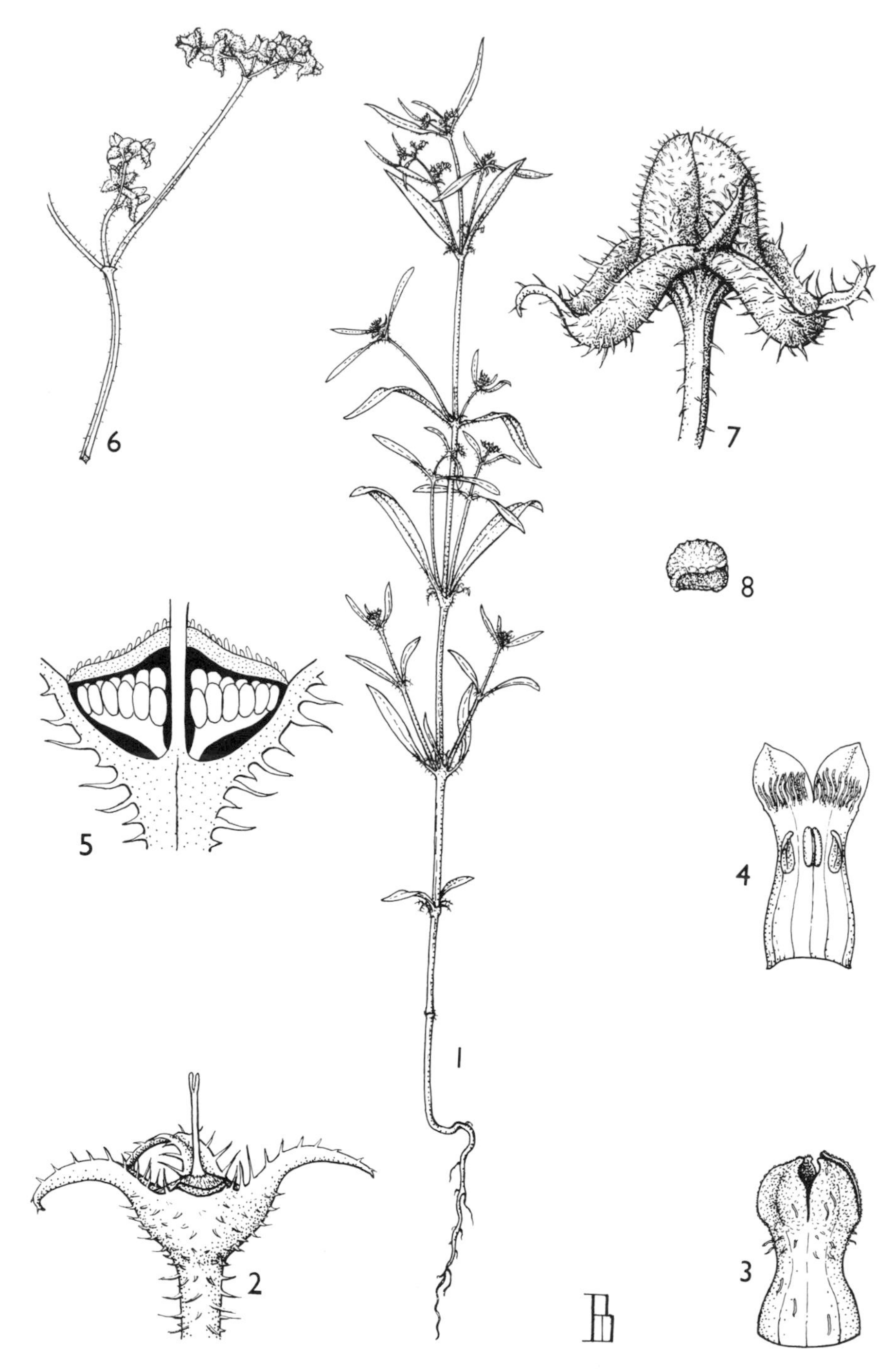

Tab. 23. MITRASACMOPSIS QUADRIVALVIS. 1, habit (×$\frac{2}{3}$), from *Richards* 5121; 2, calyx, with one lobe removed, and style (× 20), from *Haarer* 2182; 3, corolla (× 20), from *Robinson* 4999; 4, corolla, longitudinal section (× 20); 5, ovary, longitudinal section (× 30); 6, infructescence (× 2), 4–6 from *Haarer* 2182; 7, capsule (× 14); 8, seed (× 30), 7–8 from *Richards* 14972. From F.T.E.A.

not dimorphic, in small, usually few-flowered terminal dichasial inflorescences. Calyx tube ovoid; lobes 4, triangular-lanceolate. Corolla tube very short; lobes ovate, throat hairy. Ovary 2-locular, each locule with few ovules; style not exserted; stigma bifid, the lobes linear. Capsule of very characteristic shape, both loculicidal and septicidal, cordate and bilobed at the base, with the fertile part produced into a very distinct compressed obtuse central beak. Seeds few, compressed subglobose, bluntly angular, distinctly reticulately pitted.

A monotypic genus occurring in tropical Africa and Madagascar.

Mitrasacmopsis quadrivalvis Jovet in Archiv. Mus. Nat. Hist. Paris, sér. 6, **12**: 590, fig. 1–14 (1935). —Verdc. in Kew Bull. **30**: 291 (1975); in F.T.E.A., Rubiaceae 1: 248, fig. 32 (1976). TAB. **23**. Types from Madagascar.

Diotocranus lebrunii in Verh. K. Nederl. Akad. Wet. Afd. Natuurk., ser. 2, **48**(2): 48 (1952). —Tennant in Kew Bull. **22**: 438 (1968). Type from Zaire.

Diotocranus lebrunii var. *sparsipilus* Bremek. in Verh. K. Nederl. Akad. Wet. Afd. Natuurk. ser. 2, **48**(2): 49 (1952). Type from Tanzania.

Herb 8–40 cm. tall, with glabrous to sparsely hairy square stems. Leaf blades 0.8–3.2 × 0.1–1.1 cm., acute or subacute but mucronulate at the apex, cuneate at the base, margin often narrowly revolute, glabrescent to scabridulous above, glabrous beneath or the midvein and margins ciliate; petiole obsolete or very short; stipular fimbriae up to 1.5–2 mm. long from a base 0.2–1.5 mm. long, hairy. Pedicels obsolete or up to 3 mm. long in the fruiting stage. Calyx puberulous to densely scabrid pubescent; tube shallow, 0.5 mm. long or less, 1–1.1 mm. wide; lobes 0.5–0.8 × 0.3 mm. Corolla white, pink or pale mauve, tube 0.6–1 mm. long, widest at the base; lobes 0.4–0.7 × 0.4 mm., ovate, papillate inside. Style 0.3–0.6 mm. long; stigmas linear, 0.2–0.3 mm. long. Capsule 2 mm. wide, beak 1 mm. long and wide, puberulous to densely scabrid pubescent. Seeds brown c. 20 per capsule, c. 0.3 mm. long.

Zambia. B: Mongu, fl. & fr. 20.iii.1966, *Robinson* 6893 (K). N: Mpika Distr., Serenje-Mpika road, fl. & fr. 5.iv.1961, *Richards* 14972 (K). W: Mwinilunga, "L. River", fr. 17.v.1969, *Mutimushi* 3465 (K; NDO). C: Mkushi Distr., Lunsemfwa R., fl. & fr. 5.iv.1961, *Richards* 14950 (K).

Also in Zaire, Burundi, Tanzania, Angola and Madagascar. Sandy places in rocky areas, woodland, dambo edges, mossy banks beneath falls; 1110–?1500 m.

21. HEDYTHYRSUS Bremek.

Hedythyrsus Bremek. in Verh. K. Nederl. Akad. Wet., Afd. Natuurk., ser. 2, **48**(2): 149 (1952).

Small erect shrubs, invariably turning blackish on drying and with distinctly discolorous leaves. Leaves very shortly petiolate, elliptic to lanceolate, rather thick, closely placed, the lateral nerves invisible or very obscure; stipule sheath usually hairy, divided into several subulate fimbriae. Flowers heterostylous, small in many-flowered dense terminal corymbs or panicles. Calyx lobes 4, triangular or lanceolate, often with fimbriae between. Corolla shortly subcylindrical, slightly widened above; lobes 4; throat glabrous or sparsely hairy inside. Stamens well exserted in short-styled flowers. Ovary 2-locular; ovules few on peltate placentas; style filiform; stigma lobes subglobose. Capsule depressed hemispherical, produced into a conical beak which splits loculicidally and septicidally into 4 diverging valves. Seeds few, much compressed, elliptic or oblong, sometimes subangular and often slightly winged at both ends or all round, reticulate.

A genus of 2 very closely allied species confined to upland areas in tropical Africa. It might be advisable to treat these taxa as subspecies since the differences are rather trivial.

Inflorescences many-flowered, wider, over 3 cm. wide; corolla tube in short-styled flowers 1.8 mm. long, in long-styled flowers 0.7–1 mm. long; leaves mostly about 2.5 × 0.8 cm.; calyx lobes shorter than fruit beak - - - - - - - - - - - - - - 1. *spermacocinus*

Inflorescences fewer-flowered, narrower, 1–2.5 cm. wide; corolla tube in short-styled flowers 2.8–4 mm. long, in long-styled flowers c. 2 mm. long, leaves mostly 1.5 × 0.8 cm.; calyx lobes often equalling or slightly exceeding fruit beak - - - - - - - - 2. *thamnoideus*

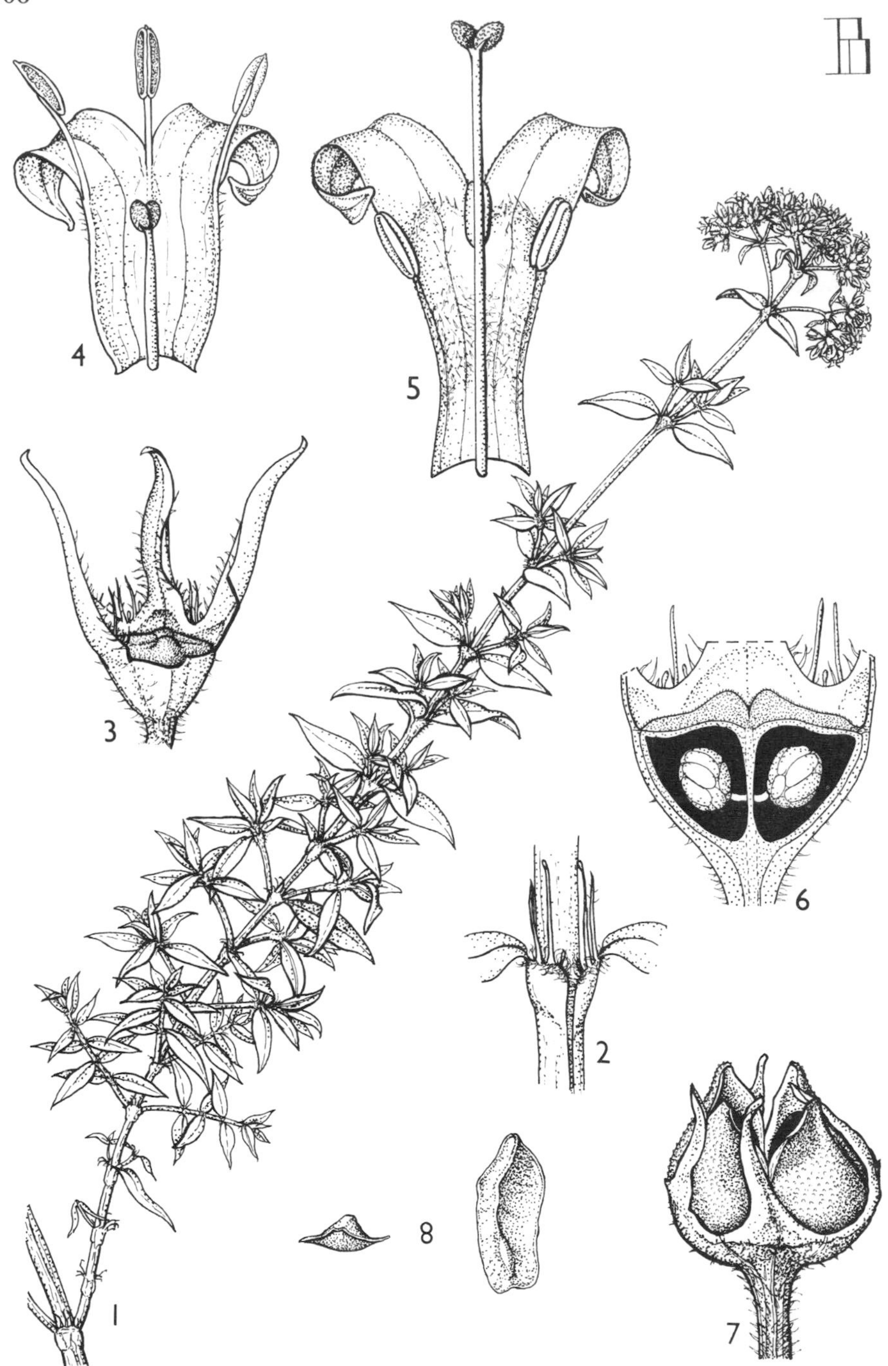

Tab. 24. HEDYTHYRSUS THAMNOIDEUS. 1, flowering branch (× ⅔); 2, node showing stipules (× 6); 3, calyx with one lobe removed (× 10); 4, short-styled flower, longitudinal section (× 10), 1–4 from *Rounce* 614; 5, long-styled flower, longitudinal section (× 10), from *Richards* 7782; 6, ovary, longitudinal section (× 20), from *Rounce* 614; 7, capsule (× 12); 8, seed, two views (× 20), 7–8 from *Greenway & Eggeling* 8687. From F.T.E.A.

1. **Hedythyrsus spermacocinus** (K. Schum.) Bremek. in Verh. K. Nederl. Akad. Wet., Afd. Natuurk., ser. 2, **48**(2): 151 (1952). —F. White, F.F.N.R.: 409 (1962) (excl. spec. cit.). Type from Angola.
Oldenlandia spermacocina K. Schum. in Engl., Bot. Jahrb. **23**: 418 (1897). —Hiern in Cat. Afr. Pl. Welw. **1**: 443 (1898).

Subshrub 0.4–1.2 m. tall; stems drying black, glabrous or with slight pubescence; internodes short. Leaf blades drying black above and usually yellow-green beneath or at least ± pale, 1–3.3 × 0.3–1.2 cm., narrowly ovate-elliptic, acute at the apex, cuneate at the base, glabrous or with few scattered hairs on midrib beneath; petiole about 1 mm. long or ± obsolete; stipule sheath pubescent, 1 mm. long with 3–5 fimbriae up to 5 mm. long. Inflorescence corymbiform with up to 100 flowers, 3.5–4 cm. across becoming 5 in fruit or with aggregate; inflorescence up to 8 cm. wide; peduncle 0–2.5 cm. long; secondary peduncles up to 1 cm. long; pedicels 0–1 mm. long; all axes shortly spreading pubescent. Calyx tube 0.8–1 mm. long; limb tube 0.2 mm.; lobes 0.5–1 mm. long. Corolla white, pale violet or mauve. Long-styled flowers: corolla tube 0.7–1 mm. long; lobes 1.8 mm. long; style 2–2.5 mm. long; filaments 0.2 mm. long. Short-styled flowers: corolla tube 1.8 mm. long; lobes 1.8 × 1.2 mm.; style 0.8–1.5 mm. long filaments 1.3–1.8 mm. long. Capsule 1–1.8 × 1.6–2 mm. including 0.9–1.2 mm. long beak.

Zambia. N: Mpika, fl. 2.ii.1955, *Fanshawe* 1950 (K; NDO). W: Mwinilunga, source of R. Zambezi, fl. 13.xii.1963, *Robinson* 5993 (K).
Also in Angola. Mushitu fringes and dambo/mushitu edges; 1380 m.

2. **Hedythyrsus thamnoideus** (K. Schum.) Bremek. in Verh. K. Akad. Wet. Afd. Natuurk. ser. 2, **48**(2): 151 (1952). —Verdc. in F.T.E.A., Rubiaceae 1: 250, fig. 33 (1976). TAB. **24**. Type from Tanzania.
Oldenlandia thamnoidea K. Schum. in Engl., Bot. Jahrb. **28**: 56 (1899); **28**: 485 (1900).
Hedythyrsus spermacocinus sensu F. White, F.F.N.R.: 409 (1962) quoad spec. cit. non (K. Schum.) Bremek. sensu stricto.

Subshrub 0.15–2 m. tall; stems drying blackish, ± glabrous or with only a few lines of hairs on young parts; internodes short. Leaf blades discolourous, green above (blackish-green when dry), whitish or silvery beneath, 0.8–2(2.5) cm. × 2.5–8 mm., narrowly elliptic to lanceolate, acute at the apex, cuneate at the base, revolute at the margins, glabrous or with a few hairs on the margins and main nerve beneath; petiole short, scarcely 1(–1.5) mm. long; stipule sheath hairy, 1–3 mm. long, with 3–7 fimbriae 1–4 mm. long, often ciliate. Inflorescences (1)1.5–2.5 cm. wide, ± less than 30-flowered; peduncle 0–1.5 cm. long, sometimes bifariously hairy; pedicels 0–1 mm. long. Calyx tube 0.5–1.5 mm. long, ovoid, glabrous, pubescent or with a few hairs at top between the lobes; lobes 1.3–2.5 mm. long, glabrous or with some hairs on margin at the base. Corolla white, pink or pale lilac; tube 1.8–4 mm. long, longest in the short-styled flowers, glabrous or hairy outside; lobes 1.5–3 × 1–1.3 mm., oblong-ovate. Style 1.7–2.8 mm. long in short-styled flowers, 3–5.8 mm. long in long-styled flowers; stigma lobes 0.2–0.5 mm. long. Capsule 1–1.2 mm. tall excluding the 1–1.6 mm. long beak, 1.8–2 mm. wide. Seeds chestnut brown, 1–1.2 mm. long.

Zambia. N: Mafinga Mts., fl. & fr. 24.v.1973, *Chisumpa* 78 (K; NDO). **Malawi**. N: Nyika Plateau, Kasaramba View Road, fl. 6.vii.1971, *Pawek* 5029 (K; MAL).
Also in Tanzania. *Podocarpus–Ochna* forest on steep rock slopes, *Myrica–Hagenia–Philippia* mist forest, above upper limit of evergreen ravine forest in bracken or *Protea–Ericaceae* scrub and also in grassland; 1950–2400 m.
F. White (F.F.N.R.: 409 (1962)) unites the two taxa but although undoubtedly close and with distributions coming near each other in Zambia, N. they are certainly not identical in either morphology or ecology. Subspecific rank is probably indicated but I have maintained the status I used in F.T.E.A. — there is no difficulty in naming the two.

22. AGATHISANTHEMUM Klotzsch

Agathisanthemum Klotzsch in Peters, Reise Mossamb., Bot. **1**: 294 (1861). —Bremek. in Verh. K. Nederl. Akad. Wet., Afd. Natuurk., ser. 2, **48**(2): 152 (1952). —Lewis in Ann. Missouri Bot. Gard. **52**: 182 (1965).

Perennial herbs or small shrubs with simple or branched stems. Leaves sessile or nearly so, narrowly lanceolate to ovate or oblong; lateral nerves mostly strong; stipule sheath produced into a 3–15-fimbriated lobe. Flowers often heterostylous, small, in many-flowered cymes or corymbs or in subglobose heads. Calyx lobes 4, lanceolate to ovate-triangular, keeled. Corolla shortly tubular, densely hairy at the throat; lobes 4, often with some additional filamentous lobes between. Ovary 2(3)-locular; ovules numerous on peltate placentas; style filiform, shortly hairy, usually undivided but sometimes bifid at the apex; stigma lobes subglobose or ovoid. Disk farinose or very shortly pubescent. Capsule subglobose, produced into a conical beak which opens both loculicidally and septicidally. Seeds numerous, angular.

A small genus of probably only 3 species confined to tropical Africa and the Comoro Is., seemingly closely allied to *Hedyotis* and possibly best considered a section of that genus. Pollen aperture morphology indicates that it may be nearer to the American *Hedyotis* L. subgen. *Edrisia* (Raf.) Lewis rather than to the Asiatic section *Diplophragma* Wight & Arn. as Bremekamp suggested.

Calyx lobes c. 2(3) mm. long; inflorescences mostly lax but sometimes flowers in collections of dense subcapitulate clusters, usually distinctly heterostylous; disk shortly pubescent; typical subsp. in coastal areas - - - - - - - - - - - - - - - - - - 1. *bojeri*
Calyx lobes 3.3–9 mm. long; inflorescences not lax, the flowers in large capitula, not distinctly heterostylous; disk mostly finely papillate; mostly inland areas - - - 2. *globosum*
(if intermediate see note at end)

1. **Agathisanthemum bojeri** Klotzsch in Peters, Reise Mossamb. Bot. **1**: 294 (1861). —Bremek. in Verh. K. Nederl. Akad. Wet. Afd. Natuurk. ser. 2, **48**(2): 154 (1952). —Verdc. in F.T.E.A. Rubiaceae 1: 255, fig. 35 (1976). Types from Zanzibar.
Oldenlandia bojeri (Klotzsch) Hiern in F.T.A. **3**: 53 (1877). —Brenan T.T.C.L.: 507 (1949). —R.O. Williams, Useful & Ornamental Pl. Zanzibar: 384, fig. (1949).

Branched herb or subshrub 0.25–1.5 m. tall; young stems densely pubescent, becoming glabrescent with age and often the brown bark peeling off. Leaves subsessile, pale, sometimes drying bright yellow-green; blades 1.2–5.8 × 0.15–1(2.1) cm., narrowly lanceolate to lanceolate, narrowly elliptic or ovate-oblong, subacute to acute and sometimes apiculate at the apex, cuneate at the base, glabrous or slightly scabridulous above, pubescent on the nerves beneath; petiole 0–1.2 mm. long; leaves often appearing whorled owing to presence of short undeveloped leafy shoots present at practically every node; stipule sheath pubescent, 0.5–4.5 mm. long with (2)5 fimbriae 1–5.8 mm. long. Inflorescences laxly corymbose, 0.9–7 cm. across or sometimes quite dense subcapitulate clusters; peduncles 0.2–4 cm. long; pedicels 0–1.5(3) mm. long; flowers distinctly heterostylous. Calyx tube 0.9–1.7 mm. long, glabrous or pubescent, the limb tube 0.2–0.8 mm. long; lobes 1.4–2(3.2) mm. long (4.6 mm. in an atypical isostylous specimen), lanceolate. Corolla white, creamy white or greenish white; tube 1–2.8 mm. long, glabrous or very sparsely hairy outside; lobes 1.5–2.7 × 0.65–1.8 mm., ovate-oblong, glabrous or very sparsely hairy outside. Disk very shortly pubescent. Style 0.7–2 mm. long in short-styled flowers, 3–4(6 *fide* Bremekamp) mm. long in long-styled flowers, usually filiform and undivided but sometimes bifid at the apex; stigma lobes 0.15–0.5 mm. long. Capsule (1)1.5–2.8 × (1.3)1.8–2.5 mm. Seeds black, 0.32 mm. long, ovoid-trigonous, reticulate.

Subsp. **bojeri**
Agathisanthemum bojeri subsp. *bojeri* var. *glabriflorum* Bremek. in Verh. K. Nederl. Akad. Wet. Afd. Natuurk., ser. 2, **48**(2): 156 (1952). TAB. **25**. Type: Mozambique, Pungoe Valley, *Vasse* 280 (P, holotype).
Agathisanthemum bojeri subsp. *australe* Bremek. in Verh. K. Nederl. Akad. Wet. Afd. Natuurk., ser. 2, **48**(2): 156 (1952). Type from S. Africa (Transvaal).
Agathisanthemum bojeri subsp. *australe* var. *glabriflorum* Bremek. in Verh. K. Nederl. Akad. Wet. Afd. Natuurk., ser. 2, **48**(2): 157(1952). Type: Mozambique, Beira railway, *Cecil* 10 (K, holotype).

Leaf blades with nerves mostly well-spaced. Inflorescences often but not always laxer. Calyx lobes mostly short and under 2 mm. long. Mostly coastal.

Zimbabwe. C: Marondera, fl. 15.i.1938, *Stent* in GHS 5465 (K; SRGH). E: Sabi-Chipinge road, fl. 1.ii.1948, *Wild* 2448 (K; SRGH). S: Ndanga, hills just N. of Chipinda Pools, fl. 17.i.1960, *Goodier* 822 (K; LISC; SRGH). **Mozambique**. N: Messalo R., fl. 20.iii.1912, *Allen* 146 (K). Z: Morrumbala, fl. & fr. x.1887, *Scott* (K). MS: Shupanga, fl. 23.ii.1862, *Meller* (K). GI: between Magul and Macia, fl. 1.vi.1959, *Barbosa & Lemos* 8552 (K; LISC; LMA). M: Maputo, between Peter and Costa do Sol, Golf course, fl. 30.iii.1960, *Balsinhas* 145 (K; LISC; LMA).

Also in Somali Republic, Kenya, Tanzania, S. Africa, Comoro Is. (Mayotte) and Madagascar. Grassland, *Brachystegia* woodland usually on sandy soil, also riverine forest and as a weed in old cultivations; 0–1140 m.

I have not retained the subsp. *australe*, distinguished by leaf shape, since the characters are far from constant and specimens occur in Kenya with the ovate leaves of subsp. *australe*. The glabrous-

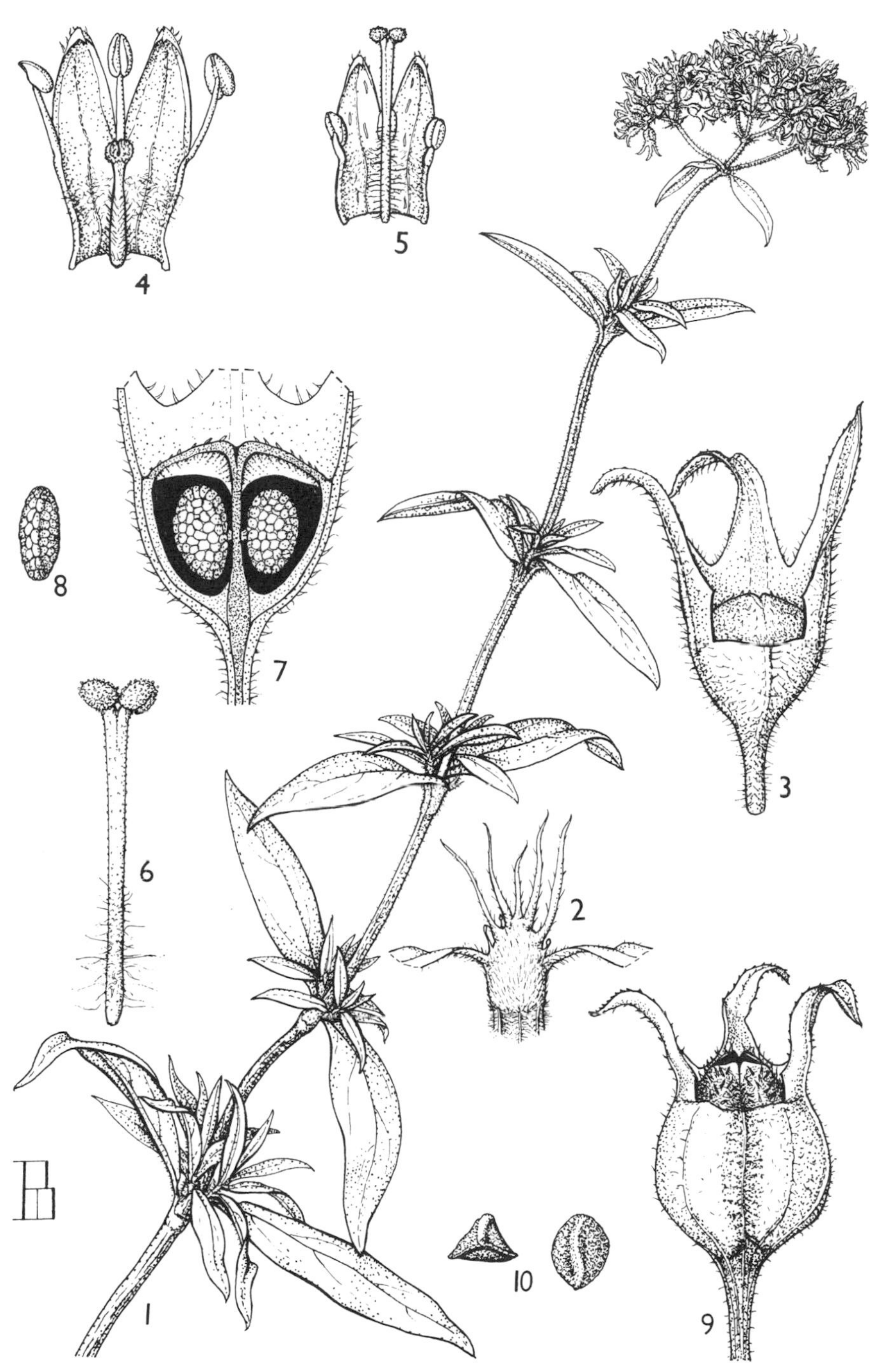

Tab. 25. AGATHISANTHEMUM BOJERI subsp. BOJERI. 1, habit ($\times \frac{2}{3}$), from *Tweedie* 1243; 2, node showing stipule (× 4), from *Faulkner* 1205; 3, calyx with one lobe removed (× 10); 4, short-styled flower, longitudinal section (× 8), 3–4 from *Tweedie* 1216; 5, long-styled flower, longitudinal section (× 8), from *Rawlins* 731; 6, style and stigma lobes (× 14); 7, ovary, longitudinal section (× 16); 8, ovules on one placenta (× 16), 6–8 from *Tweedie* 1216; 9, capsule, with one calyx lobe removed (× 10); 10, seed, two views (× 20), 9–10 *Tweedie* 1243. From F.T.E.A.

flowered varieties are more distinctive but specimens which are very slightly pubescent are frequent.

Subsp. **angolense** (Bremek.) Verdc. comb. et stat. nov. Type from Angola.
Agathisanthemum angolense Bremek. in Verh. K. Nederl. Akad. Wet. Afd. Natuurk. ser. 2, **48**(2): 154 (1952).

Leaf blades with nerves mostly basal. Inflorescences mostly quite dense subcapitulate clusters. Calyx lobes slightly longer, 2–3.5 mm. long. Inland localities.

Var. **angolense**
Agathisanthemum quadricostatum var. pubescens Bremek. in Verh. K. Nederl. Akad. Wet. Afd. Natuurk. ser. 2, **48**(2): 161 (1952) pro parte exclud. typum.

Leaf blades mostly ± ovate-elliptic, about 2.5–3.5 × 1.1–1.5 cm.

Zambia. B: Kalabo, fl. 11.vi.1983, *Fanshawe* 7824 (K; NDO). C: Chakwenga Headwaters, 100–129 km. E. of Lusaka, fl. 14.ii.1965, *Robinson* 6381 (K). S: Kalomo, v.1909, *Rogers* 8216 (K). **Zimbabwe**. N: Trelawney, Tobacco Research Station, fl. 29.iii.1944, *Jack* 247 (K; SRGH). W: Bulawayo, Burnside, Beacon Hill, fl. 24.iv.1966, *Best* 478 (K; LISC; SRGH).
Also in ?Zaire and Angola. Grassland, open *Colophospermum* and *Isoberlinia* woodland, also a weed in old cultivations; 900–1500 m.

Var. **linearifolia** Verdc. var. nov.* Type: Zambia, Mbala District, Muswilo Stream, fl. 4.ii.1965, *Richards* 19600 (K, holotype).

Leaf blades linear to linear-lanceolate, 3.5–5 × 0.5–0.7 cm.

Zambia. N: Mbala, above Kasulo House, fl. 1.ii.1952, *Richards* 618 (K).
Endemic. Open bushland and woodland; 1350–1500 m.
Although much of the coastal subsp. *bojeri* has narrowly lanceolate leaves this is an isolated population clearly distinct from the various forms of *Agathisanthemum globosum* also with narrow leaves which are common in the area. Rather similar variants have been seen from Zaire, Upemba. Subsp. *angolense* and *bojeri* are scarcely separable and certainly not maintainable as separate species. In Eastern Zimbabwe the two completely merge. Some specimens of subsp. *angolense* certainly seem to be isostylous.

2. **Agathisanthemum globosum** (Hochst. ex A. Rich.) Bremek. in Verh. K. Nederl. Akad. Wet. Afd. Natuurk. ser. 3, **48**(2): 161 (1952). —Agnew, Upland Kenya Wild Flowers: 401, fig. (1974). —Verdc. in F.T.E.A., Rubiaceae 1: 258 (1976). —Gonçalves in Garcia de Orta, Sér. Bot. **5**(2): 186 (1982). Types from Ethiopia.
Hedyotis globosa Hochst. ex A. Rich., Tent. Fl. Abyss. **1**: 360 (1847).
?*Agathisanthemum petersii* Klotzsch in Peters, Reise Mossamb. Bot. **1**: 295 (1861). Type: Mozambique, Querimba (B, holotype†).
Oldenlandia globosa (Hochst. ex A. Rich.) Klotzsch ex Hiern in F.T.A. **3**: 54 (1877); in Cat. Afr. Pl. Welw. **1**: 440 (1898).
Agathisanthemum globosum var. *subglobosum* Bremek. in Verh. K. Nederl. Akad. Wet. Afd. Natuurk. ser. 2, **48**(2): 163 (1952). Type from Tanzania.
Agathisanthemum quadricostatum Bremek., tom. cit. 160 (1952). —Verdc. in F.T.E.A., Rubiaceae 1: 256 (1976). Type from Zaire.
Agathisanthemum quadricostatum var. *pubescens* Bremek., tom. cit. 161 (1952). Type: Malawi, Zomba, *Purves* 114 (K, holotype).

Rather strictly erect herb 0.1–1.2 m. tall, with (1) several usually sparsely branched stems from a woody rootstock; branches densely pubescent with short spreading hairs; whole plant often drying the peculiar yellow-green of an aluminium accumulator. Leaves subsessile, (1)2.5–8 × 0.2–2.5(3) cm., elliptic, oblong-elliptic or oblong-lanceolate, acute at the apex, cuneate at the base, glabrescent to pubescent or scabrid on both surfaces but always some hairs on the margins; petiole less than 1 mm. long; aborted leafy shoots are present in most axils; stipule sheath triangular, 1–9 mm. long, bearing (0)3–15 linear fimbriae 3–10 mm. long or sometimes produced as a practically unfimbriated triangular lobe. Flowers typically not heterostylous with anthers and stigmas touching in bud, (3)4(5)-merous, in dense subglobose inflorescences 1–3(3.5 in fruit) cm. in diam., sessile, supported by a pair of leaves, terminal and also terminating slender axillary branches from the upper part of the stems; pedicels 0.2–3 mm. long. Calyx tube 1–2.1 mm. long, glabrous or pubescent; limb tube 0.4–1.8 mm. long; lobes (2.5)3.3–7(9) mm. long, linear-lanceolate, shortly ciliate on margins and usually costa or densely pubescent. Corolla very

* Var. *angolense* foliis linearibus differt.

variable in colour, white, cream, pale yellow, blue, lilac, pink, purple or mauve, the lobes sometimes coloured at least outside and tube whitish, glabrous, puberulous or minutely papillate; tube (1.6)3–4 mm. long; lobes 1.7–4.1 × 1–1.5 mm., triangular. Disk mostly minutely papillate but sometimes with short hairs as well. Anthers exserted, often dark. Style white to yellow, 3.7–6.5 mm. long, exserted; stigma lobes 0.4–0.7 mm. long. In heterostylous flowers of intermediates style 1.4 mm. long in short-styled flowers and 2.6–4 mm. in long-styled flowers. Capsule 1.5–3 mm. tall and wide, compressed-ellipsoid, grooved between the loculi, the beak c. 1 mm. long, crowned by the persistent calyx lobes, glabrous or pubescent. Seeds dark purplish brown, 0.6 × 0.5 mm., ovoid-trigonous, reticulate.

Zambia. B: Kaoma, near Luena R., fl. 20.xi.1959, *Drummond & Cookson* 6694 (K; SRGH). N: Mbala, Pans, fl. 18.ii.1955, *Richards* 4576 (K). W: Mwinilunga, slope E. of Matonchi Farm, fl. 29.xii.1937, *Milne-Redhead* 3867 (K). C: 45 km. SW. of Kabwe, fr. 13.vii.1930, *Hutchinson & Gillett* 3628 (K). E: near Katete, fl. 10.ii.1957, *Angus* 1503 (K; SRGH). S: 12.8 km. S. of Mapanza, 27.ii.1954, *Robinson* 560 (K). **Zimbabwe**. N: Miami, Mahobohobo area of K34 Exper. Farm, fl. 4.iii.1947, *Wild* 1707 (K; SRGH). C: St. Triashill Mission, fl. & fr. iii.1917, *Anon.* SRGH 3204 (K; SRGH). **Malawi**. N: Mzimba Distr., 4.8 km. SW of Chikangawa, fl. & fr. 11.ii.1978, *Phillips* 4220 (K; MO). C: Chitedze, Chirikanda Estate, fl. 6.ii.1959, *Robson* 1488 (BM; LISC; K). S: Zomba Plateau, fl. 6.vi.1946, *Brass* 16279 (K; NY). **Mozambique**. N: Mandimba, *Hornby* 3534 (K). Z: Namagoa, Mocuba, fl., *Faulkner* 354 (K; PRE).

Also in Gabon, Zaire, Burundi, Ethiopia, Uganda, Kenya and Angola. Grassland, *Brachystegia* woodland and derived bushland, roadsides, dambo edges, cultivations etc.; (60 (Mozambique)) 975–1800 m.

Basically there are only two species of *Agathisanthemum* in tropical Africa which typically are easy to separate but even in undoubted *Agathisanthemum globosum* there is much variation in flower-colour, indumentum and degree of exsertion of the anthers showing that the populations are far from uniform. Much of what Bremekamp has annotated as *Agathisanthemum quadricostatum* is clearly not distinguishable from other material named *globosum* in fact from exactly the same place e.g. Zomba he has named material as one or the other indiscriminately. There is, however, a grey area between *globosum* and *bojeri* and very probably a hybrid swarm has developed between the two. The coastal subsp. *bojeri* with mostly quite open inflorescences and heterostylous flowers is clearly very distinct from the more upland *globosum* with large compact inflorescences and isostylous flowers but there are many specimens not so clear cut (e.g. *Norlindh & Weimarck* 4787 (Inyanga, Cheshire)) and inland races of *bojeri* may have produced hybrid populations with *globosum*. Other specimens with *globosum* characters even occur at low altitudes in Mozambique. Studies on the distribution of heterostyly in as many populations as possible would throw light on the problem.

23. AMPHIASMA Bremek.

Amphiasma Bremek. in Verh. K. Nederl. Akad. Wet., Afd. Natuurk., ser. 2, **48**(2): 168 (1952).

Perennial herbs or subshrubs, with rather strict rush-like stems, usually glabrous or glabrescent. Leaves sessile, opposite; blades linear or filiform, rather rigid and usually erect or nearly so, acute and hard-pointed, the margins revolute; stipule sheath mostly tubular, truncate or sometimes with 2 small cusps. Flowers 4-merous, isostylous or heterostylous, in small rather dense terminal and axillary capituliform, umbelliform or cymose inflorescences, sometimes combining to form a more ample inflorescence; bracts connate in pairs. Calyx lobes triangular. Corolla mostly white; tube cylindrical or somewhat funnel-shaped; throat ± densely hairy; lobes ovate or ovate-oblong, hairy in lower half inside. Stamens included in the tube in long-styled flowers, exserted in isostylous and short-styled flowers; filaments glabrous; anthers dorsifixed. Ovary 2-locular; ovules fairly numerous on peltate slightly stipitate placentas affixed just below the middle of the septum; style glabrous, included in short-styled flowers, exserted in long-styled and isostylous flowers; stigma lobes filiform, densely covered with long papillae. Disk cushion-shaped, farinose. Capsule globose, only slightly beaked, the beak dehiscing loculicidally. Seeds dark brown, not very numerous, dorsiventrally compressed, oblong, smooth.

A small genus of probably 5–6 species in south-central and SW. Africa. I am unable to distinguish some of the 8 species listed by Bremekamp. Lewis — see Ann. Missouri Bot. Gard. **52**: 183, 202 (1965) —would not separate this genus from *Oldenlandia*. Two species occur in the Flora Zambesiaca area.

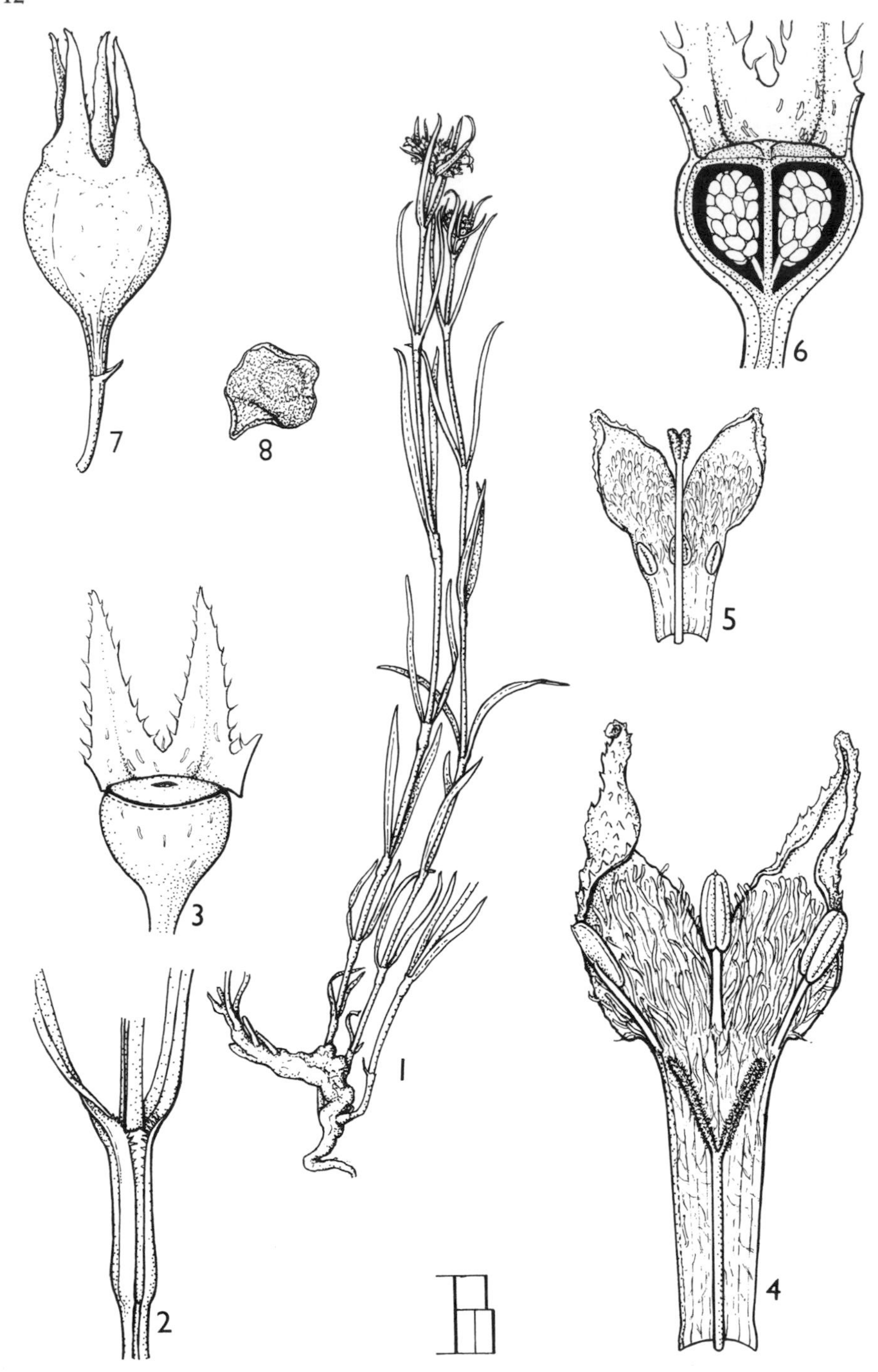

Tab. 26. AMPHIASMA LUZULOIDES. 1, habit (× ⅔); 2, node showing stipule and leaf-bases (× 4), 1–2 from *Whyte* 128; 3, calyx with two lobes removed (× 14); 4, short-styled flower, longitudinal section (× 14), 3–4 from *Pawek* 2195; 5, long-styled flower, longitudinal section (× 8), from *Whyte* 128; 6, ovary, longitudinal section (× 20), from *Pawek* 2195; 7, capsule (× 8); 8, seed (× 20), 7–8 from *Pawek* 2195. From F.T.E.A.

Leaf blades rather wider, mostly 1.5–3.5 mm. wide; pedicels 0–3 mm. long, mostly very short; corolla tube 2–4 mm. long, mostly short; stems rather shorter (N. Malawi) - - 1. *luzuloides*
Leaf blades very slender, up to c. 1 mm. wide; pedicels c. 3 mm. long; corolla tube 4 mm. long; stems graceful (W. Zambia) - - - - - - - - - - - - - 2. *redheadii*

1. **Amphiasma luzuloides** (K. Schum.) Bremek. in Verh. K. Nederl. Akad. Wet., Afd. Natuurk., ser. 2, **48**(2): 170 (1952). —Verdc. in F.T.E.A., Rubiaceae 1: 261, fig. 37 (1976). TAB. **26**. Type: Malawi, Nyika Plateau, *Whyte* (B, holotype †; K).
Oldenlandia luzuloides K. Schum. in Engl., Bot. Jahrb. **28**: 55 (1899).
Oldenlandia luzuloides K. Schum. in Engl., Bot. Jahrb. **30**: 411 (1901) nom. illegit. Type from Tanzania.
Amphiasma assimilis Bremek. tom. cit.: 172 (1952). Type: Malawi, Kondowe to Karonga, *Whyte* (K, holotype).

Perennial herb, with erect stems 14–50 cm. tall from a many-headed rhizome; stems strict, simple or sparsely branched, glabrous. Leaves held erect; blades 1–4 cm. × 1–2(3.5) mm., linear, scabridulous above and ± ciliolate at the base of the midvein beneath; stipular sheath 2–5 mm. long, truncate or with 2 minute teeth c. 0.2 mm. long. Inflorescences 2–5, situated at the apex and at the first 2 nodes beneath, capitulate, c. 6 mm. long and wide; peduncles 2–8 mm. long; pedicels 0–1.5(3) mm. long; bracts leaf-like, 3–9 × 1–2 mm.; flowers heterostylous ? or isostylous. Calyx tube glabrous, 1–1.2 mm. long; lobes 1–1.6 mm. long, with ciliolate margins. Corolla white or tinged lavender, glabrous or papillate outside; tube 2–4 mm. long, slightly funnel-shaped, pilose inside; lobes 2.8–3.3 × 1.1–1.6 mm. the lower half pilose inside. Style 4 mm. long in long-styled flowers, 2 mm. long in short-styled flowers; stigma 0.6–1 mm. long. Capsule 2 × 2.5 mm. Seeds not seen.

Malawi. N: Nyika Plateau, Chosi, fl. 21. iii.1976, *Phillips* 1509 (K; MO).
Also in S. Tanzania. Grassland, often above upper forest edges, sometimes rocky; also in grassland with *Protea* scrub etc.; 1670–2280 m.
This species was presumably based on material at Berlin sent as a duplicate from Kew. K. Schumann curiously used the same name twice, so the second is a later homonym, but fortunately the two types belong to the same species. Bremekamp distinguished *Amphiasma assimilis* by having papillate isostylous flowers but the corolla varies greatly in indumentum in *Amphiasma luzuloides* and specimens with the anthers exserted and stigmas ± exserted have been seen; whether these are variants of brachystylous forms or genuinely isostylous could only be found out in the field. Perhaps more important the pedicels are distinctly longer in *A. assimilis* on a few lower flowers. Certainly there are not two species in N. Malawi.

2. **Amphiasma redheadii** Bremek. in Verh. K. Nederl. Akad. Wet. Afd. Natuurk., ser. 2, **48**(2): 170 (1952). Type: Zambia, Mwinilunga Distr., R. Warnibobo, *Milne-Redhead* 859 (K, holotype).

Perennial herb with erect graceful stems 45 cm. tall from a several-headed rhizome; stems ± strict, not or sparsely branched, glabrous. Leaves held erect; blades 3–4 cm. × 0.5–1 mm., linear-filiform, ciliolate on the margins and scabridulous above. Stipular sheath ± 5 mm. long, truncate. Inflorescences subumbellate cymes restricted to apex of stem or with few flowers at next lower axil; pedicels all 2–3 mm. long. Calyx tube c. 1.3 mm. long, ovoid, glabrous; limb 0.3 mm. long with triangular lobes 0.8–1.1 mm. long, the margins slightly ciliolate. Corolla white, glabrous outside; tube somewhat funnel-shaped, 4 mm. long, pilose inside; lobes 2.5–7.5 × 1 mm., oblong-elliptic, the lower half pilose inside. Styles 5.5 mm. long in long-styled flowers, the stigmatic lobes 1.3 mm. long. Short-styled flowers and fruits not known.

Zambia. W: Mwinilunga Distr., R. Warnibobo, fl. 7.viii.1930, *Milne-Redhead* 859 (K).
Not known elsewhere. *Brachystegia* woodland. Very similar to *Amphiasma luzuloides* but with a different more graceful facies.

24. MANOSTACHYA Bremek.

Manostachya Bremek. in Verh. K. Nederl. Akad. Wet. Afd. Natuurk., ser. 2, **48**(2): 145 (1952).

Woody-based herbs. Leaves sessile, linear or subulate, the upper gradually passing into bracts. Stipular sheath truncate, sometimes shortly sparsely ciliate. Flowers sessile or shortly pedicellate, 4-merous, heterostylous, solitary or in axillary cymes arranged at many well-separated upper nodes in very open elongate spike-like inflorescences or each apparent cyme actually an inflorescence terminating a brachyblast; bracteoles present. Calyx tube ovoid; limb tube not developed; teeth narrowly triangular. Corolla tube short, ± cylindric or funnel-shaped, the lobes about as long or slightly longer. Anthers well-included in long-styled flowers. Ovary 2-locular; placentas peltate, attached to middle of septum; ovules several, immersed in the placenta. Style glabrous, exserted in long-styled flowers, included in short-styled flowers; stigmatic lobes 2, ovoid or filiform. Capsule subglobose, dehiscing loculicidally. Seeds few, brown, oblong, dorsiventrally compressed, not becoming glutinous when wet, the testa cells with thick cell walls and large perforations forming a reticulate network in the lumen.

A genus of 3 species in Zaire, Malawi, Zambia, Angola and just reaching S. Tanzania.

Leaves opposite, rather sparse and never appearing densely verticillate - - 1. *staelioides*
Leaves mostly in whorls of 3–4 but, due to abbreviated shoots within axils, appearing densely verticillate - - - - - - - - - - - - - - - - 2. *ternifolia*

1. **Manostachya staelioides** (K. Schum.) Bremek. in Verh. K. Nederl. Wet. Afd. Natuurk., ser. 2, **48**(2): 146 (1952). —Verdc. in Kew Bull. **35**: 322 (1980). Type from Angola.
Oldenlandia staelioides K. Schum. in Engl., Bot. Jahrb. **23**: 418 (1897). —Hiern, Cat. Afr. Pl. Welw. **2**: 446 (1898).
Oldenlandia staelioides f. *major* De Wild. in Ann. Mus. Congo, sér. 4, **2**: 150, t. 4 (1913). Type from Zaire.

Woody-based *Thesium*-like herb with several erect glabrous branched stems (15)20–45 cm. tall, the stems below the lateral branches essentially leafless. Leaves 2–6(15) × scarcely 0.5 mm., linear-subulate, glabrous or with margins ± ciliolate towards the base. Stipular sheath very short. Flowers distinctly bracteolate, solitary or few in the axils, sessile or pedicels to about 0.5 mm. or sometimes falsely pedicellate by reduction and stalk up to 1 cm. long; the nodes well separated and forming spike-like inflorescences 10–20 cm. long. Calyx tube glabrous; lobes 0.5–1.5 mm. long. Corolla white sometimes with pink-tipped lobes; tube 1.5–2 × 1.5 mm., shortly cylindrical, glabrous inside; lobes ovate, 1–2 × 0.9–1.2 mm., densely pilose inside. Style 2–2.5 mm. long in long-styled flowers, 0.5 mm. long in short-styled flowers; stigmatic lobes ovoid, 0.3 mm. long. Capsule subglobose, about 2 mm. long including the elevated beak which is about $\frac{1}{3}$ the total. Seeds 0.6 mm. long.

Zambia. W: 1.6 km. SE. of Solwezi, fr. 4.xi.1962, *Lewis* 6128 (K; MO). **Malawi**. N: Viphya, 65.6 km. SW. of Mzuzu, Chimpyai View, fr. 15.xii.1970, *Pawek* 4095 (K; MAL).
Also in Zaire, S. Tanzania and Angola, grassland, chipya woodland on dambo margins and on limestone; ?1500–2250 m.

2. **Manostachya ternifolia** Martins in Bol. Soc. Brot. sér 2, **55**: 5, t. 1 (1981). TAB. **27**. Type from Angola.

Strictly erect perennial herb 40–90 cm. tall with 1-several stems, woody at the base, 3–4-angled, glabrescent or finely shortly pubescent when young. Leaves in whorls of 3–4 or opposite, but appearing densely verticillate due to abbreviated axillary shoots, 0.5–2.5 cm. × 0.4–1 mm., subulate, rigid, the margins and midribs scabrid-ciliate; stipular sheath c. 1 mm. long, mostly shortly sparsely ciliate. Inflorescences 1–7-flowered contracted cymes, densely bracteate, axillary in the top 7–30 spaced nodes, the whole spike-like, up to 30 cm. long; bracts similar to leaves, 2–5 mm. long; pedicels 0.5–1 mm. long. Calyx tube 1.2 mm. long, ovoid, glabrous but with obvious surface raphides; lobes 4, 1.3–1.5 mm. long, subulate, scabrid-ciliate. Corolla white, glabrous outside; tube 1–1.3 × 0.8 mm., cylindric-funnel-shaped, with a ring of hairs inside at the throat; lobes 1.1–1.7 × 0.7–1 mm., triangular-ovate, papillate inside, the tip inflexed. Anthers exserted in short-styled flowers, just included in long-styled flowers. Style 1.8 mm. long in long-styled, 0.5 mm. in short-styled flowers. Capsule 1.7 × 1.3 mm. ellipsoid. Seeds 1–2(3) per loculus, brown, oblong, dorsiventrally flattened, 0.9 × 0.38 mm., the testa reticulate.

Zambia. C: Serenje, Kaombi, fl. iv.1930, *Lloyd* s.n. (BM).
Also in S. Zaire and Angola.

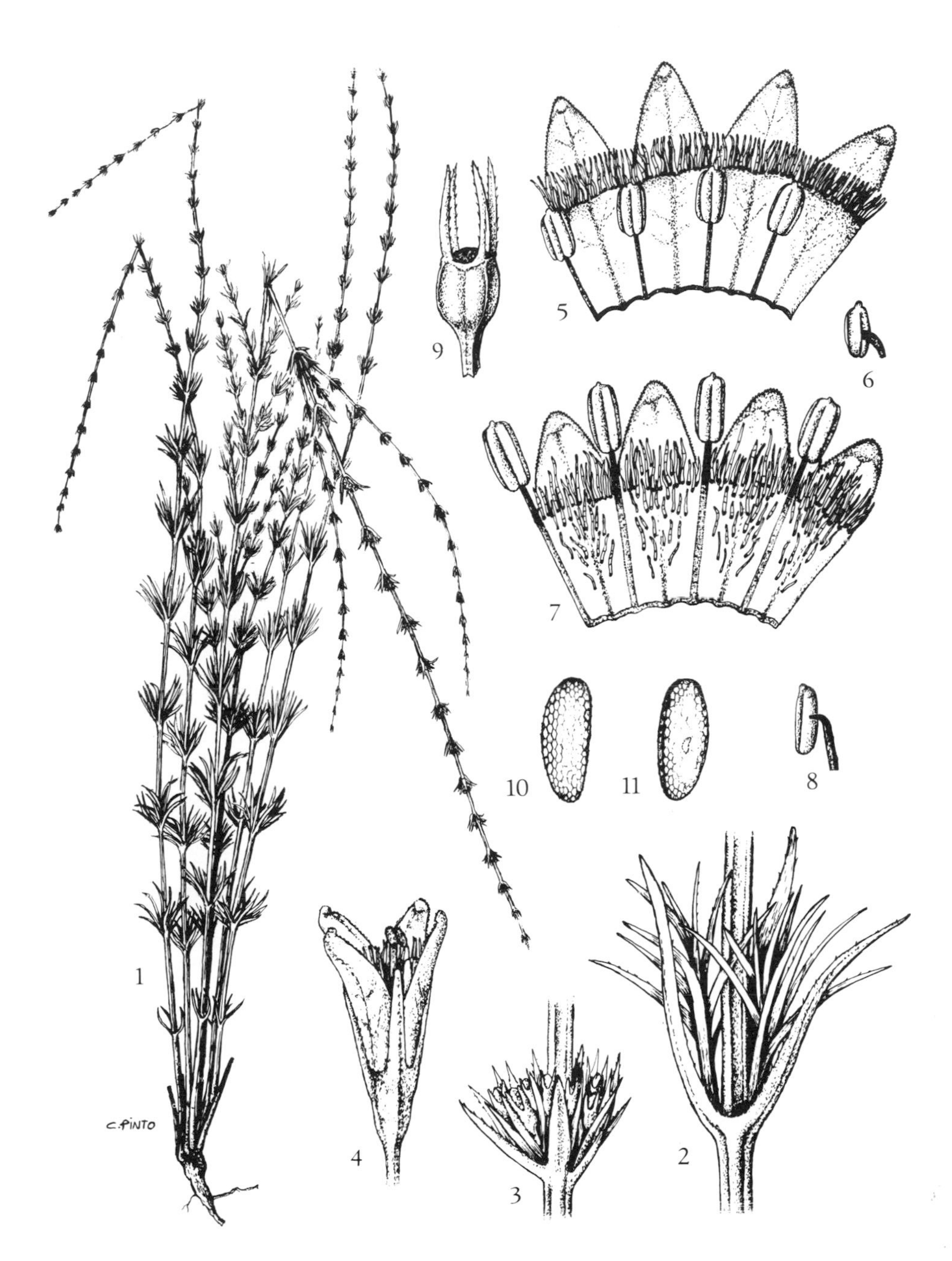

Tab. 27. MANOSTACHYA TERNIFOLIA. 1, habit (× ½); 2, node showing axillary brachyblasts (× 2½); 3, floriferous node (× 2½); 4, long-styled flower (× 7½); 5, corolla of long-styled flower opened out (× 7½); 6, part of stamen of long-styled flower (× 7½); 7, corolla of short-styled flower opened out (× 7½); 8, part of stamen of short-styled flower (× 7½), 1–8 from *Mendes* 2662; 9, fruit (× 7½); 10, seed, dorsal view (× 20); 11, seed, ventral view (× 20), 9–11 from *Bamps & Martins* 4410.

25. PENTODON Hochst.

Pentodon Hochst. in Flora **27**: 551 (1844). —Bremek. in Verh. K. Nederl. Akad. Wet., Afd. Natuurk., ser. 2, **48**(2): 175 (1952).

Glabrous subsucculent annual or short-lived perennial herbs with a slender rootstock. Leaves paired, sessile, penninerved but lateral nerves obscure; stipules with the short sheath divided into (1)2–5 narrow fimbriae. Flowers small, hermaphrodite, dimorphic or not, in few several-flowered very lax pedunculate terminal or apparently axillary (solitary at the nodes) inflorescences which appear to be verticillate, with 1–4 flowers from 1–4 widely spaced nodes. Calyx tube obconic or campanulate; lobes 5, equal, very narrowly triangular. Corolla tube narrowly funnel-shaped, throat hairy; lobes 5, ovate-triangular. Stamens exserted in short-styled forms, included in long-styled and equal-styled forms. Ovary 2-locular; placentas peltate with numerous ovules; style filiform, glabrous; stigma 2-lobed, the lobes filiform, included in short-styled forms, often exserted in equal-styled forms, exserted in long-styled forms. Fruit capsular, campanulate or oblong, the beak only slightly raised, loculicidally dehiscent. Seeds small, numerous, brown, angular.

A genus of probably only 2 species, one restricted to the Somali Republic (S.), the other widespread in Africa and extending to Arabia and Madagascar. A third species has been described from America, namely *P. halei* (Torrey & Gray) A. Gray; this occurs in the U.S.A. from Florida to Texas, Cuba and Nicaragua. Bremekamp suggests that it is conspecific with *P. pentandrus* and other writers (e.g. Correll & Johnston, Man. Vasc. Pl. Texas: 1490 (1970)) have accepted this. Clearly it cannot be specifically distinct. The material is rather uniform in facies, having small elliptic leaves and the inflorescences shorter than in most African material; curiously it is with some coastal E. African material that it can be most easily matched in the characters just mentioned; but the flowers appear to be more like those of var. *pentandrus*, although unfortunately no material has been seen which throws light on the heterostyly or otherwise of the New World material. It seems possible that the American populations are due to very few introductions of atypical material from Africa quite possibly long ago. Studies on the presence of heterostyly in these populations would be very interesting.

Pentodon pentandrus (Schumach. & Thonn.) Vatke in Oest. Bot. Zeitschr. **25**: 231 (1875). —K. Schum. in Engl. Pflanzenw. Ost-Afr. **C**: 377 (1895) (as *pentander*). —Bremek. in Verh. K. Nederl. Akad. Wet. Afd. Natuurk. ser. 2, **48**(2): 176 (1952) (as *pentander*). —Hepper, F.W.T.A. ed. 2, **2**: 213 (1963). —Lind & Tallantire, Fl. Pl. Uganda ed. 2: 160 (1972). —Agnew, Upland Kenya Wild. Fl.: 401 (1974). —Verdc. in F.T.E.A., Rubiaceae 1: 263 (1976). Type from Ghana.

Hedyotis pentandra Schumach. & Thonn. in Kongel. Danske Vid. Selsk. Skr. **3**: 71 (1827).

Oldenlandia pentandra (Schumach. & Thonn.) DC., Prodr. **4**: 427 (1830) non Retz.

Usually annual or short-lived perennial herb, with weak decumbent, procumbent or rarely suberect, often single stems 4–90 cm. long. Leaf blades (1.3)1.5–8 × (0.3)0.45–2.5 cm., linear-lanceolate, elliptic-lanceolate or elliptic, subacute to sharply acute at the apex, rounded to cuneate at the base, sessile; stipule base 0.5–3(5) mm. long; fimbriae 0.5–3 mm. long. Inflorescences 0.7–9 cm. long; peduncles (0.3)0.6–6.5 cm. long; pedicels 0.2–1.5 cm. long, spreading. Calyx tube 0.5–1.5 mm. long; limb tube 0.25–1 mm. long; lobes 0.5–1(1.5) mm. long. Corolla white, pink, pale to deeper blue or pale mauve, very often with the limb blue and the tube pale yellow; tube (1.5)2–4.5 mm. long, narrowly to widely funnel-shaped, widest in heterostylous forms, (1)2–4 mm. wide; lobes 1–3 × 0.8–2 mm., ovate-triangular, the throat hairs extending up over the inside. Style in long-styled, short-styled and equal-styled forms 2–3.5 mm., 1–1.5 mm. and 1 mm. long respectively; stigma lobes in short-styled forms short, c. 0.8–1.5 mm. long, in long-styled forms longer, 1.5–2 mm. long. Capsule (2)3–4 × (2)2.8–3.5 mm.; beak 0.5 mm. tall, crowned by the persistent calyx lobes. Seeds black, angular, 0.3 mm. long.

Var. **pentandrus**. —Bremek. in Verh. K. Nederl. Akad. Wet., Afd. Natuurk., ser. 2, **48**(2): 177 (1952). —Verdc. in F.T.E.A., Rubiaceae 1: 265 (1976).

Oldenlandia macrophylla DC., Prodr. **4**: 427 (1830). —Hiern in F.T.A. **3**: 63 (1877). Type from the Gambia.

Pentodon abyssinicus Hochst. in Flora **27**: 552 (1844). Type from Ethiopia.

Flowers isostylous or with style exserted and anthers included but never with flowers truly heterostylous; corolla usully small, the tube c. 2 mm. long, rather more cylindrical than funnel-shaped.

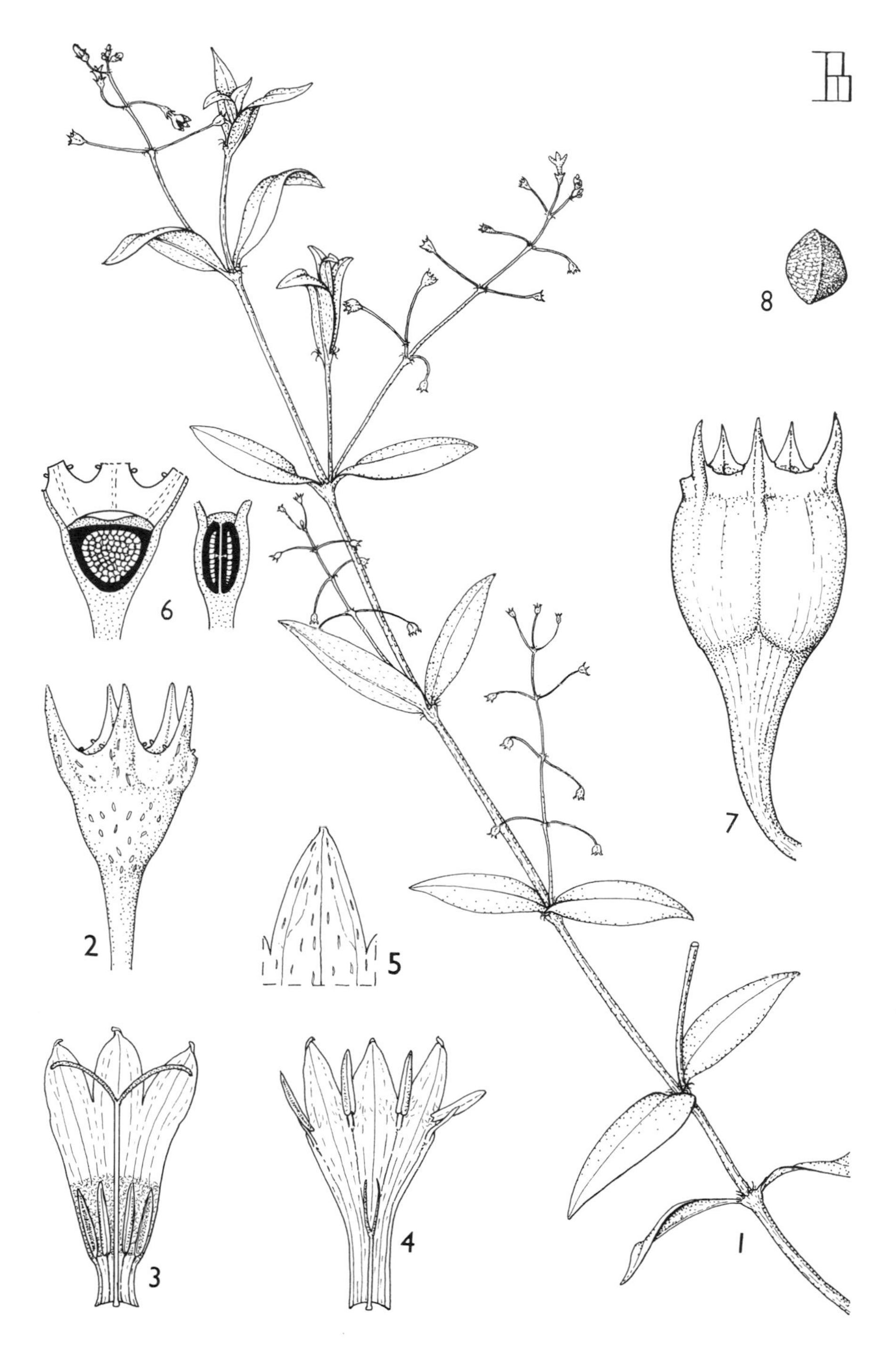

Tab. 28. PENTODON PENTANDRUS var. MINOR. 1, habit (× $\frac{2}{3}$), from *Milne-Redhead & Taylor* 8020; 2, calyx (× 10), from *Rawlins*; 3, long-styled flower, longitudinal section (× 6), from *Tanner* 2905; 4, short-styled flower, longitudinal section (× 6); 5, corolla lobe, outer side (× 8); 6, ovary, longitudinal sections (× 10), 4–6 from *Rawlins*; 7, capsule (× 6); 8, seed (× 20), 7–8 from *Tanner* 2630. From F.T.E.A.

Caprivi Strip. Linyanti R., Namatanga, fl. 29.x.1972, *Biegel et al.* 4101 (LISC; SRGH). **Botswana**. N: Zibadianja Lagoon, fl. 19.x.1972, *Biegel et al.* 4008 (K; LISC; SRGH). **Zambia**. B: Mongu, Lake Lutende, fr. 20.iii.1966, *Robinson* 6887 (K). N: Chibaya, fl. & fr. 15.x.1949, *Bullock* 1281 (K). W: Mwinilunga Distr., R. Lunga, below R. Mudjanyama, fl. & fr. 25.xi.1937, *Milne-Redhead* 3395 (K). E: Sasare, fl. & fr. 8.xii.1958, *Robson* 863 (BM; LISC; K). **Zimbabwe**. N: Mwenezi, Lundi R., near Fishan, fl. 28.iv.1962, *Drummond* 7787 (LISC; SRGH) (var. doubtful.) W: Hwange Distr., Victoria Falls, fl. 27.wi.1978, *Mshasha* 121 (K; SRGH). **Malawi** N: Nyika Plateau, Mwanemba, fr. & fl. ii.1903, *McClounie* 74 (K). C: Dedza, Unisensu Ntata-taka, fl. 22.x.1968, *Salubeni* 1172 (K; SRGH). S: Blantyre, *Buchanan* ? 146 (K).

Widespread in tropical Africa from Cape Verde Is., W. Africa (Senegal-Angola), Arabia, Sudan, Ethiopia and Somali Republic to East & South West Africa, Madagascar; also in U.S.A. (Florida & Texas), Cuba, Nicaragua and Brazil. Riverine grassland, *Phragmites* and *Papyrus* swamps, water-hole and lake edges, rocky places in rivers, wet mud in swamp-forest and rain-forest etc.; 750–900 m.

Specimens from Malawi, Nyika Plateau must be from a much higher altitude but the only actual figure 8000 ft. (McClounie) is suspect.

Var. **minor** Bremek. in Verh. K. Nederl. Akad. Wet. Afd. Natuurk. ser. 2, **48**(2): 179 (1952). —Verdc. in F.T.E.A., Rubiaceae 1: 265, fig. 38 (1976). TAB. **28**. Type from S. Africa.
Pentodon decumbens Hochst. in Flora **27**: 552 (1844). Type as for var. *minor*.
Hedyotis pentamera Sond. in Harv. & Sond., F.C. **3**: 12 (1865). Type as for var. *minor*.

Flowers heterostylous, some with style included and anthers well-exserted and others exactly vice versa; corolla usually longer, the tube c. (2.5)3–4 mm. long, usually distinctly funnel-shaped and often with a wide throat up to 4 mm. wide.

Zimbabwe. E: Mutare, Dura Ranch, fl. 10.iii.1957, *Chase* 6361 (K; SRGH). S: Ndanga Distr., Umtilikwe R., fl. 26.i.1949, *Wild* 2741 (K; SRGH). **Malawi**. S: Zomba, Lake Chilwa, fl. 28.i.1959, *Robson* 1330 (BM; LISC; K). **Mozambique**. N: Niassa, Marrupa, road to Nungo, margin of R. Messalo between Montanhas Mirenge and Mucuwango, fl. & fr. 20.ii.1981, *Nuvunga* 664 (K; LMU). Z: R. Zambezi, Kongone Mouth, fl. i.1861, *Kirk* " Fig. No. 331" (K). MS: near Sena, sand-bank in R. Zambezi, xi.1859, *Kirk* s.n. (K). GI: Inhambane, Inharrime, Ponta Zavora, fl. & fr. 4.iv.1959, *Barbosa & Lemos* in *Barbosa* 8494 (K; LISC; LMA). M: Maputo, Costa do Sol, fl. & fr. 4.xi.1963, *Balsinhas* 654 (K; LISC; LMA).

Uganda, Kenya, Tanzania (including Zanzibar and Pemba), South Africa (Transvaal, Natal), Swaziland and also in Seychelles (material mostly too poor to be certain of variety). By pools etc. in grass and sedge associations, river banks and beds, dambos, seasonal bogs, damp sand dunes etc.; ± sea-level–1050 m.

There is no doubt that Bremekamp is correct in dividing this species into two taxa. Since var. *pentandrus* has long-styled forms certain specimens can be very difficult to name. Studies of heterostyly in populations of this species, particularly the inland ones are needed. Bremekamp also uses the size of the capsule as a distinguishing character but I have not found it constant.

26. LELYA Bremek.

Lelya Bremek. in Verh. K. Nederl. Akad. Wet. Afd. Natuurk., ser. 2, **48**(2): 181 (1952). —W.H. Lewis in Ann. Missouri Bot. Gard. **52**: 189 (1965).

Prostrate perennial herb. Leaves paired, small, shortly petiolate; stipules in lower part of stem with sheath produced into an undivided triangular lobe, in the upper part deeply bifid into narrowly triangular fimbriae or less often with 3 fimbriae. Flowers small, hermaphrodite, dimorphic, in terminal triads or sometimes solitary. Calyx tube ovoid or ellipsoid; lobes 4, equal, oblong-elliptic or oblong-spathulate, rather thick, exceeding the tube, joined at the base to form a free tube. Corolla tube shortly cylindrical or narrowly funnel-shaped, sparsely pilose inside; lobes 4, elliptic-oblong to oblong-lanceolate. Stamens completely exserted in short-styled forms. Ovary 2-locular, with peltate placentas and 8 to fairly numerous ovules; style stoutly filiform, minutely bifid at apex, the stigma lobes short, subglobose or ellipsoid. Fruit capsular with bony walls, ellipsoid, produced into a solid beak; at length apically loculicidally dehiscent. Seeds few per locule, blackish, angular, pitted.

A monotypic genus occurring in widely separated areas of tropical Africa and scarcely separable from *Oldenlandia*, but apart from the bony fruits the pollen differs from all known species of that genus (see Lewis loc. cit.: 191 (1965) for details).

Lelya prostrata (Good) W.H. Lewis in Ann. Missouri Bot. Gard. **52**: 189 (1965). —Verdc. in F.T.E.A., Rubiaceae 1: 266 (1976). Type from Angola.

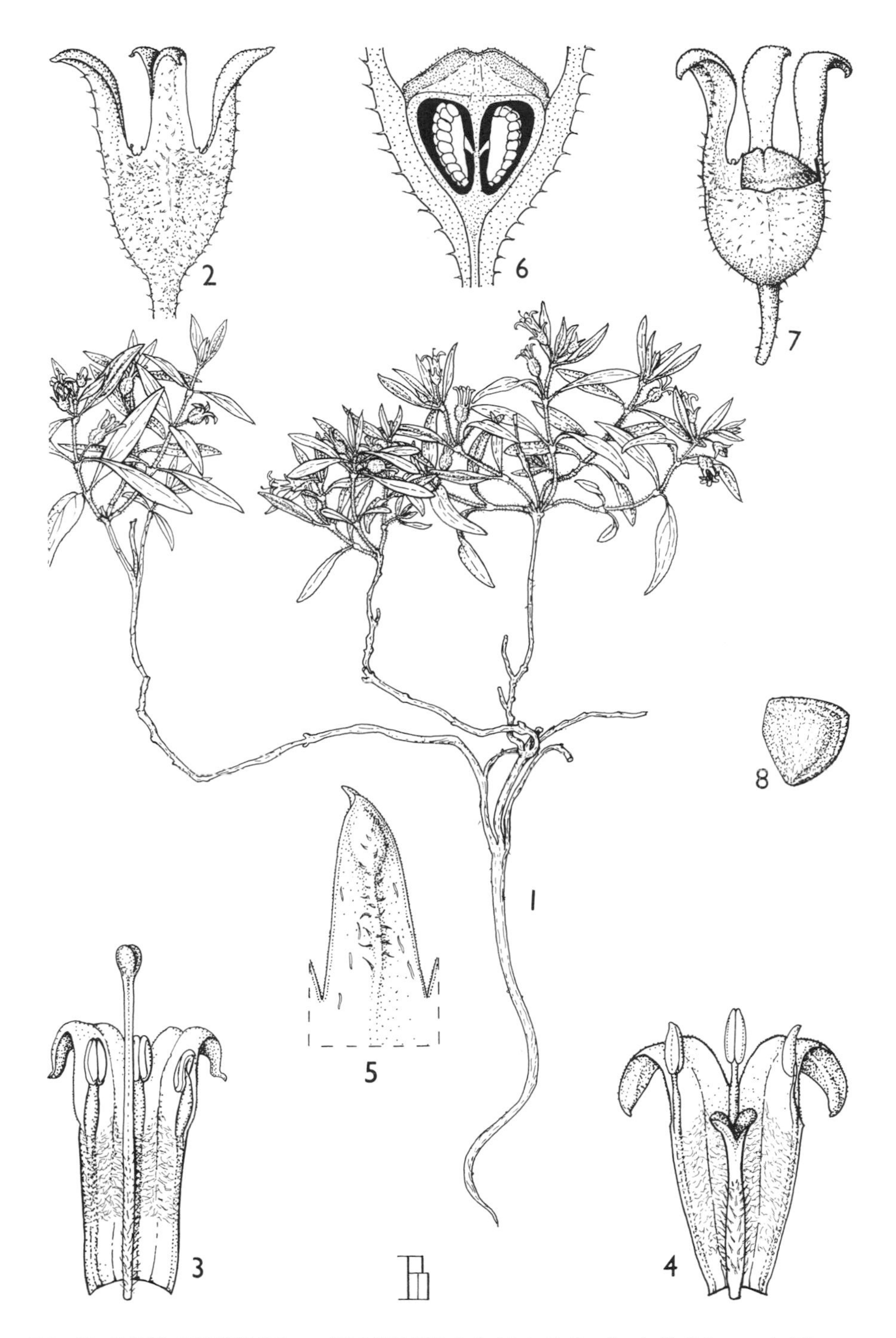

Tab. 29. LELYA PROSTRATA var. PROSTRATA. 1, habit (× 1); 2, calyx (× 8); 3, long-styled flower, longitudinal section (× 8), 1–3 from *Lewis* 6067; 4, short-styled flower, longitudinal section (× 8), from *Robson & Angus* 221; 5, corolla lobe, outer side (× 10); 6, ovary, longitudinal section (× 14); 7, capsule with one calyx lobe removed (× 6), 5–7 from *Lewis* 6067; 8, seed (× 20), from *Salubeni* 1202. From F.T.E.A.

Perennial herb, with many stems 2.5–10 cm. long radiating from a woody rootstock, much branched at the base; sometimes rooting at nodes *fide Staples* 153; often forming small mats; branchlets usually spreading pubescent at first, later glabrous or all subglabrescent. Leaf blades 0.5–1.5 cm. × 1–10 mm., elliptic to almost linear, acute, apiculate or somewhat obtuse at the apex, cuneate at the base; the margins often recurved, rather thick, frequently drying the curious yellow-green of an aluminium-accumulating plant, glabrous or scabridulous above, pubescent or glabrous beneath with 2–3 distinct lateral nerves; petiole 1–2.5 mm. long, united with the stipule sheath in lower half; stipule sheath 1–2.5 mm. long, triangular-ovate; lobes 3–5, 0.8–3 mm. long. Pedicels 1–4 mm. long, mostly pubescent. Calyx tube 0.4–1.8 mm. long, pubescent or glabrous, the limb 2.5–3.5 mm. long, the tubular part up to 0.3–1 mm. long, the lobes 1–2.5 mm. long, up to 1 mm. wide, acute or subobtuse, keeled. Corolla white or blue, glabrous or sparsely hairy outside; tube 1.6–3 mm. long; lobes 2–5 × 1.2 mm. Style in short-styled flowers 2 mm. long, in long-styled flowers 3–5.5 mm. long; stigma lobes 0.4–0.6 mm. long. Capsule 2 × 2–2.3 mm., globose or ovoid, with short spreading hairs or glabrescent, the beak 0.8 mm. long. Seeds round in outline, 0.7 mm. in diam.

Var. **prostrata**. —Verdc. in F.T.E.A., Rubiaceae 1: 268, fig. 39 (1976). TAB. **29**.
Spermacoce prostrata Good in Journ. Bot., Lond. **65**. Suppl. **2**: 42 (1927).
Lelya osteocarpa Bremek. in Verh. K. Nederl. Akad. Wet., Afd. Natuurk., ser. 2, **48**(2): 181 (1952). —Hepper, F.W.T.A. ed. 2, **2**: 212 (1963). Type from Nigeria.

Stems and petioles densely pubescent; leaf blades elliptic to lanceolate, pubescent beneath; corolla pubescent outside; calyx tube pubescent.

Zambia. W: Solwezi Dambo, fl. 24.ix.1930, *Milne-Redhead* 1197 (K). **Malawi**. N: Nyika Plateau, fl. 11.xi.1967, *Richards* 22487 (K).
Also in Nigeria, Zaire and Angola. Dry grassy hillside, in grass tussocks, open burnt ground; 1500–2250 m.

Var. **angustifolia** (Bremek.) W.H. Lewis in Ann. Missouri Bot. Gard. **52**: 189 (1965). —Verdc. in F.T.E.A., Rubiaceae 1: 268 (1976). Type from Tanzania.
Lelya osteocarpa var. *angustifolia* Bremek. in Verh. K. Nederl. Akad. Wet., Afd. Natuurk. ser. 2, **48**(2): 183 (1952).

Stems and petioles ± glabrescent or pubescent; leaf blades narrowly to very narrowly elliptic, glabrous; corolla and calyx tube glabrous.

Zambia. N: 1.6 km. S. of Mbala, fl. 30.x.1962, *Lewis* 6119 (K; MO). W: Ndola, fl. 21.x.1953, *Fanshawe* 361 (K; NDO). **Malawi**. N: Nyika Plateau, 3.2 km. SW. of Rest House, fl. 21.x.1958, *Robson & Angus* 221 (BM; LISC; K).
Also in S. Tanzania. Recently burnt open woodland, bushland, grassland or bare ground, mostly sandy, also dambos etc.; 1500–2150 m.

27. OLDENLANDIA L.

Oldenlandia L., Sp. Pl.: 119 (1753); Gen. Pl. ed. 5: 55 (1754). —Bremek. in Verh. K. Nederl. Akad. Wet., Afd. Natuurk., ser. 2, **48**(2): 183 (1952).

Annual or perennial herbs or rarely subshrubs, the stems erect or prostrate, simple or branched, or rarely cushion-herbs. Leaves opposite; stipules with 1–several fimbriae from a short base which is often adnate to the leaf base. Flowers mostly small, hermaphrodite, heterostylous or isostylous, in terminal or axillary lax or dense inflorescences or sometimes fasciculate or solitary at the nodes. Calyx lobes 4 (rarely more), mostly small, equal, narrowly to broadly triangular. Corolla tube usually short, cylindrical; lobes 4; throat often hairy. Stamens enclosed or exserted. Ovary 2-locular; ovules mostly numerous on peltate placentas; style short or long, included or exserted, filiform; stigma lobes linear to subglobose. Capsule subglobose to oblong, usually with a loculicidally dehiscent beak. Seeds mostly numerous, angular or subglobose, smooth or alveolate, often becoming viscid when moistened; testa cells smooth to distinctly punctate, granular or tuberculate.

A large genus, sometimes estimated at 300 species (but probably nearer a third that number)in the tropics of both the Old and New Worlds. Many botanists consider that *Oldenlandia* L. itself (let alone the many segregate genera proposed by Bremekamp) should be merged with *Hedyotis* L. e.g. Torrey & Gray, Fl. N. Amer. **2**: 37–43 (1841), Fosberg in Va. Journ. Sci. **2**: 106–111 (1941) & in Castanea **19**: 25–37 (1954), Lewis in Southwest. Nat. **3**: 204–207 (1959) & in Rhodora **63**: 216–223 (1961), Fukuoka in Tonan Ajia Kenkyu **8**: 305–336 (1970), etc. Shinners in Field & Lab. **17**: 166–169 (1949) and Lewis *loc. cit.* (1961) merge *Houstonia* L. (based on N. American species) with *Hedyotis*. In other papers, however, Lewis continued to use the genus *Oldenlandia* (Lewis in Grana Palynologica **5**: 330–341 (1964)) and in a more extensive paper, Cytopalynological Study of African Hedyotideae (Rubiaceae), Ann. Missouri Bot. Gard. **52**: 182–211 (1965), he upheld many of Bremekamp's segregate genera as well as *Oldenlandia*. If all the genera closely related to *Hedyotis* are sunk into it, then it forms an unwieldy unit covering a very wide range of structure and habit. It is true that Bremekamp's segregate African genera were based mainly on a study of the African taxa only and he himself predicted that an extension of this study on a worldwide basis would result in the necessity to erect numerous new genera. The African genera mostly have distinct facies although the technical characters are frequently trivial; no-one can doubt the distinctness of genera such as *Kohautia* or *Mitrasacmopsis*, which I would certainly keep distinct even if *Oldenlandia* and *Hedyotis* were merged. For the purposes of the African floras I have maintained nearly all Bremekamp's genera.

Bremekamp has divided the genus into a number of subgenera and a key to these will be found on pages 184-6 of his monograph. Those occurring in the Flora Zambesiaca area are listed below (one being new) showing which species belong to them. I have not entirely followed Bremekamp's order for this account.

1. *Anotidopsis* (Hook.f.) K. Schum. (species 1–5)
2. *Orophilum* Bremek. (species 6–10)
3. *Hymenophyllum* Bremek. (species 11)
4. *Trichopodium* Bremek. (species 12)
5. *Octoneurum* Bremek. (species 13)
6. *Aneurum* Bremek. (species 14)
7. *Phymatesta* Verdc. subgen. nov.* (species 15, 16)
8. *Cephalanthium* Bremek. (species 17)
9. *Oldenlandia* (*Euoldenlandia*) (species 18–23)
10. Unnamed groups (species 24–25)

Whether the plants are heterostylous or isostylous is a useful character in this genus. The following are heterostylous, i.e. they exist in two very distinct forms, one with flowers having a short included style and anthers well-exserted, and the other with the anthers included and the style well-exserted: 5, 6, 8, 10, 13 and 17. The following are isostylous, i.e. with the anthers and style of more or less equal lengths or at least at constant relative heights or dolichostylous with the style long but the opposite form with the anthers exserted missing: 1–4, 7, 9, 11, 12 (in Flora Zambesiaca area), 14, 15?, 16?, 18 (in Flora Zambesiaca area), 19, 20 (in Flora Zambesiaca area), 21–24, ?25.

1. Calyx lobes basically 4 - - - - - - - - - - - - - - - - - - 2
– Calyx lobes 5–8 - - - - - - - - - - - - 21. *capensis* var. *pleiosepala*
2. Inflorescences terminal or terminal and axillary, capitate to very lax - - - - - 3
– Inflorescences axillary (only occasionally appearing falsely terminal due to shortness of young terminal shoots and their overtopping by inflorescences from the upper axils or rarely to the suppression of some of the upper leaves) - - - - - - - - - - - - 20
3. Flowers in dense terminal and sometimes axillary clusters or if inflorescences laxer or compound then at least pedicels very short, up to 1.5(3) mm. long - - - - - 4
– Flowers in rather dense to extremely lax terminal and lateral inflorescences or at least the pedicels well-developed and clearly evident - - - - - - - - - - - 11
4. Slender, short, little-branched or unbranched, erect ephemeral herbs with linear to linear-lanceolate leaves - - - - - - - - - - - - - - - - - - - 5
– More robust herbs of if ephemeral then with broader leaves - - - - - - 6
5. Inflorescences ± condensed heads; corolla tube 2–3.3 mm. long; not drying a characteristic yellow-green - - - - - - - - - - - - - - 17. *nematocaulis*
– Inflorescences dichasially cymose with distinct central flower and somewhat elongated lateral several-flowered branches; corolla tube just under 1 mm. long; drying a characteristic yellow-green colour - - - - - - - - - - - - - - 25. *robinsonii*
6. Corolla tube narrowly cylindrical 3.2–4.5 mm. long erect; lateral nerves very evident 11. *echinulosa*
– Corolla tube more shortly broadly cylindrical; erect or decumbent; lateral nerves evident or ± obscure 7
7. Stigmata ± subglobose; true wet-land plants (widespread) - - - - - - - - 8

* Subgen. *Stachyanthi* Bremek. similis ob semina madefacta glutinosa, folia angusta, inflorescentias terminales sed cellulis testae proprie reticulatis et tuberculatis, floribus probabiliter isostylis differt. Typus: *Torre* 5046A (LISC) *Oldenlandia verrucitesta* Verdc.

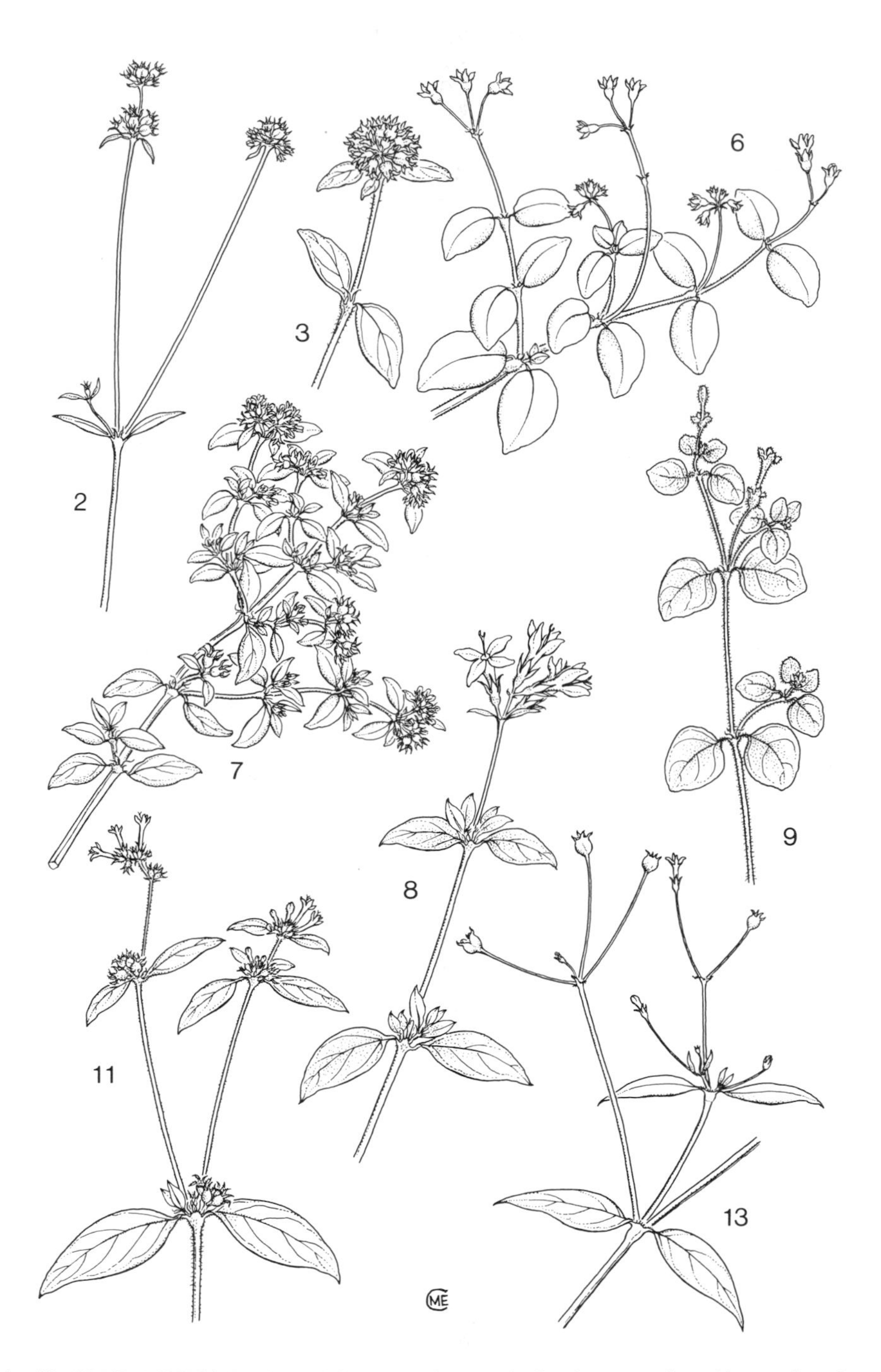

Tab. 30. OLDENLANDIA. Portions of flowering shoot, (× 1). Species as numbered in text. 2, —O. ANGOLENSIS, from *Brummitt, Chisumpa and Polhill* 14094; 3, —O. GOREENSIS, from *Richards* 13732; 6, —O. HOCKII, from *Harley* 9486; 7, —O. GEOPHILA, from *Fanshawe* 1079; 8, —O. RUPICOLA, from *Pawek* 3894; 9, —O. CANA, *Grosvenor* 373; 11, —O. ECHINULOSA var. ECHINULOSA, from *Pawek* 8257; 13, —O. AFFINIS subsp. FUGAX, from *Pawek* 4471.

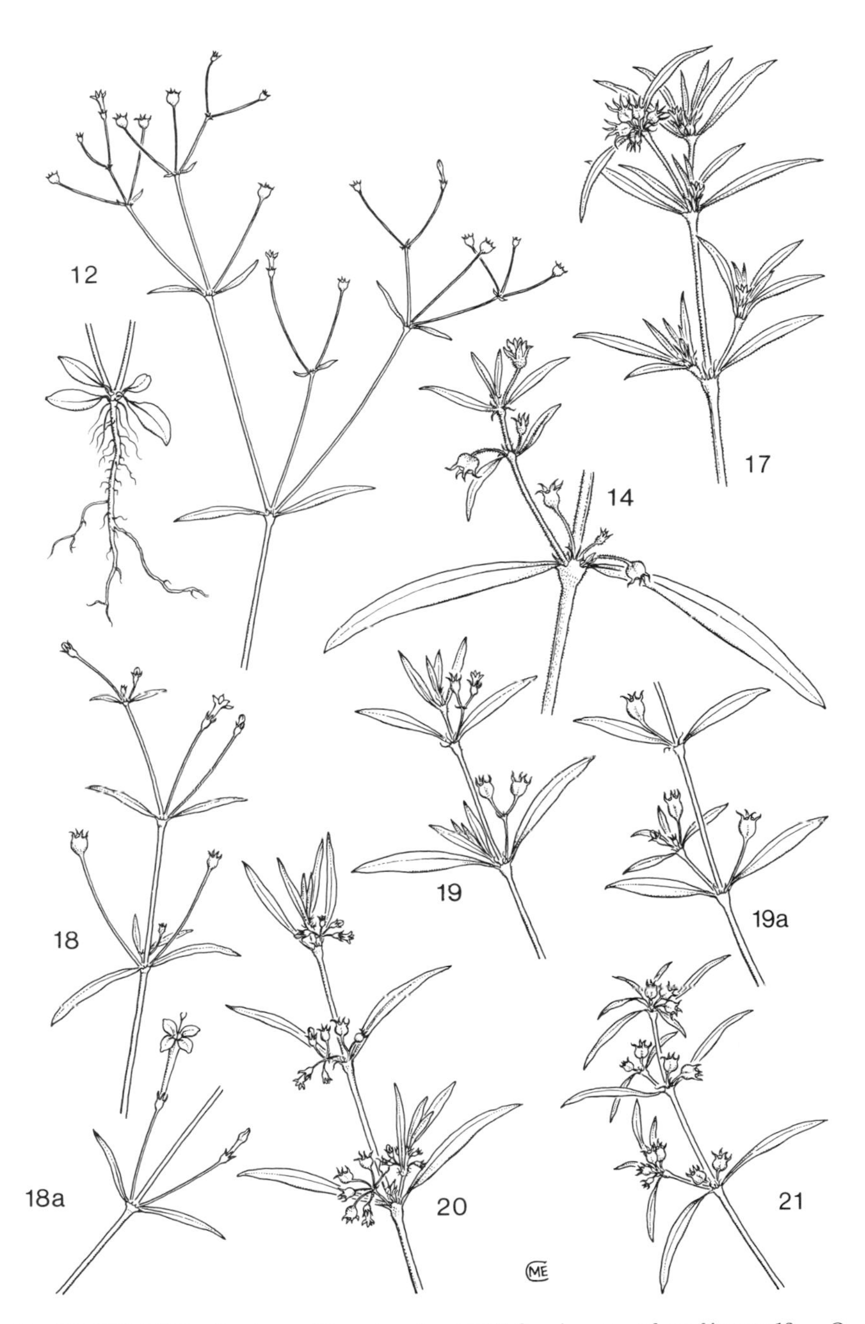

Tab. 31. OLDENLANDIA. Portions of flowering shoot, (× 1). Species as numbered in text. 12, — O. ROSULATA, from *Phillips* 1630; 14, —O. LANCIFOLIA var. SCABRIDULA, from *Richards* 8139; 17, —O. NEMATOCAULIS var. NEMATOCAULIS, from *Richards* 10990; 18, —O. HERBACEA var. HERBACEA, from *Best* 84; 18a, —O. HERBACEA var. GOETZII, from *Richards* 5126; 19, —O. CORYMBOSA var. CORYMBOSA, from *Best* 75; 19a, —O. CORYMBOSA var. CAESPITOSA, from *Pawek* 7094; 20, —O. FASTIGIATA, from *Lemos & Macuacua* 26; 21, —O. CAPENSIS, from *Mutimushi* 2333.

- Stigmata elongate; prostrate plant of sandy dambo margins and banks (Zambia, W) - - - - - - - - - - - - - - - - - -7. *geophila*
8. Erect plant with narrowly elliptic to linear-lanceolate leaves - - - - 2. *angolensis*
- Erect to prostrate plants with broader lanceolate to ovate leaves - - - - - - - 9
9. Leaves shortly scabrid-pubescent on upper surface; stipules with several fimbriae (known from one specimen) - - - - - - - - - - - - - - - - - 4. sp. A.
- Leaves glabrous or with pubescence only on margins and venation; stipules mostly bifid or with 2 lobes - 10
10. Usually erect herb with narrowly oblong-elliptic sessile leaves; stipule lobe bifid to about middle (Mozambique, M) - - - - - - - - - - - - - - - 1. *cephalotes*
- Prostrate or decumbent herb with broadly ovate to narrowly oblong-elliptic ± petiolate leaves; stipules with often more developed or separate divisions or only bifid as above (very widespread) - - - - - - - - - - - - - - - - - 3. *goreensis*
11. Flowers in dense fascicles but pedicels 1–2 becoming 7–14 mm. long; stigmatic lobes subglobose - - - - - - - - - - - - - - - - - - 5. *oxycoccoides*
- Flowers in laxer inflorescences; stigmatic lobes mostly elongate - - - - - - - 12
12. Slender annual herbs; corolla tube 0.8–3.6 mm. long; not heterostylous - - - - 13
- Perennial herbs; corolla tube 2–10 mm. long; flowers often heterostylous - - - - 14
13. Slender erect herb with basal leaves rosulate, spathulate to elliptic, often persistent and contrasting with the filiform to linear-lanceolate stem leaves; pedicels very slender 0.4–3.8 cm. long; corolla tube 1.2–3.6 mm. long - - - - - - - - - - - 12. *rosulata*
- Probably decumbent herb with all leaves ovate and without rosulate basal leaves; pedicels or apparent pedicels 0.2–1.7 cm. long; corolla tube 0.8 mm. long (Malawi, S) 24. *machingensis*
14. Testa cells reticulate and with characteristic tuberculation; leaves narrowly elliptic to lanceolate; corolla white (Mozambique) - - - - - - - - - - - - - - 15
- Testa cells not tuberculate; leaves more ovate or if narrow then flowers bright blue 16
15. Inflorescences elongate with leaves reduced or suppressed at upper nodes; corolla tube c. 2 mm. long - - - - - - - - - - - - - - - - 15. *verrucitesta*
- Inflorescences more condensed; leaves present at all nodes; corolla tube more slender 3-4 mm. long - - - - - - - - - - - - - - - - - 16. sp. B.
16. Flowers bright blue in lax divaricate paniculate inflorescences; leaves narrowly lanceolate, tapering acuminate; capsule ± exactly globose - - - - - - - - 13. *affinis*
- Flowers white or lilac in less spreading inflorescences; leaves more ovate; capsule not so perfectly globose - - - - - - - - - - - - - - - - - - - 17
17. Very intricately branched herb with very slender stems; corolla tube glabrous inside, 4–7 mm. long (Zimbabwe, E) - - - - - - - - - - - - - - - 10. *tenella*
- Less intricately branched herb with coarser stems; corolla tube hairy at the throat 18
18. Leaves and stems ± densely covered with short white spreading hairs; leaves broadly ovate, obtuse or rounded at apex, rounded-truncate at base (Zimbabwe, E) - - - - 9. *cana*
- Leaves and stems glabrescent or more sparsely pubescent; leaves not so broadly ovate, usually acute - 19
19. Leaves not fleshy, the lateral nerves invisible or clearly visible; pedicels 0.5–2.5(7) mm. long, mostly ± short; plants mostly over 10 cm. long; corolla tube 2–10 mm. long 8. *rupicola*
- Leaves somewhat fleshy, the lateral nerves not visible; pedicels 0.2–1 cm. long; plants usually small, 10 cm. long or less; corolla tube c.2 mm. long - - - - - - - 6. *hockii*
20. Flowers 1-several at the nodes, each solitary on its peduncle or only 2–3 per peduncle due to suppression of leaves near apex of the shoot - - - - - - - - - - 21
- Flowers in 2–several-flowered inflorescences or in fascicles at least at some nodes; at other nodes the flowers may be solitary - - - - - - - - - - - - - 26
21. Capsule distinctly broader than long, rather saccate at the base; spreading herb with long shoots - - - - - - - - - - - - - - - - - - 14. *lancifolia*
- Capsule ± subglobose or oblong, not at all saccate at the base - - - - - - 22
22. Pedicels longer (0.2)0.4–2.6 cm. long, the flowers never appearing ± sessile when young 24
- Pedicels mostly very short, 1–2(4) mm. long, the flowers sometimes ± sessile at first 23
23. Calyx lobes about 1 mm. long or less - - - - - - - - - - 21. *capensis*
- Calyx lobes 3–3.5 mm. long (Zimbabwe, E/S) - - - - - - - - 22. *geminiflora*
24. Capsule markedly beaked; corolla narrowly tubular 0.2–1.1 cm. long (according to variety); testa cells very strongly punctate; plant usually erect (seeds developing a distinct layer of mucilage when soaked in water*) - - - - - - - - - - - - - 18. *herbacea*
- Capsule much less beaked; testa cells in sp. 19 not strongly punctate - - - - - 25
25. Plant covered with short white spreading pubescence; corolla tube 1.5–2 mm. long; erect (Zambia, N) - - - - - - - - - - - - - - - - - - 23. sp. C
- Plant glabrous or obscurely pubescent or scabridulous, not densely covered with spreading pubescence; corolla tube shortly cylindrical, 0.6–1 mm. long; mostly prostrate - - - - - - - - - - - - - - 19. *corymbosa* var. *caespitosa*

* This character can be useful for distinguishing this species from *Oldenlandia rosulata* some specimens of which are deceptively similar.

26. Corolla 2–2.5 × 1.5 mm.; leaves elliptic or elliptic-ovate, 0.8–1.3 cm. × 5–8 mm., rather thick - - - - - - - - - - - - - - - - see 6. *hockii*
- Corolla shorter and much narrower; leaves thinner - - - - - - - - 27
27. Flowers in pedunculate 2-several-flowered cymes sometimes mixed with solitary flowers - - - - - - - - - - - - - - - 19. *corymbosa*
- Flowers in sessile to very shortly pedunculate fascicles, often 10–20 flowers at a node - - - - - - - - - - - - - - - 20. *fastigiata*

1. **Oldenlandia cephalotes** (Hochst.) Kuntze, Rev. Gen. Pl. **1**: 292 (1891). —Bremek. in Verh. K. Nederl. Akad. Wet. Afr. Natuurk., ser. 2, **48**(2): 238 (1952). Type from Natal.
Hedyotis cephalotes Hochst. in Flora **27**: 553 (1844). —Sond. in F.C. **3**: 9 (1865).
Oldenlandia sphaerocephala Schinz in Mem. Herb. Boiss. **1**: 65 (1900). Type: Mozambique, Delagoa Bay, *Junod* 400 (Z, holotype).

Annual or ?perennial herb with 1–several erect, ascending or decumbent stems to 20–50 cm. long; stems simple or sparsely branched above, glabrous or with sparsely pilose nodes and some sparse hairs beneath the inflorescence. Leaves 1.2–3.5 × 0.3–1.5 cm., linear-oblong, oblong, narrowly ovate or elliptic-oblong, acute at the apex, cuneate at the base, rather rigid, glabrous, subsessile. Stipular sheath very short, produced into a bifid lobe 2–2.5 mm. long, including the c. 1 mm. long triangular lobes. Flowers isostylous in 20–30-flowered capitate inflorescences c. 1 cm. diam., terminal and sometimes also from upper axils, sessile, if foliar organs beneath inflorescences are treated as reduced leaves; pedicels up to 1 mm. long. Calyx tube scarcely 1 mm. long, ovoid, sparsely pubescent; lobes 1.2 mm. long, ovate-triangular, with very sparsely ciliolate margins. Corolla white; tube 0.5 m. long, the lobes 1.2 mm. long. Staminal filaments and style 0.3 mm. long. Capsule broadly subglobose, 1.5–2 mm. long including 0.5 mm. beak, 2–2.3 mm. wide, the pedicels up to 2 mm. long.

Mozambique. GI: Inhambane, Homoine, Machongo Inhaliave, fl. & fr. 18.ix.1948, *Myre & Carvalho* 233 (LISC; LMA). M: Ponta d'Ouro, fl. & fr. 7.vii.1980, *Jansen & de Koning* 7305 (K).
Also in S. Africa. Grassland, in damp places; 15 m.
There is at Kew a specimen collected by Lt. J.H. Speke in 1860 at Delagoa Bay. I am not entirely convinced that this taxon is distinct from *Oldenlandia goreensis*. Bremekamp's characters erect v. decumbent and rooting at nodes; stipule lobe bifid v. bipartite (!) are very feeble and, moreover, both species are recorded from Magaia in Mozambique, *Schlechter* 12049 as *O. cephalotes* and *Schlechter* 12045 as *O. goreensis*. Study in the field is needed to confirm that two species do occur. One of the cited specimens is described as "ca. creeping herb".

2. **Oldenlandia angolensis** K. Schum. in Engl., Bot. Jahrb. **23**: 412 (1897). —Bremek. in Verh. K. Nederl. Akad. Wet. Afd. Natuurk., ser. 2, **48**(2): 195 (1952). —Verdc. in F.T.E.A., Rubiaceae 1: 278 (1976). TAB. **30**. Type from Angola.

Short-lived perennial (or ? sometimes annual) herb 15–60 cm. tall; stems erect or suberect, mostly sparsely branched, slender, glabrous (save at the nodes) or rarely hairy. Leaf blades (0.3)1–2 cm. × 1–4.5 mm., narrowly elliptic to linear-lanceolate, subacute at the apex, cuneate at the base, rather rigid and held erect, mostly drying a distinct yellowish-green and with surface finely reticulate, glabrous but margins very slightly scabridulous or rarely hairy all over; petiole obsolete or very short; stipule sheath c. 1 mm. long, usually with a few hairs, rarely densely hairy, produced into a 2-fid lobe 1.5–3 mm. long. Inflorescences dense, subcapitate, 5–15-flowered, 5–8 mm. in diam., terminal and also in leaf axil below the terminal head and also sometimes on axillary side branches from terminal and lower nodes, usually sessile but peduncles up to 5 cm. long; pedicels 0.5–1.5(3) mm. long. Flowers not dimorphic, equal-styled. Calyx tube 1 mm. long, subglobose, glabrous to sparsely or rarely spreading hairy; ± 1–1.2 mm. long, lobes deltoid, glabrescent or with rather long white marginal hairs or rarely more densely covered with spreading hairs. Corolla white to dull mauve; tube 0.5 mm. long; lobes 1–1.3 mm. long, ovate-lanceolate. Capsule purple, subglobose, 1.5 × 2 mm.; beak slightly raised, 0.5 mm. long, glabrescent to spreading hairy. Seeds black, angular, 0.4–0.5 mm. long.

Var. **angolensis**. —Verdc. in F.T.E.A., Rubiaceae 1: 278, fig. 40/1 (1976).

Plant glabrescent or with sparse indumentum only.

Botswana. N: Okavango Swamp, Xhere Lediba, fl. 15.ii.1973, *Smith* 391 (K; SRGH). **Zambia**. B: 16 km. E. of Mongu, fl. 24.x.1965, *Robinson* 6690 (K). N: Lake Chila, fl. & fr. 24.ix.1949, *Bullock* 1072 (K). W: 37 km. W. of Mwinilunga on Matonchi road, fl. & fr. 24.i.1975, *Brummitt et al.* 14094 (K).

Zimbabwe. N: 4.8 km. SW. of Gokwe, Chimvuri Vlei, fl. 12.xi.1963, *Bingham* 893 (K; SRGH). C: Harare, fl. & fr. 20. ii.1927, *Eyles* 4725 (K; SRGH). E: Chimanimani, upper Haroni, fl. 31.i.1957, *Phipps* 253 (K; SRGH). **Malawi**. N: Rumphi Distr., Nyika Plateau, Chelinda Bridge, fl. & fr. 29.iii.1970, *Pawek* 3424 (K).

Also in Cameroon, Zaire, Uganda, Tanzania and Angola. *Miscanthidium* swamp edges, boggy grassland, muddy dambos, riverine marshes, etc.; 1160–2250 m.

3. **Oldenlandia goreensis** (DC.) Summerh. in Kew Bull. **1928**: 392 (1928). —Bremek. in Verh. K. Nederl. Akad. Wet., Afd. Natuurk. ser. 2, **48**(2): 196 (1952). —Hepper, F.W.T.A. ed. 2, **2**: 211 (1963). —Lind & Tallantire, Fl. Pl. Uganda ed. **2**: 159 (1972). —Agnew, Upland Kenya Wild. Fl.: 400 (1974). —Verdc. in F.T.E.A., Rubiaceae 1: 279 (1976). TAB. **30**. Type from Senegal.

Hedyotis goreensis DC., Prodr. **4**: 421 (1830).

Oldenlandia trinervia sensu Hiern in F.T.A. **3**: 63 (1877); Cat. Afr. Pl. Welw. **2**: 449 (1898). —K. Schum. in Engl., Pflanzenw. Ost-Afr. **C**: 372 (1895) non Retz.

Annual or possibly sometimes a short-lived perennial herb with prostrate, decumbent or ascending stems (7)10–40(90) cm. long, branched at the base, sometimes rooting at the nodes, glabrescent to sparsely hairy. Leaf blades 0.5–2.5(4.5) × 0.3–1.5(2) cm., elliptic to ovate, obtuse or subacute at the apex, rounded or cuneate at the base, glabrous save for the scabridulous to ciliate margins and appressed hairs on the midvein and some nerves; petiole 0.5–2 mm. long, glabrous or ciliate; stipule sheath 0.7–1(2) mm. long, produced into a 2-fid lobe 1–2 mm. long, ciliate. Inflorescences terminal, sessile, 9–25-flowered, subglobose, those on reduced side branches in lower axils appearing like axillary fascicles; pedicels 1–3 mm. long, glabrous to pilose. Flowers not dimorphic, 4-merous or rarely 5–6-merous. Calyx tube c. 1 mm. long, subglobose, glabrous to spreading pubescent; lobes 1–1.5 × c. 0.5 mm., narrowly triangular, ciliate. Corolla white or less often pink or red, glabrous to pilose outside; tube 0.3 mm. long, sparsely pilose inside; lobes 0.8–1.2 × 0.7 mm., ovate-triangular. Capsule subglobose, 1.5 × 2 mm., glabrescent to pilose; beak slightly raised, puberulous. Seeds black, angular, 0.4 mm. long.

Var. **goreensis** Bremek, in Verh. K. Nederl. Akad. Wet., Afd. Natuurk., ser. 2, **48**(2): 197 (1952). —Verdc. in F.T.E.A., Rubiaceae 1: 279 (1976).

Stems glabrous or glabrescent.

Botswana. N: Okavango Swamp, Kwando R., fl. 5.xi.1980, *Smith* 3556 (K; SRGH). **Zambia**. B: Kalabo, fl. & fr. 14.x.1962, *Robinson* 5489 (K). N: Mbala, Chilongowelo, fl. & fr. 11.xi.1954, *Richards* 2206 (K). W: Ndola, fl. & fr. 2.i.1960, *Robinson* 3278 (K). C: Great E. Road, 38 km. E. of Lusaka — Chinkomba, fr. 2.x.1971, *Kornaś* (K; KRA). **Zimbabwe**. N: Sengwa Research Station, fl. & fr. 18.x.1968, *Jacobsen* 258 (K; LISC; SRGH). W: Hwange Distr., near Gwaai R. Hotel, fl. & fr. 9.vii.1970, *Rushworth* 2520 (K; SRGH). E: Vumba, Burma Valley, Bomponi, fl. & fr. 3.xii.1961, *Wild & Chase* 5547 (K; LISC; SRGH). S: Chibi Distr., 6.4 km. N. of Lundi R. Bridge, near Madzivire Dip, fl. & fr. 3.v.1962, *Drummond* 7881 (K; LISC; SRGH). **Malawi**. N: Livingstonia, Kaziweziwe R., fl. & fr. 8.i.1959, *Robinson* 3118 (K). C: 6.4 km. S. of Dedza, fl. & fr. 2.ii.1959, *Robson* 1425 (BM; LISC; K). S: Mulanje Mt., foot of Great Ruo Gorge, fl. & fr. 18.iii.1970, *Brummitt & Banda* 9202 (K; MAL). **Mozambique**. N: Marrupa to Lichinga, km. 25, fr. 10.viii.1981, *Jansen et al.* 165 (K; WAG). MS: Inyamadzi Valley, fl. & fr. 8.ix.1906, *Swynnerton* 1534 (BM; K).

Also in W. Africa from Senegal to Angola, Zaire, Burundi, Sudan, Ethiopia, East Africa to South Africa (Transvaal), Madagascar, Seychelles and Mascarene Is. Permanently damp areas, riverine marshland, *Miscanthidium* swamp, seasonally flooded sandy areas by lakes, water-holes, etc., muddy banks, irrigation ditches, etc.; 650–1850 m.

Var. **trichocaula** Bremek. in Verh. K. Nederl. Akad. Wet., Afd. Natuurk., ser. 2, **48**(2): 198 (1952). —Verdc. in F.T.E.A., Rubiaceae 1: 279 (1976). Type from Togo.

Stems sparsely to fairly densely spreading hairy.

Zimbabwe. N: Mwenezi, SW. Mateke Hills, Malangwe R., fl. & fr. 6.v.1958, *Drummond* 5593 (K; LISC; SRGH). **Malawi**. N: Mzimba Distr., 4.8 km. W. of Mzuzu, Katoto, fl. & fr. 3.viii.1973, *Pawek* 7296 (K; MO). **Mozambique**. Z: Mocuba, Namagoa, fl. & fr. vii.1945, *Faulkner* 239 (K; PRE). M: Magala, fr. 17.v.1898, *Schlechter* 12045 (K).

Also in W. Africa, Kenya and Tanzania. Sandy boggy areas, moist streamsides, etc.; 15–1350 m.

Bremekamp (op. cit.: 200 (has cited *Faulkner* 239 as *Oldenlandia verticillata* Bullock ex Brem. var. *trichocarpa* Brem. (i.e. *O. bullockii* Brem.) but the K specimen at least is certainly *O. goreensis*. No material of *O. bullockii* has been seen from the Flora Zambesiaca area. *Mogg* 29081 (Mozambique. GI: Bazaruto Is., fr. 6.xi.1958 (LISC)) a small annual ± 8 cm. tall with small leaves with margins appearing ± serrate due to marginal hairs is probably only a depauperate form of this variety but further collections are needed.

4. **Oldenlandia** sp. A.

Stems weak, unbranched save for a few short apical side-shoots, square with marked edges, glabrous save for very slight pubescence in grooves above, c. 35 cm. tall. Leaves 0.7–1.4 × 0.3–0.9 cm., elliptic to narrowly ovate, acute at the apex, narrowed into a short petiole up to 4 mm. long, slightly scabrid with short hairs above, on margins and on the few lateral nerves beneath; stipules with 3–5 fimbriae to 3 mm. from a short base. Flowers in small terminal heads; peduncles 0.3–1.5 cm. long; pedicels obsolete; bracts filiform-lanceolate, up to 1.5 mm. long. Calyx tube c. 1 mm. long, obconic, glabrous save for few short emergences; lobes about 1.2 mm. long, lanceolate, ciliate with short hairs. Corolla similar to *O. goreensis* (DC.) Summerh. Capsule unknown.

Zambia. S: Lukanda Valley, fl. *Allen* 458 (K).

Only one specimen known. 'Wetland'.

The locality is given on the label as "Rhodesia, Lukanda Valley"; there is a Lukampa Valley in Nkai District of Zimbabwe W. but this seems unlikely. Lukanda has not been traced. Bremekamp has annotated the sheet "This is doubtless a new species differing from *O. goreensis* in the greater length of the corolla, much longer filaments and longer anthers". He did not mention the specimen in his revision. As he states on the label it is best left undescribed until more material turns up; apparently it has never been recollected.

5. **Oldenlandia oxycoccoides** Bremek. in Kew Bull. **13**: 382 (1959). —Verdc. in F.T.E.A., Rubiaceae 1: 281, fig. 40/4 (1976). Type from Tanzania.

Prostrate short-lived perennial mat-forming herb, with ± glabrous or pubescent stems (2)10–30 cm. long rooting at the nodes. Leaf blades 4–8 × 2–5 mm., ovate, subacute to rounded at the apex, rounded at the base, often rather thick with only midrib visible, shortly ciliate around the margins but otherwise glabrous; petiole 0.5–1 mm. long; stipule sheath and lobe 0.4–1 mm. long, produced into 2 fimbriae 0.2–0.5 mm. long, ciliate. Flowers heterostylous, 1–5, terminal or axillary, in fascicles without a common peduncle; pedicels 1–2 mm. lengthening to 7(14) mm. after flowering. Calyx tube 0.8 mm. long, obconic, glabrous or sparsely pubescent; lobes 1 × 0.5 mm., oblong-triangular. Corolla white with pinkish-mauve limb or reddish green or pink; tube 0.2 mm. long; lobes 1.5 × 1 mm., triangular, glabrous. Style in short-styled flowers 0.1 mm. long, in long-styled flowers 1.2 mm. long; stigma lobes 0.1–0.2 mm. in diam., subglobose. Capsule 1 × 1.7 mm., ± glabrous, not beaked. Seeds dark brown.

Var. **A**.

Stems spreading pubescent. Some leaves at least more narrowly ovate and more acute.

Zambia. W: 2 km. along Kanyama road from Mwinilunga–Solwezi road, fl. & fr. 25.ii.1975, *Hooper & Townsend* 369 (K).

Also in S. Tanzania. Wet lake-edges; 1400 m.

Despite having some characters close to *O. bullockii* Bremek. there is no doubt that this matches some specimens of *oxycoccoides* from Tanzania. Too little material of the species is available to assess variation — altogether only 5 specimens have been collected. *Ricardo* 123 (Zambia N: Shiwa Ngandu, fr.17.i.1937) annotated *O. rupicola* var. *psilogyna* by Bremekamp seems to be nearer this species.

6. **Oldenlandia hockii** De Wild. in Fedde. Repert. **13**: 108 (1914). —Bremek. in Verh. K. Nederl. Akad. Wet. Afd. Natuurk., ser. 2, **48**(2): 203 (1952). —Verdc. in F.T.E.A., Rubiaceae 1: 285, fig. 49/9 (1976). TAB. **30**. Type from S. Zaire.

Small presumably decumbent perennial herb 4–9 cm. long; stems very shortly scabridulous-pubescent. Leaves blue-green or purplish-green, 0.8–1.3 cm. × 5–8 mm., elliptic or elliptic-ovate, subacute at the apex, ± rounded at the base, somewhat fleshy, revolute, glabrous or minutely scabrid above, glabrous beneath, discolorous, the rhaphides particularly evident on the lower surface; midvein evident but no lateral veins visible; petioles c. 1 mm. long; stipule sheath c. 1 mm. long, divided into c. 3 fimbriae 0.8 mm. long. Flowers heterostylous, in several-flowered erect axillary ± dichasial cymes; peduncles (0.1)1–2.5 cm. long; pedicels 0.2–1 cm. long, glabrous or minutely pubescent. Calyx tube 1.5 mm. long, campanulate; lobes 1–1.8 × 0.5–0.8 mm., ovate, mostly minutely scabrid-ciliate on the margins. Corolla white; tube 2.1–2.5 mm. long, hairy at the throat; lobes 1.5–1.8 × 1.3 mm., ovate, apiculate, densely papillate inside; style hairy, 1.3 mm. long in short-styled flowers, 3 mm. long in long-styled flowers; stigma lobes oblong or elliptic, 0.6–0.8 mm. long, densely papillate-hairy. Ripe capsule not seen.

Malawi. C: Dedza, Chongoni Forest Reserve, fl. 6.xi.1967, *Salubeni* 866 (K; SRGH).
Also in S. Zaire and W. Tanzania. Ecology not known, elsewhere it occurs in burnt ground in *Brachystegia* woodland and open bushland.
I am not certain about the naming of this specimen but the obscure nervation and long pedicels suggest it belongs to this species and in general facies it is very similar.

7. **Oldenlandia geophila** Bremek. in Verh. K. Nederl. Akad. Wet., Afd. Natuurk., ser. 2, **48**(2): 20 (1952). TAB. **30**. Type: Zambia, Mufulira, *Eyles* 8131 (K, holotype; SRGH).

Prostrate matted much branched herb 10–45 cm. long; stems ± glabrous, 4-angled. Leaves 6–12 × 3.5–7 mm., ovate or elliptic, acute at the apex, cuneate at the base, scabrid above and less obviously so beneath on the midrib and nerves which are distinctly visible; petiole c. 1 mm. long; stipule sheath scarcely 1 mm. long, usually with two subulate fimbriae 1–1.5 mm. long. Flowers in triads or few-flowered cymes at the apex of the stems or rarely solitary; peduncle up to 1.5 cm. long; pedicels c. 1 mm. long. Calyx tube c. 1 mm. long, obovoid, glabrous; lobes 1–1.8 × 0.7 mm., linear to narrowly ovate, glabrous or margins appearing minutely serratulate due to short emergences. Corolla greyish-blue or purple; tube 0.8–1 mm. long, pilose inside at base and near throat; lobes 1 × 1 mm. long, broadly ovate. Flowers apparently isostylous; style glabrous, 1.6 mm long, the anthers shortly exserted. Capsule 1–1.5 × 1.5–2 mm., depressed subglobose, scarcely beaked, glabrous.

Zambia. W: Luano, fl. & fr. 9.v.1967, *Mutimushi* 1969 (K; NDO).
Not known elsewhere. Sandy dambo margins, drain-banks; 1400 m.
Bremekamp assumed the flowers were heterostylous but I believe they are isostylous with the stigmatic lobes either just below or just above the anther tips; this needs examining in the field. Only 3 specimens have been seen.

8. **Oldenlandia rupicola** (Sond.) O. Kuntze, Rev. Gen. Pl. **1**: 293 (1891). —Bremek. in Verh. K. Nederl. Akad. Wet., Afd. Natuurk., ser. 2, **48**(2): 208 (1952). —Verdc. in F.T.E.A., Rubiaceae 1: 284, fig. 40/8 (1976). TAB. **30**. Type from South Africa (Natal).
Hedyotis rupicola Sond. in Harv. & Sond., F.C. **3**: 12 (1865).

Perennial procumbent usually mat-forming herb, with glabrous to hairy much-branched stems often rooting at the nodes, or scrambling to suberect, 0.8–1.4 m. long, or prostrate with erect flowering shoots (2.5)7–10 cm. tall. Leaf blades 0.3–1.2(3.7) cm. × 3–9(19) mm., elliptic-ovate to ovate, acute at the apex, rounded to cuneate at the base, sometimes rather thick, scabridulous to hairy above, glabrous beneath or sometimes slightly hairy near the margin, rarely pubescent on midrib; petiole 0.5–2(5) mm. long, glabrous to sparsely hairy; stipule sheath 0.5–1 mm. long, glabrous to sparsely hairy, with 5–7 lobes 1–3 mm. long. Flowers heterostylous in terminal triads or few- to several-flowered cymes, less often solitary; peduncle 0–1.5(2) cm. long; pedicels 0.5–2.5(7) mm. long, glabrous. Calyx tube 1–1.5 mm. long, obconic, glabrous, hairy or with sparse to dense subglobose papillae, sometimes purple-tipped; lobes 1–4 × 0.7–1 mm., usually broadly triangular, less often ovate or lanceolate, keeled, glabrous or sometimes hairy on the midvein and margins, free or connate at the base. Corolla white, pale blue, lilac or pink; tube 2–7(10) mm. long, hairy in the throat; lobes 1.5–5.5 × 1.2–2.5 mm., triangular to oblong-lanceolate, hairy inside. Style 2–3(6) mm. long in short-styled forms, 5–8(12) mm. long in long-styled forms; stigma lobes 0.3–0.7 mm. long. Capsule 2–2.5 × 2.2–3 mm., subglobose, glabrous, papillate or hairy, beak slightly raised. Seeds blackish, c. 0.5(0.8) mm. long, elliptic, flat on one side, somewhat conically raised on the other, strongly reticulate.

Var. **rupicola**. —Bremek. in Verh. K. Nederl. Akad. Wet., Afd. Natuurk., ser. 2, **48**(2): 209 (1952). —Verdc. in F.T.E.A., Rubiaceae 1: 284 (1976).
Oldenlandia oliveriana K. Schum. in Engl., Hochgebirgsfl. Trop. Afr.: 397 (1892); in Engl., Pflanzenw. Ost-Afr. **C**: 375 (1895). Type: Mozambique, Namuli Plateau, Makua Country, *Last* (B, holotype †; K).
Oldenlandia junodii Schinz in Viert. Nat. Ges. Zürich **52**: 431 (1907) non Schinz (1900) nom. illegit. Type from S. Africa (Transvaal).
Oldenlandia rogersii S. Moore in Journ. Bot., Lond. **59**: 229 (1921). Type from S. Africa (Transvaal).
Oldenlandia rupicola var. *psilogyna* Bremek. loc. cit. Type as for *O. oliveriana*.
Oldenlandia rupicola var. *parvifolia* Bremek. loc. cit. Type as for *O. junodii*.
Oldenlandia greenwayi Bremek. tom. cit. 207 (1952). Type from Tanzania.

Stems glabrescent. Ovary glabrous to densely covered with short or subglobose papillae.

Zimbabwe. C: Shurugwe Peak, fl. 19.iii.1964, *Wild* 6436 (K; LISC; SRGH). E: Chimanimani Mts., fl. 20.viii.1954, *Wild* 4586 (K; SRGH). **Malawi**. S: Mulanje Mt., path to Luchenza Plateau, fl. 5.vi.1962, *Richards* 16517 (K). **Mozambique**. Z: Namuli, Makua Country, *Last* (K). MS: Chimanimani Mts., fl. 7.vi.1949, *Wild* 2935 (K; SRGH).

Also in Kenya, Tanzania and South Africa. Woodland edges, *Widdringtonia* forest clearings, etc., also in grassland and on wet rocks; 425–2250 m.

There is a good deal of variation in the indumentum of the ovary, etc., and certainly a great deal in the size of the leaves and corolla; I have not been able to retain Bremekamp's varieties cited in synonymy above; var. *hirtula* (Sond.) Bremek. which occurs in S. Africa has hairy stems, leaves, calyx tubes and capsules, and often has much larger leaves and flowers. The species as a whole is very variable and extremes can appear markedly different, e.g. *Wild* 4586 with leaves c. 2.5 × 1.5 cm. (cited above) and *Loveridge* 476 from Domboshawa Rock near Harare with leaves c. 7 × 5 mm. *Torre & Paiva* 11374, Mozambique, N, Serra de Ribáuè (Mepaluè), fl. & fr. 23.iii.1964 (LISC) differs from other material in its facies, more spicate inflorescences, leaves puberulous on both sides and papillate fruits but all these characters can be found in some specimens throughout the range of this very variable species and it is at best a distinctive local variant.

9. **Oldenlandia cana** Bremek. in Kew Bull. **25**: 186 (1971). TAB. **30**. Type: Zimbabwe, Chimanimani Mts., below Digby's Pool, *Grosvenor* 373 (K; SRGH; U, holotype).

Decumbent branched herb with slender ± unangled stems 15–25 cm. long, densely covered with short spreading white hairs. Leaves 3–10 × 2.5–8 mm., broadly ovate, rounded but very shortly obscurely apiculate at the apex, rounded-truncate at the base, densely pubescent with ± adpressed white hairs; petiole 1–2 mm. long; stipular-sheath short with few filiform fimbriae 1 mm. long. Flowers isostylous, truly terminal and solitary, solitary in the axils or sometimes in terminal pairs or threes; pedicels 1–4 mm. long, becoming up to 1 cm. long in fruit, pubescent like the stems. Calyx tube similarly pubescent, 1 mm. long, ovoid; lobes foliaceous, about 1 mm. long and wide, enlarging to 1.4 mm. after flowering, ovate to lanceolate, acuminate, pubescent. Corolla densely pubescent outside; tube just over 4 mm. long, glabrous inside; lobes 2.5 × 1.5 mm., ovate. Style 2.5 mm. long; stigmatic arms 0.6 mm. long. Capsule 1.8 × 2.2 mm., depressed obovoid, densely covered with spreading white hairs. Seeds black with reticulate testa.

Zimbabwe: E: Chimanimani Mts., just below Digby's Pool, fl. & fr. 10.iv.1967, *Grosvenor* 373 (K; SRGH; U).

Endemic. Habitat not known; 1650 m.

10. **Oldenlandia tenella** (Hochst.) O. Kuntze, Rev. Gen. Pl. **1**: 293 (1891). —Bremek. in Verh. K. Nederl. Akad. Wet., Afd. Natuurk. ser. 2, **48**(2): 211 (1952). Type from S. Africa (Natal).
Hedyotis tenella Hochst. in Flora **27**: 553 (1844). —Sond. in Harv. & Sond. F.C. **3**: 13 (1865).

Decumbent or hanging, very intricately branched plant, 30 cm. long; stems very slender, ± 4-angled, glabrous. Leaves 0.3–1.4 cm. × 2.5–9.5 mm., elliptic or ovate, obtuse or subacute at the apex, rounded then abruptly narrowed at the base into a petiole up to 3 mm. long, mostly very thin, glabrous; stipule sheath 0.5 mm. long with 2 or 3 very short lobes or fimbriae under 1 mm. long. Flowers heterostylous, 1–2(3) terminal and also truly axillary ones; pedicels slender, 1.5–9 mm. long becoming up to 11 mm. long in fruit. Calyx glabrous; tube ± 0.7 mm. long, obovoid; lobes foliaceous, 0.7–1.1 mm. long, accrescent in fruit to 1.7 mm., lanceolate to ovate, ± acuminate. Corolla white or pale violet; tube 4–5 mm. long, ± slender, glabrous inside; lobes c. 2.2–5 × 2.3 mm. Filaments 0.2 mm. long in long-styled flowers, 2 mm. long in short-styled flowers and style 5 and 2.5 mm. long respectively. Capsule 1.8–2 × 1.8–2.2 mm., obovoid, glabrous, crowned by the ± venose calyx lobes. Seeds black with carunculate testa.

Zimbabwe. E: Chimanimani, Silver Streams, fl. & fr. 29.v.1957, *Whellan* 1287 (K; LISC; SRGH).
Also in South Africa (Natal and Transvaal). In rock crevices, in forest; c. 1200 m.

11. **Oldenlandia echinulosa** K. Schum. in Pflanzenw. Ost-Afr. **C**: 375 (1895). —Bremek. in Verh. K. Nederl. Akad. Wet., Afd. Natuurk., ser. 2, **48**(2): 213 (1952). —Verdc. in Kew Bull. **32**: 608 (1978). Type: S. Malawi, without locality, *Buchanan* 498 (B, holotype †; K).
Oldenlandia pellucida var. *echinulosa* (K. Schum.) Verdc. in Kew Bull. **30**: 292 (1975); in F.T.E.A., Rubiaceae 1: 286 (1976) non rite publ.

Erect branched or unbranched annual herb 2–40 cm. tall; stems glabrous or scabrid-papillate on the ribs. Leaf blades 1.2–6.5 × 0.6–2.8 cm., elliptic-lanceolate, acute to acuminate at the apex, cuneate at the base, thin, sparsely very shortly strigose-pubescent above, later glabrescent, margin usually pubescent, glabrescent beneath save for the midvein; petiole 2–5(10) mm. long, glabrous or sparsely ciliate; stipule sheath 0.5–1 mm. long, usually produced into a lobe 1.5 mm. long, bearing 2–5 fimbriae 1–3 mm. long. Flowers not heterostylous, arranged in small heads 4 mm. wide, terminal or axillary, the heads often in threes; peduncle 0.2–1.5 cm. long; pedicels 0–1.2 mm. long. Calyx tube 0.5 mm. long, subglobose, densely covered with short spreading hairs; lobes 0.7–2 × 0.5 mm., lanceolate, ciliolate on the margins and midvein. Corolla white or tube green and lobes pale lilac, sometimes with a double mauve spot on each; tube 3.2–4.5 × 0.4–0.9 mm., very narrowly cylindrical; lobes 0.4–1.8 × 0.4–1.2 mm., broadly elliptic. Style 1.7–5 mm. long; stigmatic lobes 0.4–1.2 mm. long. Capsule ± biglobose, 1.5 × 2.2 mm., densely covered with short spreading bristly hairs. Seeds black, c. 0.5 mm. long, ellipsoid, strongly reticulate.

Var. **echinulosa**

Oldenlandia nesaeoides Hiern in Cat. Welw. Afr. Pl. **1**: 448 (1898). TAB. **30**. Type from Angola.

Plants often more robust and branched, up to 40 cm. tall, but sometimes unbranched; leaves less tenuous; calyx lobes up to 2 mm. long; corolla tube up to 4.5 mm. long; anthers usually included.

Zambia. N: Mbala, Ndundu to Kawimbe road, fl. & fr. 27.iv.1966, *Richards* 21460 (K). W: Solwezi, fl. & fr. 9.iv.1960, *Robinson* 3475 (K). S: Mumbwa, *Macaulay* 614 (K). **Zimbabwe**. N: Miami, fl. iv.1926, *Rand* 85 (BM). E: Mutare, Odzani R. Valley, *Teague* 526 (BOL; K). **Malawi**. N: Nkhata Bay Distr., 8 km. E. of Mzuzu, Rose Falls, fl. & fr. 2.iv.1969, *Pawek* 1911 (K). C: Dedza, Chongoni Forest, fl. & fr. 17.iii.1969, *Salubeni* 1267 (K; SRGH). S: Mt. Zomba Plateau, *Whyte* (K).

Also in Zaire, Burundi, Sudan, Tanzania and Angola. *Brachystegia* woodland, grassland, road verges, stream banks, dambos, etc., also in gardens and cultivations and sometimes on termite mounds, rocky gorge sides; (?510)900–1800(?2100) m.

Var. **pellucida** (Hiern) Verdc. in Kew Bull. **32**: 608 (1978). Type from Angola.

Oldenlandia pellucida Hiern in Cat. Afr. Pl. Welw. **1**: 448 (1898). —Bremek. in Verh. K. Nederl. Akad. Wet., Afd. Natuurk. ser. 2, **48**(2): 212 (1952). —Verdc. in F.T.E.A., Rubiaceae 1: 286 (1976).

Oldenlandia golungensis Hiern in Cat. Afr. Pl. Welw. **1**: 451 (1898). —Bremek. tom. cit.: 214 (1952). Type from Angola.

Oldenlandia pellucida var. *robustior* Bremek. tom. cit.: 213 (1952). Type from Tanzania.

Oldenlandia echinulosa sensu Hepper, F.W.T.A. ed. 2, **2**: 211 (1963) non K. Schum. sensu stricto.

Oldenlandia pellucida var. *pellucida* —Verdc. in F.T.E.A., Rubiaceae 1: 286, fig. 40/11 (1976).

Plants mostly weak and relatively unbranched 4–35 cm. tall; leaves very tenuous; calyx lobes 0.7–1.2 mm. long; corolla tube 1.5–3 mm. long; anthers usually very slightly exserted.

Zambia. N: Kasama Distr., Chishimba Falls, fl. & fr. 31.iii.1955, *Exell et al.* 1359 (BM; LISC). W: 3 km. S. of Solwezi, Chifubwa R. Gorge, fl. 20.iii.1961, *Drummond & Rutherford-Smith* 7112 (K; SRGH). **Zimbabwe**. N: Darwin, Umsengezi Swamp, fr. 9.vi.1955, *Whellan* 878 (K; SRGH). C: Harare, Rumani, fr. 10.v.1948, *Wild* 2528 (K; SRGH). **Malawi**. N: Mzuzu, Marymount toward Tung Estate, fl. & fr. 14.v.1972, *Pawek* 5371 (K).

Also in Sierra Leone, Nigeria, Cameroon, Sudan, Tanzania and Angola. Damp rocks by rivers and streams, damp humus on granite cave floors, mossy banks and also roadsides; 600–1800 m.

Although extremes of the two taxa mentioned above are very distinctive, intermediates occur; the floral characteristics mentioned by Bremekamp do not always work. There is a tendency for var. *pellucida* to occur in wild habitats and var. *echinulosa*, the more robust weedy variant, to occur in disturbed habitats. Most W. African material seems to be nearer var. *pellucida*.

12. **Oldenlandia rosulata** K. Schum. in Engl., Bot. Jahrb. **23**: 416 (1897). —Hiern, Cat. Afr. Pl. Welw. **1**: 447 (1898). —Bremek. in Verh. K. Nederl. Akad. Wet., Afd. Natuurk. ser. 2, **48**(2): 225 (1952). —Verdc. in F.T.E.A., Rubiaceae 1: 290 (1976). TAB. **31**. Type from Angola.

Slender erect annual herb 2.5–35 cm. tall, with glabrous or basally scabrid pubescent stems. Basal leaves rosulate, the blades 3–7 × 1.3–3 mm., spathulate, lanceolate or elliptic, obtuse, glabrous, often eventually disappearing before maturity; stem leaves 0.3–3 cm. × 0.3–2.2 mm., filiform, linear or linear-lanceolate, glabrous; petioles obsolete; stipule sheath 0.3 mm. long with 2 short teeth 0.5 mm. long. Flowers heterostylous or isostylous, terminal or axillary, either solitary or in groups of 2 or 3, in larger plants forming a paniculate inflorescence; peduncles 0.9–2.8 cm. long; pedicels very graceful, 0.4–3.8 cm. long. Calyx tube 0.5–0.7 mm. long, subglobose, glabrous or with minute hairs; lobes 0.3–0.9 mm. long, broadly triangular. Corolla white to pale purple, sometimes with a dark spot or streaks near the throat; tube 1.2–3.6 mm. long, somewhat hairy at the throat with white or purple hairs; lobes 0.7–2 mm. long, oblong-ovate. Style 1.2 mm. long in short-styled forms and 2.9 mm. in long-styled forms; stigma lobes 0.6–0.8 mm. long; isostylous forms have the style of intermediate length and the stigma lobes are at the same height as or higher than the anthers. Capsule 1–1.8 × 2 mm., subglobose, glabrous, beaked, the valves rounded and well separated after dehiscence. Seeds black, angular, c. 0.4 mm. long, reticulate.

Var. **parviflora** Bremek. in Verh. K. Nederl. Akad. Wet., Afd. Natuurk. ser. 2, **48**(2): 225 (1952). —Verdc. in F.T.E.A., Rubiaceae 1: 290, fig. 40/14 (1976). Type from Rwanda.

Flowers isostylous, often small, with corolla tube 1.2–2(3) mm. long and lobes 0.7 mm. long.

Zambia. N: Mbala, fl. & fr. 3.iv.1963, *Richards* 18060A (K). C: Kabwe, fl. & fr. 23.ix.1947, *Brenan & Greenway* 7926 (EA; K). **Malawi**. N: Nyika Plateau, Businande road, fl. & fr. 3.iv.1976, *Phillips* 1630 (K; MO). C: Dedza Distr., Chongoni Forest Reserve, foot of Chencherere Hill, fl. 25.iv.1970, *Brummitt* 10146a (K; MAL). S: Zomba Plateau, road to summit just before turning to Chingwe's Hole, fl. 16.ii.1970, *Brummitt* 9173 (K; MAL). **Mozambique**. N: Amaramba, W. side of Serra Mitucué, 20 km. from Nova Freixo (Cuamba) fl. & fr. 15.ii.1964, *Torre & Paiva* 10614 (C; COI; LISC; LMU; SRGH; WAG); T: Macanga, Mt. Furancungo, fl. & fr. 15.iii.1966, *Pereira et al.* 1681 (BR; LMU).

Also in Rwanda, Burundi, Uganda and Tanzania. Grassland (sometimes with *Protea* etc.), dambos, stream-banks, wet flushes on hillsides, shallow soil on rocks, *Brachystegia–Uapaca* woodland, also by paths in old cultivations, etc.; 1260–2300 m.

I am not at all certain of the validity of Bremekamp's varieties; certainly most of the East African material is isostylous. Some populations may, however, be uniformly long-styled; *Milne-Redhead & Taylor* 9007 (Tanzania, Songea District, Matengo Hills, about 1.5 km. N. of Miyau by R. Utili, 2 Mar. 1956) shows 9 such plants on one sheet and a further sheet from the same area is similar. This may correspond to var. *rosulata*; I have not seen two forms in any collection examined from other areas – all plants have been dolichostylous. *O. rosulata* and *O. herbacea*, particularly weak forms of the latter which may even have some broader lowermost leaves (? cotyledons) preserved, have been confused. They can always be distinguished if ripe seeds are available. Those of *O. herbacea* immersed in water for an hour develop a thick coating of mucilage whereas those of *O. rosulata* do not; furthermore the testa cells of *O. herbacea* are very coarsely granulate whereas those of *O. rosulata* are smooth, the differences being so marked that they can be seen by viewing the seed under high magnification without making microscopical preparations. Some robust large-leaved specimens have been referred to *Kohautia confusa* (Hutch. & Dalz.) Bremek.

13. **Oldenlandia affinis** (Roem. & Schult.) DC., Prodr. **4**: 428 (1830). —Bremek. in Verh. K. Nederl. Akad. Wet., Afd. Natuurk. ser. 2, **48**(2): 226 (1952). —Hepper F.W.T.A. ed. 2, **2**: 212 (1963). —Verdc. in F.T.E.A., Rubiaceae 1: 291 (1976). Type from India.
Hedyotis affinis Roem. & Schult., Syst. Veg. **3**: 194 (1818).

Perennial herb with erect scrambling or trailing stems 0.2–1.2 m. tall radiating from a woody rootstock; stems glabrous, with 2 raised lines on each internode below the stipules. Leaf blades 1.2–8 × 0.1–1.6 cm., elliptic-oblong to elliptic- or linear-lanceolate, acute to tapering acuminate at the apex, cuneate to rounded at the base, scabridulous above particularly near the margins, glabrous beneath; petiole very short or not developed; stipule sheath 1 mm. long, with 1–5 short fimbriae or colleters 0.2–0.5 mm. long, sometimes with bristly hairs. Flowers heterostylous, in lax terminal or pseudo-axillary lax dichasia often running together to form ample panicles; pedicels 0.2–1(1.5) cm. long (up to 3 cm. in inflorescences reduced to 1 flower). Calyx tube 0.6–1 mm. long, subglobose, glabrous; lobes 0.5–1 mm. long, triangular, keeled, glabrous or with ciliate margins. Corolla dark blue, or blue-purple to deep violet; tube 3–4.5 mm. long, hairy inside; lobes 1.8–3.2 × 0.8–1.5 mm., elliptic or oblong. Style 3 mm. long in short-styled flowers, 4.5 mm. long in long-styled flowers; stigma lobes 1–3 mm. long. Capsule 2–2.8 × 2.2–2.5 mm., globose, glabrous, obscurely 8-ribbed, scarcely beaked, crowned by the calyx lobes. Seeds brown, angular, c. 0.5 mm. long, hilum borne on one of the keels, reticulate.

Subsp. **fugax** (Vatke) Verdc. in Kew Bull. **30**: 293 (1975); in F.T.E.A., Rubiaceae, 1: 292, fig. 41/15 (1976). TAB. **30**. Type from Zanzibar.
Hedyotis decumbens Hochst. in Flora **27**: 552 (1844). — Sond. in Harv. & Sond. F.C. **3**: 11 (1865). Type from S. Africa (Natal).

Hedyotis (Kohautia?) fugax Vatke in Oest. Bot. Zeitschr. **25**: 232 (1875).

Oldenlandia decumbens (Hochst.) Hiern in F.T.A. **3**: 54 (1877). —K. Schum. in Engl., Planzenw. Ost-Afr. **C**: 376 (1895). —Hiern, Cat. Afr. Pl. Welw. **1**: 442 (1898). —Hutch. & Dalz., F.W.T.A. ed. 1, **2**: 132 (1931) non Spreng. 1815 nom. illegit.

Oldenlandia affinis Bremek. in Verh. K. Nederl. Akad. Wet., Afd. Natuurk. ser. 2, **48**(2): 226 (1952) pro parte. —Hepper, F.W.T.A. ed. 2, **2**: 212 (1963) non (Roem. & Schult.) DC. sensu stricto.

Plant more robust with coarser shorter pedicels, more evident bracts, rather longer calyx lobes and a thicker-walled more globose capsule.

Zambia. B: Masese, fr. 12.v.1962, *Fanshawe* 6826 (K; NDO). N: Chipili, fl. & fr. 7.vi.1957, *Robinson* 2248 (K). C: 48 km. E. of Kapiri Mposhi, Gt. N. Road/Lunsemfwa R. crossing, fl. & fr. 17.i.1959, 3232 (K). S: Machili, fl. & fr. 18.i.1961, *Fanshawe* 6149 (K; NDO). **Zimbabwe**. W: Victoria Falls, Big Tree, fl. & fr. 21.iii.1979, *Mshasha* 188 (K; SRGH). E: Vumba, Norseland, fl. & fr. 9.ii.1949, *Chase* 1528 (BM; K; LISC; SRGH). S: Bikita, fl. & fr. *Wild* 4389 (K; SRGH). **Malawi**. N: Mzimba, Marymount, Mzuzu, fl. & fr. 3.iii.1971, *Pawek* 4471 (K; MAL). S: N. of Zomba on road up to Plateau, fl. & fr. 24.ii.1977, *Grosvenor & Renz* 897 (K; SRGH). **Mozambique**. N: Niassa, 38.4 km. E. of Ribáuè, fl. & fr. 17.v.1961, *Leach & Rutherford-Smith* 10903 (K; LISC; SRGH). Z: Lugela, Moebeda road, fl. & fr. 6.iv.1948, *Faulkner* 228 (K). T: Between planalto do Songo and river, fl. & fr. 23.ii.1972, *Macêdo* 4891 (LISC; LMA). MS: N. of Moribane, 5.vi.1971, *Pope* 455 (K; LISC; SRGH). M: Maputo, 4 km. from Catembe, R. Tembe, fr. 6.v.1981, *de Koning & Boane* 8675 (BM; K; LMU).

Widespread in tropical Africa from Liberia to Angola, Gabon, Sudan and southwards to S. Africa (Natal and Transvaal); also in Madagascar and Comoro Is. Deciduous thicket and woodland including *Brachystegia*, dry rocky hillsides, grassland, also roadsides and abandoned cultivations, etc.; 0–1500 m.

Subsp. *affinis* occurs in India and the Malay Peninsula.

14. **Oldenlandia lancifolia** (Schumach.) DC., Prodr. **4**: 425 (1830). —Hiern in F.T.A. **3**: 61 (1877). —K. Schum. in Engl., Pflanzenw. Ost-Afr. **C**: 375 (1895). —Bremek. in Verh. K. Nederl. Akad. Wet., Afd. Natuurk. ser. 2, **48**(2): 230 (1952). —Hepper, F.W.T.A., ed. 2, **2**: 212 (1963). —Agnew, Upland Kenya Wild Fl.: 400 (1974). —Verdc. in F.T.E.A., Rubiaceae 1: 292 (1976). Type from Ghana.

Hedyotis lancifolia Schumach. in Schumach. & Thonn., Beskr. Guin. Pl.: 72 (1827).

Perennial (rarely annual) straggling or prostrate herb, often much branched near the base into almost simple stems, (5)20–60(90) cm. long, which sometimes root at the nodes, glabrous or in some variants scabridulous when young; sometimes forming loose mats. Leaf blades 1–6 × 0.2–1.2 cm., linear to linear-lanceolate, less often elliptic or lanceolate, acute at the apex, cuneate at the base, scabridulous above near the margins, glabrous beneath or with midvein scabridulous; petioles not developed, very short and adnate to the stipule sheath which is 1 mm. long and bears 2–5 linear fimbriae 1.5 mm. long. Flowers not heterostylous, often solitary at the nodes (pseudo-axillary) or sometimes several at the nodes but then these actually borne on very reduced axillary shoots; pedicels slender, 0.5–3 cm. long, glabrous or scabridulous. Calyx tube 0.8 mm. tall, 1.5 mm. wide, bowl-shaped, glabrous or with scattered very short hairs; lobes 1–1.8 mm. long, triangular, acuminate at the apex, glabrous or sparsely scabridulous. Corolla white, sometimes tinged pink or purple; tube 1 mm. long, glabrous inside; lobes 1–2 mm. long, triangular. Style slightly longer than the tube; stigma lobes 0.7–1.4 mm. long. Capsule 2.2–3 mm. tall including the 1 mm. tall beak, 3.2–5 mm. wide, depressed subglobose, grooved at the middle. Seeds pale brown, angular, 0.3–0.4 mm. long, strongly reticulate.

Var. **scabridula** Bremek. in Verh. K. Nederl. Akad. Wet., Afd. Natuurk. ser. 2, **48**(2): 232 (1952). —Verdc. in F.T.E.A., Rubiaceae 1: 293, fig. 41/16 (1976). TAB. **31**. Type from Tanzania.

Young stems, pedicels and leaf blades scabridulous; calyx tube often puberulous. Leaf blades attaining maximum dimensions.

Botswana. N: 9 km. E. of Makalambedi, Botletle R., fl. & fr. 21.iii.1965, *Wild & Drummond* 7211 (K; LISC; SRGH). **Zambia**. B: Mongu, Lake Lutende, fl. & fr. 20.iii.1966, *Robinson* 6888 (K). N: Mbala, Lake Chila, fl. & fr. 25.iv.1952, *Richards* 1514 (K). W: 12 km. W. of Mwinilunga, Kanjima dambo, fl. & fr. 24.i.1975, *Brummitt et al.* 14054 (K). C: Chiwefwe, fl. & fr. 1.v.1957, *Fanshawe* 3239 (K; NDO). S: Kalomo, fl. & fr. v.1909, *Rogers* 8241 (K; SRGH). **Zimbabwe**. N: Sebungwe Distr., spring at headwaters of R. Masumo below Chizarira Hills, fl. & fr. 10.xi.1958, *Phipps* 1447 (K; SRGH). W: Matobo, Farm Besna Kobila, fl. & fr. ii.1959, *Miller* 5768 (K; SRGH). C: Harare, Cleveland Dam, fl. & fr. 3.iii.1950, *Wild* 3236 (K; SRGH). E: Chimanimani, Tarka Forest Reserve, fl. & fr. ii.1971, *Goldsmith* 5/71 (K; LISC; SRGH). **Malawi**. N: Mzimba Mt., Mzuzu, Marymount Dambo, fl. & fr. 15.i.1971, *Pawek* 4317 (K; MAL). C: Lilongwe–Chipata road, Namitete R., fl. & fr. 5.ii.1959, *Robson* 1459 (BM; LISC; K). S: Mulanje Mt., foot of Great Ruo Gorge, fl. & fr. 18.iii.1970, *Brummitt & Banda* 9205 (K; MAL). **Mozambique**. N: Niassa, 20 km. from Marrupa towards Lichinga, R. Mussoro, fl. & fr. 19.ii.1981, *Nuvunga* 620 (K; LMU). Z: Lower Shire, illegible, fl. & fr. ii.1888, *L. Scott* (K).

W. Africa from Sierra Leone to Angola and Sudan, Zaire, East Africa southwards to the Transvaal; Madagascar.

In various aquatic and semi-aquatic situations on mud or peat or sometimes in water even to a depth of 1.5 m., e.g. *Phragmites* and *Miscanthidium–Papyrus* swamps, rocky edges of rivers, spray-zone bogs near waterfalls, irrigation ditches, rock-pool edges, also in drier places, dambos, etc.; 600–1560 m.

Robinson 6720 (Barotseland, Mongu) has been referred to var. *lancifolia* but the pedicels, much shorter than usual, are distinctly scabridulous; it is akin to var. *sesseensis* Bremek. but differs in foliage.

15. **Oldenlandia verrucitesta** Verdc. sp. nov.* Type: Mozambique, Zambezia, Montes do Ile, fl. & fr. 2.iv.1943, *Torre* 5046A (COI; EA; K; LD; LISC holotype & isotype; LMU; SRGH).

Annual herb according to collector but probably perennial, with weak probably straggling glabrous quadrangular stems branched from the base. Leaves subsessile, 1–2.5 cm. × 1–6 mm., lanceolate, acute at the apex, attenuate at the base, scabrid puberulous above, ± glabrous beneath; stipule sheath c. 0.5 mm. long with 5–7 unequal fimbriae up to 1.5 mm. long. Flowers probably isostylous in narrow rather spike-like cymes due to suppression of leaves and bracts at flowering nodes on the branches of the dichasium which well overtop the true terminal flower; 1–2 flowers at each node; pedicels 2–3 mm. long but when single flowers top reduced branches apparently up to 1.5 cm. Calyx tube c. 0.7 mm. long, subglobose, distinctly papillate; lobes c. 1 mm. long, narrowly triangular or oblong, somewhat decurrent in fruit. Corolla white; tube scarcely 2 mm. long, narrowly funnel-shaped, glabrous outside and throat not bearded; lobes c. 1 mm. long, ovate oblong. Anthers sessile, about half-exserted; style exserted, the stigmatic lobes about 1 mm. long. Capsule depressed, usually broader than long, 1.5–3 × 1.5–3 mm., distinctly papillate, the beak slightly raised, the valves c. 0.5 mm. tall. Seeds black, ellipsoid, the longest dimension about 0.8 mm., the testa cells polygonal with straight walls and characteristic rather coarsely tuberculate reticulate lumen.

Mozambique. Z: Montes do Ile, fl. & fr. 2.iv.1943, *Torre* 5046A (COI; EA; K; LD; LISC; LMU; SRGH). Endemic. Thin soil over rock.

Although collected long before Bremekamp's revision he unfortunately never saw this specimen. It seems closest to *Oldenlandia flosculosa* Hiern but differs markedly in its characteristic testa and needs to be placed in a separate subgenus.

16. **Oldenlandia** sp. B.

Rhizomatous subshrubby herb with very woody base, much branched at the base with many prostrate to ascending quadrangular glabrous stems c. 25 cm. long arranged in a rosette. Leaves ± subsessile, 0.5–1 cm. × 1–3.5 mm., narrowly elliptic to lanceolate, acute at the apex, cuneate at the base, scabridulous-puberulous above, glabrescent beneath; stipule sheath 0.5 mm. long with 5–7 unequal fimbriae up to c. 1.5 mm. long. Cymes terminal, few-flowered, contracted with a terminal flower overtopped by short branches but all nodes usually with leaves save one below terminal flower; true pedicels obsolete or very short but up to 3 mm. long in fruit. Calyx tube 0.7 mm. long, subglobose, with papillae on the nerves; lobes c. 1 mm. long, triangular, slightly decurrent in fruit, with minutely papillate keels. Corolla white; tube slender, 3–4 mm. long, glabrous but throat with ± sparse hairs; lobes c. 1.5 mm. long, elliptic. Filaments exserted 0.5 mm., the anthers just reaching tips of lobes; style included, bifid at apex, the true stigmatic lobes about 1 mm. long, just exserted. Capsule depressed, wider than tall, 2.5 × 3 mm., the beak under 1 mm. tall. Seeds similar to last species.

Mozambique. N: 15 km. from Murrupala on road to Nampulo, Mt. Namuato, fl. & fr. 30.i.1968, *Torre & Correia* 17482 (COI; K; LD; LISC; LMU).

Endemic. 'Rupideserta' on soil over rocks, with *Entada* etc. on rocky mountain; 650 m. Possibly conspecific with the last species but only one specimen of each has been seen.

* *Oldenlandia flosculosae* Hiern ob structuram inflorescentiae similis sed fructibus majoribus papillatis, cellulis testae reticulatis valde tuberculatis differt.

17. **Oldenlandia nematocaulis** Bremek. in Verh. K. Nederl. Akad. Wet., Afd. Natuurk. ser. 2, **48**(2): 222 (1952). —Verdc. in Kew Bull. **30**: 295, fig. 8 (1975); in F.T.E.A., Rubiaceae 1: 302 (1976). Type from Tanzania.

Slender strict annual herb 2–20 cm. tall, with either unbranched or single branched stems with some of the branchlets basal and almost as long as or overtopping the main stem; stems angled, densely to sparsely scabridulous with short papilla-like hairs, particularly on the angles. Leaf blades 0.25–2.2 cm. × 0.1–1.5 mm. wide, linear to linear-lanceolate, acute at the apex, scarcely narrowed at the base, scabridulous-pubescent or glabrous or glabrous above and scabridulous on margins and midrib beneath, revolute; petiole not developed; stipule sheath 0.7–2 mm. long, divided into (1)2–5 filiform pubescent setae 0.5–3.5 mm. long. Flowers heterostylous, in sessile mostly subglobose 1-many-flowered inflorescences supported by the 2 terminal leaves or sometimes the apical internodes suppressed and lower axillary inflorescences joined with the terminal one and apparently supported by a whorl of 4 leaves, up to 4–9 mm. in diameter in fruit but minute specimens may bear only 1 fruit; pedicels often purplish, 0.2–1.5 mm. long, sometimes papillate. Calyx tube 0.6–0.8 mm. long, ovoid or subglobose, scabridulous; lobes often purplish, 1–3 × mostly 0.1–0.2 mm., linear-lanceolate, probably attaining the maximum length only in fruit, the margins scabridulous-ciliate. Corolla white, pink, bluish white or pale mauve; tube slender, 2–3.3 mm. long, widened at the throat, upper half hairy inside in short-styled forms, throat densely barbate in long-styled forms; lobes 1–1.5 × 0.7–0.9 mm., oblong-elliptic. Filaments exserted 1 mm. in short-styled flowers, included in the throat in long-styled flowers. Style 2 mm. long in short-styled flowers, the stigma lobes 0.8 mm. long, 3.3–4.3 mm. long in long-styled flowers, the stigma lobes 0.3–1 mm. long. Capsule 0.9–1.8 mm. tall (including the beak), 1.2–1.8 mm. wide, subglobose to transversely oblate, papillate, slightly scabridulous or glabrous, the beak very rounded, scarcely raised c. 0.5 mm. tall, the persistent calyx lobes slightly decurrent. Seeds brown, 0.25–0.35 mm. long, ellipsoid, reticulate, the testa cells granular.

Var. **nematocaulis.** —Verdc. in F.T.E.A., Rubiaceae 1: 304, figs. 40/29, 41/29 and 44 (1976). TAB. **31**.

Flowers with pedicels 0.2–1.5 mm. long.

Zambia. N: Mbala Distr., 4.8 km. from Kawimbe on Sumbawanga Road, fl. & fr. 25.ii.1959, *Richards* 10990 (K). **Malawi**. N: Nkhata Bay Distr., Viphya Plateau, 59.2 km. SW. of Mzuzu, fl. & fr. 27.iii.1976, *Pawek* 10931 (K; MAL; MO; SRGH; UC).

Also in Burundi and Tanzania. Damp pockets between flat rocks, sandy gravel over rock slab; 1650 m.

Var. **pedicellata** Verdc. in Kew Bull. **30**: 296 (1975). —Verdc. in F.T.E.A., Rubiaceae 1: 304 (1976). Type from W. Tanzania.

Flowers with pedicels 2–4 mm. long.

Zambia. N: Kasama Distr., Mungwi–Kasama Road, fl. & fr. 26.ii.1962, *Richards* 16191 (K).

Also in Tanzania. Probably roadside; 1320 m.

18. **Oldenlandia herbacea** (L.) Roxb., Hort. Bengal.: 11 (1814); in Fl. Indica, ed. Carey & Wallich **1**: 445 (1820). —DC., Prodr. **4**: 425 (1830). —Bremek. in Verh. K. Nederl. Akad. Wet., Afd. Natuurk. ser. 2, **48**(2): 244 (1952). —Wild, Common Rhod. Weeds: fig. 58 (1955). —Hepper in F.W.T.A. ed. 2, **2**: 212 (1963). —Lind & Tallantire, Fl. Pl. Uganda ed. 2: 160, fig. 101 (1972). —Agnew, Upland Kenya Wild Fl.: 401 (1974). —Verdc. in F.T.E.A., Rubiaceae 1: 305, fig. 41/31 (1976). Type from Sri Lanka.

Hedyotis herbacea L., Sp. Pl.: 102 (1753).

Annual or perennial erect, decumbent or spreading branched herb 1.5–60 cm. tall, with glabrous 4-ribbed stems. Leaf blades 0.6–5.5 cm. × 1–3.5(5) mm., linear to linear-lanceolate, acute at the apex, cuneate at the base, glabrous or with a few setae at the margins; petiole not developed; stipule sheath short, rarely exceeding 0.5 mm., truncate, with a few setae c. 0.3 mm. long but not fimbriate. Flowers usually isostylous but in one variety markedly heterostylous, solitary or paired at the nodes; pedicels graceful, spreading (0.3)0.8–3.5 cm. long. Calyx tube 0.5–1 mm. long, ovoid, glabrous, papillate or shortly hairy; lobes 0.5–2.5 mm. long, narrowly triangular, scabridulous on the margins. Corolla white, lilac or mauve or tube green and lobes mauve with purple marks; tube 0.2–1.1 cm. long, cylindrical; lobes 1–3 mm. long, ovate. Stigma lobes filiform, 0.7–0.9 mm.

long. Capsule pale straw-coloured, 2.2–5 × 1.5–2 mm., subglobose, crowned by the blackish calyx lobes, glabrous, papillate or shortly hairy; beak 0.3–1 mm. long. Seeds brown, 0.2–0.4 mm. long, ovoid to ellipsoid, angular, reticulate.

Key to infraspecific variants

1. Corolla tube short, 2–3.7(4) mm. long; flowers not heterostylous, the anthers never well exserted - - - - - - - - - - - - - - - - 2
– Corolla tube longer, (3.5)7–9(11) mm. long - - - - - - - - - - - 3
2. Calyx tube and capsule glabrous - - - - - - - - - - var. *herbacea*
– Calyx tube and capsule papillate or shortly hairy - - - - - - var. *papillosa*
3. Flowers distinctly heterostylous, the anthers well-exserted in short-styled plants; mostly perennial decumbent herbs with less slender corolla tube (no material yet seen from Flora Zambesiaca area) - - - - - - - - - - - - - - - var. *holstii*
– Flowers not distinctly heterostylous, the anthers either just included or only with the tips protruding from the throat; mostly erect annuals with very slender corolla tube var. *goetzei*

Var. **herbacea**. —Bremek. in Verh. K. Nederl. Akad. Wet., Afd. Natuurk. ser. 2, **48**(2): 245 (1952). —Verdc. in F.T.E.A., Rubiaceae 1: 305, fig. 41/31 (1976). TAB. **31**.

Oldenlandia heynii G. Don, Gen. Syst. **3**: 531 (1834). —Oliv. in Trans. Linn. Soc. **29**: 84 (1873). —Hiern in F.T.A. **3**: 59 (1877) pro parte. —K. Schum. in Engl., Pflanzenw. Ost-Afr. **C**: 376 (1895). Type from India.

Hedyotis trichopoda A. Rich., Tent. Pl. Abyss. **1**: 360 (1847). Type from Ethiopia.

Hedyotis dichotoma A. Rich., Tent. Fl. Abyss. **1**: 361 (1847) non Roth. Type from Ethiopia.

Hedyotis heynii (G. Don) Sond. in Harv. & Sond. F.C. **3**: 10 (1865).

Mostly erect annual herbs; flowers small, isostylous; corolla tube 2–3.7 mm. long; ovary and capsule glabrous.

Botswana. N: Gubatsha Hill, fr. 21.v.1977, *Smith* 2076 (K; SRGH). SE: 48 km. SE. of Gaborone, Ootse Hill, fl. & fr. 6.iv.1974, *Mott* 217c (K; UBLS). **Zambia**. B: near Senanga, 29.vii.1952, *Codd* 7230 (K; PRE). N: Mbala, Uningi, Pans, fl. & fr. 23.iii.1966, *Richards* 21378 (K). W: Ndola, fl. & fr. iv.1961, *Wilberforce* A/59 (K). C: Chakwenga Headwaters, fl. & fr. 27.iii.1965, *Robinson* 6484 (K; LISC). S: Mazabuka, Siamambo Forest Reserve, 13.iii.1960, *White* 7752 (FHO; K). **Zimbabwe**. N: Urungwe National Park, 20 km. from Makuti on road to Kariba, fl. & fr. 18.ii.1981, *Philcox et al.* 8724 (K). W: Matobo, Besner Kobila, fr. i.1953, *Miller* 1452 (K; SRGH) (atypical). C: Harare, Cleveland Dam, fr. 3.ix.1925, *Peter* 30683 (B; K). E: Chimanimani, Ngorima Reserve (E), near Lusitu R., fl. & fr. 23.xi.1967, *Simon & Ngoni* 1280 (K; SRGH). S: 6.4 km. E. of Zimbabwe, fl. & fr. 1.vii.1930, *Hutchinson & Gillett* 3357 (BM; K). **Malawi**. N : Mzimba, 4.8 km. SW. of Mzuzu, Katoto, fl. & fr. 17.iv.1973, *Pawek* 6547 (K; MAL; MO; UC). C: 20 km. SW. of Lilongwe, 2 km. NE. of Malingunde, fl. & fr. 25.iii.1977, *Brummitt et al.* 14919 (K; MAL). S: Kirk Range, Zangano Hill, fl. & fr. 31.i.1959, *Robson* 1390 (BM; K; LISC). **Mozambique**. N: Nampula, fl. & fr. 6.v.1937, *Torre* 1457 (COI; LISC). Z: Between Ile and Nampevo, 9.4 km. from Ile, fl. & fr. 18.ix.1949, *Barbosa & Carvalho* 4139 (LISC; LMA). MS: Manica, Mavita no cimo do Rutando, fl. & fr. 28.iv.1948, *Barbosa* 1623 (LISC).

Widespread in tropical and southern Africa, also in Madagascar etc. and Asia. Catholic in ecology, dry woodland, roadsides, rocky areas, laterite, dry waterholes but also in marshy grassland and seepage areas; occasionally a weed; 40–1850 m.

I have not formally retained var. *suffruticosa* Bremek. and var. *flaccida* Bremek. (see Bremek. loc. cit.: 248). Both *Johnson* 158 cited above and *Eyles* 4868 (from Zimbabwe, Harare) were referred to the latter by its author.

Var. **papillosa** (Chiov.) Bremek. in Verh. K. Nederl. Akad. Wet., Afd. Natuurk. ser. 2, **48**(2): 249 (1952). —Verdc. in F.T.E.A., Rubiaceae 1: 306 (1976). Type from Somalia.

Oldenlandia dichotoma var. *papillosa* Chiov. in Result. Sci. Miss. Stef.-Paoli Coll. Bot.: 89 (1916).

Mostly erect annual herbs; flowers small, isostylous with corolla similar to that of var. *herbacea*; ovary and capsule distinctly papillose or shortly densely hairy.

Zambia. S: Mumbwa, *Macaulay* 613 (K). **Malawi**. N: Rumphi, Chelinda R., fr. 4.v.1974, *Pawek* 8616 (K; MAL; UC). **Mozambique**. N: Amaramba, W. side of Serra Mituquê 20 km. from Nova Freixo (Cuamba). fl. & fr. 15.xi.1964, *Torre & Paira* 10614 A (LISC; LMU).

Also in East Africa, Rwanda and Somalia. Grassy thicket; 1080–1200 m.

Var. **goetzei** Bremek. in Verh. K. Nederl. Akad. Wet., Afd. Natuurk. ser. 2, **48**(2): 249 (1952). —Verdc. in F.T.E.A., Rubiaceae 1: 307 (1976). TAB. **31**. Type from Tanzania.

Annual or perennial mostly erect herb; flowers longer, isostylous or at least the anthers never long-exserted (but style often long-exserted); corolla tube 0.7–1.1 cm. long; ovary and capsule glabrous.

Zambia. B: 59 km. WSW. of Kabompo on Zambezi road, fl. & fr. 24.iii.1961, *Drummond & Rutherford-Smith* 7302 (K; SRGH). N: Mbala Distr., Sumbawanga road, 8 km. from Kiwonki, fl. & fr. 10.11.1955, *Richards* 4434A (K). W: N. of Mwinilunga, Luakera Falls, fl. & fr. 25.i.1938, *Milne-Redhead* 4340 (BM; K). C: near Kanona on Great North Road, Musha Hills, fl. & fr. 8.iv.1932, *Clair-Thompson* 1278 (K). S: Kalomo Distr., Mochipapa, near Choma, fl. & fr. 10.iii.1962, *Astle* 1493 (K). **Zimbabwe**. (all intermediate). N: Lomagundi Distr., between Kirkdale Pass and Umvukwes, 37 km. N. of Kildonan, fl. & fr. 10.iii.1982, *Brummitt & Drummond* 15839 (K; SRGH). W: Matopos, fl. & fr. iii.1918, *Eyles* 997 (BM; K; SRGH). C: Rusape, Makoni, St. Faith's Mission Farm, fl. & fr. 26.xii.1968, *Norman* R32 (K) (see note). E: Chimanimani, Gungunyana Forest Reserve, fl. & fr. xi.1961, *Goldsmith* 108/61 (K; SRGH). **Malawi**. N: Mzimba, 4.8 km. W. of Mzuzu, fl. & fr. 16.i.1971, *Pawek* 4324 (K; MAL). **Mozambique**. N: Marrupa, road towards Lichinga, km. 20, area Missor, fl. 16.ii.1982, *Nuvunga* 541 (K; LMU).

Also in S. Tanzania. Sandy areas, grassland, woodland, also rocky places but often in damp or swampy places; 450–1680 m.

The Zimbabwe material is intermediate between var. *herbacea* and var. *goetzei* and some specimens also have relatively more elongate calyx lobes. *Norman* R32 cited above is said to have dark crimson flowers although this is certainly not apparent from the dry material and needs confirmation that such a variant exists.

Bremekamp has annotated *Benson* 1199 (Malawi, Livingstonia), but not cited it, as a long-styled form of var. *holstii* (K. Schum.) Bremek. but I see no reason why it is not var. *goetzei* which commonly occurs in the area. No short-styled specimens have been seen from within the Flora Zambesiaca area. It has, however, been retained in the varietal key.

19. **Oldenlandia corymbosa** L., Sp. Pl.: 119 (1753). —Hiern in F.T.A. **3**: 62 (1877) pro parte. —K. Schum. in Pflanzenw. Ost-Afr. **C**: 375 (1895) pro parte. —Hutch. & Dalz., F.W.T.A. ed. 1, **2**: 132 (1931) pro parte. —Bremek. in Verh. K. Nederl. Akad. Wet., Afd. Natuurk. ser. 2, **48**(2): 254 (1952). —Hepper in F.W.T.A., ed. 2, **2**: 211 (1963). —Lewis in Grana Palynologica **5**: 330 (1964); in Missouri Bot. Gard. 53: 257 (1966). —Hallé, Fl. Gabon, 12 Rubiacées: 99, fig. 17/3 (1966). —Agnew, Upland Kenya Wild Fl.: 401, fig. (1974). —Verdc. in F.T.E.A., Rubiaceae 1: 308 (1976). Type from West Indies.

Hedyotis corymbosa (L.) Lam., Tab. Encycl. **1**: 272 (1792).

Annual herb, sparsely to very densely branched near the base; stems prostrate to ± erect, 1.5–30 cm. long, ridged, glabrous or scabridulous or pubescent on the ribs. Leaf blades 0.6–3.5(5.3) cm. × 0.5–7 mm., linear to narrowly elliptic, acute and apiculate at the apex, narrowed to the base, glabrous to sparsely scabridulous above and on margins and also beneath, particularly on the main nerve; petioles not developed; stipule sheath 0.5–2(3) mm. long, produced at the middle with (2)3–5 unequal fimbriae, 0.5–1(2.5) mm. long. Flowers not heterostylous, variously arranged, either 1–several single flowers in the axils or in 2–5(6)-flowered pedunculate umbel-like inflorescences, both kinds present on one branch or even at one node, the peduncles and pedicels mostly long and slender but rarely the flowers are fasciculate; peduncles (0)0.5–1.8(2.3) cm. long; pedicels (1.8)3–6(13) mm. long. Calyx tube 0.7–1 mm. long, ellipsoid; lobes 0.5–1.8 mm. long, triangular, setulose on the margins. Corolla white or tinged blue, pink or purple or with 2 pink stripes on each lobe; tube 0.6–1 mm. long; lobes 0.5–1.2 mm. long, ovate to oblong. Style 0.5–1.5 mm. long. Capsule 1.2–2.2 × (1)1.8–2.8 mm., ovoid, the beak scarcely raised. Seeds pale brown, c. 0.3 mm. long, ellipsoid or very obtusely depressed conic, reticulate.

Key to infraspecific variants

1. Inflorescences mostly all 1-flowered - - - - - - - - var. *caespitosa*
– Inflorescences nearly all 2–5(6)-flowered - - - - - - - - - 2
2. Leaves mostly elliptic, up to 7 mm. wide - - - - - - - var. *corymbosa*
– Leaves mostly linear, 1–1.5(3) mm. wide; plants more strictly erect with leaves held more erect than in other variants - - - - - - - - - - - - - 3
3. Plants mostly over 10 cm. tall - - - - - - - - - - var. *linearis*
– Plants very small, mostly 1.5–7 cm. tall - - - - - - - - var. *nana* (not yet found in Flora Zambesiaca area)

Var. **corymbosa**. —Verdc. in F.T.E.A., Rubiaceae 1: 309 (1976). TAB. **31**.

Oldenlandia caespitosa var. *major* sensu Bremek. in Verh. K. Nederl. Akad. Wet., Afd. Natuurk. ser. 2, **48** (2): 263 (1952) pro parte

Oldenlandia caespitosa var. *lanceolata* Bremek. sensu Bremek. op. cit. 264 pro parte

Plants erect or prostrate. Lamina mostly narrowly elliptic. Inflorescences nearly all 2–5(6)-flowered. Style glabrous or with few hairs.

Zambia. N: Mbala Distr., Lake Tanganyika, beyond Kasakalwa, fl. & fr. 8.ii.1964, *Richards* 18957 (K). W: Nkana/Kitwe Sewage Disposal Works, fl. & fr. 19.iii.1959, *Shepherd* 29A (K). C: Lusaka, fl. & fr. 19.iii.1955, *Best* 75 (K). E: Chipata (Fort Jameson), Chizombo, fl. & fr. 20.ii.1969, *Astle* 5486 (K; SRGH). S: Mumbwa, *Macaulay* 331 (K). **Zimbabwe**. W: Victoria Falls, fl. & fr. 8.vii.1930, *Hutchinson & Gillett* 3437 (K) (intermediate). **Malawi**. N: Nkhata Bay, White Fathers, fl. & fr. 9.v.1971, *Pawek* 4797* (K; MAL). C: Nkhota Kota, fl. & fr. 16.vi.1970, *Brummitt* 11477* (K; MAL). S?: *Buchanan* 421 p.p. (BM). **Mozambique**. N: Malema, Mutuali, fl & fr. 24.iv.1961, *Balsinhas & Marrime* 434 (LISC; LMA) (population with *corymbosa–capensis* intermediates). Z: Kongone mouth of Zambesi, fl. & fr. 29.i.1861, *Kirk* (K). T?: Between Tete and the sea coast, Mazzaro, fl. & fr. 27.iii.1860, *Kirk* (K). MS: Chemba, about 19 km from Tambara, towards Lupata, fl. 16.v.1971, *Torre & Correia* 18498 (LISC; LMA; LMU). GI: Outskirts of Chibuto near Lake Seozola, fl. & fr. 30.vi.1950, *Myre* 895 (LISC; LMA) (*sensu lato*).

Widespread throughout Africa with similar forms throughout tropics and subtropics of the World. Bare sand and laterite, grassland, woodland, also flooded lake shores, etc.; often occurs as a weed; (c. 0)450–1450 m.

Var. **linearis** (DC.) Verdc. in Kew Bull. **30**: 296 (1975); in F.T.E.A., Rubiaceae 1:309, fig. 41/33 (1976). Type from Senegal.
Oldenlandia linearis DC., Prodr. **4**: 425 (1830). —Bremek. in Verh. K. Nederl. Akad. Wet., Afd. Natuurk. ser. 2, **48**(2): 258 (1952). —Hepper in F.W.T.A. ed. 2, **2**: 211 (1963).

Plants mostly strictly erect. Leaves held more erect than in other variants, the blades usually linear (only 1–1.5(3) mm. wide) and often longer. Inflorescences 1–2(3)-flowered. Style glabrous. Capsules usually smaller than in var. *corymbosa*.

Botswana. N: Moremi, Tsetse Fly camp, fr. iv.1968, *Lambrecht* 529 (K; SRGH) (doubtful). SE: Morale Research Station, fl. & fr. 3.iii.1978, *Hansen* 3360 (BM; BR; C; EA; GAB; K, PRE; SRGH; UPS; WAG). **Zambia**. B: Sesheke, *Gairdner* 508 (K). W: Mwinilunga, slope E. of Matonchi R., fl. 7 fr. 13.x.1937, *Milne-Redhead* 2750 (K). S: Namwala, fl. & fr. 17.iv.1963, *van Rensburg* 2016 (K). **Zimbabwe**. N: Mtoko, Chitora R. road, fl. & fr. iv.1956, *Davies* 1916 (K; SRGH). W: Matobo, Besna Kobila Farm, fl. & fr. iii.1961, *Miller* 7740 (K; SRGH). C: Harare, Twentydales, fl. & fr. 28.iv.1948, *Wild* 2519 (K; SRGH). S: Chibi Distr., near Madzivire Dip, c. 6.5 km. N. of Lundi R. Bridge, fl. & fr. 3.v.1962, *Drummond* 7919 (K; SRGH). **Malawi**. N: Mzimba Distr., Campira, Katete Mission, fl. & fr. 19.iv.1974, *Pawek* 8391A (K). C: Dedza, Mphunzi, fl. & fr. 20.iii.1971, *Salubeni* 1535 (K; SRGH).
Also in Senegal, Sudan, Ethiopia, Burundi, Zaire, Uganda, Kenya and Tanzania. Sandy pans, grassland, shallow soil in rocky places, damp stream edges, also as a weed in gardens and lawns; 500–1350 m.
Although nominally long retained as a distinct species and, when typical, easily recognised there is no clear dividing line between this and *Oldenlandia corymbosa*; it may in fact be an ecotype of dryish places but it is absent from many areas and one suspects some genetic basis. *Richards* 15138 (Zambia, Mbala, Kalambo R., Sansia Falls, fl. & fr. 12.v.1961, 1500 m.) has tentatively been referred here but has a distinctive facies.

Var. **caespitosa** (Benth.) Verdc. in Kew Bull. **30**: 298 (1975); in F.T.E.A., Rubiaceae 1: 310 (1976). TAB. **31**. Type from Liberia.
Oldenlandia herbacea Roxb. var. *caespitosa* Benth. in Hook., Niger Fl.: 403 (1849).
Oldenlandia caespitosa (Benth.) Hiern in F.T.A. **3**: 61 (1877). —Bremek. in Verh. K. Nederl. Akad. Wet., Afd. Natuurk. ser. 2, **48**(2): 262 (1952). —Hepper in F.W.T.A., ed. 2, **2**: 212 (1963).
Oldenlandia tenuissima Hiern in F.T.A. **3**: 61 (1877). Type: Zambia, "Island at Victoria Falls" *Kirk* (K, holotype).
Oldenlandia corymbosa var. *subpedunculata* O. Kuntze, Rev. Gen. Pl. **3**: 121 (1893). Type: Mozambique, without locality, *O. Kuntze* (K, isotype).
Oldenlandia caespitosa var. *subpedunculata* (O. Kuntze) Bremek. tom. cit.: 263 (1952).
Oldenlandia caespitosa var. *major* Bremek. tom. cit. Type from S. Africa (Natal).

Typically a small herb 2–30 cm. tall with several prostrate stems, but sometimes erect and unbranched. Leaf blades typically small and narrowly elliptic, 0.7–2(2.8) cm. × 1.5–3(4.5) mm. Flowers typically all solitary at the nodes, but often also a few in 2-flowered cymes or sometimes, where reduced branchlets are leafless, simulating 2–3-flowered cymes. Style often sparsely hairy at the middle**.

* These are absolutely typical *corymbosa* with 3–4-flowered inflorescences.
** Bremekamp uses this as a character to separate *O. caespitosa*, which he maintains as a species, from *O. capensis*, etc. On the type sheet, however, he has added style glabrous in pencil against his drawing.

Zambia. N: Kasama–Mpika road, near Chambeshi pontoon, fl. & fr. 29.iv.1962, *Richards* 16392 (K). C: Kabwe, fl. & fr. v.1914, *Rogers* 7688 (K). S: Livingstone, Victoria Falls, Knife Edge, fl. & fr. 3.viii.1947, *Brenan & Greenway* 7132 (EA; FHO; K). **Zimbabwe**. N: Mweneze, Malangwe R., SW. Mateke Hills, fl. & fr. 6.v.1958, *Drummond* 5603 (K; LISC; SRGH). W: Victoria Falls, between rain-forest and chasm, fl. & fr. 14.ix.1935, *Galpin* 14988 (K; PRE). C: Harare, Golf course, fl. & fr. 25.iii.1949, *Wild* 2818 (K; SRGH) (depauperate). E: Mutare, Odzani R. Valley, *Teague* 168 (BOL; K). **Malawi**. N: Mzimba Distr., 4.8 km. W. of Mzuzu, Katoto, fl. & fr. 10.ii.1974, *Pawek* 8076 (K; MAL; MO; SRGH; UC). C: Dedza, Ntaka-taka, fl. & fr. 10.iv.1969, *Salubeni* 1294 (K; SRGH). S: Chikwawa Distr., Lengwe Game Reserve, fl. & fr. 5.iii.1970, *Brummitt* 8887 (K). **Mozambique**. N: Malema, Mutuáli, CICA Expt. Station, fl. & fr. 24.iv.1961, *Balsinhas & Marrime* 434 (BM; K; LMA). Z: Quelimane, Namagoa, fl. & fr. 3.vii.1949, *Faulkner* K442 (K). T: between Mutarara and Dóvo, 18.7 km. from Mutarara, fl. & fr. 16.vi.1949, *Barbosa & Carvalho* 3116 (LISC; LMA). MS: Spungabera Catholic Mission, fl. & fr. 14.vi.1942, *Torre* 4329 (LISC).

Widespread from Cape Verde Is., N. & W. Africa to Somalia, Natal and Angola, Madagascar, Mascarene Is. and Middle East. Damp often sandy places, forest edges, woodland, grassland, rocks by streams, often by roadsides and as a weed in cultivations, rice fields etc; 0–1560 m.

20. **Oldenlandia fastigiata** Bremek. in Verh. K. Nederl. Akad. Wet., Afd. Natuurk. ser. 2, **48**(2): 260 (1952). —Agnew, Upland Kenya Wild Fl.: 401 (1974). —Verdc. in F.T.E.A., Rubiaceae 1: 311 (1976). TAB. **31**. Type from Tanzania.

Annual or perennial herb with stems mostly branched at the base, erect, straggling or ascending, (6)15–60 cm. long, ribbed, glabrous or slightly to rarely densely scabridulous on the ribs. Leaf blades 1.2–4.5(6) cm. × 1–5(8) mm., linear to linear-lanceolate, acute at the apex, narrowed to the base, entirely glabrous or with very small obscure papilla-like hairs above, particularly when very young, and sometimes on margins and main nerve beneath; petioles not developed; stipule sheath 1–2 mm. long, scabrid-papillose, bearing 3–5 fimbriae of varying length, 0.5–2.5 mm. long. Flowers isostylous or heterostylous, in 3–7(10)-flowered sessile or pedunculate axillary cymes or fascicles; peduncle 0–3 cm. long, usually very much shorter than the subtending leaf, typically $\frac{1}{4}$ the length; pedicels 1.5–4 mm. long, glabrous or rarely scabridulous. Calyx tube 0.5–0.8 mm. long, subglobose, glabrous; lobes 1–2 mm. long, narrowly triangular, scabridulous on the margins, keeled. Corolla white or tinged blue or lilac-pink; tube 1–1.6 mm. long, the throat densely hairy; lobes 0.7–2 × 0.6 mm., ovate-oblong, papillate inside. Style in short-styled flowers 0.6 mm. long, in long-styled flowers 2.8 mm. long, in isostylous flowers 0.6–1.2 mm. long; stigma lobes 0.3–1 mm. long. Capsule 1–2 × 1.5–2.5 mm. depressed globose, glabrous, the beak scarcely raised. Seeds pale brown, c. 0.2 mm. long, angular, bluntly conical, strongly reticulate.

Var. **fastigiata**. —Verdc. in F.T.E.A., Rubiaceae 1: 312 (1976)
Oldenlandia fastigiata var. *longifolia* Bremek. in Verh. K. Nederl. Akad. Wet., Afd. Natuurk. ser. 2, **48**(2): 261 (1952). Type from Tanzania.

Inflorescence mostly sessile or only shortly pedunculate, the peduncles 2 mm. long, rarely longer, up to 1 cm. Flowers not heterostylous, mostly smaller, the tube of the corolla c. 1 mm. long.

Malawi. N: Likoma I., fr. viii.1887, *Bellingham* (BM.) (intermediate with *O. corymbosa*). S: Chikwawa Distr. Lower Mwanza R., fl. & fr. 6.x.1946, *Brass* 18013 (K; NY). **Mozambique**. N: Erati, between Namapa and Ocua near bridge over R. Lúrio, fl. & fr. 9.iii.60, *de Lemos & Macúacua* 26 (K; LISC; LMA).

Also in Zaire, Sudan, Ethiopia, Somalia, Kenya and Tanzania. Sandy beach, riverine; 180 m.

In Zambia and Zimbabwe many specimens ± intermediate between *O. fastigiata* and *O. capensis* occur but in East Africa the species are rather more distinctive and simple sinking is probably not the best solution.

21. **Oldenlandia capensis** L.f., Suppl. Pl.: 127 (1781). —DC., Prodr. **4**: 424 (1830). —Hiern in F.T.A. **3**: 62 (1877). —K. Schum., in Engl., Pflanzenw. Ost-Afr. **C**: 375 (1895). —Bremek. in Verh. K. Nederl. Akad. Wet., Afd. Natuurk. ser. 2, **48**(2): 265 (1952). —Hepper in F.W.T.A. ed. 2, **2**: 211 (1963). —Verdc. in F.T.E.A., Rubiaceae 1: 313 (1976). TAB. **31**. Type from South Africa.

Prostrate or ascending annual herb (2)8–22 cm. tall or long, sometimes forming mats; stems branched, glabrescent or with minutely setulose ribs. Leaf blades 0.7–3.4 cm. × 0.5–4.5 mm., linear to narrowly elliptic, subobtuse to usually acute at the apex, cuneate at the base, scabrid-papillose above, on the inrolled margins and often on the main nerve beneath but mostly glabrous beneath; petiole absent or apparent petiole 1–2 mm. long;

stipule sheath 1.2–2 mm. long, with 3–7 long and short fimbriae up to 1.5 mm. long. Flowers not heterostylous, sometimes solitary at the nodes or few in subsessile axillary fascicles; peduncles obsolete or up to 2 mm. long; pedicels 1–3 mm. long. Calyx tube c. 1 mm. long, obconic to campanulate, glabrous or sparsely pubescent; lobes 4–8, 0.6–1.1 × 0.2–0.3 mm., triangular or linear-subulate, margins setulose-ciliolate. Corolla white, deep lilac, dull reddish, pink or lilac; tube 0.6–0.9 mm. long, cylindrical; lobes 4, 0.5–0.6 mm. long. Anthers included. Style 0.2 mm. long, glabrous; stigma lobes 0.2 mm. long, adjacent to the anthers. Capsule 1.75–2 mm. long and white, obconic, glabrous or shortly pubescent, the beak straight, slightly raised, 0.3–0.4 mm. tall. Seeds brown, 0.3 mm. long, ellipsoid-angular or ± cone-shaped with flat base.

Var. **capensis**. —Bremek. in Verh. K. Nederl. Akad. Wet., Afd. Natuurk. ser. 2 **48**(2): 266 (1952). —Hepper in F.W.T.A., ed. 2, **2**: 211 (1963). —Verdc. in F.T.E.A., Rubiaceae 1: 313 (1976).

Calyx lobes 4.

Botswana. N: 17 km. SE. of Maun at Samadupe Drift, Botletle R., fl. & fr. 23.i.1972, *Biegel & Gibbs-Russell* 3739 (K; LISC; SRGH). **Zambia**. B: near Kalabo resthouse, fl. & fr. 13.xi.1959, *Drummond & Cookson* 6408 (K; SRGH) (pubescent variant). W: Mwinilunga, 8–10 km. SE. of Zambia/Angola border, Mudileji (Mujileshi) R., 7.xi.1962, *Lewis* 6173 (K; MO). E: Chipata, fr. 13.x.1967, *Mutimushi* 2333 (K; NDO). S: Victoria Falls, Livingstone I., fl. & fr. 21.xi.1949, *Wild* 3127 (K; SRGH). **Zimbabwe**. N: Mtoko, Mkota Reserve, Mazoe R., fl. & fr. 1.x.1948, *Wild* 2692 (K; SRGH) (intermediate). W: Bulawayo, Hillside Dam, fl. & fr. iv.1959, *Miller* 5886 (K; LISC; SRGH). C: Makoni, Rusape, fl. & fr. 7.i.1931, *Norlindh & Weimarck* 4170 (K; LD; SRGH). S: Beitbridge, 9.6 km. NW. of junction of Limpopo with Shashi R., fl. & fr. 6.v.1959, *Drummond* 6095 (K; SRGH). **Malawi**. N: Lake Malawi, Umbaka R., xii.1887, *Scott* s.n. (K). S: Mangochi, Nampingudya Stream, fl. & fr. 23.iv.1955, *Banda* 81 (K). **Mozambique**. N: Cabo Delgado, about 4 km. from Montepuez towards Nmuno, fl. & fr. 25.xii.1963, *Torre & Paiva* 9679 (LISC; LMU; SRGH). Z: Quelimane, fl. & fr. 14.x.1965, *Mogg* 32267 (J; LISC). T: Right bank of R. Zambezi, near Quartel, fl. & fr. 21.x.1965, *Neves Rosa* 113 (LISC; LMA). MS: Gorongosa, Parque Nacional de Caca margem direita do R. Urema proximo do batclão a 26 km. du Acampamento de Chitengo, fl. & fr. 8.xi.1963, *Torre & Paiva* 9120 (BR; EA; LISC; LMA; PRE; WAG). GI: Aldeia da Barragem a 7 km. a montante da Barragem, edge of R. Limpopo, fl. & fr. 20.xi.1957, *Barbosa & Lemos* 8221 (LISC; LMA). M: Maputo, Jardim Vasco da Gama, fl. & fr. 9.xi.1971, *Balsinhas* 2254 (K; LISC).

More or less throughout Africa from Marocco and Egypt to the Cape Peninsula, also in Madagascar and Yugoslavia. River-banks in alluvium and cracks of rocks, grassland, roadsides, etc.; 0–910 m.

Several specimens from N. Malawi, Mzimba, Mzuzu, e.g. *Pawek* 8103 & 8258 have 2-flowered cymes and are intermediate with *O. corymbosa*. *Smith* 2732 from N. Botswana and many other specimens are intermediates between *O. corymbosa* var. *caespitosa* and *O. capensis*. Bremekamp has annotated *Greatrex* in GHS13971 from Harare as *O. geminiflora* (Sond.) O. Kuntze but it seems to me to be *capensis;* in his revision he cites it with some doubt.

Var. **pleiosepala** Bremek. in Verh. K. Nederl. Akad. Wet., Afd. Natuurk, ser. 2, **48**(2): 267 (1952). —Verdc. in F.T.E.A., Rubiaceae 1: 313 (1976). Type from Russia (Azerbaydzhan).

Karamyschewia hedyotoides Fisch. & Mey. in Bull. Soc. Nat. Mosc. **1838**: 266 (1838). Type as for var. *pleiosepala*.

Theyodis octodon A. Rich., Tent. Fl. Abyss. **1**: 364 (1847). Type from Ethiopia.

Oldenlandia hedyotoides (Fisch. & Mey.) Boiss., Fl. Orient. **3**: 11 (1875). —Hiern in F.T.A. **3**: 64 (1877).

Calyx lobes 5–8.

Botswana. N: Banks of Okavango R. at 18°48.75'S, 22°20.6'E, fr. 27.i.1976, *Smith* 1563 (K; SRGH). **Zambia**. N: Mbereshi – Luapula R., swamp, fl. & fr. 14.i.1960, *Richards* 12353 (K). W: Solwezi, fl. & fr. 8.i.1969, *Mutimushi* 2936 (K; NDO). C: Lusaka, near Chinkuli, Nyanshishi R., fr. 10.xii.1972, *Kornaś* 2794 (K; KRA). E: Lundazi R., above dam. fl. & fr. 19.xi.1958, *Robson* 669 (BM; K; LISC; SRGH). **Malawi**. S: Mulanje, fl. & fr. xii.1893, *Scott Elliot* 8654 (BM; K). **Zimbabwe**. W: Victoria Falls, Cataract I., fl. & fr. 23.xi.1949, *Wild* 3164 (K; SRGH). **Mozambique**. MS: Gorongosa National Park, margin of R. Urema, fl. & fr. 27.x.1965, *Torre & Pereira* 12608 (LISC; LMU; SRGH; WAG).

Widespread in Africa and Algeria and Egypt to Zimbabwe and Angola, also Madagascar, Transcaucasia and Iran. Seasonally flooded river banks. dam-edges, dambos etc., bare mud, grassland, riverine rocks and open woodland; also as a weed in moist gardens; 900–1200 m.

This is a very distinctive plant the status of which could stand more investigation in the field and breeding experiments. The flowers are more truly sessile and the completely prostrate habit distinctive. I have, however, refrained from restoring it to specific rank.

The classification of *O. corymbosa*, *O. fastigiata* and *O. capensis* is most unsatisfactory and probably all should be considered components of a broad 'superspecies'; in particular *O. corymbosa* var.

caespitosa merges with *O. capensis* which itself when several flowers occur in each axil grades into *O. fastigiata*.

22. **Oldenlandia geminiflora** (Sond.) O. Kuntze. Rev. Gen. Pl. **1**: 292 (1891). —Bremek. in Verh. Fl. Nederl. Akad. Wet. Afd. Natuurk. ser. 2, **48**(2): 268 (1952). Type from S. Africa (Transvaal).
Hedyotis geminiflora Sond. in Linnaea **23**: 51 (1850); in Harv. & Sond., F.C. **3**: 10 (1825).

Herb of sandy river beds very close to *O. capensis*, with several presumably prostrate stems 7.5–20 cm. long from a probably ± woody base; shoots shortly hispidulous. Leaf blade 2–4 × 0.2 cm., with strongly revolute margins and similar indumentum to stems. Fruiting pedicels 1–4 mm. long. Calyx lobes distinctly longer than in *O. capensis*, 3–3.5 mm. long.

Zimbabwe. E/S: Sabi R., Birchenough Bridge, fr. 12.ix.1949, *Chase* 1776 (BM; SRGH).
This is probably no more than a variety of *O. capensis* but little material has been seen including only one specimen from the Flora Zambesiaca area. *Greatrex* in G.H.S. 13971 (Harare, Arthur's Seat, 17.xi.1945) annotated by Bremekamp as *O. geminiflora* but cited with some doubt in his revision appears to me to be true *O. capensis*.

23. **Oldenlandia** sp. C.

Annual unbranched or branched herb 8–11 cm. tall with slender ± 4-angled stems covered with spreading short white pubescence; roots very slender. Leaves distinctly ascending, 0.7–1.7 cm. × 0.3–1 mm., linear, apiculate, with similar pubescence to stem, sessile; stipule sheath 1.5 mm. long joining the two 'petioles', with 3–5 filiform setae 0.5–2 mm. long. Flowers essentially solitary on pedicels c. 4 mm. long, lengthening to 1.1 cm. in fruit; occasionally 2–3 flowers occur in the same axil, at least sometimes due to very abbreviated shoots and also a few apparently pedunculate 2-flowered inflorescences occurring, the 'peduncles' up to 1.5 cm. long but judging by stipule remnant subtending the pedicels they are actually leafless branches. Calyx with short spreading white hairs; tube c. 1 mm. long, ± obovoid; lobes 1.2 mm. long, lanceolate. Corolla white; tube 1.5–2 mm. long; lobes 1.2 × 0.5 mm. long. All flowers seen on several different plants have stigmata exserted and anthers just included. Capsules compressed, c. 1.5 × 2 mm., ± hemispherical, the beak scarcely raised.

Zambia. N: Lunzua Electricity Station, Upper Falls, fl. & young fr. 19.iii.1966, *Richards* 21365 (K).
Endemic. Sandy ground; 1200 m.
This had been named as *Oldenlandia linearis* DC. and I originally considered it might be this or some similar variant of *Oldenlandia corymbosa* but many characters of corolla, calyx lobes, indumentum and inflorescence suggest it is a distinct species but this needs confirmation from further material. It is clearly an ephemeral of short duration.

24. **Oldenlandia machingensis** Verdc. sp. nov.* Type: Malawi, Machinga Distr., Munde Hill Saddle, T.A. Nkhokwe, *Patel* 884 (K, holotype; MAL).

Small weak annual herb about 20 cm. long or tall; stems slender, 4-angled, glabrous. Leaves 0.5–2 × 0.2–1.1 cm., narrowly ovate, acute at the apex, rounded then abruptly narrowed at the base into a short petiole 1–2 mm. long, glabrous; stipule sheath c. 1 mm. long with c. 5 often purplish fimbriae 2–3 mm. long. Flowers solitary in the leaf axils or apparently in 2–3-flowered scorpioid cymes due to suppression of leaves and bracts and actually some branches with a truly terminal flower; pedicels or apparent pedicels 2 mm. long, lengthening in fruit up to 1.7 cm. long. Calyx tube 0.5 mm., with short acute swollen-based hairs; lobes up to c. 1.5 mm. long in fruit, 0.5 mm. wide at the base, narrowly triangular. Corolla white; tube 0.8 mm. long; lobes 1.3 × 0.5 mm., oblong. Anthers about $\frac{3}{4}$ exserted; style and stigmata together about 1.2 mm. long, the latter c. 0.4 mm. and just reaching anther tips, the flowers obviously not heterostylous. Capsule 2.5–3 × 2–2.5 mm. including the low rounded beak, broadly subglobose. Seeds compressed, dark brown, c. 0.25 mm. long; testa cells densely coarsely verrucose, with straight walls.

Malawi. S: Machinga Distr., Munde Hill Saddle, T.A. Nkhokwe, fl. & fr. 30.iv.1982, *Patel* 884 (K; MAL).

* Ab omnibus speciebus characteribus sequentibus simul combinatis, foliis tenuibus ovatis, inflorescentiis laxe cymosis terminalibus, floribus isostylis parvis, tubo corollae vix 1 mm. longo, cellulis testae dense verrucosis differt.

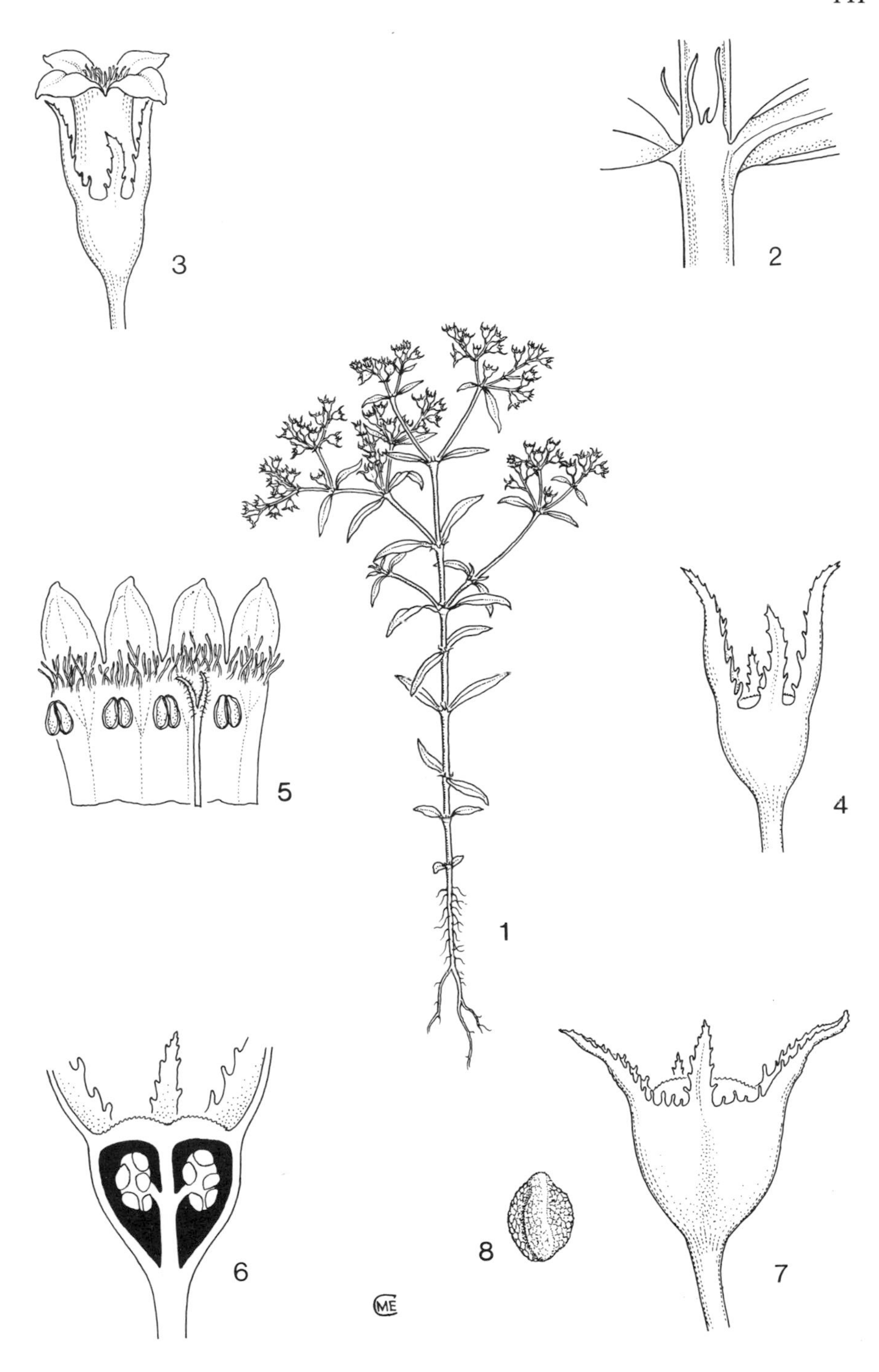

Tab. 32. OLDENLANDIA ROBINSONII. 1, habit (× 1); 2, stipules (× 8); 3, flower (× 24); 4, calyx (× 20); 5, corolla opened out (× 20); 6, ovary, longitudinal section (× 30); 7, capsule (× 12); 8, seed (× 50), all from *Robinson* 3632.

Not knowm from elsewhere. "Under rocks".
I have been unable to fit this into any of Bremekamp's subgenera; it does not seem likely to be an introduced species and presumably needs a new subgenus.

25. **Oldenlandia robinsonii** Verdc. sp. nov.* TAB. **32**. Type: Zambia, Mwinilunga, Matonchi, *Robinson* 3632 (K, holotype).

Branched or ± unbranched slender annual herb 4–10 cm. tall, drying yellow-green, glabrous; stems 4-angled, the angles almost winged when dry. Leaves 3–12 × 0.3–1.8 mm., linear-lanceolate, narrowly acute, sessile. Stipular sheath joined to the leaf bases c. 1 mm. long with 2–several minute lobes, fimbriae 0.1–0.6 mm. long. Flowers in terminal several-flowered dichasial cymes with diverging branches, the flowers closely placed; one truly terminal flower present between each pair of branches; rhachis almost winged in dry state. Calyx tube 0.5 mm. long, obovoid or semi-globose; lobes linear-lanceolate, equal or 2 long and 2 short, the longest 0.7–1 mm. long, becoming slightly longer in fruit with minute linear colleters between the lobes which are persistent in fruit. Corolla blue (tube looks blue and lobes yellowish in dried plant); tube just under 1 mm. long, hairy at the throat; lobes 0.8 × 0.4 mm., oblong, ± acute. Anthers blue, small, situated just below the throat, 0.25 mm. long. Style and 2 short minutely hairy stigmatic lobes together about 1 mm. long. Capsule 1.2 × 1.5 mm., compressed obovoid, grooved between the locules in dry state; beak low and rounded, emarginate, the valves distinctly granulate. Seeds black, several per locule, c. 0.25 mm. long; testa cells smooth with wavy walls.

Zambia. Mwinilunga, Matonchi, fl. & fr. 16.iv.1960, *Robinson* 3632 (K).
Not known from elsewhere. Lateritic gravel drying out after flooding; 1400 m.

28. OTIOPHORA Zucc.**

Otiophora Zucc. in Abh. Akad. München. **1**: 315 (1832). —Verdc. in Journ. Linn. Soc. Bot. **53**: 383 (1950); in Garcia de Orta, Sér. Bot. **1**: 25 (1973); in F.T.E.A., Rubiaceae 1: 315 (1976).
Mericocalyx Bamps in Bull. Jard. Bot. Brux. **29**: 147 (1959).

Annual or perennial herbs, subshrubs or sometimes dwarf shrubs. Stems erect to prostrate, ± terete to obscurely 4-angled, covered with yellowish to whitish, ± spreading or curled multicellular hairs above, often becoming glabrescent below. Leaves decussate or occasionally in whorls of 3(4), often seemingly in much larger numbers at nodes ('pseudo-verticillate'; due to the presence of leafy, much-contracted short shoots); blades variable in size and shape, ± ovate to linear, ± acute to obtuse at apex, membranaceous to coriaceous, with 2–4 usually ± indistinct lateral veins on either side of midvein, glabrous or hairy on margins, veins or surfaces, margins ± flat to slightly revolute; petioles short to subobsolete; stipular sheaths ± cup-shaped, with 3–5(7) ± linear setae or fimbriae on either side, the median often (much) longer than the others. Inflorescence head- or spike-like, usually elongated when in fruit. Flowers paired, on very short, persistent pedicels, hermaphrodite, protandrous, 4–5-merous. Calyx: lobes unequal, 1(2; very rarely 3) enlarged, foliaceous, other lobes minute, setiform. Bud with limb abruptly expanded into a subcylindrical to ovoid head, or limb and tube not distinct from each other. Corolla glabrous or throat and tube hairy inside, lobes lanceolate to oblong, ± acute at apex. Stamens exserted, filaments inserted at or near throat, thin, ± erect and stiff, glabrous or hairy near base; anthers oblong. Ovary bicarpellate and biovulate (but one carpel sometimes smaller and ± reduced or in one species completely sterile), mostly shortly hairy; style glabrous, filiform; stigma bifid, lobes filiform. Fruit crowned by persistent calyx lobes, dehiscing into two mericarps, each convex on dorsal side and plane or slightly concave on ventral side. Seed ± oblong to ± ovoid, granulate, dorsal side convex, ventral side flat, concave or ± hollowed out, with a median longitudinal ridge.

A genus of 17 species confined to Africa south of the Sahara and to Madagascar. 10 species occur in the Flora Zambesiaca area.

* Ob inflorescentias ± dense dichasialiter cymosas terminales, habitum ephemerum, colorem in siccitate flavoviridem, flores minutos isostylos, folia lineari-lanceolata perdistincta et nulla affinitate arcte obvia in sectione nova probabiliter ponenda.
** Tribal position doubtful.

Otiophora was previously included in the Anthospermeae but detailed investigations [Puff in Bothalia **14**: 185–188 (1983)] showed that the genus must be excluded from that tribe. *Otiophora* agrees with the Hedyotideae in several characters but also approaches the Spermacoceae in certain other features. Verdcourt disagrees with its position in the Hedyotideae.

1. Throat of corolla distinctly hairy or at least filaments hairy at base - - - - - 2
– Throat of corolla glabrous or nearly so, filaments glabrous - - - - - - 5
2. Bud with limb abruptly expanded into a subcylindrical head - - - 5. *pycnostachys*
– Bud claviform, limb and tube not distinct from each other - - - - - - - - 3
3. Leaves ovate to ovate-lanceolate, (7)10–33 × (3.8)4.5–13 mm., decussate or in whorls of 3(4) - - - - - - - - - 4a. *inyangana* subsp. *inyangana*
– Leaves linear-lanceolate to lanceolate, 10–40(47) × 1.6–6(7.5) mm., strictly decussate 4
4. Leaves 10–18 × 1.6–3.5(4) mm.; corolla tube (4)5.5–9 mm. long 4b. *inyangana* subsp. *parvifolia*
– Leaves 23–40(47) × 3–6(7.5) mm.; corolla tube (3.2)3.8–4.5(5) mm. long 3. *lanceolata*
5. Flowering and fruiting inflorescence lax, spike-like (1.5)4–25 cm. long - - - - 6
– Flowering and fruiting inflorescence condensed, head-like or only very slightly elongated in fruit - 8
6. Dwarf shrub or woody herb; leaves linear, (0.8)1–2 mm. wide, glabrous 2. *angustifolia*
– Usually an annual; leaves broadly lanceolate to linear-lanceolate, 2–20(22) mm. wide, often hairy - 7
7. Plant usually erect; leaves (10)15–50(60) × 2–8(11) mm.; petioles subobsolete - - - - - - - - - - - - 1a. *scabra* subsp. *scabra*
– Plant usually straggling; leaves (25)30–90 × (6)8–20(22) mm.; petioles (2)4–8 mm. long - - - - - - - - - - - - - 1b. *scabra* subsp. *diffusa*
8. Plant with several erect stems to c. 45 cm. from a woody base; corolla tube (5)7–12 mm. long, lobes (4)5–6 mm. long - - - - - - - - - - - - - - - 6. *caerulea*
– Tufted, cushion- or mat-forming many-stemmed herb, stems to c. 20 cm. long; corolla tube c. (2)2.5–6 mm. long, lobes to 3 mm. long - - - - - - - - - - - 9
9. Plant with numerous (c. 20–50) stems from a very woody rootstock; leaves small, 9–12 × 3.5–4.5 mm., ovate to ovate-lanceolate - - - - - - - - - - - 8. *multicaulis*
– Plant with fewer (to c. 25) stems from a less woody rootstock; leaves longer, (7)10–26(34) × 1.3–12(15) mm., variable - - - - - - - - - - - - - - 10
10. Leaves 1.3–6 mm. wide, elliptic to linear-lanceolate, mostly ± glabrous (in Flora Zambesiaca area) - - - - - - - - - - - - - - - - - - - 7. *stolzii*
– Leaves mostly wider, 4–12(15) mm., ovate to elliptic, densely hairy or at least with hairs on the midvein - - - - - - - - - - - - - - - - - - 11
11. Corolla glabrous - - - - - - - - - - - - - - - 9. *villicaulis*
– Corolla with scattered long hairs on the outside - - - - - - - 10. *parviflora*

1. **Otiophora scabra** Zucc. in Abh. Akad. München. **1**: 315 (1832). —Drake in Grandidier, Hist. Phys. Madagascar **36** [Hist. Nat. Pl. **6**]: t. 412 (1898). —K. Schum. in Engl. & Prantl, Pflanzenfam. **4**, 4: 133, fig. 42/F (1891). —Verdc. in Journ. Linn. Soc., Bot. **53**: 391, fig. 5, 3/h & 4/s (1950); in F.T.E.A., Rubiaceae 1: 317, fig. 45 (1976). —Puff in Bull. Jard. Bot. Nat. Belg. **51**: 120, fig. 1. (1981). —Gonçalves in Garcia de Orta, Sér. Bot. **5**: 199 (1982). Type from Madagascar.

Annual or rarely perennial herb or subshrub, erect or straggling. Stems c. 15–100 cm. long, slender, ± much-branched, densely hairy above, glabrescent below or rarely subglabrous. Leaves decussate, often pseudo-verticillate; blades (10)15–90 × 2–20(22) mm., ovate to lanceolate to linear-lanceolate, glabrous to densely hairy, particularly above; petioles 0–8 mm. long; stipular sheath with 3–4 setae, c. 1–3 mm. long. Inflorescence many-flowered, spike-like, to c. 25 cm. long when in fruit. Calyx: enlarged, foliaceous lobe c. 3–5 × 1–2 mm., lanceolate, margins usually hairy, setiform lobes small or minute, unequal, densely hairy. Bud with limb abruptly expanded into an ovoid head. Corolla lilac, mauve, blue, reddish or white, tube (2)3.5–6 mm. long, filiform, glabrous, lobes 2–2.5 × 0.5–0.8 mm., lanceolate, usually hairy outside; filaments c. 1.5–2.5 mm. long, glabrous, anthers c. 1 mm. long; style (3)5–8 mm. stigma lobes 1–1.5 mm. long; ovary c. 1 × 0.6–0.8 mm., densely hairy. Fruit yellowish brown to dark purplish brown; mericarps 2 × 0.7–0.9 mm. Seed blackish, c. 1.4–1.6 × 0.8 mm., ovoid, dorsal side convex, slightly keeled, ventral side flat.

Subsp. **scabra**

Pentanisia spicata S. Moore in Journ. Bot. **46**: 38, 76 (1908). TAB. **33**. Type: Zimbabwe, Mazoe, Iron Mask Hill, *Eyles* 522 (BM, holotype; BOL; SRGH).

Otiophora scabra var. *glabra* Verdc. in Journ. Linn. Soc., Bot. **53**: 395 (1950). Type from Angola.

Otiophora "sp. 3" [Zambia, 48 km. S. of Kabwe, *Anton-Smith* in GHS 185736] sensu Verdc. in Garcia de Orta, Sér. Bot. **1**: 27 (1973).

Longest internodes 10–40(50) mm.; fruiting spike to c. 15 cm. long.

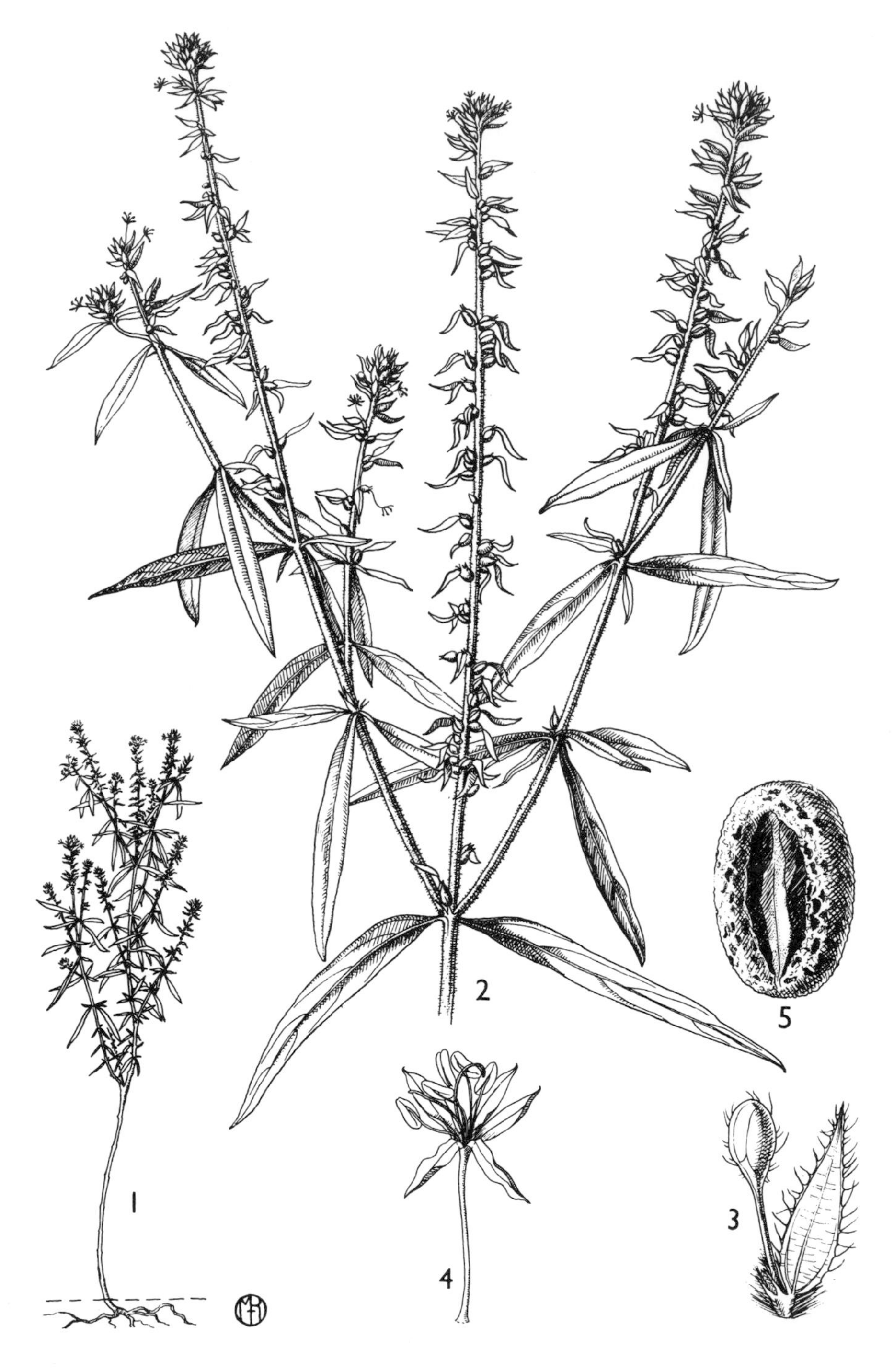

Tab. 33. OTIOPHORA SCABRA subsp. SCABRA. 1, habit (× $\frac{1}{6}$); 2, upper portion of plant (× 1); 3, flower bud (× 7); 4, flower with calyx removed (× 7); 5, seed (× 20), all from *Schlieben* 1951. From F.T.E.A.

Zambia. B: 32 km. NE. of Mongu, fl. & fr. 10.xi.1959, *Drummond & Cookson* 6296 (E; K; LISC; PRE; SRGH). N: Mbala Distr., Kalombe gorge, 1500 m., fl. & fr. 12.ii.1965, *Richards* 19656 (BR; K). W: Mwinilunga Distr., near source of Matonchi R., fl. & fr. 7.x.1937, *Milne-Redhead* 2629 (BR; K; PRE; SRGH). C: Serenje Distr., Kundalila Falls, fl. i., *Williamson* 2137 (SRGH). S: Mazabuka Distr., c. 48 km. N. of Choma, fl. 5.i.1957, *Robinson* 2021 (K; SRGH). **Zimbabwe**. N: Mazoe, Iron Mask Hill, c. 1520 m., fl. i.1905, *Eyles* 604 (BM; K; SAM; SRGH). C: Harare, fl. & fr. 30.i.1979, *Puff* 790130-1/1 (WU). E: Chimanimani Mts., below Mountain Hut, c. 1650 m., fl. & fr. 28.xii.1959, *Goodier & Phipps* 318 (K; PRE; SRGH). S: Mberengwe Distr., Mt. Buhwa, c. 1400–1450 m., fl. & fr. 4.v.1973, *Biegel, Pope & Simon* 4304 (K; LISC; PRE; SRGH). **Malawi**. N: Mzimba Distr., Mzuzu, Marymount, c. 1300–1370 m., fl. & fr. 11.xi.1965, *Pawek* 61 (K; MAL; SRGH). C: Lilongwe Distr., Dzalanyama Forest Reserve, 6 km. SE. of Chaulongwe Falls, 1230–1290 m., fr. 22.iii.1970, *Brummitt* 9266 (K; SRGH). S: Mt. Mulanje, Chambe path, c. 1600 m., fl. 10.ii.1978, *Puff* 780210-1/2 (J; WU). **Mozambique**. N: Serra de Ribáuè (Mepaluè), c. 900 m., fl. & fr., *Torre & Paiva* 11341 (LISC). T: Mt. Furancungo, 1265–1450 m., fl. & fr. 17.iii.1966, *Pereira, Sarmento & Marques* 1868 (LMU). MS: Báruè, Serra de Choa, 24 km. W. of Vila Gouveia, c. 1400 m., fl. & fr., *Torre & Correia* 15424 (LISC).

Also in Tanzania, Burundi, Zaire, Angola and Madagascar. Grassland, scrub, bush- or woodland; also in disturbed ground.

Peculiar pathological forms (sterile flowers with modified, leaf-like floral organs; 'organoid galls'?) occur occasionally (Malawi, Mozambique).

Subsp. **diffusa** (Verdc.) Puff in Bull. Jard. Bot. Nat. Belg. **51**: 126 (1981). Type: Zambia, at source of Matonchi R., *Milne-Redhead* 3740 (K, holotype; BR; PRE).

Otiophora scabra var. *diffusa* Verdc. in Journ. Linn. Soc., Bot. **53**: 395 (1950).

Longest internodes 20–60(70) mm.; fruiting spike to c. 25 cm. long.

Zambia. N: Mbala Distr., road to Mlefu, c. 1520 m., fl. & fr. 23.i.1969, *Sanane* 411 (K; SRGH). W: Zambezi R., c. 48 km. NW. of Mwinilunga, fl. & fr. 24.v., *Angus* 2317 (FHO; K).

Also in Zaire and Angola. Apparently mostly in moist to wet habitats such as mushitos, vleis and marshes.

2. **Otiophora angustifolia** Verdc. in Garcia de Orta, Sér. Bot. **1**: 27, TAB. I (1973). Type: Zambia, Luapula Distr., Mbereshi–Kawambwa road, near Timnatushi turning, *Richards* 12406 (K, holotype; SRGH).

Dwarf shrub or woody herb, several- to many-stemmed, usually ± erect. Stems c. 10–30 cm. long, ± much-branched, with hairs in 2 rows. Leaves decussate, often pseudo-verticillate; blades ascending to ± erect, c. (12)15–32 × (0.8)1–2 mm., linear, glabrous, somewhat coriaceous; petioles subobsolete; stipular sheath with usually 3 minute setae. Inflorescence spike-like, c. 8–15 cm. long. Calyx: enlarged, foliaceous lobe(s) 1(2), c. 1–6 × 0.5–1.2 mm., narrowly oblong-lanceolate, glabrous, the others minute, c. 0.2–4 mm. long. Bud with limb abruptly expanded into an oblong-elliptic head. Corolla white, pale blue or pale pink, glabrous altogether, tube c. 4–6 mm. long, filiform, lobes 2.5–3 × 0.5–0.8(1) mm., oblong-elliptic; filaments c. 1.5–2.5 mm. long, glabrous, anthers c. 0.8–1.2 mm. long; style 5.5–8 mm. long, stigma lobes c. 1–1.8 mm. long; ovary c. 0.8–1.2 × 0.4–0.6 mm., glabrous. Fruit brownish; mericarps c. 1.8–2.2 × 0.5–0.7 mm. Seed dark brown, c. 1.5–2 × 0.6 mm., ± ovoid, dorsal side convex, slightly keeled, ventral side concave.

Zambia. N: Mungwi–Kasama road, 1260 m., fl. & fr. 30.iv.1962, *Richards* 16417 (K; SRGH).
Known only from Zambia. Rocky outcrops.

3. **Otiophora lanceolata** Verdc. in Garcia de Orta, Sér. Bot. **1**: 28, TAB. II (1973). — Puff in Journ. S. Afr. Bot. **47**: 327 (1981). Type: Mozambique, Manica e Sofala, Makurupini R., c. 5 km. from its confluence with Haroni R., E. bank, *Biegel* 3582 (K, holotype; BR; COI; EA; LD; PRE; SRGH).

Subshrub, many-stemmed, straggling. Stems slender, up to c. 1 m. long, usually much-branched, with hairs in 2 rows; internodes (much) longer than leaves. Leaves decussate but sometimes pseudo-verticillate; blades spreading to ascending, 23–40(47) × 3–6(7.5) mm., lanceolate or linear-lanceolate, glabrous except for a few short hairs on margins and midvein above; petioles c. 0–1(1.5) mm. long; stipular sheath with usually 3 small setae. Inflorescence c. 20–30-flowered; ± head-like when in flower, often much elongated, ± spike-like, up to c. 9(12) cm. long when in fruit. Calyx: enlarged, foliaceous lobe c. 2.5–4(5) × 0.8 mm., ± lanceolate, glabrous or sometimes margins shortly hairy, setiform lobes c. 0.2–1 mm. long. Bud claviform, limb and tube not distinct from each other. Corolla white, glabrous outside, throat and upper third of tube a little hairy inside, tube (3.2)3.8–4.5(5) mm. long, narrowly infundibuliform, lobes (2.8)3.2–3.8 × 0.6–1 mm., ± lanceolate; filaments c. 2.8–3.5 mm. long, glabrous, anthers c. 0.6–0.9 mm. long; style

6–9(10.5) mm. long, stigma lobes c. 0.8–1.5 mm. long; ovary c. 0.7–1 × 0.5–0.7 mm., ± densely covered with short, ± curled hairs, the carpel crowned by the enlarged calyx lobe sometimes less hairy than the other and occasionally smaller, ± reduced. Fruit brownish; mericarps 1.5–2 × (0.4)0.6–1 mm., the mericarp crowned by the enlarged calyx lobe sometimes without seed. Seed brownish-black, c. 1.1–1.3 × 0.5–0.9 mm., ± ovoid, dorsal side convex, often ± keeled, ventral side strongly concave.

Zimbabwe. E: Chimanimani, 1.6 km. below Haroni Gorge, 450 m., fl. & fr. 10.i.1969, *Macdonald* 24 (K; SRGH). **Mozambique**. MS: 3.2 km. S. of Makurupini Falls, 390 m., fl. & fr. 25.xi.1967, *Simon & Ngoni* 1312 (SRGH).

Known only from the Haroni-Makurupini area. Mostly in damp to wet grassy areas, below 500 m.

4. **Otiophora inyangana** N.E. Br. in Kew Bull. **1906**: 107 (1906). —Verdc. in Journ. Linn. Soc., Bot. **53**: 397 (1950). Type: Zimbabwe, Inyanga Mts., *Cecil* 203 (K, holotype).

Perennial herb, subshrub or dwarf shrub, usually many-stemmed, ± erect, sometimes forming cushions or mats, with an often branched, massive woody base, distinctly yellow in cross section. Stems 10–90 cm. long, unbranched to ± much-branched, with hairs in 2 rows or hairy all over. Leaves decussate or in whorls of 3(4), sometimes pseudo-verticillate; blades erect, ascending or spreading, (7)10–33 × 1.6–13 mm., ovate to ovate-lanceolate, broadest at lower third or near middle, lanceolate or linear-lanceolate, glabrous or hairy on midvein, lateral veins and (occasionally) upper surface; petioles 0–1.5(2) mm.; stipular sheath with 3(4–5) small setae. Inflorescence several- to many-flowered, head-like, (1)2–4 cm. in diam. when in flower, often much elongated, spike-like, 3–5(12) cm. long when in fruit. Calyx: enlarged, foliaceous lobe (3.5)4–8 × 0.8–2.2(3) mm., ovate-lanceolate to linear-lanceolate, often glabrous, setiform lobes c. 0.2–1 mm. long. Bud claviform, limb and tube not distinct from each other. Corolla whitish, pale pink, pink, pale blue, mauve or purplish, usually glabrous outside, sometimes a little hairy near apex, throat and at least upper half of tube ± densely hairy inside, tube (3.5)4–9 mm. long, narrowly infundibuliform, lobes (2.5)3–5.5 × (0.4)0.6–1.2 mm., lanceolate; filaments 2–5.5(6) mm. long, mostly hairy near base, anthers 0.5–0.9 mm. long; style 6.8–11 mm. long, stigma lobes (0.4)0.6–1.8 mm. long; ovary 0.7–1.2 × 0.5–0.8 mm., sparsely to densely covered with short hairs, the carpel crowned by the enlarged calyx lobe often less hairy than the other. Fruit greenish to brownish; mericarps 1.5–2.5 × 0.6–1.2 mm. Seed black, (1)1.2–1.8 × 0.4–0.8 mm., ± ovoid, dorsal side convex, ± keeled, ventral side flat or concave.

Subsp. **inyangana**

Leaves decussate or in whorls of 3(4), (7)10–33 × (3.8)4.5–13 mm., ovate to ovate-lanceolate, broadest at lower third or near middle. Plants mostly not densely leafy, internodes (15)20–40(60) mm. long (shorter only in low, ± cushion-forming plants). Inflorescence usually many-flowered. Corolla tube (3.5)4–6.8(8) mm. long.

Zimbabwe. E: Inyanga Distr., Mare R., fl. & fr. 21.x.1946, *Wild* 1555 (K; SRGH). **Mozambique**. MS: Tsetserra, 2140 m., fl. & fr. 7.ii.1955, *Exell, Mendonça & Wild* 237 (BM; LISC; SRGH).

Known only from the E. highlands of Zimbabwe and neighbouring parts of Mozambique. Open grassland or sometimes at the edge of scrub or forest or on rocky outcrops; 1200–2580 m.

Highly variable in growth habit (plants erect to low and cushion- or mat-forming; stems woody and ± much-branched or flowering shoots thin and unbranched), indumentum, size, shape and arrangement of the leaves, size of the inflorescence and size and shape of the foliaceous calyx lobe. Environmental factors (exposure or non-exposure to veld fires; altitude) contribute to the variability; cf. Puff, op.cit. for details.

Subsp. **parvifolia** (Verdc.) Puff in Journ. S. Afr. Bot. **47**: 325 (1981). Type: Zimbabwe, Chimanimani Mts., *Munch* 87 (K, holotype; SRGH).

Otiophora inyangana var. *parvifolia* Verdc. in Journ. Linn. Soc., Bot. **53**: 399 (1950).

Leaves strictly decussate, 10–18 × 1.6–3.5(4) mm., mostly linear-lanceolate to lanceolate. Plants quite densely leafy, internodes only (2)7–25 mm. long. Inflorescence fewer-flowered. Corolla tube (4)5.5–9 mm. long.

Zimbabwe. E: Chimanimani Mts., Long Gulley, c. 1650 m., fl. & fr., *Noel* 2150 (SRGH). **Mozambique**. MS: Chimanimani Mts., unnamed triple falls on tributary below Martin Falls, c. 1200 m., fl. & fr., *Taylor* 1796 (E; NU; SRGH).

Endemic to the Chimanimani Mts. Open, sometimes rocky, grassland areas or at the edge of scrub; possibly confined to quartzite areas; c. 1200–1800(2300) m.

5. **Otiophora pycnostachys** K. Schum. in Engl., Pflanzenw. Ost-Afr. **C**: 393 (1895). —Verdc. in Journ. Linn. Soc., Bot. **53**: 402, fig. 4/G, M, Q (1950); in F.T.E.A., Rubiaceae 1: 319 (1976). —Puff in Bull. Jard. Bot. Nat. Belg. **51**: 129 (1981). Type from Tanzania.

Perennial herb or subshrub with a woody base, several-stemmed, erect or straggling. Stems 15–100 cm. long, branched or sometimes unbranched, with 2 rows of hairs above, glabrescent below. Leaves decussate, pseudo-verticillate; blades (10)15–60 × (1)2–10(12) mm., lanceolate, linear-lanceolate or linear, less commonly ovate-lanceolate, glabrous or a little hairy on margins and midvein above; petioles (sub)obsolete; stipular sheath with 1–5 filiform setae, 1–3(4) mm. long. Inflorescence many-flowered, head-like or subcylindrical and to c. 20 mm. long when in flower, often somewhat elongated and more spike-like when in fruit. Calyx: enlarged, foliaceous lobe c. 3–6(8) × 0.5–1(2) mm., linear-lanceolate, glabrous or hairy on margins, setiform lobes minute, hairy. Bud with limb abruptly expanded into a subcylindrical head. Corolla white, pinkish or crimson, tube (3)4–8 mm. long, filiform, glabrous, lobes 3–5.5 × 0.8–1 mm., lanceolate, often a little hairy near apex; filaments 3–5 mm. long, with long white hairs near base, anthers 0.8–1.2 mm. long; style 7–10 mm. long, stigma lobes c. 1 mm. long; ovary 1–1.5 × 1 mm., mostly rather densely hairy. Fruit yellowish brown; mericarps 2.5–3(4) × 0.7–1 mm. Seed brownish, c. 2–2.5(3) × 0.6–0.8 mm., oblong, dorsal side convex, somewhat keeled, ventral side ± flat.

Zambia. N: c. 80 km. S. of Mbala, fl. & fr. 18.vii.1930, *Hutchinson & Gillett* 3835 (BM; K; SRGH). W: Kasompe R., W. of Mwinilunga, fl. 31.x.1937, *Milne-Redhead* 3031 (BR; K; PRE).

Also in Uganda, Tanzania, Zaire (mainly Shaba) and Angola. Marshy or swampy places, bogs.

Variable in leaf size and shape and indumentum; rather densely hairy forms, however, have so far only been recorded from Shaba.

6. **Otiophora caerulea** (Hiern) Bullock in Kew Bull. **1933**: 471 (1933). —Verdc. in Journ. Linn. Soc., Bot. **53**: 404, fig. 3/B, C, fig. 4/E, R (1950); in Garcia de Orta, Sér. Bot. **1**: 29 (1973); in F.T.E.A., Rubiaceae 1: 321 (1976). —Puff in Bull. Jard. Bot. Nat. Belg. **51**: 131, fig. 2 (1981). Type from Angola.

Pentanisia caerulea Hiern, Cat. Afr. Pl. Welw. **1**, 2: 471 (1898). Type as above.

Otiophora pulchella De Wild. in Ann. Mus. Congo, Sér. Bot. **4**: 230 (1903). Type from Zaire.

Perennial herb, several-stemmed, erect, with an often branched, massive woody rootstock. Stems 10–45 cm. long, often unbranched, with 2 rows of hairs or, rarely hairy all over. Leaves decussate, sometimes pseudo-verticillate; blades often ascending or erect, (18)25–55(80) × 2–12(15) mm., elliptic-lanceolate to linear, sometimes ovate, mostly glabrous; petioles (sub)obsolete; stipular sheath with 1–3 lanceolate to filiform setae c. 1–3(5) mm. long. Inflorescence several- to many-flowered, head-like, 1–2.5(4) cm. in diam. Calyx: enlarged, foliaceous lobe(s) 1(2), 5–10 × 1–2.5 mm., linear to (elliptic-) lanceolate, mostly glabrous, setiform lobes c. 0.3–1(2) mm. long. Bud with limb abruptly expanded into an ovoid or subcylindrical head. Corolla blue to mauve or pinkish-mauve, tube (5)7–12 mm. long, filiform, glabrous, lobes (4)5–6 × 0.7–1 mm., lanceolate, sometimes a little hairy outside; filaments (3)4–5 mm. long, glabrous, anthers c. 1 mm. long; style 8–14 mm. long, stigma lobes 1–3 mm. long; ovary c. 1 × 1 mm., glabrous. Fruit yellowish brown to dark brown, subglobose; mericarps 1.5–3 mm. in diam. Seed dark brown, 1.2–2.5 mm. in diam., subglobose, dorsal side somewhat keeled.

Zambia. N: Mbala Distr., escarpment above Inono R., c. 1220 m., fl. 12.iii.1955, *Richards* 4907 (BR; K; SRGH). W: Mwinilunga Distr., Sinkabolo Dambo, fl. & fr. 29.9.1938, *Milne-Redhead* 4388 (BR; K; PRE). **Malawi**. N: Nyika plateau, c. 1830–2130 m., fl. vii.1896, *Whyte* 193 (K).

Also in Tanzania, Burundi, Zaire (mainly Shaba) and Angola. Open or wooded grassland, sometimes in (seasonally) damp areas; 1220–2130 m.

Very variable in leaf size and shape but fairly uniform in floral characters and growth form. Unusual hairy mutants have been recorded from the Mwinilunga area (and Shaba); they appear to be rare and occur next to 'typical' glabrescent specimens.

7. **Otiophora stolzii** (Verdc.) Verdc. in Kew Bull. **14**: 351 (1960); in F.T.E.A., Rubiaceae 1: 320 (1976). Type from Tanzania.

Otiophora parviflora var. *stolzii* Verdc. in Journ. Linn. Soc. **53**: 407 (1950); in Bull. Jard. Bot. Brux. **23**: 63 (1953). Type as above.

Perennial tufted or cushion-forming herb with up to c. 25 stems from a somewhat woody rootstock. Stems 10–20 cm. long, erect or suberect, often unbranched, mostly with 2 rows of hairs. Leaves decussate; blades (7)10–20 × 1.3–6 mm., elliptic to linear-lanceolate, glabrous or sometimes a little hairy above; petioles subobsolete or to c. 1 mm. long; stipular sheath with 3 filiform setae to c. 1 mm. long. Inflorescence several-flowered, head-like, c. 1 cm. in diam. when in flower, somewhat elongated when in fruit. Calyx: enlarged, foliaceous lobe(s) 1(2), 3–6 × 1–2 mm., elliptic to linear-lanceolate, often glabrous, setiform lobes less than 1 mm. long. Bud with limb abruptly expanded into an ovoid head. Corolla white, pink, lilac or pale blue, tube 3.5–5.5 mm. long, filiform, glabrous, lobes 2–3 × 0.8–1 mm., lanceolate, sometimes a little hairy outside, filaments c. 1–2 mm. long, glabrous, anthers to c. 0.7 mm. long; style 5–6 mm. long, stigma lobes 1–1.5 mm. long; ovary c. 1 × 1 mm., densely hairy. Fruit yellowish brown to dark brown; mericarps 2–2.5 × 1–1.5 mm. Seed brown, c. 1.5–2 × 0.8–1.2 mm., oblong, dorsal side convex, slightly keeled, ventral side ± flat.

Malawi. N: Nyika Plateau, fl. & fr. 14.iii.1961, *Robinson* 4502A (SRGH).
Also in Tanzania and Zaire (Shaba). Grassland; rocky places.
Verdcourt (in F.T.E.A.) records this species from both the Malawi and Zambia side of the Nyika Plateau, but I have seen no specimen from Zambia.
O. stolzii and the following three species (*O. multicaulis*, *O. villicaulis* and *O. parviflora*) are very closely allied ['*Otiophora villicaulis* complex' —cf. Puff in Bull. Jard. Bot. Nat. Belg. **51**: 134 (1981)]. *O. stolzii* comes closest to *O. multicaulis;* the cited collection, the only one seen from the Flora Zambesiaca area, somewhat approaches *O. multicaulis*. The above description was drawn up from the cited collection and specimens from Tanzania and Zaire (Shaba).

8. **Otiophora multicaulis** Verdc. in Journ. Linn. Soc., Bot. **53**: 405 (1950). Type: Zambia, Mwinilunga Distr., SW. of Dobeka Bridge, *Milne-Redhead* 2739 (K, holotype; BR; PRE).

Perennial tufted or cushion- or mat-forming herb with 20–50 stems from a ± massive woody rootstock. Stems c. 8–15 cm. long, erect, branched or unbranched, hairy all over or with 2 rows of hairs. Leaves decussate; blades 9–12 × 3.5–4.5 mm., ovate to ovate-lanceolate, often a little hairy on margins and veins, especially below, or glabrescent; petioles subobsolete or to c. 1 mm. long; stipular sheath with 1–3 minute setae. Inflorescence few- to several-flowered, head-like, to c. 1 cm. in diam. Calyx: enlarged, foliaceous lobe 3–5 × 1–1.5(2) mm., ovate-lanceolate to lanceolate, often a little hairy, setiform lobes minute. Bud with limb abruptly expanded into an ovoid head. Corolla lilac, tube 2–4 mm. long, filiform, glabrous, lobes 2–2.5 × 0.5–0.8 mm., lanceolate, glabrous or with a few short hairs outside, filaments c. 1–1.7 mm. long, glabrous, anthers c. 0.5 mm. long; style 3–3.5 mm. long, stigma lobes 1–1.5 mm. long; ovary 1 × 0.8–1 mm., hairy. Fruit brownish; mericarps c. 2 × 1.5 mm. Seeds brown, c. 1.5 × 1 mm., (sub)globose to ± oblong, dorsal side convex, slightly keeled, ventral side ± flat.

Zambia. W: Mwinilunga Distr., 16 km. along road from Matonchi Farm, 1200 m., fl. & fr. 17.xi.1962, *Richards* 17268 (K; LISC).
Also in Zaire (Shaba). Grassland.
Seed-shape, used to distinguish this species from *O. stolzii* (above) by Verdcourt [in Garcia de Orta, Sér. Bot. **1**: 26 (1973)], proved to be an unreliable character (too variable).

9. **Otiophora villicaulis** Mildbr. in Notizbl. Bot. Gart. Berlin **15**: 637 (1941). —Verdc. in Journ. Linn. Soc., Bot. **53**: 411 (1950); in F.T.E.A., Rubiaceae 1: 322 (1976). —Puff in Bull. Jard. Bot. Nat. Belg. **51**: 137 (1981). Type from Tanzania.

Perennial tufted or cushion- or mat-forming herb with several to c. 20 stems from a somewhat woody rootstock. Stems 4–12(20) cm. long, mostly (sub)procumbent, often branched, (densely) hairy all over or sometimes with hairs in 2 rows. Leaves decussate; blades 10–20(25) × 5–12(15) mm., ovate to round, ovate-lanceolate or elliptic, slightly to densely hairy, or with a few hairs on midvein, at least above, or (rarely) ± glabrous; petioles subobsolete to c. 1 mm. long; stipular sheath with 3(5) filiform setae to c. 2 mm. long. Inflorescence ± few- to several-flowered, head-like, 1–1.5 cm. in diam. Calyx: enlarged, foliaceous lobe(s) 1(2), 2–5 × 1–2 mm., ovate-lanceolate to lanceolate, hairy or glabrescent, setiform lobes minute. Bud with limb abruptly expanded into an ovoid head. Corolla (bluish) lilac, mauve or red, glabrous, tube (4)5–6 mm. long, filiform, lobes 1.5–3 × 0.5–1 mm., (ovate)lanceolate; filaments c. 1.5–2 mm. long, glabrous, anthers 0.7–1 mm. long, style 5.5–8 mm. long, stigma lobes 1.5–2 mm. long; ovary c. 1 × 1 mm., densely hairy. Fruit yellowish brown; mericarps c. 1.5–2 × 1–1.5 mm. Seed dark brown, 1–1.5 × 0.8–1.2

mm., (sub)globose to ± oblong, dorsal side convex, somewhat keeled, ventral side ± concave.

Leaves mostly ovate to round, 6–12(15) mm. wide, slightly to densely hairy var. *villicaulis*
Leaves mostly ovate-lanceolate to elliptic, 5–8 mm. wide, mostly with a few hairs on midvein - - - - - - - - - - - - - - - - var. *iringensis*

Var. **villicaulis**

Otiophora latifolia Verdc. in Journ. Linn. Soc., Bot. **53**: 406 (1950). Type from Zaire.
Otiophora latifolia var. *villosa* Verdc. in Journ. Linn. Soc., Bot. **53**: 407 (1950). Type: Zambia, Mwinilunga Distr., Kalenda Plain, *Milne-Redhead* 4410 (K, holotype; BR; PRE).
Otiophora villicaulis var. *villosa* (Verdc.) Verdc. in F.T.E.A., Rubiaceae 1: 323 (1976). Type as above.
Otiophora latifolia var. *bamendensis* Verdc. in Journ. Linn. Soc., Bot. **53**: 407 (1950). Type from Cameroon.

Stems (densely) hairy all over.

Zambia. N: Mbala Distr., Kambole, c. 1520 m., fl. 1896, *Nutt* s.n. (K). W: Mufulira, c. 1220 m., fl. & fr. 13.ii.1949, *Cruse* 483 (K). **Malawi**. N: Nyika Plateau, near Nganda Peak, c. 2440 m., fl. & fr. 9.i.1974, *Pawek* 7912 (K; MO). C: Ntchisi Forest Reserve, 1950 m., fl. & fr. 26.iii.1970, *Brummitt* 9427 (K; SRGH; approaching var. *iringensis*). S: Zomba Plateau, Queen's View, fl. & fr., *Salubeni* 227 (MAL; SRGH).
Also in Tanzania, Burundi, Zaire (Shaba), Angola, Cameroon and Nigeria. Open or wooded grassland, often in disturbed areas, bare patches, fire-breaks, etc.

Var. **iringensis** (Verdc.) Puff in Bull. Jard. Bot. Nat. Belg. **51**: 139 (1981). Type from Tanzania.
Otiophora parviflora var. *iringensis* Verdc. in Journ. Linn. Soc., Bot. **53**: 408 (1950); in F.T.E.A., Rubiaceae 1: 323 (1976). Type as above.

Plants generally less densely hairy than those of var. *villicaulis*.

Malawi. N: Nkhata Bay Distr., Mzuzu junction to Chikwina, km. 10, c. 1520 m., fl. & fr., *Pawek* 8832 (MO). C: Ntchisi Mt., 1540 m., fl. & fr. 20.ii.1959, *Robson* 1679 (K; LISC; MO; SRGH).
Also in Tanzania. Open or wooded grassland.
Some collections from Malawi (Nyika and Zomba Plateau, Ntchisi Mt.) are ± intermediate between var. *villicaulis* and var. *iringensis*.

10. **Otiophora parviflora** Verdc. in Journ. Linn. Soc., Bot. **53**: 407 (1950); in F.T.E.A., Rubiaceae 1: 323 (1976). Type: Zambia, just NE. of Mufulira, *Cruse* 161 (K, holotype).

Perennial tufted herb with few to c. 20 stems from a ± woody rootstock. Stems 6–15(20) cm. long, ± erect, often unbranched, hairy all over or with 2 rows of hairs. Leaves decussate; blades (7)10–26(34) × 4–10(12) mm., elliptic or sometimes ovate, with short hairs above and longer hairs below, at least on midvein, seldom ± glabrous above; petioles c. 1–2 mm. long; stipular sheath with 3 subulate setae to 1 mm. long. Inflorescence several-flowered, head-like, c. 1 cm. in diam. Calyx: enlarged, foliaceous lobe(s) 1(2), 3–6.5 × 1–2.5 mm., lanceolate, hairy or glabrescent, other lobes unequal, some occasionally obsolete. Bud with limb abruptly expanded into an ovoid head. Corolla white, lilac-blue or reddish, hairy; tube 2.5–5(6) mm. long, filiform, lobes 2–2.5 × 0.5–0.8 mm., lanceolate, filaments 1.5–2 mm. long, glabrous, anthers 0.5–0.7 mm. long; style 3.5–6 mm. long, stigma lobes 1.5–1.8 mm. long; ovary c. 1 × 1 mm., densely hairy. Fruit yellowish-brown to dark brown; mericarps c. 2 × 1.5 mm. Seed brownish, 1–1.5 × 1 mm., (sub)globose, dorsal side convex, ± keeled, ventral side concave.

Zambia. W: just NE. of Mufulira, fr. 3.ii.1949, *Cruse* 475 (K). C: Serenje Distr., 8 km. SE. of Kanona, on road to Kundalila Falls, fl. 16.ii.1970, *Drummond & Williamson* 9629 (K; SRGH).
Also in Zaire (Shaba) and ?Tanzania. Open or wooded grassland, sometimes in seasonally damp areas.

Tribe 10. **ANTHOSPERMEAE**

Shrubs, dwarf shrubs, short-lived subshrubs or perennial herbs. Flowers ⚥, ♂ or ♀, sexes variously distributed, plants quite often dioecious; corolla cylindrical to subcampanulate, in ⚥ and ♂ always larger than in ♀; filaments inserted at base or at least below middle of corolla tube; disk absent; ovary mostly 2- or 1-locular, each locule with a single basally attached anatropous ovule; style 0 or very short; stigmas long, hairy. Fruit dry or fleshy, dehiscing into 2 mericarps or into exocarp-valves and endocarp plus seed, or indehiscent. Seed with membranous testa and copious endosperm, remaining enclosed in endocarp.

A tribe centred in tropical, subtropical and temperate regions of the S. hemisphere. All African and Madagascan taxa belong to subtribe *Anthosperminae,* which is distinguished by *dry* fruits and unisexual and/or *protandrous* hermaphrodite flowers.

Wind-pollination (flowers small and inconspicuous, odourless, without disk, often unisexual; anthers dangling on long, slender filiform filaments; stigmas long exserted, hairy all around, etc.) is characteristic for the entire tribe. See Puff in Journ. Linn. Soc., Bot. **89**: 357 (1982) for further details.

Inflorescence made up of axillary, ± sessile flower clusters; fruit supported by a carpophore - - - - - - - - - - - - - - **29. Anthospermum**
Inflorescence terminal, paniculate to thyrso-paniculate, fruit never supported by a carpophore - - - - - - - - - - - - - - - **30. Galopina**

29. ANTHOSPERMUM L.

Anthospermum L., Sp. Pl. **1**: 1058 (1753); Gen. Pl., ed. 5: 479 (1754).

Large shrubs, dwarf shrubs, short-lived subshrubs or perennial herbs. Leaves decussate or occasionally in whorls of 3 (rarely 4), often seemingly in much larger numbers at nodes*, blades ± broad and large to ± ericoid and small, mostly narrowed to base, acute to acuminate (seldom ± mucronate or ± obtuse) at apex, shortly petiolate to sessile, with ± cup-shaped stipular sheaths bearing one to many setae** or fimbriae on either side. Inflorescence frequently leafy and inconspicuous, made up of mostly subsessile, many- to very few-flowered cymes, in dioecious taxa often sexually dimorphic (♀ inflorescence contracted, ± cylinder-like). Flowers mostly subsessile, subtended by a pair of leafy bracts, ♂, ⚥ or ♀, 4–5-merous. Calyx: lobes large, conspicuous to small, indistinct or ± lacking. ⚥, ♂: corolla tube ± cylindrical, broadly funnel-shaped to subcampanulate, lobes recurved, ± lanceolate; anthers yellowish to whitish, exserted, dangling on long slender filiform filaments. ♀: corolla much smaller; tube cylindrical, lobes mostly erect, linear to ± lanceolate. Ovary bicarpellate and biovulate [in 2 SW. Cape species one carpel reduced]; style 0 or very short; stigmas 2 [only in one SW. Cape species 1] long exserted, hairy, greyish to greenish white, seldom purplish red. Fruit crowned by persistent calyx lobes, supported by a ± U-shaped carpophore (cf. TAB. **34**, figs. B5 and D3.), dehiscing into two mericarps, each convex on dorsal side, plane to concave or sometimes hollowed out and with a prominent to inconspicuous median longitudinal ridge on ventral side.

A genus of 39 species widely distributed in Africa south of the Sahara and in Madagascar; the highest species concentration is found in S. Africa. 9 species occur in the Flora Zambesiaca area.
The species are difficult. Some of the reasons for this are:
— Environment-induced morphological variability. Individuals exposed to fire and capable of resprouting from a woody base often differ markedly from shrubby plants of the same taxon sheltered from fire.
— The occurrence of characters which are constant in some but variable in other taxa. Notably the arrangement of the leaves — e.g., decussate *or* in whorls of 3 or 4 in *A. ternatum* (variable even within populations), or either strictly decussate or strictly in whorls of 3 (e.g., *A. welwitschii* vs. *A. whyteanum*). For this reason (and also in view of the variable sex distributions within a taxon: see below) taxa should be studied with care in the field and representative samples should be collected.
— The occasional occurrence of flowers 'transitional' between ⚥ and 'pure' ♀. Such 'transitional' flowers have corollas intermediate in size between ⚥ and ♀ and are characterized by the presence of small but clearly discernible, (±) pollenless anther rudiments; 'pure' ♀ with small corollas have no rudimentary anthers. In some essentially dioecious taxa, ♂ flowers do occasionally revert to ⚥; in some it is only odd ⚥ flowers on a single ♂ plant of a population, in others a considerable number of individuals may show this phenomenon. Also the reverse (transitions

* Due to the presence of leafy much-contracted short shoots (cf. TAB. **34**, fig. A.); this situation is referred to as 'pseudo-verticillate' in the key and in the descriptions.
** Single median setae are often flanked by a pair of minute gland-tipped setae. As the latter are often no longer discernible on older leaves/stipular sheaths, no mention is made of them in the following descriptions.

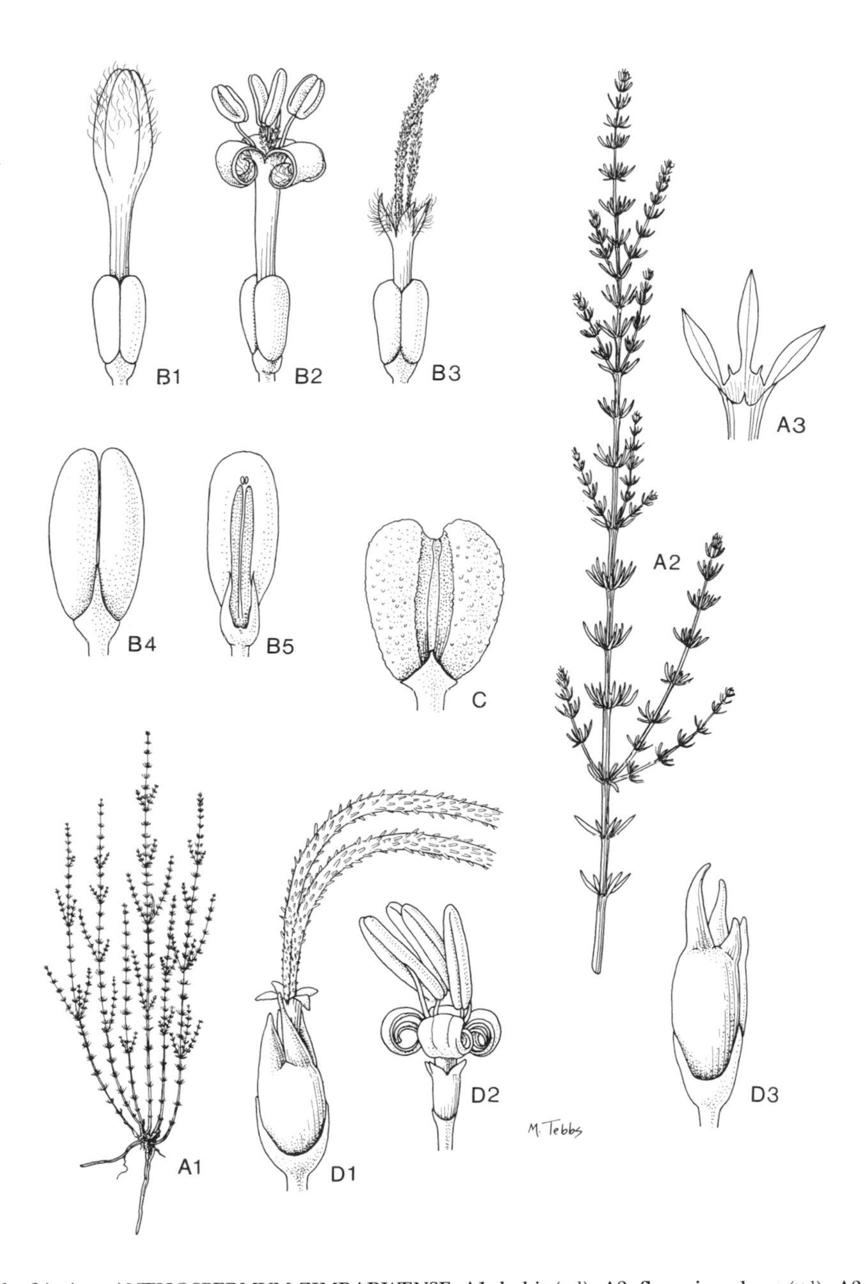

Tab. 34. A. —ANTHOSPERMUM ZIMBABWENSE. A1, habit (× $\frac{1}{4}$); A2, flowering shoot (× $\frac{1}{2}$); A3, stipular sheath (× 20), A1–3 from *Puff* 790125-3/1. B. —ANTHOSPERMUM TERNATUM subsp. RANDII. B1, hermaphrodite bud (× 15); B2, hermaphrodite flower (× 15); B3, female flower (× 15), B1–3 from *Puff* 780215-3/1; B4, fruit (two mericarps), lateral view (× 20); B5, mericarp, dorsal view, with attached carpophore (× 20), 4–5 from *Bingham* 1211. C. —ANTHOSPERMUM HERBACEUM. Fruit (two mericarps), lateral view, note broad, longitudinal groove between mericarps (× 25), from *Puff* 810916-3/1. D. —ANTHOSPERMUM VALLICOLA. D1, female flower (× 25); D2, male flower (× 25); D3, fruit with attached carpophore (× 25), all from *Puff* 790125-1/1.

from ⚥ to ♂) is possible (*Note*: the flowers are always protandrous; ⚥ flowers at anthesis — i.e., with stigmas which have not yet fully elongated — should, therefore, not be confused with such transitional flowers). In both transitions ♂→⚥ and ⚥→♂ there are, however, no marked corolla size and shape differences.

— Sex dimorphism in dioecious taxa. Differences between the sexes may go beyond corolla size and shape differences; dimorphic inflorescences are rather common (contracted and often quite conspicuous in ♀); occasionally ♂ and ♀ differ somewhat in leaf size and shape. It is recommended that both ♂ and ♀ are collected whenever possible.

Hybridization represents an additional problem. Experiments carried out in the greenhouse proved that (not even so closely allied) species of *Anthospermum* can be crossed and form viable hybrids. Thus, virtually anywhere where two species happen to grow sympatrically and flower at the same time, hybrids can be expected to occur. Especially in the Eastern Highlands of Zimbabwe (and adjacent parts of Mozambique), where no less than 7 taxa (out of a total of 10 in Flora Zambesiaca area) may occur together, the frequency of putative hybrids appears to be quite high. Because of the variability of numerous taxa and the presence of only few 'good' and useable distinguishing characters between taxa, definite proof for the hybrid nature of an individual or an entire population is not easily obtained. One can, thus, not always be certain whether an 'atypical' specimen still belongs to one taxon (i.e. represents a geno- and/or phenotypic 'extreme') or is of hybrid origin.

1. Erect, mostly single-stemmed shrubs c. 1–3 m. tall - - - - - - - - 2
– Several- to many-stemmed dwarf shrubs, subshrubs, 'woody herbs' or perennial herbs 6
2. Leaves decussate - - - - - - - - - - - - - - - - - - 3
– Leaves in whorls of 3 - - - - - - - - - - - - - - - - - 4
3. Leaves 10–35 × (1)1.5–3.5 mm., strictly decussate - - - - - - - 1. *welwitschii*
– Leaves (2)3–10(15) × 0.5–1.5(2) mm., occasionally decussate, more commonly in whorls of 3; plants mostly densely leafy above, 'ericoid' in appearance - - - 2. *usambarense*
4. Leaves (2)3–10(15) mm. wide, surfaces very shortly hairy - - - 4. *ammannioides*
– Leaves 0.5–2(2.5) mm. wide, mostly glabrous - - - - - - - - - - 5
5. Leaves (2)3–10(15) mm. long; stipular sheaths with 3–7(8) setae between two neighbouring leaves; persistent calyx lobes 0.2–0.6 mm. long - - - - - - - 2. *usambarense*
– Leaves (10)15–35(55) mm. long; stipular sheaths mostly with one seta between two neighbouring leaves; persistent calyx lobes (0.5)0.8–1.8(2.1) mm. long - - - - 5. *vallicola*
6. Plants dioecious; leaves always in whorls of 3, glabrous, to 8(12) mm. long; ♂ flowers with (± broadly) funnel-shaped corolla tubes 0.5–1 mm. long; mericarps 1.4–1.7 mm. long - - - - - - - - - - - - - - - 3. *zimbabwense*
– Plants not dioecious; leaves decussate or in whorls of 3(4), if in whorls of 3 either variously hairy or glabrous and to 25(30) mm. long; ♂ and ⚥ flowers with cylindrical to (narrowly) funnel-shaped corolla tubes to 2.2(3.7) mm. long; mericarps to 3 mm. long - - - - - 7
7. Leaves on both surfaces densely covered with spreading hairs 0.2–0.5(0.9) mm. long; mostly distinct dwarf shrubs; mericarps usually crowned by calyx lobes (0.2)0.3–0.7 mm. long - - - - - - - - - - - - - - - - - 6. *whyteanum*
– Leaves glabrous or variously hairy (but not as above); perennial herbs or subshrubs; mericarps not crowned by distinct calyx lobes or calyx lobes minute, to 0.3 mm. long - - 8
8. Leaves in whorls of 3, decussate or occasionally in whorls of 4; stipular sheaths with one hairy, subulate seta (0.7)1–3(3.7) mm. long between 2 neighbouring leaves; mericarps 2–3 × 0.8–1.2 mm. - - - - - - - - - - - - - - - 7. *ternatum*
– Leaves strictly decussate; stipular sheaths with one shorter, glabrous seta or (3)5(7) setae between 2 neighbouring leaves; mericarps smaller and broader - - - - - - - 9
9. Often scrambling, straggling or trailing perennial herbs; leaves 5–55 × (1)2–25 mm., ovate to lanceolate; ♂ and ⚥ flowers with cylindrical corolla tubes (1.5)2.3–7 mm. long; fruits with ± broad, conspicuous longitudinal grooves between the mericarps - - 8. *herbaceum*
– Mostly subshrubs with numerous short, mostly unbranched stems; leaves (4)6–12(22) × (0.5)0.8–1.5(2) mm., linear-lanceolate; ♂ and ⚥ flowers with narrowly funnel-shaped corolla tubes (0.5)0.7–1.4(1.7) mm. long; fruits without ± broad, conspicuous longitudinal grooves between the mericarps - - - - - - - - - - 9. *rigidum* subsp. *pumilum*

1. **Anthospermum welwitschii** Hiern, Cat. Afr. Pl. Welw. **1**, 2: 500 (1898). —Brenan in Mem. N.Y. Bot. Gard. **8**: 455 (1954). —Verdc. in F.T.E.A., Rubiaceae 1: 330 (1976). —Puff in Fl. Southern Afr. **31**, 1(2): 12 (1986); in Pl. Syst. Evol., Suppl. 3: 190 (1986). Type from Angola.
Anthospermum cliffortioides K. Schum. in Engl., Bot. Jahrb. **30**: 416 (1901). Type from Tanzania.

Dioecious shrub, single- or (seldom) several-stemmed, ± erect, 1–3 m. tall. Stems mostly much-branched above, branches usually ± regular, paired, often ascending, glabrous to very shortly hairy. Leaves decussate, mostly pseudo-verticillate; blades 10–35 × (1)1.5–3.5 mm., (narrowly) oblanceolate, elliptic to ± linear-lanceolate, mostly glabrous; margins ± flat; petioles 0–1 mm. long; stipular sheath with 3–5(7,8) setae, the longest (0.5)0.7–4.5(5.6) mm. Flowers subsessile to shortly pedicellate (pedicels to 0.7 mm. in ♂), in clusters of many (very many: ♀) at nodes, inflorescences dimorphic, in ♀ often quite contracted, dense, ± cylindrical inflorescence zones; corolla 4-merous, greenish-yellow to pale yellow, occasionally purplish tinged outside, mostly glabrous. ♂: tube (0.5)0.7–1.2 mm. long, funnel-shaped, lobes (1.2)1.5–2.2(2.7) × (0.4)0.6–0.9(1.1) mm.; anthers 1–2 mm. long; small rudimentary ovary with 4 minute calyx lobes. ♀: tube 0.2–0.5 mm. long, lobes 0.2–0.7 × 0.1–0.2 mm.; style 0–1 mm. long, stigmas 3–7.5(10) mm. long; ovary 0.6–1 × 0.3–0.6 mm., with 4 sometimes unequal calyx lobes. Fruit mostly reddish-brown, shiny; mericarps 1.5–2.7 × 0.7–1.2 mm., oblong, elliptic to ± obovate, glabrous or (seldom) ± sparsely shortly hairy, with 2 ± triangular calyx lobes (0.3)0.5–1(1.2) × 0.2–0.5 mm., one occasionally longer than the other.

Zambia. W: Solwezi, Luamisamba R., fl. 14.iv.1969, *Mutimushi* 3119 (K; SRGH). E: Nyika Plateau, fr. 31.xii.1962, *Fanshawe* 7385 (K). **Malawi**. N: Misuku Hills, Mughesse Forest, fl. 19.xii.1978, *Puff* 781219-1/4 (WU). S: Mt. Mulanje, 2400 m., fl. & fr. 28.vi.1946, *Brass* 16495 (BR; K; MO; SRGH; US).

Widely but disjunctly distributed mostly in mountain areas in Kenya, Tanzania, Zaire (Shaba), Angola and S. Africa (Transvaal). At the edge of forest or scrub or in streambank vegetation; (?)2000–2400 m.

The cited specimen from Zambia W, the only collection known from that region, is rather atypical.

See also *A. usambarense*, below, and 6., *A. whyteanum* for details on the suspected occurrence of hybrids with these species.

2. **Anthospermum usambarense** K. Schum. [in Engl., Bot. Jahrb. **17**: 165 (1893); in Abhandl. Preuss. Akad. Wiss. **1894**: 69 (1894): in observ.] in Engl., Bot. Jahrb. **28**: 112 (1899). —Orth in Walter, Vegetationsbilder **25**(8): t. 43a & b (1940). —Robyns, Fl. Sperm. Parc Nat. Albert **2**: 372, t. 37, fig. 15 (1947). —Brenan in Mem. N.Y. Bot. Gard. **8**: 455 (1954). —Hedberg, Afroalp. Vasc. Pl.: 176, 328 (1957). —Dale & Greenway, Kenya Trees & Shrubs: 425 (1961). —Verdc. in F.T.E.A., Rubiaceae 1: 331 (1976). —Puff in Pl. Syst. Evol., Suppl. 3: 196 (1986). Syntypes from Tanzania.

Anthospermum leuconeuron K. Schum. in Engl., Bot. Jahrb. **30**: 416 (1902). Type from Tanzania.

Anthospermum prittwitzii K. Schum. & K. Krause in Engl., Bot. Jahrb. **39**: 570 (1907). Type from Tanzania.

Anthospermum keilii K. Krause in Engl., Bot. Jahrb. **48**: 431 (1912). Type from Burundi.

Anthospermum aberdaricum K. Krause in Notizbl. Bot. Gart. Berlin **10**: 609 (1929). Syntypes from Kenya.

Dioecious shrub, single-stemmed, erect, 1–3 m. tall. Stems mostly much-branched above, branches ± regular, arising in threes (less commonly paired), often ascending to ± erect, usually shortly hairy, ± densely leafy. Leaves in whorls of 3 or, less often, decussate, pseudo-verticillate; blades (2)3–10(15) × 0.5–1.5(2) mm. (on new growth occasionally to 20 × 3 mm.), narrowly obovate, oblanceolate, oblong-elliptic, lanceolate to ± linear, mostly glabrous; margins often (strongly) revolute; petioles subobsolete; stipular sheath shortly hairy, with 3–7(8) setae, the longest 1–4.1 mm. long. Flowers subsessile, in clusters of many at nodes, inflorescences rather dimorphic, in ♀ often quite contracted, dense, ± cylindrical inflorescence zones; corolla 4-merous (very seldom 5-merous), greenish to creamy yellow, often dark purplish-red or -brown tinged outside, glabrous or with a few odd hairs. ♂: tube 0.7–1.6 mm. long, funnel-shaped, lobes 1.4–2.4(2.7) × 0.6–1(1.4) mm.; anthers (0.7)1–1.7 mm. long; small rudimentary ovary mostly with 4 small calyx lobes, occasionally also rudimentary stigmas present. ♀: tube 0.3–0.7 mm. long, lobes 0.2–0.7 × 0.1–0.3 mm.; style 0–0.5 mm. long; stigmas 3–6.1 mm. long; ovary 0.7–0.8 × 0.5–0.6 mm., with 4 subequal or 2 longer and 2 shorter calyx lobes. Fruit reddish-brown, shiny or greyish-brown; mericarps 1.2–2.2(2.4) × 0.7–1 mm., ± obovate, oblong to elliptic, mostly glabrous, with 2 ± triangular calyx lobes 0.2–0.6 × 0.3–0.4 mm., one often larger than the other.

Zambia. E: Nyika Plateau, fl. & fr. 26.vi.1966, *Fanshawe* 9764 (K; SRGH). **Malawi**. N: Nyika Plateau, c. 23 km. SE. on Kasaramba road, fl. & fr. 23.iv.1976, *Pawek* 11206 (K; MO; PRE; SRGH; WAG).

Centred in the tropical E. African mountains (S. Sudan to S. Tanzania). Mostly in forest edge-grassland border in 'ericaceous' vegetation; c. 1850–2400 m.

There is some evidence that there are ± extensive hybrid populations *A. usambarense* × *A. welwitschii* in N. Malawi (Misuku Hills; e.g. *Puff* 781221-2/3; WU) and adjacent parts of Tanzania; suspected hybrids are ± intermediate in their characters and will either key out to one or the other species. The types of *A. leuconeuron*, *A. prittwitzii* and *A. uwembae* Gilli [in Ann. Naturhist. Mus. Wien **77**: 18 & fig. 1 (1973)] (all from S. Tanzania) may represent putative hybrid collections. The first two more closely approach *A. usambarense* in their characters; *A. uwembae* is morphologically closer to *A. welwitschii*.

There are also strong indications that *A. usambarense* forms hybrids with *A. whyteanum* in areas where the two species grow in immediate vicinity, e.g. on the Nyika Plateau (*Fanshawe* 9747; K).

3. **Anthospermum zimbabwense** Puff in Pl. Syst. Evol., Suppl. 3: 201 (1986). TAB. **34**, fig. A. Type: Zimbabwe, Rhodes Inyanga National Park, Mt. Inyangani, *Puff* 790125–3/1 (BM; SRGH; WU, holotype;).

Dioecious subshrub or dwarf shrub, several- to many-stemmed, erect, 0.2–0.6(1) m. tall, often with a quite thick, ± disk-like woody base. Stems ± unbranched to much-branched, branches often more numerous in ♂ than ♀, ascending to ± erect, papillate to very shortly hairy. Leaves in whorls of 3, pseudo-verticillate; blades (3)4–8(12) × 0.5–1.5(3) mm., linear to linear-lanceolate or narrowly oblanceolate, glabrous; margins mostly revolute; petioles subobsolete; stipular sheath with a small, narrow seta, often flanked by a smaller seta on either side. Flowers subsessile, in clusters of several (♂) to many (♀) at nodes, ♀ inflorescences often more conspicuous than ♂, ± compact and cylindrical; corolla 4-merous, greenish-yellow to yellow, in ♂ usually a little hairy outside. ♂: tube 0.5–1 mm. long, (± broadly) funnel-shaped, lobes 1.3–2 × 0.5–0.8 mm.; anthers (0.8)1–1.3 mm. long; rudimentary ovary minute, hardly discernible. ♀: tube 0.2–0.4 mm. long, lobes 0.3–0.5 × 0.1–0.2 mm.; style ± 0, stigmas 3.4–9.1 mm. long; ovary 0.7–0.8 × 0.6 mm., with 4 indistinct calyx lobes or calyx lobes lacking. Fruit reddish-brown; mericarps 1.4–1.7 × 0.7–0.9 mm., ± obovate, ± glabrous or with a few whitish hairs, with 2 indistinct, ± triangular calyx lobes hardly longer than 0.1 mm. or calyx lobes ± lacking.

Zimbabwe. E: Mutare Distr., Banti Forest, fl. 4.ii.1955, *Exell, Mendonça & Wild* 174 (BM; LISC; SRGH). **Mozambique**. MS: Tsetserra, fl. 8.ii.1955, *Exell, Mendonça & Wild* 293 (BM; LISC; SRGH).

Known only from Zimbabwe and Mozambique. Mostly in fire-prone grassland; 1650–2150(2500) m.

See also 5., *A. vallicola* and 6., *A. whyteanum* for comments and details on occasional hybrids with these species.

4. **Anthospermum ammannioides** S. Moore in Journ. Linn. Soc., Bot. **40**: 102 (1911). ("ammanioides"). —Puff in Pl. Syst. Evol., Suppl. 3: 209 (1986). Type: Zimbabwe, "Gazaland", Chimanimani, *Swynnerton* 2156 (BM, holotype; K).

Dioecious or ± dioecious (♂ or ♀ plants occasionally with odd ⚥ flowers) shrub, single-stemmed, erect, 1–3 m. tall. Stems much-branched above, branches ascending to ± erect, densely covered with greyish-white hairs 0.2–0.7 mm. long. Leaves in whorls of 3; blades 18–45(65) × (2)3–10(15) mm. (in ♂ often somewhat smaller than in ♀), narrowly (ob)-lanceolate to oblong-lanceolate, both surfaces sparsely to ± densely covered with very short whitish spreading hairs, but lower surfaces often more densely hairy, particularly on mid-vein, and upper surfaces sometimes ± glabrous; margins usually ± flat; petioles subobsolete or 2–6 mm. long on largest leaves; stipular sheath hairy, with a hairy seta (0.7)1.5–4(6.5) mm. long, sometimes flanked by a smaller seta on either side. Flowers subsessile (♀) or with pedicels c. 0.7–1.5(2) mm. long (♂), in clusters of many (very many:♀) at nodes, inflorescences dimorphic, in ♀ compact, ± cylindrical inflorescence zones 20–45(70) mm. long and 15–20 mm. in diam.; corolla 4-merous (very rarely 5-merous), greenish-yellow, sometimes buds purplish tinged, at least in ♂ hairy outside. ♂ (odd ⚥): tube (0.5)0.7–1 mm. long, (± narrowly) funnel-shaped, lobes 1.5–2.4 × 0.6–1 mm.; anthers 1.2–1.7 mm. long; ♂: minute, hardly discernible rudimentary ovary. ♀: tube 0.3–0.5(0.8) mm. long, lobes 0.2–0.5 × 0.1–0.2 mm.; style (0.5)2–5(6.5) mm. long, stigmas (4.8)6–14 mm. long (much shorter in odd ⚥); ovary 0.7–1 × 0.6–0.8 mm., with 4 indistinct calyx lobes or calyx lobes ± lacking. Fruit reddish-brown; mericarps (1.5)1.8–2.2(2.4) × 0.8–1 mm., ± oblong to obovate, glabrous, with 2 indistinct, ± triangular calyx lobes 0.1–0.2 mm. long or calyx lobes ± lacking.

Zimbabwe. E: Chimanimani Distr., Farm Kasipiti, fl. 1.vi.1966, *Loveridge* 1579 (BR; K; SRGH). **Mozambique**. MS: Báruè, serra de Choa, 16 km. from Vila Gouveia, fl. 26.v.1949, *Torre & Correia* 18685 (LISC; LMU).

Known only from Zimbabwe and Mozambique. At forest edges, in streambank scrub or in bush clumps amongst rocks; (1100)1400–2300(2500) m.

Phipps 409 (K; SRGH) from the Chimanimani Mts., Stonehenge Plateau (fl. 2.ii.1957) may be *A. ammannioides* × *A. whyteanum*; see also 6., *A. whyteanum*.

5. **Anthospermum vallicola** S. Moore in Journ. Linn. Soc., Bot. **40**: 103 (1911). — Puff in Pl. Syst. Evol., Suppl. 3: 212 (1986). TAB. **34**, fig. D. Type: Zimbabwe, "Gazaland", Chimanimani Mts., *Swynnerton* 2155 (BM, holotype).

Dioecious shrub, often single-stemmed, erect, 1–3 m. tall. Stems much-branched above, branches ascending, ± papillate, densely leafy. Leaves in whorls of 3, pseudo-verticillate; blades (10)15–35(55) × (0.5)1–2(2.5) mm., linear to linear-lanceolate, glabrous, upper epidermis shiny, with large, conspicuous epidermis cells; margins revolute to ± flat; petioles subobsolete; stipular sheath hairy, with a narrowly triangular seta (0.3)0.7–1(1.5) mm. long, sometimes flanked by a shorter seta on either side. Flowers subsessile (♀) or with peduncles and pedicels to c. 1 mm. long (♂), in clusters of many (very many: ♀) at nodes, inflorescences dimorphic, in ♀ ± dense, cylindrical inflorescence zones; corolla 4-merous, greenish-yellow to pale yellow, sometimes reddish-tinged outside, glabrous. ♂: tube 0.6–0.8(1) mm. long, (broadly) funnel-shaped, lobes 1.7–2.4(2.8) × 0.7–1 mm.; anthers (1.2)1.5–1.7(2) mm. long; rudimentary ovary sometimes hardly discernible, minute, with 4 ± linear to linear-lanceolate calyx lobes 0.1–0.4 mm. long, and occasionally also rudimentary stigmas present. ♀: tube 0.7–1.4 mm. long, lobes 0.4–0.8 × 0.1–0.2 mm.; style 0.5–1.5 mm. long, stigmas 5–9.5(14) mm. long; ovary 1–1.5 × 0.8–1.3 mm., with 4 long calyx lobes. Fruit reddish-brown, shiny; mericarps (1.8)2.2–3(3.5) × 1.3–1.7(1.8) mm., oblong, glabrous, with 2 narrowly triangular to ± deltoid calyx lobes (0.5)0.8–1.8(2.1) mm. long, often unequal in size, one sometimes only half as long as the other.

Zimbabwe. E: Chimanimani Distr., Mt. Peni summit, fl. 14.iv.1957, *Chase* 6415 (B; K; LISC; PRE; SRGH). **Mozambique**. MS: Gorongosa Mt., Gogogo summit area, fl. v.1969, *Tinley* 1838 (K; LISC; SRGH).

Known only from Zimbabwe and Mozambique. In mixed sclerophyll scrub and 'heath' communities dominated by *Philippia*, usually in rocky areas; c. 1700–2600 m.

In the Chimanimani mountains, *A. vallicola* appears to hybridize occasionally with *A. zimbabwense*; putative hybrids: e.g. *Grosvenor* 196 (LISC; SRGH) or *Wild* 4578 (LISC; MO; SRGH).

6. **Anthospermum whyteanum** Britten in Trans. Linn. Soc., Bot. ser. 2, **4**: 16 (1894). —Brenan in Mem. N.Y. Bot. Gard. **8**: 455 (1954). —Verdc. in F.T.E.A., Rubiaceae 1: 331 (1976). —Puff in Pl. Syst. Evol., Suppl. 3: 268 (1986). Type: Malawi, Mt. Mulanje, *Whyte* 48 (BM, holotype).

Anthospermum albohirtum Mildbr. in Notizbl. Bot. Gart. Berlin **15**: 637 (1941). Type from Tanzania.

Non-dioecious (♂, ⚥ + ♀ or ♀) dwarf shrub, several- to many-stemmed, suberect or ± rounded and low. Stems 20–80(125) cm. long, erect or ascending, mostly much-branched, branches ± erect or ascending, densely covered with whitish or reddish-brown hairs 0.2–0.9(1.2) mm. long. Leaves in whorls of 3, pseudo-verticillate; blades spreading to ascending, (2)3–11(15) × 0.8–3(5) mm., oblanceolate, narrowly ovate-lanceolate to linear-lanceolate, mostly both surfaces densely covered with whitish spreading hairs 0.2–0.5(0.9) mm. long, sometimes less hairy below except for midrib; margins mostly strongly revolute; petioles subobsolete; stipular sheath hairy, with a seta (0.6)1–2.5 mm. long. Flowers subsessile, in clusters of 9–3 at nodes, ♀ inflorescences occasionally somewhat congested and ± cylindrical; corolla 4-merous (very rarely also 5-merous in ♂ or ⚥), greenish-yellow to creamy yellow, at least near tips of lobes hairy outside. ♂, ⚥ : tube (0.7)1.2–2.2 mm. long (in ♂ often somewhat smaller than in ⚥), (narrowly) funnel-shaped, lobes (1.7)2–2.7(3.4) × (0.5)0.7–0.9 mm.; anthers 1.3–1.9 mm. long (in ♂ often smaller than in ⚥); ♂: small rudimentary ovary, sometimes also rudimentary stigmas present; ⚥: stigmas 2–3.4 mm. long, ovary 0.7–1.2 × 0.5–0.8 mm., hairy, with 4 calyx lobes. ♀: tube 0.3–0.7 mm. long, lobes 0.3–0.7(1) × 0.2–0.3 mm.; style 0.3–0.8 mm. long; stigmas 2.7–12.2 mm. long; ovary as in ⚥. Fruit reddish-brown; mericarps (1.6)1.8–2.2 × 0.7–1 mm., ± oblong, quite densely covered with whitish, often upwardly directed hairs 0.2–0.7 mm. long, usually with 2 small, ± triangular calyx lobes (0.2)0.3–0.7 × 0.2–0.3 mm., often hidden amongst hairs.

Zambia. N: Danger Hill, 29 km. NNE. of Mpika, fl. 18.ii.1970, *Drummond & Williamson* 9708 (K; SRGH). E: Mafinga Hills, 23.v.1973, *Chisumpa* 38 (K). **Zimbabwe**. N: Mazoe, Iron Mask Hill, *Eyles* 343 (BM; BOL; SRGH; atypical). C: Mt. Wedza, fl. 27.ii.1964, *Wild* 6352 (K; LISC; SRGH). E: Inyanga Distr., Juliasdale, 2 km. W. of Punch Rock, fl. 23.i.1973, *Biegel* 4162 (B; BR; K; LISC; MO; PRE; SRGH). **Malawi**. N: Nyika Plateau, at Chelinda Bridge, fl. 10.i.1974, *Pawek* 7935 (K; MAL; MO; SRGH). C: Dedza Distr., Chongoni Forest Reserve, Chiwau (=Ciwao) Hill, fl. 16.v.1960, *Chapman* 707 (BM; K; MAL; PRE; SRGH). S: Blantyre Distr., Ndirande Mt., fl. 2.v.1970, *Brummitt* 10325 (K; SRGH). **Mozambique**. Z: Gúruè, Nàmuli, *Andrada* 1858 (COI; LISC). T: Angónia, N. part of Serra Dómuè, *Macuáca & Mateus* 1099 (LMA). MS: Báruè, Serra de Choa, 26 km. from Vila Gouveia to the border, *Torre & Correia* 15401 (LISC).

Also in S. Tanzania. Mostly on rocky outcrops; 1250–2500 m.

A. whyteanum seems to be capable of forming viable hybrids with any other shrubby or dwarf shrubby *Anthospermum* species, notably *A. usambarense* (Nyika Plateau; see 2., above), *A. welwitschii* (Mt. Mulanje; putative hybrids fairly common in some areas), *A. rigidum* subsp. *pumilum* (E. Zimbabwe; putative hybrids infrequent) and *A. zimbabwense* (E. Zimbabwe; putative hybrids probably rather common). Putative hybrids often seem to be more similar in appearance to *A. whyteanum* (hairy, but less densely so than 'pure' *A. whyteanum*); they are normally found together with the parental species in the same general area.

Hutchinson & Gillett 4082 (K; SRGH) from Zambia (C), 32 miles NE. of Serenje Corner, 5200 ft., fl. 25.vii.1930, and *Torre & Paiva* 10309 (LISC) from Mozambique (N), Ribáuè, Serra de Ribáuè (Mepalué), c. 1600 m., fl. 28.i.1964, may represent an odd state of the species (rather glabrescent foliage); as mature fruits are absent, their true identity cannot be established with certainty.

7. **Anthospermum ternatum** Hiern, Cat. Afr. Pl. Welw. **1**, 2: 499 (1898). —Verdc. in F.T.E.A., Rubiaceae 1: 328 (1976). —Puff in Pl. Syst. Evol., Suppl. 3: 281 (1986). Syntypes from Angola.

Non-dioecious (⚥, ♀, occasionally also ⚥ + ♀, ⚥ + ♂ or ♂) short-lived (sometimes biennial*) herb, 'woody herb' or subshrub with a somewhat woody base, erect to ± straggling. Stems 20–100(150) cm. long, few- to ± much-branched, branches often ascending to ± erect, shortly hairy. Leaves in whorls of 3, decussate or, occasionally, in whorls of 4, often pseudo-verticillate; blades spreading to ± erect, (4.5)6–45(55) × (0.5)0.8–5(8) mm., ovate- or oblong-lanceolate to ± linear, glabrous or variously hairy, but most commonly with short hairs (<0.1–0.3 mm. long) above and with longer hairs, to c. 0.5 mm., at least on midrib below; upper surface sometimes with conspicuous large epidermal cells; margins ± flat to revolute; petioles subobsolete or to 3 mm. on largest leaves; stipular sheath shortly hairy, with a hairy subulate seta (0.7)1–3(3.7) mm. long, often flanked by a much shorter seta on either side. Flowers in subsessile to slightly elongated axillary clusters of several to few at nodes; corolla 4-merous, yellowish to yellowish-green, at least near tips of lobes hairy outside, rarely glabrous altogether. ⚥, ♂: tube (1.7)1.9–2.5 mm. long, cylindrical, lobes 2–2.7 × 0.5–0.7 mm.; anthers 1–2.2 mm. long. ♂: small rudimentary ovary, sometimes also rudimentary stigmas present; ⚥: stigmas 1.9–4.7 mm. long, ovary 0.8–1.4 × 0.5–0.7 mm., glabrous or hairy, without or seldom with indistinct minute calyx lobes. ♀: tube 0.3–0.7 mm. long (1.2–1.7 mm. long if rudimentary anthers present), cylindrical, lobes 0.4–1 × 0.1–0.3 mm. (1–1.7 × 0.2–0.5 mm. if rudimentary anthers present); style c. 0–1 mm. long; stigmas 4–10.2 mm. long; ovary as in ⚥. Fruit reddish-brown; mericarps 2–3 × 0.8–1.2 mm., oblong to narrowly obovate, covered with whitish spreading hairs (0.1)0.3–0.7 mm. long at least near apex or entirely glabrous, without distinct calyx lobes.

Leaves ± linear and sometimes ± needle-like due to strongly revolute margins, (4.5)6–25(30) × (0.5)0.8–1.2(1.5) mm.; fruits mostly hairy, at least near apex - - - subsp. *ternatum*
Leaves ovate- or oblong-lanceolate to ± linear-lanceolate, (10)12–45(55) × 1.5–5(8) mm.; fruits glabrous or hairy - - - - - - - - - - - - subsp. *randii*

Subsp. **ternatum**

Plant mostly erect, sparsely branched; stems to c. 70(100) cm. long. Leaves mostly in whorls of 3 but occasionally decussate; blades glabrous altogether or sometimes scabrous and/or with longer hairs below (leaves then usually smaller and narrower than in subsp. *randii*); petioles subobsolete.

Zambia. N: Mbala Distr., Chilongowelo escarpment, fl. 6.iv.1962, *Richards* 16255 (K; LISC; SRGH). E: Mafinga, fl. 23.v.1973, *Fanshawe* 11937 (K). **Zimbabwe**. N: Mt. Darwin Distr., Mt. Banirembizi, fl. 14.v.1955, *Watmough* 136 (SRGH). **Malawi**. N: Mzimba Distr., Marymount, towards Lunyangwa, fl. & fr. 14.vi.1972, *Pawek* 5453 (K; MO; SRGH). S: Zomba Rock, fl. & fr. 1896, *Whyte* s.n. (K). **Mozambique**. N: Lichinga, fl. 25.iv.1934, *Torre* 68 (LISC).

Also in Angola, Zaire (Shaba) and S. Tanzania. Primarily in miombo or related woodlands, often in open, grassy or rocky areas; c. 1050–2000 m.

Subsp. **randii** (S. Moore) Puff in Pl. Syst. Evol., Suppl. 3: 285 (1986). TAB. **34**, fig. B. Type: Zimbabwe, Harare, *Rand* 475 (BM, holotype).

Anthospermum randii S. Moore in Journ. Bot., Lond. **40**: 253 (1902). Type as above.

Anthospermum erectum Suesseng. in Trans. Rhod. Sci. Assoc. **43**: 54 (1951). Type: Zimbabwe, Marondera, 18.iii.1941, *Dehn* 55 (M, holotype).**

* Young plants usually come into flower in their first growing season and may thus look like annuals; unless burnt they will continue growth in the following season.

** In SRGH, there is another *Dehn* 55 collection from Marondera, collected, however, on May 7, *1942*. This is not mentioned by Suessenguth and must, therefore, not be considered type material. The *Dehn* 55 collections from Rusape, made in *1952* "55/52" (K; LISC; MO; S; SRGH) and "55'/52" (BR; K; M), also must not be confused with the type.

Plant often more branched than in subsp. *ternatum* and not uncommonly straggling; stems to 150 cm. long. Leaves in whorls of 3, decussate or occasionally in whorls of 4; blades variously hairy but most commonly with short hairs above and longer hairs on midrib below, less often subglabrous (leaves then larger and broader than in subsp. *ternatum*); largest leaves with petioles to 3 mm. long.

Zambia. C: Lusaka Distr., Livingstone-Lusaka Road, fr. 29.iii.1961, *Richards* 14907 (K). S: c. 8 km. E. of Choma, fl. & fr. 28.iii.1955, *Robinson* 1215 (K; SRGH). **Zimbabwe**. N: Mazoe Distr., Umvukwes, Horseshoe Mine, fl. & fr. 16.iv.1960, *Leach & Brunton* 9859 (K; SRGH). W: Bulawayo, fl. & fr. iv.1911, *Rogers* 13681 (BOL; J; PRE). C: c. 12 km. S. of Gweru, fl. 25.ii.1967, *Biegel* 1948 (K; MO; SRGH). E: c. 6 km. from Inyanga village on Troutbeck road, fl. & fr. 26.iv.1967, *Rushworth* 737 (MO; SRGH). S: Masvingo Distr., Kyle National Park, fr. 21.v.1971, *Grosvenor* 646 (SRGH). **Malawi**. N: Mzimba Distr., Hora Mt., fl. & fr. 30.xii.1978, *Puff* 781230–1/1 (BR; J; W; WU). C: Dedza Distr., Chongoni Forest Reserve, fr. 16.v.1969, *Salubeni* 1339 (MAL; SRGH).

Also in Angola (and Tanzania?). Habitats as for subsp. *ternatum*; c. 950–2100 m.

A. ternatum is very variable. The two subspecies are easily separable in Zambia and N. Malawi but specimens from elsewhere may be difficult to place.

Especially in E. Zimbabwe, the delimition of subsp. *randi* and *A. whyteanum* may be problematic.

8. **Anthospermum herbaceum** L.f., Suppl.: 440 (1781). —Murr., Syst. Veg.: 919 (1784). —Brenan in Mem. N.Y. Bot. Gard. **8**: 455 (1954). —Agnew, Upl. Kenya Wild Fl.: 407, fig., p. 406 (1974). —Verdc. in F.T.E.A., Rubiaceae 1: 325, fig. 46 (1976). —Puff in Fl. Southern Africa **31**, 1(2): 22 (1986); in Pl. Syst. Evol., Suppl. 3: 300 (1986). TAB. **34**, fig. C. Type from S. Africa.

Anthospermum lanceolatum Thunb., Prodr. **1**: 32 (1794); Fl. Cap., ed. Schultes: 157 (1823). —Cruse, Rub. Cap.: 12 (1825); in Linnaea **6**: 12 (1831). —Sond. in Harv. & Sond., F.C. **3**: 30 (1865). —De Wild., Pl. Bequaert. **2**: 301 (1923). Type from S. Africa.

Anthospermum ferrugineum Eckl. & Zeyh., Enum. Pl. Afr. Austr. Extratrop.: 366 (1836). Type from S. Africa.

Anthospermum nodosum E. Mey. in Drège in Flora **26**, Bes. Beigabe: 164 (1843) nom nud.

Anthospermum muriculatum Hochst. ex A. Rich., Tent. Fl. Abyss. **1**: 345 (1848). —Hiern in F.T.A. **3**: 229 (1877). Syntypes from Ethiopia.

Anthospermum hedyotideum Sond. in Harv. & Sond., F.C. **3**: 30 (1865). Syntypes from S. Africa.

Anthospermum lanceolatum var. *latifolium* Sond. in Harv. & Sond., F.C. **3**: 30 (1865) [*Anthospermum latifolium* E. Mey. in Drège in Flora **26**, Bes. Beigabe: 164 (1843) nom. nud.] Lectotype from S. Africa.

Anthospermum lanceolatum var. *hedyotideum* (Sond.) O. Kuntze, Rev. Gen. Pl. **3**: 117 (1898).

Anthospermum mildbraedii K. Krause in Wiss. Ergebn. Deutsch. Zent.-Afr. Exped. **2**: 341 (1914). Type from Rwanda.

Non-dioecious (⚥, ⚥ + ♀, ♀, ⚥ + ♂, ♂ + ⚥ + ♀, less commonly ♂ or ♂ + ♀) perennial herb or ± sub shrub, several- to many-stemmed, scrambling, straggling or trailing, sometimes ± erect, occasionally low and ± mat- or cushion-forming, somewhat woody near base and often with ± thick, woody root; sometimes rather short-lived. Stems slender, 7.5–250(300) cm. long, unbranched to much-branched; branches often ± regular, arising in pairs, frequently with short branches of a higher order, papillate, shortly hairy or sometimes glabrous, often with short shoots bearing rather small leaves. Leaves decussate; blades 5–55 × (1)2–25 mm., ± ovate, ovate-lanceolate, lanceolate to ± linear-lanceolate, cuneate to rounded at base, glabrous, ± densely papillate or shortly hairy, often distinctly discolourous; petioles 0.7–6.5 mm. long; stipular sheath mostly with (3)5(7) filiform setae, the longest (0.3)0.7–6.1 mm. Flowers mostly on short lateral branches, in ± sessile to somewhat elongated clusters of many to c. 6 at nodes; corolla 4-merous, greenish to yellow or yellowish, sometimes reddish-purplish tinged, mostly papillate or shortly hairy at least near tip. ⚥, ♂: tube (1.5)2–3.7 mm. long (in ⚥ often longer than in ♂), cylindrical, lobes (1.5)2–2.7(3.4) × 0.3–0.7 mm.; anthers (0.9)1.2–2(2.5) mm. long; ♂: small rudimentary ovary and stigmas present, the latter hidden in corolla tube; ⚥: stigmas (2)2.4–5(6.4) mm. long, ovary 0.5–1.2 × 0.4–0.7 mm., sometimes with 4 often indistinct calyx lobes. ♀: tube 0.3–0.7(1.2) mm. long, lobes 0.3–0.7(1) × 0.1–0.3 mm.; style 0–0.8 mm. long; stigmas (2)4–10.2 mm. long; ovary as in ⚥. Fruit yellowish-brown or reddish-brown, mostly with ± broad, conspicuous longitudinal grooves between mericarps; mericarps (1.5)1.7–2.5(2.8) × 0.9–1.6 mm., elliptic, oblong to ± obovate, ± densely covered with ± tuberculate structures, shortly hairy, papillate or subglabrous, occasionally with 2 obscure, ± triangular to rounded calyx lobes 0.2–0.3 × 0.2–0.3 mm.

Zambia. N: Mbala, fl. & fr. 10.ix.1969, *Fanshawe* 10620 (BR; K; SRGH). E: Nyika Plateau, below Rest House, towards N. Rukuru waterfall, fl. & fr. 27.x.1958, *Robson* 394 (BM; BR; K; LISC; PRE; SRGH). **Zimbabwe**. N: Mazoe Distr., Suri Suri dam, fl. 16.vi.1971, *Gibbs Russell* 1144 (SRGH).

W: Matobo, Farm Besna Kobila, fl. & fr. 1.ix.1956, *Miller* 3665 (K; LMA; SRGH). C: Makoni Distr., Rusape, Chiduku Sacred Forest, fl. & fr. 9.xii.1964, *West* 6182 (SRGH). E: Stapleford Forest Reserve, fl. & fr. 13.vi.1934, *Gilliland* 309 (BM; K: MO; PRE; SRGH). **Malawi**. N: Chitipa Distr., around Chisenga, foot of Mafinga Mts., fl. & fr. 12.vii.1970, *Brummitt* 12036 (K; SRGH). C: Ntchisi Mt., fl. & fr. 2.viii.1946, *Brass* 17107 (BM; BR; GH; K; MO; PRE; SRGH; US). S: Zomba Plateau, fl. & fr. 3.vi.1946, *Brass* 16189 (BM: BR; GH; K; MO; PRE; SRGH; US). **Mozambique**. N: Metónia, Lichinga, fl. & fr. 1.vi.1934, *Torre* 196 (COI; LISC). Z: Namuli Peaks, fl. & fr. 26.vii.1962, *Leach & Schelpe* 11477 (K; SRGH). T: NE. of Vila Couthino, Mozambique/Malawi border, fl. & fr. 16.ii.1978, *Puff* 780216–1/1 (WU). MS: Manica, Mavita, on M'chere R., fl. & fr. 30.i.1948, *Barbosa* 913 (LISC).

In E. Africa from Ethiopia S. to S. Africa; also in SW. Arabia. In forest edge vegetation, scrub, riverine thicket, at the edge of marshes, or occasionally in grassland; 450–2300 m.

Very variable in leaf size and shape, indumentum and habit.

Wild 4324 (K; LISC; MO; SRGH) from Zimbabwe, Mberengwa, Mt. Buhwa (fl. 10.xii.1953), described as "tufted bush to 1 foot", is probably *A. herbaceum* × *A. rigidum* subsp. *pumilum*.

9. **Anthospermum rigidum** Eckl. & Zeyh., Enum. Pl. Afr. Austr. Extratrop.: 367 (1836). Type from S. Africa.

Subsp. **pumilum** (Sond.) Puff, stat. nov. *

Anthospermum pumilum Sond. in Harv. & Sond., F.C. **3**: 31 (1865). Type from S. Africa.

Anthospermum humile N.E.Br. in Kew Bull. **1895**: 145 (1895). Type from S. Africa.

Anthospermum ericoideum K. Krause in Engl., Bot. Jahrb. **39**: 570 (1907). —Engler, Pflanzenw. Afr. **1**: 574, fig. 508 (1910). —Launert & Roessler in Merxm. Prodr. Fl. SW. Afr. **115**: 8 (1966). Type from SW. Africa/Namibia.

Anthospermum pumilum var. *pilosum* Phillips in Ann. S. Afr. Mus. **16**: 112 (1917). Type from Lesotho.

Anthospermum spicatum Suesseng. in Trans. Rhod. Sci. Assoc. **43**: 55 (1951). Type: Zimbabwe, Marondera, *Dehn* 547 (M, holotype).

Non-dioecious (⚥, ⚥ + ♀, ♀, occasionally ♂ or ⚥ + ♂) subshrub with an often massive, ± rosette-like woody base. Stems numerous, mostly unbranched, (5)8–20(30) cm. long, papillate to very shortly hairy and mostly ± densely leafy above. Leaves decussate, pseudo-verticillate; blades (4)6–12(22) × (0.5)0.8–1.5(2) mm., linear, linear-(ob-)lanceolate or narrowly lanceolate, ± membranous; flat to somewhat revolute margins mostly papillate, midrib below often reddish-brown and prominent; petioles subobsolete; stipular sheath with 1 (rarely 3) small seta(e). Flowers subsessile, in clusters of 6–2 (occasionally more) at nodes; corolla 4-merous, greenish to yellowish, often papillate near tip. ⚥, ♂: tube (0.5)0.7–1.4(1.7) mm. long, narrowly funnel-shaped, lobes 1.2–1.9(2.5) × 0.3–0.7 mm.; anthers 1–1.8(2) mm. long; ⚥: stigmas often shorter than in ♀; ovary 0.5–0.9 × 0.3–0.8 mm., with 4 ± indistinct calyx lobes. ♀: tube 0.2–0.5 mm. long, lobes 0.2–0.5 × 0.1–0.2 mm.; style 0–0.5 mm. long; stigmas 2.4–9.8 mm. long; ovary as in ⚥. Fruit reddish-brown, shiny; mericarps (1.5)1.8–2.4 × 1–1.5 mm., elliptic to obovate, mostly ± glabrous, often with 2 ± broadly triangular to rounded calyx lobes 0.1–0.3 × 0.3–0.4 mm.

Botswana. SE: c. 14 km. S. of Ramotswa (Ramoutsa), fr. 19.i.1960, *Leach & Noel* 215 (K; LISC; SRGH). **Zambia**. S: Mazabuka Distr., Kalomo, fr. 1.i.1958, *Robinson* 2559 (K; M; SRGH). **Zimbabwe**. W: Matobo, Farm Besna Kobila, fl. & fr. xi.1955, *Miller* 3118 (LISC; PRE; SRGH). C: Garch Farm, Harare South, fr. 16.i.1964, *Strang* 2324 (SRGH). E: Mutare Distr., Penhalonga, fl. & fr. 2.xi.1956, *Robinson* 1840 (K; MO; SRGH). S: Bikita, fl. & fr. 20.x.1930, *Fries, Norlindh & Weimarck* 2122 (BM; K; MO; PRE; SRGH). **Mozambique**. T: Angónia, around Ulongue, fl. & fr. *Stefanesco & Nyongani* 270 (LMA).

Also in S. Tanzania, Angola, SW. Africa/Namibia and S. Africa. Usually in (rocky) grassland or open woodland; c. 1050–1850 m.

Subsp. *rigidum* is confined to the drier parts of S. Africa.

* In Fl. Southern Afr. **31**, 1(2): 25, 26 (1986) and in Pl. Syst. Evol., Suppl. 3: 329, 342 (1986) I united *A. rigidum* with *A. pumilum* and considered *A. rigidum* as a subsp. of *A. pumilum*. As this was incorrect the above change becomes necessary.

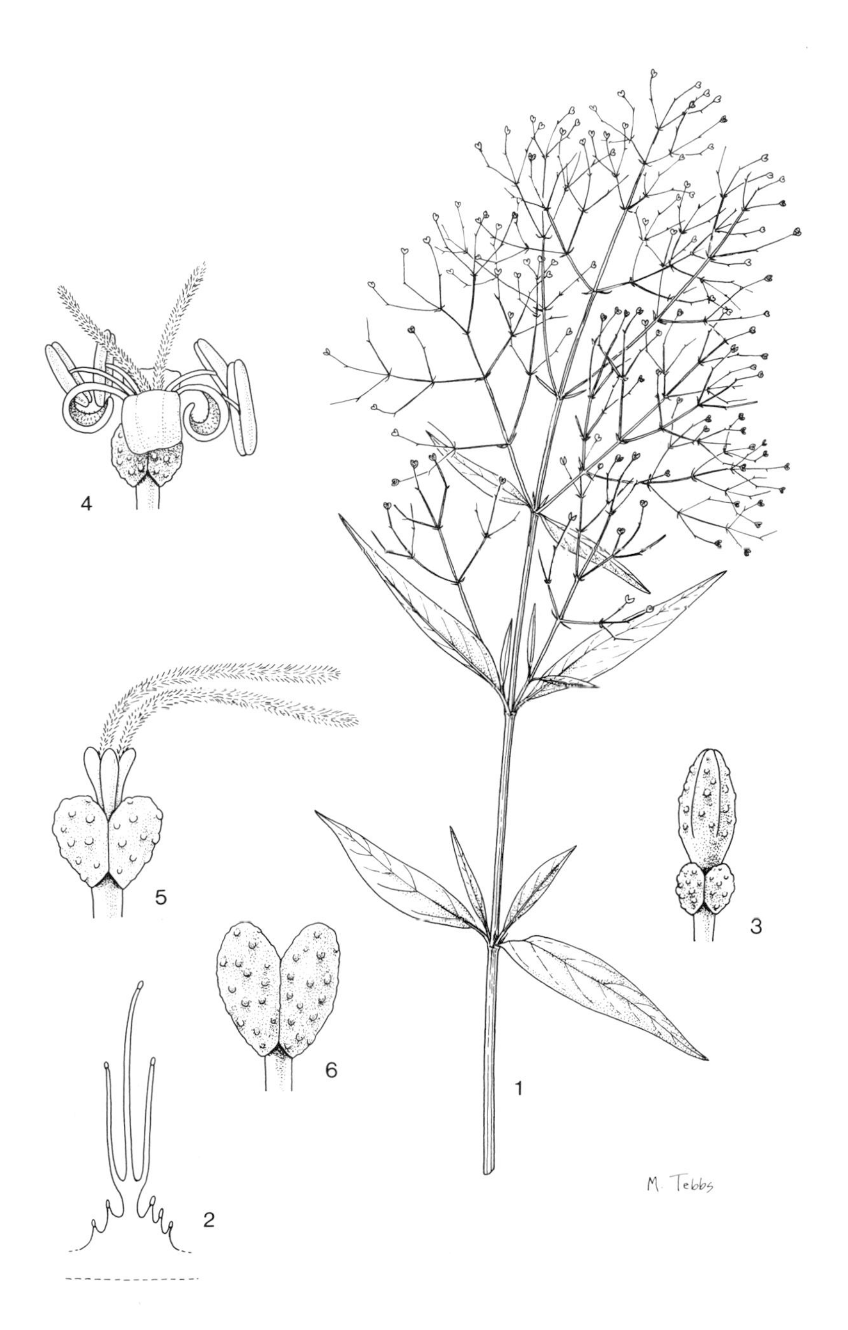

Tab. 35. GALOPINA CIRCAEOIDES. 1, part of plant with fruiting inflorescence (× ½); 2, part of stipular sheath with setae (× 9), 1–2 from *Jacobs* 1598; 3, hermaphrodite bud (× 9); 4, hermaphrodite flower (× 9); 5, female flower (× 9), 3–5 from *Puff* 781203-1/1; 6, fruit (two mericarps) in lateral view, from *Puff* 780326-2/1.

30. **GALOPINA** Thunb.

Galopina Thunb., Nov. Gen. Pl. **1**: 3 (1781).
Oxyspermum Eckl. & Zeyh., Enum. Pl. Afr. Austr. Extratrop.: 365 (1836).
Phyllis sensu Cruse in Linnaea **6**: 19 (1831).

Perennial herbs with branched, often ± woody rhizomes or rootstocks. Leaves decussate, broadly ovate to lanceolate, distinctly petiolate, with stipular sheaths bearing 3–5(7) setae on either side. Inflorescence terminal, paniculate to thyrso-paniculate, bracteate. Flowers ⚥, ♀ or ♂, 4(5)-merous. Calyx obsolete. ⚥, ♂: Corolla: tube (very) short, broadly funnel-shaped to campanulate, lobes recurved, ± lanceolate; anthers yellowish to whitish, exserted, dangling on long slender filiform filaments. ♀: corolla much smaller, tube cylindrical, sometimes 0, lobes erect to spreading, ± linear. Ovary bicarpellate and biovulate; style 0; stigmas 2, long exserted, in ⚥ often shorter and thinner than in ♀, hairy, greyish-white, yellowish-grey or greenish. Fruit dehiscent, not supported by a carpophore; mericarps ± glabrous, tuberculate or covered with long hairs, dorsal side convex, ventral side plane to concave.

A SE. African genus of 4 species; only 1 species extending to the Flora Zambesiaca area.

Galopina circaeoides Thunb., Nov. Gen. Pl. **1**: 3 (1781). —Cruse, Rub. Cap.: 18 (1825). —Sond. in Harv. & Sond., F.C. **3**: 26 (1865). —Brenan in Mem. N.Y. Bot. Gard. **8**: 454 (1954). —Puff in Fl. Southern Afr. **31**, 1(2): 48 (1986); in Pl. Syst. Evol., Suppl. 3: 430 (1986). TAB. **35**. Type from S. Africa.

Galopina circaeoides var. *glabra* Kuntze, Rev. Gen. Pl. **3**, 2: 120 (1898). Type from S. Africa.
Galopina circaeoides var. *pubescens* Kuntze, Rev. Gen. **3**, 2: 120 (1898). Type from S. Africa.
Anthospermum galopina Thunb., Prodr. **1**: 32 (1794). Type as for *G. circaeoides*.
Phyllis galopina (Thunb.) Cruse in Linnaea **6**: 20 (1831).

Several- to many-stemmed perennial herb. Stems 0.3–0.8(1.2) m. long, ascending to erect or occasionally decumbent and rooting at nodes, glabrous or sometimes sparsely hairy; often with much-contracted short shoots bearing small leaves. Leaf blades (30)40–70(90) × (8)10–25(30) mm., ± lanceolate to ovate-lanceolate, narrowed to base, acute at apex, glabrous or with some short whitish spreading hairs; petioles 3–10(14) mm. long; stipular sheath with 3 or 5(7) setae, free portion of the longest 4–8 mm. Inflorescence broadly pyramidal to ± spheroidal, lax, 12–30 × (8)10–28 cm.; peduncles and pedicels filiform, slender, glabrous, strongly divergent in fruit, elongating to (7)12–25(35) mm. Corolla whitish, creamy white, yellowish or greenish, occasionally reddish purplish tinged outside, glabrous or papillate. ⚥, ♂ tube 0.3–0.5 mm. long, lobes 1.2–1.8 × 0.4–0.6 mm.; anthers 1.1–1.6 mm. long. ♀: tube 0–0.3 mm. long, lobes 0.6–0.8(1) × 0.1–0.3 mm.; stigmas (2.5)3–5(6.5) mm. long; ovary 0.5–1 × (0.5)0.8–1.3 mm., ± densely warty. Fruit blackish to black; each mericarp 1.5–2 × (0.7)1–1.4 mm., oblong to ± obovate, ± densely warty.

Zimbabwe. E: Mutare Distr., Engwa, fl. 3.ii.1955, *Exell, Mendonça & Wild* 148 (BM; LISC; SRGH). **Malawi**. S: Zomba Plateau, fl. 7.vi.1946, *Brass* 16295 (K; MO; SRGH; US). **Mozambique**. MS: Manica, Macequece, E. Vumba Mts., fl. & fr. 18.iii.1948, *Barbosa* 1193 (LISC).
Also in S. Africa. Forest or scrub, often in ± shady and damp to wet places; (300)1000–2100 m.

Tribe 11. **SPERMACOCEAE**

1. Ovary 3-locular; stigmas 3; fruit with 3 cocci; straggling procumbent herbs **35. Richardia**
– Ovary usually 2-locular; stigmas 2 or 1 capitate stigma; fruit with 2 cocci, capsular or circumscissile - - - - - - - - - - - - - - - - 2
2. Fruit circumscissile about its middle, the top lifting off like a lid; flowers minute, clustered in sessile globose inflorescences at the nodes; seeds with a dorsal ± X-shaped groove - - - - - - - - - - - - - - - - - **34. Mitracarpus**
– Fruit indehiscent, with 2 cocci or a capsule splitting longitudinally - - - - - 3
3. Succulent creeping plant of the seashore, with imbricated leaves joined by quite broad sheathing stipules with very short processes; stems rooting at the nodes; fruits indehiscent - - - - - - - - - - - - - - **31. Phylohydrax**
– Plant not a littoral succulent - - - - - - - - - - - - - - 4

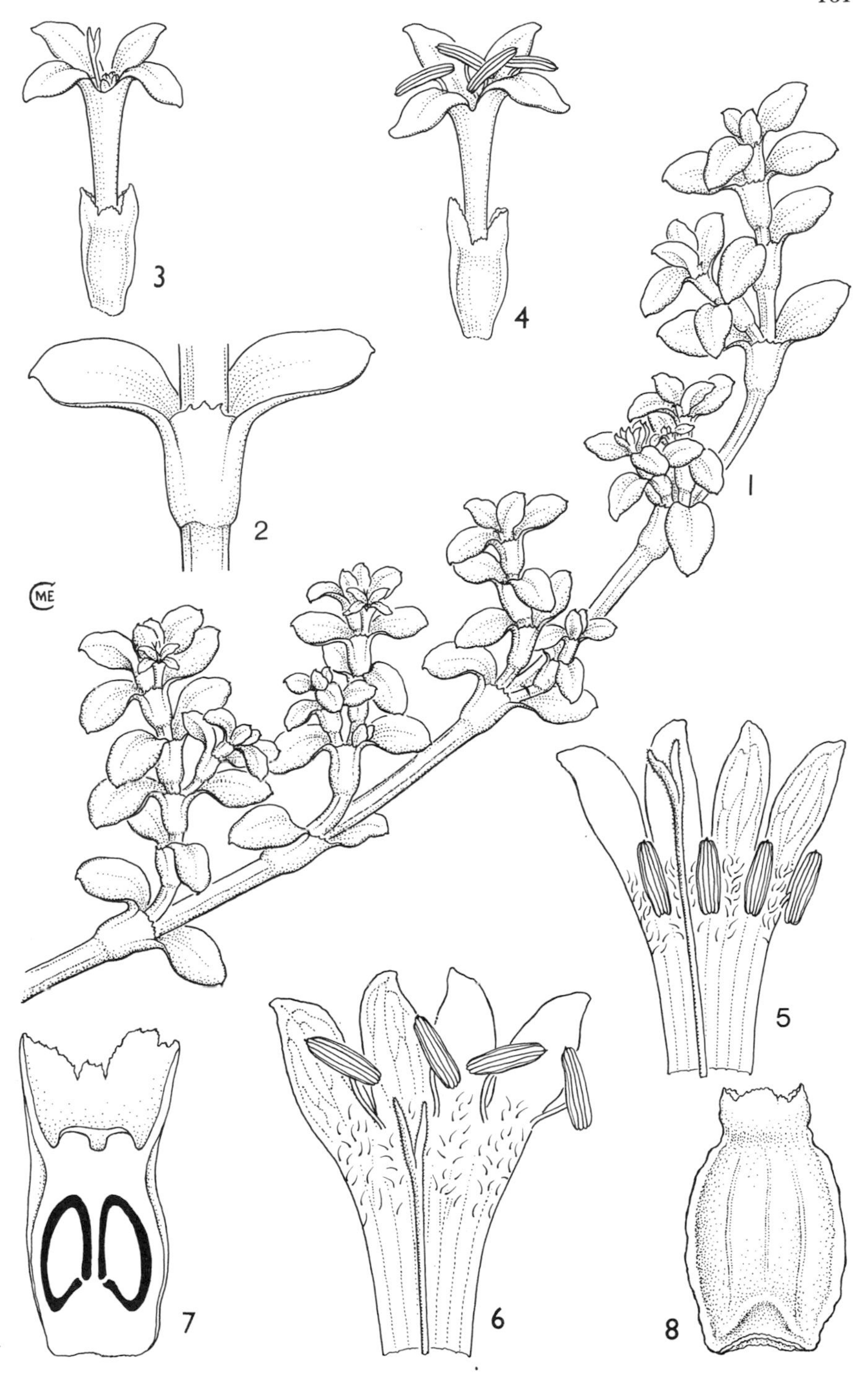

Tab. 36. PHYLOHYDRAX CARNOSA. 1, habit (× 1); 2, node showing stipule (× 2), 1–2 from *Noel* 2494; 3, long-styled flower (× 4); 4, short-styled flower (× 4); 5, corolla of long-styled flower opened out (× 6); 6, corolla of short-styled flower opened out (× 6); 7, ovary, longitudinal section (× 10); 8, fruit (× 6), 3–8 from *Wood* 11683.

4. Fruit with 2 cocci separating but ± indehiscent. - - - - - - - **32. Diodia***
– Fruit capsular with 2 valves, either dehiscing from base to apex (sect. *Arbulocarpus*) or in the reverse direction (sect. *Borreria)* or with 2 cocci, 1 or both ± dehiscent **33. Spermacoce**

31. PHYLOHYDRAX

Phylohydrax Puff in Pl. Syst. Evol. **154**: 343 (1986).

Prostrate succulent perennial herbs confined to the littoral zone. Leaves sessile, ovate-oblong to linear-lanceolate, decussate, closely imbricated on short flowering shoots; stipule sheath cupular, adnate to the petioles, with 1–2 short setae or up to 8 very small processes. Flowers solitary, axillary, hermaphrodite, heterostylous. Calyx tube oblong-ovoid, 4-angled; limb irregularly lobed, lacerate or with 2–4 indistinct triangular lobes. Corolla somewhat fleshy; tube funnel-shaped, hairy inside for upper half; lobes 4, valvate, ovate to elliptic-lanceolate. Filaments filiform; anthers linear, exserted. Disk fleshy. Ovary bilocular; ovules solitary in each locule, attached near the base of the septum; style filiform, ± glabrous or papillate; stigma lobes linear, exserted. Fruit indehiscent, small, ellipsoid to oblong-ovoid, compressed, smooth or finely ribbed when dry, 1–2-locular, 1–2-seeded. Seeds linear-oblong, dorsally convex, or ± angular, ventrally ± flat or rounded but not grooved; testa minutely granular; endosperm cartilaginous.

A genus related to *Diodia* and distributed from Madagascar to Tanzania, Mozambique and S. Africa (Natal and E. Cape Province), with 2 species, 1 of which occurs in the Flora area. Medley Wood (in litt. on *Wood* 11683) noted the heterostyly and specifically mentions that of 410 flowers of *P. carnosa* (Hochst.) Puff 180 were short-styled and 230 long-styled. Formerly included in *Hydrophylax* L.f. now restricted to one species occurring in Thailand, India and Sri Lanka.

Phylohydrax carnosa (Hochst.) Puff in Pl. Syst. Evol. **154**: 362, figs. 1b, 2a, c, d, 3b–d, 9 (1986). TAB. **36**. Type from S. Africa (Cape).
Diodia carnosa Hochst. in Flora **27**: 555 (1844).
Hydrophylax carnosa (Hochst.) Sond. in Harv. & Sond., F.C. **3**: 25 (1865). —Wood, Natal Pl. **6**: t. 539 (1910). —Ross, Fl. Natal: 337 (1973). —Gibson, Wild Fl. Natal t. 103/8 (1975).

Fleshy glabrous herb with long decumbent stems; stems branched, the branches smooth, flattened and deeply grooved one face. Upper stems mostly pale, lower and rhizomes darker, apparently with a reddish dye. Leaves 0.7–2.3 × 0.3–1 cm., oblong-elliptic to oblong, rounded to subacute at apex, usually ± mucronate, narrowed at the base but sessile on to the petiole sheath; stipule sheath up to 8 mm. long with 1 to several short lobes scarcely 0.5 mm. long. Calyx tube 2 mm. long, the limb ± 2 mm. long, erose or with triangular lobes ± 1 mm. long and wide. Corolla white or pinkish; tube 4–5 mm. long; lobes 3–4 × 1.5–1.7 mm., oblong-elliptic, ± oblique at apex. Long-styled flowers with anthers 1.7 mm. long and about $\frac{2}{3}$ exserted; style exserted 2.5 mm. the stigmatic lobes 1.2 mm. long. Short-styled flowers with anthers 2.4 mm. long, versatile, the filaments exserted 1.5 cm.; tips of stigmatic lobes exserted for about 1 mm. Fruits 3–6 × 2–4 mm.; seeds c. 3.5 mm. long.

Mozambique. MS: Beira, N. of Macuti Beach, fl. 10.ix.1962, *Noel* 2494 (K; LISC; SRGH). M: S. of Maputo, Ponta do Ouro beach, fl. 24.xii.1948, *Gomes e Sousa* 3909 (K; PRE).
Also in S. Africa (NE. Cape and Natal). Coastal sand dunes; c. 25 m.
I referred Tanzanian material of the genus to *Phylohydrax madagascariensis* (Willd. ex Roem. & Schultes) Puff; there appear to be foliage differences. The relationship between the two species will remain doubtful until better material of both, particularly fruits and seeds, are available. The Asiatic species *Hydrophylax maritima* L.f. is very distinct.

32. DIODIA L.

Diodia L., Sp. Pl.: 104 (1753); Gen. Pl., ed. 5: 45 (1754).

Annual or perennial erect or mostly decumbent or prostrate herbs; stems often 4-angled. Leaves opposite or sometimes appearing whorled due to the presence of

* Since the division into *Spermacoce* and *Diodia* is not based on very reliable characters, the key to the species of *Spermacoce* also includes those of *Diodia*.

reduced axillary shoots, ± sessile or shortly petiolate; blades linear to ovate; stipules with bases united to the petiole, divided into fimbriae. Flowers mostly small, not heterostylous, in small axillary clusters or, in one group of American species, terminal in spikes or capitula. Calyx tube ellipsoid, ovoid or obconic; lobes 2–4(rarely 5–6), sometimes with some small accessory teeth, ± persistent. Corolla tube ± funnel-shaped, glabrous or hairy at the throat; lobes 4 (rarely 5–6), valvate. Stamens 4 (rarely 5–6), exserted, the filaments inserted at the throat. Ovary 2 (rarely 3–4)-locular; ovules solitary in each cell, attached to the middle of the septum; style filiform, exserted; stigma 2-lobed or ± capitate. Fruit of 2 (rarely 3–4) indehiscent cocci. Seeds oblong, dorsally convex, ventrally longitudinally grooved, rarely with a transverse groove or in one subgenus (*Pleiaulax* Verdc.) distinctly lobed.

A genus of some 30–50 species, mostly in the New World. The circumscription of the genera *Diodia, Borreria, Spermacoce* and several satellite genera has been a very difficult matter indeed. If only the African species were involved it would seem reasonable to treat all as synonyms of *Spermacoce* but a survey of the New World material shows that *Diodia* is quite distinctive and best kept separate.

Scrambling or procumbent plant 1–4.5 m. long with usually many lateral branches from main stem; leaves scabrid but less pubescent - - - - - - - - - - 1. *sarmentosa*
Plant with more or less erect or ascending stems to 45 cm. from a several-headed woody root; leaves usually ± scabrid but distinctly more pubescent especially beneath - - 2. *flavescens*

1. **Diodia sarmentosa** Sw., Prodr. Veg. Ind. Occ.: 30 (1788); Fl. Ind. Occ. **1**: 231 (1797). —S. Moore in Fl. Pl. Jam. **7**: 117, fig. 35 (1936). —Adams, Fl. Pl. Jam.: 731 (1972). —Verdc. in F.T.E.A., Rubiaceae 1: 336, fig. 48/8–10 (1976). TAB. **37**. Type from Jamaica.
Diodia pilosa Schumach. & Thonn., Beskr. Guin. Pl.: 76 (1827). Type from Ghana.
Spermacoce pilosa (Schumach. & Thonn.) DC., Prodr. **4**: 553 (1830). —Hiern in F.T.A. **3**: 235 (1877). —Hutch. & Dalz., F.W.T.A. ed. 1, **2**: 135 (1931).
Diodia breviseta Benth. in Hook., Niger Fl.: 424 (1849). —Hiern in F.T.A. **3**: 231 (1877). Type from Fernando Po.
Diodia scandens auctt. eg. Hepper, F.W.T.A. ed. 2, **2**: 216, fig. 245 (1963) non Sw.

Straggling, scrambling or procumbent perennial or annual herb 1–4.5 m. long, often with many lateral branches from the main stem; stems 4-angular, pubescent on the angles but at length glabrous. Leaf blades often rather yellowish green, 1.8–6.3 × 0.7–2.8 cm., elliptic, acute at the apex, narrowed to the base, scabrid above with dense very short to longer tubercle-based hairs, pubescent beneath; petiole 1–5 mm. long; stipule bases 1–2 mm. long with lines of hairs, bearing 5–7 setae 1–7 mm. long. Flowers usually few in axillary clusters at most nodes, the inflorescences up to 1.2 cm. in diameter in fruiting stage. Calyx tube glabrous or pubescent, 1.5–2 mm. long, obconic; lobes 4, often unequal, 1.5–3 × 0.8 mm., oblong-lanceolate or narrowly triangular, ciliate. Corolla mauve or white; tube glabrous, 1.8 mm. long, funnel-shaped; lobes 1 mm. long and wide, triangular, with a few hairs outside. Filaments exserted 0.5 mm. Style exserted 1.5 mm., minutely papillate. Cocci 3.5–5 × 2.5 × 1.2 mm., semi-oblong-ellipsoid, or sometimes more globose, quite definitely not readily dehiscent. Seeds dark blackish red, 2–4 × 1.5 × 0.8 mm., compressed ellipsoid, with a broad ventral groove, dorsally finely rugulose, and with an apical slit represented by an impressed dorsal line extending about a third the length of the seed and on to the ventral surface.

Zambia. N: Mbala, Pansa R., fl. & fr. 6.x.1949, *Bullock* 1154 (K). W: Mwinilunga, Lisombo R. tributary, fr. 14.vi.1963, *Loveridge* 969 (K; LISC; SRGH). C: 16 km. W. of Luangwa R. Bridge, fl. & fr. 6.ix.1947, *Greenway & Brenan* 8058 (EA; K). **Zimbabwe**. E: S. end of Chimanimani Mts., Haroni-Makurupini Forest, fr. 11.vi.1971, *Pope* 475 (K; LISC; SRGH). **Mozambique**. N: between Mutivaze and Nampula, fr. 24.iv.1937, *Torre* 1452 (LISC). MS: Cheringoma, Durundi, fl. & fr. 25.v.1948, *Barbosa* 1684 (BM; LISC; LMU).
Widespread in tropical Africa and also in tropical Asia, America and the Mascarenes. Evergreen forest particularly fringing 'mushitu' edges, open riverine vegetation, bushland, also rocky places near rivers; 40–1470 m.

2. **Diodia flavescens** Hiern, Cat. Afr. Pl. Welw. **1**: 501 (1898). Type from Angola.
Diodia serrulata sensu K. Schum. in Warb., Kunene-Samb.-Exped. Baum: 393 (1903) as "(Schumach. & Thonn.) K. Schum." non (Beauv.) G. Taylor.

Erect or ascending herb 12–45 cm. tall with several (to c. 10) branched or often ± unbranched stems from a several-headed woody base; stems 4-angled and densely pubescent above, more rounded at base and often ± glabrous, the epidermis eventually

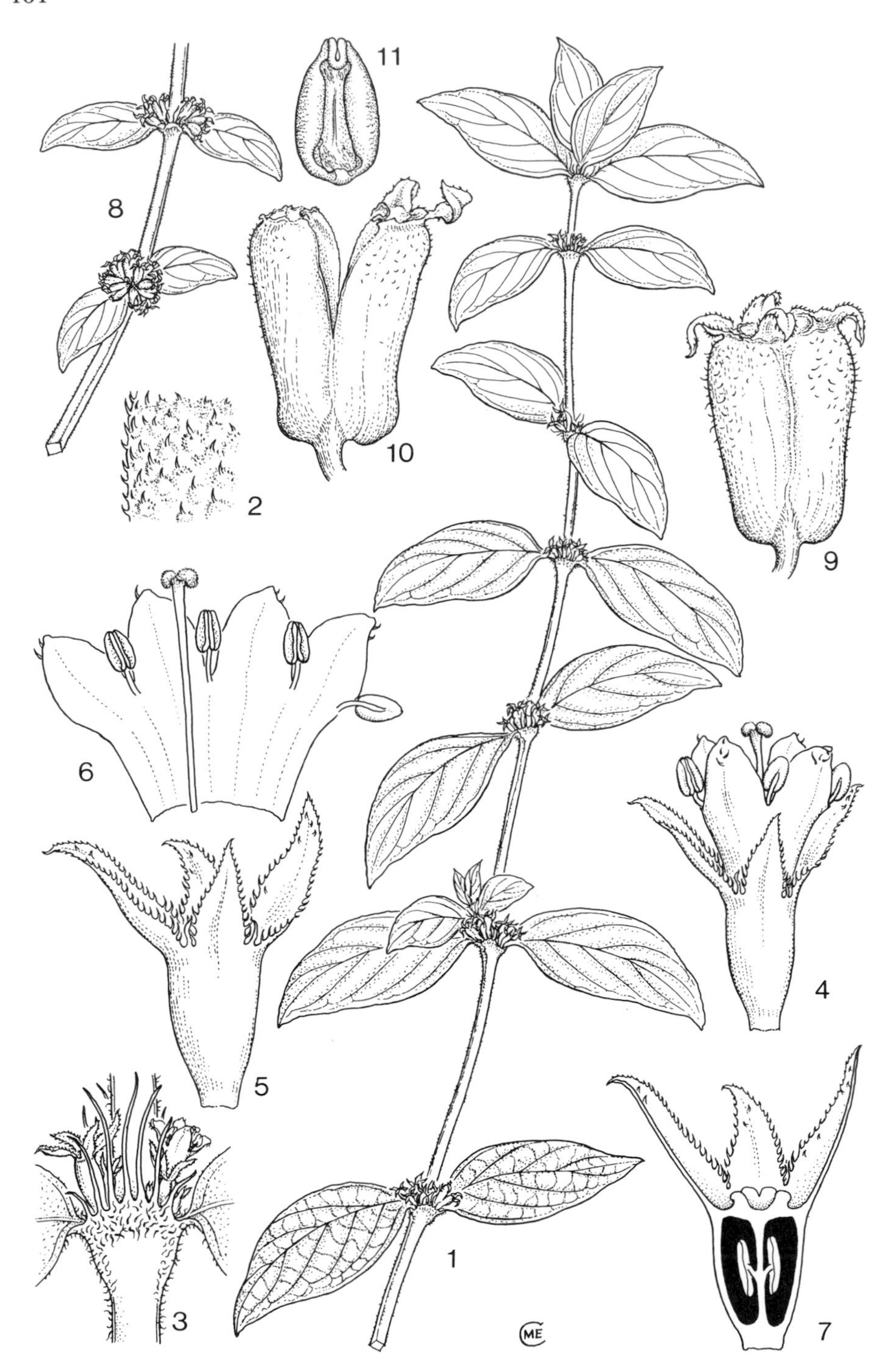

Tab. 37. DIODIA SARMENTOSA. 1, habit (×⅔); 2, part of superior leaf surface (× 8); 3, stipules (× 4), 1–3 from *Milne-Redhead* 3132; 4, flower (× 10); 5, calyx (× 12); 6, corolla opened out (× 12); 7, ovary, longitudinal section (× 12), 4–7 from *Robinson* 6082; 8, part of fruiting branch (×⅔); 9, fruit (× 8); 10, fruit splitting into two parts (× 8); 11, seed (× 8), 8–11 from *Bullock* 1154.

peeling. Leaves 1.3–3.5(4) × 0.5–1.7 cm., oblong-elliptic to elliptic, acute at the apex, cuneate to rounded at the base, somewhat to ± distinctly scabrid above, densely rather softly pubescent beneath; petiole c. 1 mm. long; stipule bases 1–2 mm. long, fringed with hairs and with 5–11 usually reddish setae 1–5 mm. long. Flowers few to several in axillary clusters at most nodes, the nodal clusters usually less than 1 cm. diam. in fruiting stage. Calyx tube pubescent, c. 1 mm. long, obconic; lobes c. 4, equal or ± unequal, 1.3–3 × 0.5–1 mm., oblong-lanceolate to lanceolate, ciliate. Corolla pale yellowish, greenish or white, tube 2.5–3 mm. long, funnel-shaped; lobes 1–2.2 × 0.8–1.2 mm. narrowly triangular, with some hairs outside. Filaments exserted 0.7–1 mm. Style exserted 2–2.5 mm.; stigmatic head 0.5 mm. wide, minutely bilobed. Cocci 2.5–3 × 1.5–2 × c. 1.3 mm., semi-ellipsoid or semi-obovoid, densely pubescent. Seeds brownish black, 2–2.8 × 1.6–2 × 1–1.3 mm., semi-oblong-ovoid, or rectangular in outline, with a ventral groove broadened at the apex, dorsally slightly rugulose and with an apical slit similar to but ± more evident than in last species.

Zambia. B: Kalabo, fl. 16.xi.1959, *Drummond & Cookson* 6506 (K; SRGH). N: Shiwa Ngandu, fl. 3.i.1937, *Ricardo* 129 (BM). W: slope E. of Matonchi Farm, fl. 4.x.1937, *Milne-Redhead* 2556 (K). C: 8 km. E. of Chiwefwe, fr. 15.vii.1930, *Hutchinson & Gillett* 3678 (K). S: Machili, fr. 24.ii.1961, *Fanshawe* 6318 (SRGH).

Also in Angola. *Brachystegia* woodland and derived clearings, wasteland on red chipya soil, grassy places on laterite gravel; 1350 m.

Although typically distinguished in habit from the last species there is practically no morphological difference of much significance, only denser indumentum, perhaps larger flowers and more ovoid fruits. The existence of this clearly derived taxon suggests that *D. sarmentosa* is truly native in Africa and perhaps introduced into the New World and not vice versa. A specimen from Mongu, Barotseland (*Drummond & Cookson* 6605 (K; SRGH)) has narrower leaves than the other material seen.

33. SPERMACOCE L.

Spermacoce L., Sp. Pl.: 102 (1753); Gen. Pl., ed. 5: 44 (1754).
Borreria G.F.W. Mey., Prim. Fl. Esseq.: 79, t. 1 (1818).
Octodon Schumach. & Thonn., Beskr. Guin. Pl.: 74 (1827).
Hypodematium A. Rich., Tent. Fl. Abyss. **1**: 348 (1848), non Kunze (1833).
Dichrospermum Bremek. in Bull. Jard. Bot. Brux. **22**: 75 (1952).
Arbulocarpus Tennant in Kew Bull. **12**: 386 (1958).

Annual or perennial herbs or small subshrubs, with glabrous, pubescent, hispid or scabrid very often 4-angled prostrate to erect stems. Leaves opposite or falsely whorled, sessile or petiolate, the petioles often united with the stipule sheath, which is mostly divided into 1-many ± filiform fimbriae. Flowers mostly small or occasionally medium-sized, hermaphrodite, not heterostylous (except in sect. *Galianthe* (Griseb.) K. Schum. which does not occur in Africa), sessile, mostly in axillary ± globose often very many-flowered clusters or less often in terminal capitula, supported by 1–2(more) pairs of leaves forming bracts, or 1–few in the axils (some extra-African species have extensive terminal inflorescences); sometimes the axillary nodal clusters run together to form a spike-like inflorescence the leaf pairs forming scattered or congested bracts; the flowers in the globose clusters are frequently intermixed with numerous ± scarious filiform bracteoles. Calyx tube obovoid, turbinate or obconic; lobes 2–4(8), mostly triangular, oblong or lanceolate, often ± persistent, sometimes with intermediate denticles. Corolla funnel-shaped or salver-shaped, the tube sometimes very slender, sometimes with a ring of hairs inside towards base; throat glabrous to hairy; lobes (3)4, mostly spreading, valvate. Stamens 4, the filaments inserted in the tube or at the throat, the linear to oblong anthers included or mostly exserted. Ovary 2-locular, the amphitropous ovules solitary in each locule, attached to the middle of the septum; style filiform, mostly exserted; stigma capitellate or with 2 short lobes. Fruit mostly a 2-valved capsule dehiscing from the apex downwards with the septum disappearing (in sect. *Borreria* (G.F.W. Mey.) Verdc.) or sometimes 2-coccous, one dehiscent but the other remaining ± closed (sect. *Spermacoce*) or in a few species (sect. *Arbulocarpus* (Tennant) Verdc.) the capsule splitting from the base upwards but valves remaining attached by the calyx limb which is not split across, the whole falling off like a lid, a persistent septum being left behind. Seeds oblong, ellipsoid or ovoid, usually shining brown with a thin often clearly reticulate testa, ventrally grooved; albumen horny or fleshy.

A large genus of worldwide distribution in the tropics and subtropics, with some 150–250 species according to various estimates, mainly American but with many species in Africa and some also in most tropical and subtropical areas; many have become weeds of cultivation. Some of the species are very similar in facies to members of the *Oldenlandia* group of genera but can always be absolutely distinguished by the solitary ovules and discoid pluricolpate pollen grains.

It has been accepted practice for a hundred years or so to split up this genus, but although the technical characters of the fruit dehiscence seem admirable on paper there is no associated habit facies and it is not possible without additional knowledge to assign flowering material to the genera concerned. I have therefore followed Hooker in Benth. & Hook., Gen. Pl. **2**: 145 (1873) in his circumscription of the genus with the addition of *Octodon* (which he kept separate) to the synonymy as suggested by Hepper (Kew Bull. **14**: 260 (1960)). Bremekamp (Rec. Trav. Bot. Néerland. **31**: 305 (1934)) who has been responsible for much generic segregation in this family states his position as follows. "The differences between *Diodia, Spermacoce* and *Borreria* are also very small and hardly of sufficient importance to justify their separation. In contradistinction with the difference between *Diodia* and *Hemidiodia* which is not only taxonomically of little value but also difficult to see the differences between *Diodia, Spermacoce* and *Borreria* are at least easily recognizable. For this admittedly purely opportunistic reason I have retained these genera". I agree entirely but I must report that Steyermark (*in litt.*) although agreeing with much of this argument, thinks the technical characters are adequate and maintains the genera, thus avoiding a great many name changes since very many S. American species have been described in *Borreria*. As far as the African species are concerned I would also unite *Spermacoce* with *Diodia* and the very close relationship can be seen by comparing *Diodia sarmentosa* with *Spermacoce princeae*, but a survey of the New World material of *Diodia* has convinced me this would not be a wise course — certainly the type species of *Diodia, D. virginiana* L. differs widely in its fruit structure.

For convenience the species of *Diodia* have also been included in the key as well as *Spermacoce assurgens* Ruiz & Pavon (= *laevis* auctt. non Lam.) which must occur as a weed eventually if not already overlooked. Some species keyed out on capsule and seed characters early in the key are keyed again on floral characters later, so if fruits are not present these couplets may be passed over starting at dichotomy 5.

1. Capsule splitting from the base, both valves lifting off, joined at the apices by the persistent calyx; annual with funnel-shaped corolla tube 5–8.5 mm. long; inflorescences mostly terminal (sect. *Arbulocarpus*) - - - - - - - - - - - - - - 20. *sphaerostigma*
– Capsule splitting from the apex to the base into 2 separate valves or fruit dividing into indehiscent or tardily dehiscent cocci - - - - - - - - - - - - - 2
2. Fruit dividing into ± indehiscent cocci - - - - - - - - - - - - - 3
– Fruit capsular with 2 valves - - - - - - - - - - - - - - 4
3. Scrambling or procumbent plant 1–4.5 m.; corolla tube about 1.5–2 mm. long - - - - - - - - - - - - - - *Diodia sarmentosa* (see p. 163)
– More or less erect or ascending stems to 45 cm.; corolla tube 2.5–3 mm. long - - - - - - - - - - - - - - *Diodia flavescens* (see p. 163)
4. Seeds provided with a very distinct cream-coloured basal conical appendage (elaiosome) (TAB. **40**, fig. 21c)* - - - - - - - - - - - - - - - 21. *congensis*
– Seeds without or with only a very small appendage (a small whitish aril may be present) or fruits not present - - - - - - - - - - - - - - - - 5
5. Flowers mostly numerous in dense terminal heads supported by 1 pair of leaves, mostly without any additional flowers at the nodes beneath or rarely at one node beneath; corolla usually blue or violet, but sometimes pale** - - - - - - - - - - - - - 6
– Flowers in all or at least several of the upper nodes, but in some cases these run together to form a spicate or even capitate collection of inflorescences which are easily distinguished from the terminal heads mentioned above by the numerous supporting leaves, which appear to radiate due to suppression of the nodes - - - - - - - - - - - - - 10
6. Corolla tube funnel-shaped, 5–8.5 mm. long; plant annual, with white, blue or purple flowers - - - - - - - - - - - - - 20. *sphaerostigma*
– Corolla tube very narrow, or if funnel-shaped then plant perennial with a woody rootstock 7
7. Calyx lobes 4; corolla tube 1–2.1 cm. long - - - - - - - - - 23. *dibrachiata*
– Calyx lobes 2 or reduced to fringe; corolla tube 0.5–1.3 cm. long - - - - - 8
8. Stems glabrous or sparsely pubescent - - - - - - - - - - - 26. *perennis*
– Stems densely covered with spreading or ascending hairs - - - - - - - - 9
9. Leaves more or less glabrous; stem indumentum ascending; corolla tube 1.3 cm. long; lobes 5.5 × 2 mm. (Lake Bangweulu) - - - - - - - - - - - - - 24. *samfya*
– Leaves glabrous to densely scabrid-pubescent; stem indumentum spreading; corolla tube 6–7 mm. long, lobes 2–4.5 × 1.2–2 mm. - - - - - - - - - 25. *phyteumoides*

* Seeds of sp. 22 not known — might be similar.
** This couplet can be difficult but is too useful to dispense with. In doubtful cases both leads should of course be tried.

10. Flowers in terminal spicate or capitate heads supported by or bearing several pairs of leaves - - - - - - - - - - - - - - - - - - 11
– Flowers not in terminal spike or heads, the flowering nodes well separated but apical one often with more than 2 leaves - - - - - - - - - - - - - - - 19
11. Flowering nodes condensed into heads supported by numerous associated leaves; corolla tube 1–5 mm. long; calyx lobes 2 or 4 - - - - - - - - - - - - - - 12
– Flowering nodes condensed into spikes; calyx lobes 2 (rarely 3 or 2 large and 2 small in *S. subvulgata*) - - - - - - - - - - - - - - - - - - 16
12. Corolla tube c. 5 mm. long; calyx lobes 2.5–5 mm. long* - - - - - - - - 13
– Corolla tube 1–1.5 mm. long; calyx lobes 1–2 mm. long (if corolla tube 0.6–1.3 cm. long and calyx lobes 2 return to couplet 9) - - - - - - - - - - - - - 14
13. Calyx lobes 2; corolla blue with funnel-shaped tube - - - - - - 28. *annua*
– Calyx lobes 4; corolla white or blue with tube filiform below, narrowly funnel-shaped above - - - - - - - - - - - - - - - - - - 9. *huillensis*
14. Very slender unbranched annual with flowering heads about 5 mm. across (Zimbabwe — Hurungwe) - - - - - - - - - - - - - - - - - 8. *sp. A*
– Coarser ± unbranched annuals; flowering heads usually much wider - - - - 15
15. Leaf blades without thick pale margins and midrib; heads with very numerous filiform bracteoles (TAB. **38**, fig. 7a) - - - - - - - - - - - - - 7. *chaetocephala*
– Leaf blades with thick pale margins and midrib; heads (TAB. **39**, fig. 10d) without very numerous narrow bracteoles - - - - - - - - - - - - - - - - 10. *radiata*
16. Calyx lobes 0.35–2 cm. long; valves of capsule almost round, opening right out; seeds with basal cream-coloured appendage (TAB. **40**, fig. 21c) - - - - - - - - 21. *congensis*
– Calyx lobes 3–7.5 mm. long; valves of capsule ± elliptic; seeds without an appendage** 17
17. Leaves linear, almost filiform, 1.5–3 cm. × 0.5 mm.; plant with c. 15 slender unbranched stems from extreme base - - - - - - - - - - - - - - - - - 22. *sp. B*
– Leaves sometimes linear but never so narrow - - - - - - - - - - - 18
18. Corolla tube 9.5 mm. long, very narrowly cylindrical; leaf blades 2–9 mm. wide 1. *arvensis****
– Corolla tube 4 mm. long; leaf blades 1–2(4) mm. wide - - - - - 3. *subvulgata*
19. Corolla tube under 3 mm. long - - - - - - - - - - - - - - - 20
– Corolla tube over 3 mm. long - - - - - - - - - - - - - - - 27
20. Leaf blades ± linear-lanceolate to filiform-subulate - - - - - - - - - 21
– Leaf blades distinctly elliptic - - - - - - - - - - - - - - - 25
21. Lower inflorescences unilateral, i.e. in only one axil at each node (TAB. **39**, fig. 13a); calyx lobes 1–1.8 mm. long; capsule and whole plant ± glabrous - - - - - - 13. *natalensis*
– Lower inflorescences forming regular verticils in both axils at each node; calyx lobes minute, or up to 2 mm. long - - - - - - - - - - - - - - - - - - 22
22. Leaves filiform-subulate, 1–8 cm. × 0.2–0.8 mm.; calyx lobes 8, minute, c. 0.5 mm. long; plant almost glabrous save for characteristic hairs on upper half of fruit - - - - 6. *filifolia*
– Leaves often narrow but never filiform; calyx lobes 2 or 4, minute or up to 2 mm. long 23
23. Inflorescence 0.5–2 cm. wide, with very numerous reddish-brown setiform bracteoles (TAB. **38**, fig. 7a); flower clusters always supported by several pairs of leaves 7. *chaetocephala*
– Inflorescences mostly smaller and under 1 cm. wide, with far fewer flowers and less conspicuous bracteoles; flower clusters supported by 1–several pairs of leaves - - - - - 24
24. Plant of drier places, with leaf blades up to 1.5–3 cm. long; calyx lobes 0.6–1.2 mm. long; capsule glabrous or pubescent - - - - - - - - - - - - - - - 11. *pusilla*
– Plant of marshy places, with leaf blades up to 2.6–6 cm. long, glabrous; calyx lobes minute, 0.25 mm. long; capsule glabrous - - - - - - - - - - - - 12. *quadrisulcata*
25. Inflorescences mostly with 10–20 or more flowers; calyx lobes 0.6 mm. long; corolla tube 1.2 mm. long; fruit distinctly capsular; erect or scrambling herb 0.3–1.2 m. tall (not yet recorded for Flora Zambesiaca area) - - - - - - - - - - - - - *assurgens* (*laevis auctt.*)
– Inflorescences with mostly less than 10 flowers - - - - - - - - - - 26
26. Weak herb, erect, decumbent or prostrate, 3–40 cm. tall, with fine fibrous roots; leaf blades (TAB. **39**, fig. 14) rounded to ± acute, mostly glabrous except for very short marginal hairs and sometimes a few on the midrib beneath; calyx lobes 2(4), 0.6–0.8 mm. long 14. *mauritiana*
– Robust, prostrate or scrambling herb with stems 1–3.6 m. long; leaf blades tapering, acute, scabrid-pubescent above; calyx lobes 4, 1.5 mm. long; corolla tube 1.8 mm. long; fruit distinctly dividing into 2 ± indehiscent cocci - - - - - - *Diodia sarmentosa* (p. 163)
27. Prostrate or ascending herbs with elliptic or ovate leaves; calyx lobes 4 - - - - 28
– More or less erect herbs or if ± prostrate (*S. kirkii* etc.) then leaves ± linear; calyx lobes 2–4 29
28. Plant of upland areas, 1180–2250 m., flowers in few- to many-flowered dense axillary clusters; corolla tube 5–10.5 mm.; capsule-lobes mostly not diverging to 180°; stipular setae usually longer, 7–8.5 mm. - - - - - - - - - - - - - - - - 15. *princeae*

* In Fl. Trop. E. Africa couplet 14/1 should read 1–5 mm. not 1–1.3 mm.

** Seeds of species 22 unknown but since believed to be closely related to species 21, appendage may be well-developed.

*** If calyx lobes large see also species 2.

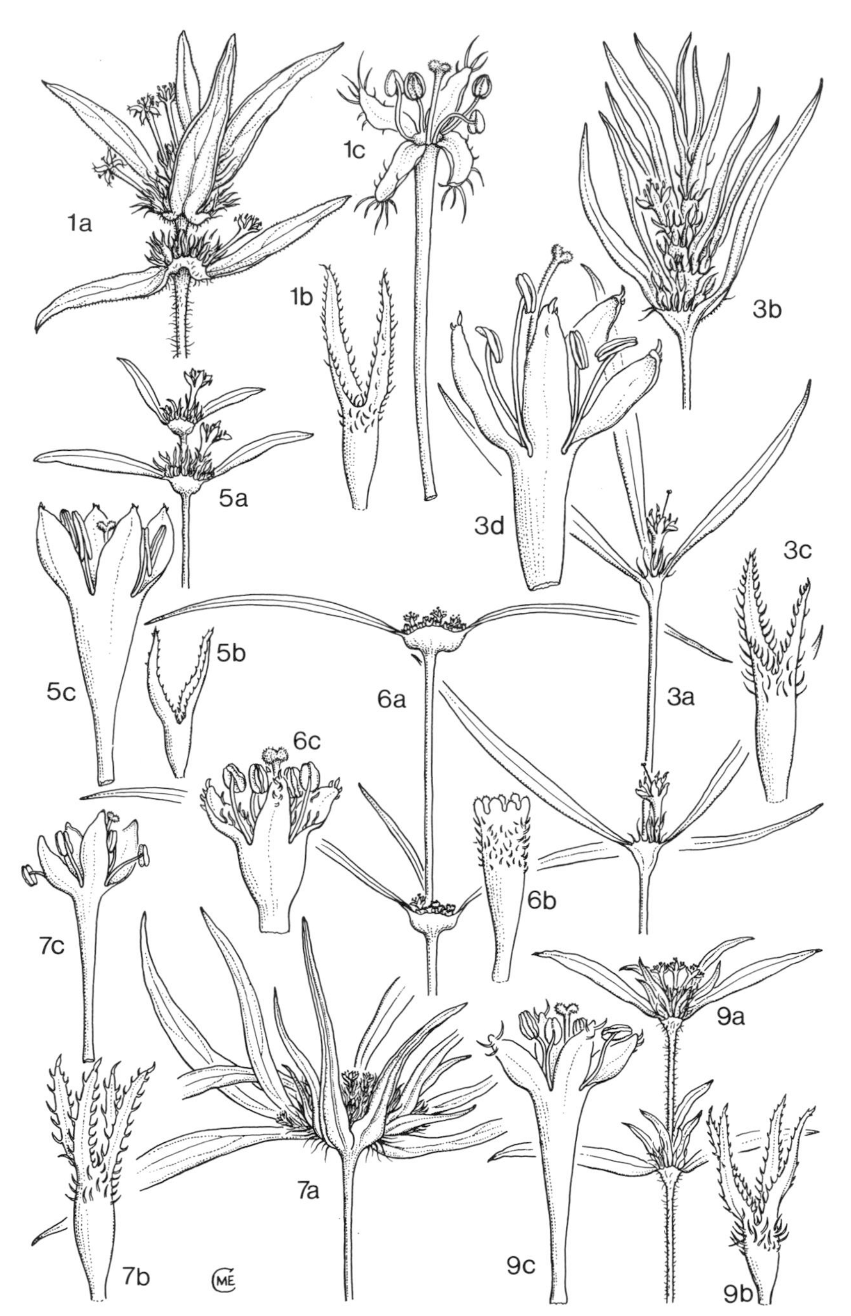

Tab. 38. SPERMACOCE. Species numbered as in text. 1. —S. ARVENSIS. 1a, inflorescence (× 1); 1b, calyx (× 6); 1c, corolla (× 6), 1a–c from *Greenway & Polhill* 11486. 3. —S. SUBVULGATA var. SUBVULGATA. 3a, inflorescence (× 1), from *Greenway & Kanuri* 14142; 3b, inflorescence (× 1); 3c, calyx (× 6); 3d, corolla (× 6), 3b–d from *Greenway & Polhill* 11461. 5. —S. BANGWEOLENSIS. 5a, inflorescence (× 1); 5b, calyx (× 4); 5c, corolla (× 4), 5a–c from *Watmough* 181. 6. —S. FILIFOLIA. 6a, inflorescence (× 1); 6b, calyx (× 10); 6c, corolla (× 10), 6a–c from *Fanshawe* 9137. 7. —S. CHAETOCEPHALA. 7a, inflorescence (× 1); 7b, calyx (× 12); 7c, corolla (× 12), 7a–c from *Bally* 10817. 9. —S. HUILLENSIS. 9a, inflorescence (× 1); 9b, calyx (× 8); 9c corolla (× 8), 9a–c from *Cruse* 302.

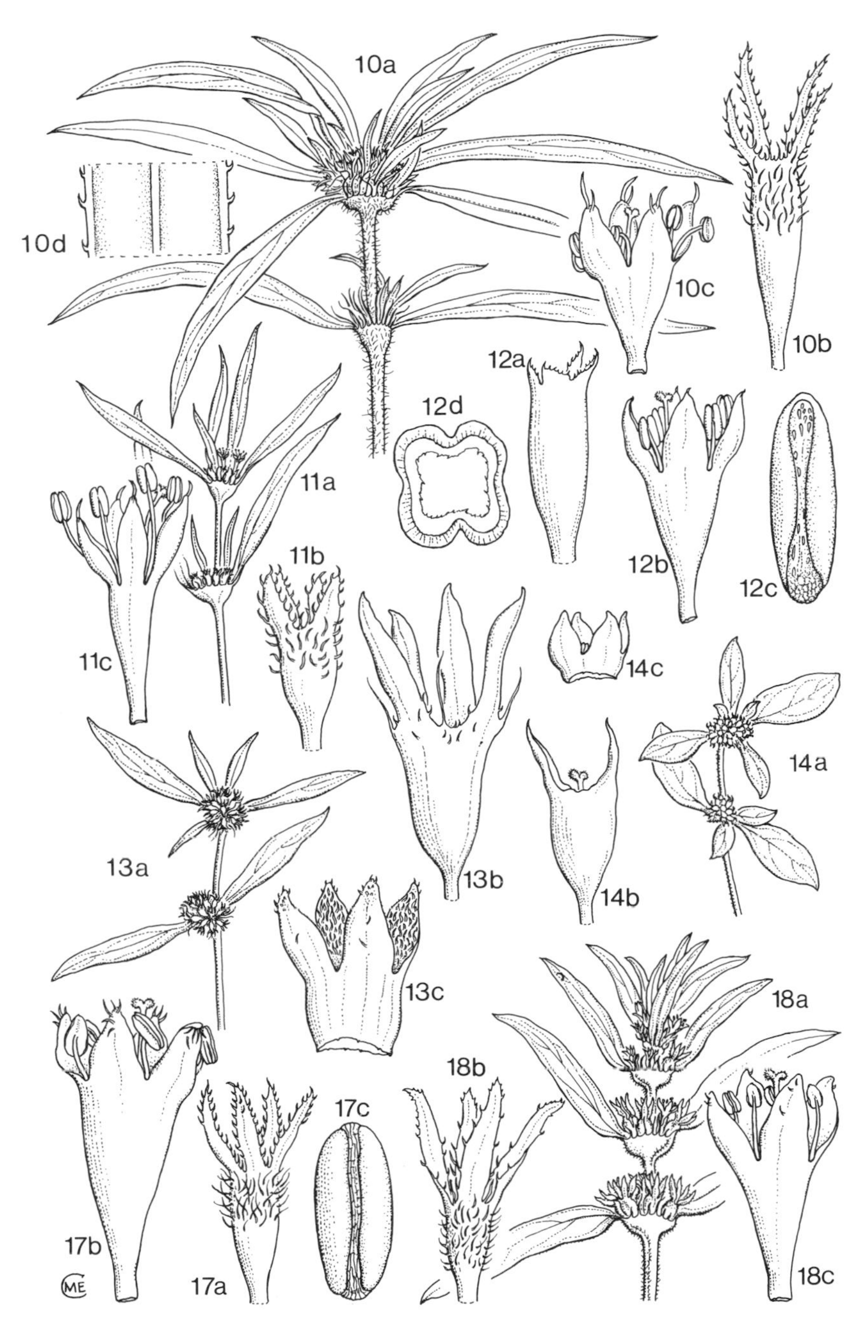

Tab. 39. 10.—S. RADIATA. 10a, inflorescence (× 1); 10b, calyx (× 12); 10c, corolla (× 12); 10d, detail of inferior leaf surface (× 12), 10a–d from *Hazel* 683. 11.—S. PUSILLA. 11a, inflorescence (× 1); 11b, calyx (× 16); 11c, corolla (× 16), 11a–c from *Norman* 94. 12.—S. QUADRISULCATA. 12a, calyx (× 12); 12b, corolla (× 12); 12c, seed (× 12); 12d transverse section of stem (× 6), 12a–d from *Milne-Redhead & Taylor* 9072. 13.—S. NATALENSIS. 13a, inflorescence (× 1); 13b, calyx (× 12); 13c, corolla (× 12), 13a–c from *Milne-Redhead & Taylor* 8761. 14. —S. MAURITIANA. 14a, inflorescence (× 1); 14b, flower with corolla removed (× 12); 14c, corolla (× 12), 14a–c from *Symes* 512. 17.—S. SENENSIS. 17a, calyx (× 6); 17b, corolla (× 6), 17a–b from *Richards* 17959; 17c, seed (× 12), from *Smith* 1117. 18.—S. DESERTI. 18a, inflorescence (× 1); 18b, calyx (× 6); 18c, corolla (× 6), 18a–c from *Skarpe* s.n.

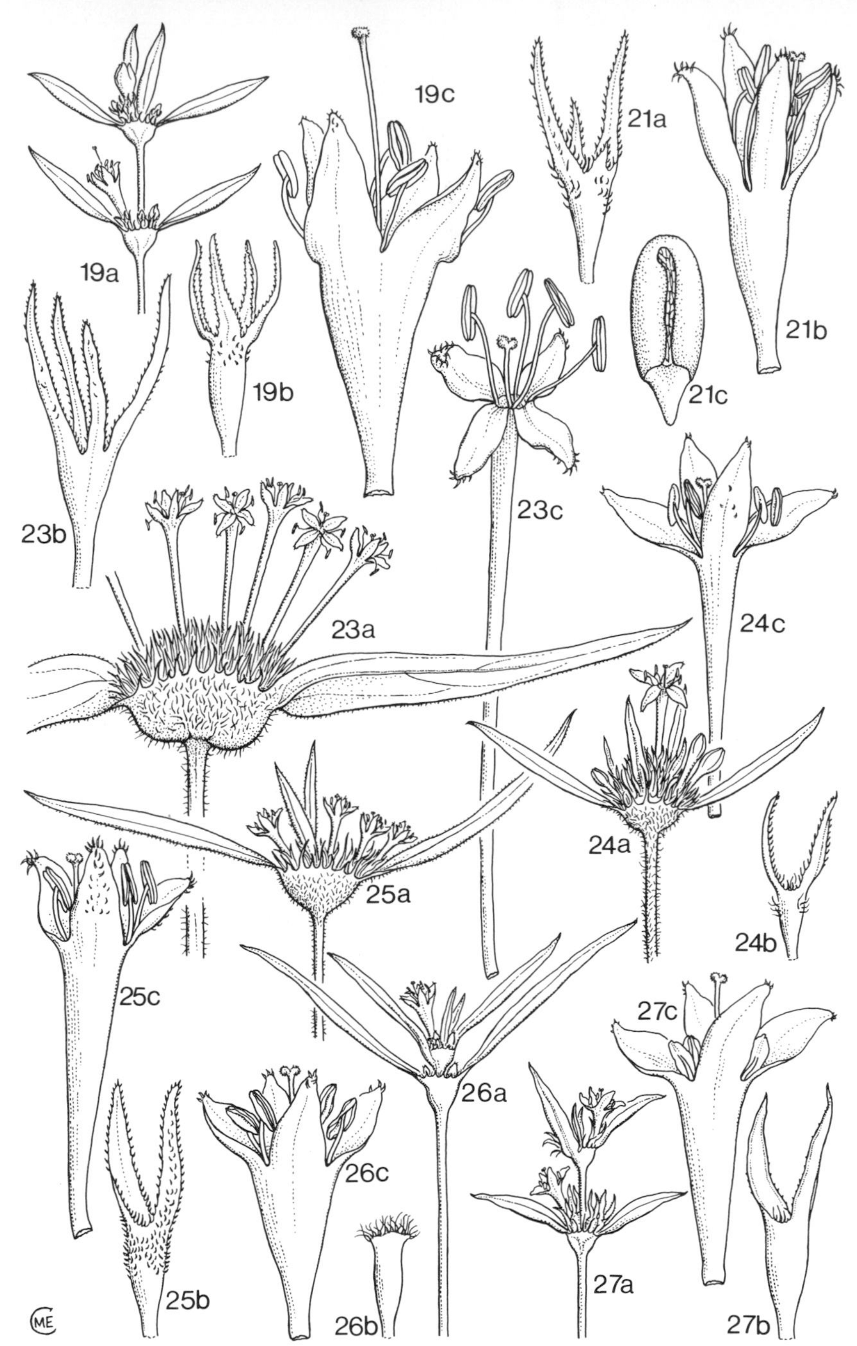

Tab. 40. 19. —S. LATITUBA. 19a, inflorescence (× 1); 19b, calyx (× 4); 19c, corolla (× 4), 19a–c from *Milne-Redhead & Taylor* 9482. 21. —S. CONGENSIS. 21a, calyx (× 3); 21b, corolla (× 3), 21a–b from *Richards* 11152; 21c, seed (× 6), from *Street* 798; 23. —S. DIBRACHIATA. 23a, inflorescence (× 1); 23b, calyx (× 4); 23c, corolla (× 4), 23a–c from *Richards* 8610; 24. —S. SAMFYA. 24a, inflorescence (× 1); 24b, calyx (× 3); 24c, corolla (× 3), 24a–c from *Chabwela* s.n. 25. —S. PHYTEUMOIDES var. PHYTEUMOIDES. 25a, inflorescence (× 1); 25b, calyx (× 5); 25c, corolla (× 5), 25a–c from *Richards* 1162. 26. —S. PERENNIS var. PERENNIS. 26a, inflorescence (× 1); 26b, calyx (× 4); 26c, corolla (× 4), 26a–c from *Wright* 352 (× 4). 27. —S. THYMOIDEA. 27a, inflorescence (× 1); 27b, calyx (× 5); 27c, corolla (× 5), 27a–c from *Milne-Redhead* 4338.

– Plant of littoral areas, 20–60 m.; flowers in 2–6-flowered clusters; corolla tube 3–4 mm. long; capsule lobes ultimately diverging at 180°; stipular setae usually short, mostly well under 7 mm. - - - - - - - - - - - - - - - - - 16. *schlechteri*
29. Calyx lobes 2 - - - - - - - - - - - - - - - - - - 30
– Calyx lobes (3)4(6) - - - - - - - - - - - - - - - - - 35
30. Leaves with margins distinctly thickened and similar to midrib; seeds sometimes with very distinct appendage - - - - - - - - - - - - - - - - 31
– Leaves with margins not so distinctly thickened; seeds without or with only a trace of an appendage - - - - - - - - - - - - - - - - - 32
31. Calyx lobes elongated, 0.8–2 cm. long; seeds with a distinct basal appendage (elaiosome) - - - - - - - - - - - - - - - - 21. *congensis*
– Calyx lobes 3–4 mm. long; ripe seeds not seen but probably without a distinct appendage - - - - - - - - - - - - - 27. *thymoidea*
32. Corolla tube very slender, almost filiform (Zimbabwe S, Masvingo) - - 2. *bisepala*
– Corolla tube more funnel-shaped - - - - - - - - - - - - 33
33. Much-branched clump-forming prostrate woody shrublet; stipule bases papillate-puberulous (Zambia, Bangweulu) - - - - - - - - - - - - - 5. *bangweolensis*
– Annual or perennial unbranched or sparsely branched herbs; stipule bases glabrous or pubescent or if prostrate then coastal in distribution - - - - - - - - - 34
34. ± Prostrate plant of coastal sandy places, sea-shore, etc.; stipular setae rather short, ± 3 mm. long; leaves subsucculent; elaiosome at base of seed truncate - - - - - - 4. *kirkii*
– ± Erect plant of mostly inland localities (to 1650 m.); stipular setae up to 6.5 mm. long; leaves thinner; elaiosome at base of seed more pointed (if descriptions do not fit return to 29) - - - - - - - - - - - - - - - 3. *subvulgata*
35. *Inflorescences supported by small leaves under half the size of the stem leaves; style and stigma well exserted from the corolla, much over-topping the anthers; stems glabrous or hairy; corolla tube distinctly funnel-shaped (TAB.* **40**, fig. 19c); leaf blades 1–4.7 cm. long 19. *latituba*
– Inflorescence leaves not markedly smaller; style and stigma not exserted much beyond the anthers; stems hairy - - - - - - - - - - - - - - - - 36
36. Leaves lanceolate, mostly under 5 mm. wide but some in *species* 3 up to 8 mm. wide 37
– Leaves elliptic or elliptic-lanceolate, 0.25–2.2 cm. wide usually some at least 5 mm. 38
37. More slender herb, with more numerous flowers in the inflorescences; corolla tube slender, usually 0.5 mm. wide near the middle - - - - - - - - - - 9. *huillensis*
– Coarser herb with fewer flowers in the inflorescences; corolla tube more cylindrical, 1.5 mm. wide near the middle - - - - - - - - - - 3. *subvulgata* var. *quadrisepala*
38. Stems sparsely to densely pubescent all over; leaves pubescent on both faces (widespread) - - - - - - - - - - - - - 17. *senensis*
– Stems pubescent or scabrid on the angles; leaves scabrid-ciliate on the margins but otherwise glabrous (SE. & SW. Botswana) - - - - - - - - - - - - 18. *deserti*
(If no satisfactory result achieved try the alternative leads at couplets 5, 10 and 11).

1. **Spermacoce arvensis** (Hiern) Good in Journ. Bot. Lond. **65**, Suppl. 2: 41 (1927). —Verdc. in F.T.E.A., Rubiaceae 1: 351, fig. 49/5 (1976). TAB. **38**, fig. 1. Types from Angola.
Tardavel arvensis Hiern, Cat. Afr. Pl. Welw. **1**: 504 (1898).
Borreria arvensis (Hiern) K. Schum. in Just's Bot. Jahresb. **26**(1): 391 (1900).

Annual herb (3)8–30(70) cm. tall, sometimes slender in habit but often coarse and appearing quite different; stems drying blackish purple, often with paler longitudinal lines, densely covered with spreading hairs, usually sparingly branched or unbranched. Leaf blades 1.3–7 cm. × 2–9 mm., lanceolate, acute at the apex, narrowed at the base into the stipule sheath, covered with long rather coarse hairs on both surfaces and also scabrid with shorter hairs near the margins; stipules with fimbriae crimson or red-brown in life but mostly blackish purple when dry; base 2 mm. long with 5–7 fimbriae 3–8 mm. long, with long white hairs. Flowers in many-flowered verticillate clusters, mostly at the upper nodes which run together to form a dense coarse strobilate inflorescence 0.7–1.5 cm. wide, but usually with 1–2 clusters at the nodes beneath and well separated; bracteoles stipule-like, of 5–6 lanceolate fimbriae 5–6 mm. long, joined at the base, ciliate along the margins. Calyx tube 1.8 mm. long, ellipsoid; lobes 2, 3 mm. long, 0.6 mm. wide at the base, lanceolate. Corolla white or with purplish tips to lobes outside or blue; tube 8–12 mm. long, very narrowly cylindrical; lobes 1.8–3 × 0.8–1.2 mm., narrowly elliptic or linear-oblong, sparsely hairy outside. Filaments exserted c. 1–2 mm. Style exserted 1.2–2 mm.; stigma slightly bifid, 0.5 mm. wide. Capsule 3–3.5 × 1.8–2 mm., ellipsoid, shortly hairy at the apex. Seeds chestnut-brown; 2.7 mm. long, 1.1–1.5 mm. wide, 0.9 mm. thick, semi-ellipsoid, shiny, flat and narrowly grooved ventrally, the groove opening out at each end, convex dorsally, rugulose.

Zambia. C: near Lusaka, 5 km. W. of Kasisi, fl. 3.iv.1972, *Kornaś* 1507 (K; KRA). E: Katete, St.

Francis' Hospital, fl. 17.iii.1957, *Wright* 183 (K). S: Namwala, fl. 16.iv.1963, *Van Rensburg* 1980 (K; SRGH). **Zimbabwe**. N: Umvukwes Mts., central area near Mutorashanga Pass, fl. 24.iv.1948, *Rodin* 4456 (K; UC). W: Bulilima-Mangwe Distr. 32 km. N. of Marula, Mananda Dam, fl. & fr. 20.iv.1972, *Grosvenor* 739 (K; LISC; SRGH). C: Chegutu, fl. 31.iii.1948, *Hornby* 2875 (K; SRGH). E: Mutare, fl. & fr. 27.iii.1955, *Chase* 5530 (K; BM; LISC; SRGH). S: 32 km. N. of Masvingo, fr. 4.v.1962, *Drummond* 7970 (SRGH). **Malawi**. N: Mzimba Distr., Champira, Kamatawo, fl. 19.iv.1974, *Pawek* 8400 (K; MAL; MO). C: Lilongwe, Chitedze, fl. 22.ii.1955, *Exell et al.* 1113 (BM; LISC; SRGH). **Mozambique**. N: about 3 km. from Mutuáli towards Malema, fl. 17.iii.1964, *Torre & Paiva* 11226 (LISC).

Also in Tanzania and Angola. Open bushland, *Colophospermum*, *Brachystegia*, *Julbernardia globiflora*, *Acacia albida–Kigelia* woodlands, also seasonally damp grasslands and cultivations; 580–1440 m.

2. **Spermacoce bisepala** Verdc. sp. nov.* Type: Zimbabwe, Masvingo. *Monro* 941 (BM, holotype; SRGH).

Herb to 20 cm., well-branched; stems probably erect or ascending, densely pubescent with spreading hairs. Leaves 2.5–3.5 × 0.3–0.8 cm., narrowly elliptic to oblong-lanceolate, acute at the apex, narrowed to sessile base, pubescent on both sides, with longer hairs on mid-rib beneath and short ones rendering margins rough; stipule bases 2.5–3 mm. long, densely pubescent, with 7–9 linear-lanceolate fimbriae 1–3.5 mm. long. Flowers in dense clusters 1–1.5 cm. wide at apical 4–5 nodes, mostly well-separated but first 2 approximate on one shoot (possibly separating later?); bracteoles c. 3 mm. long, ciliate. Calyx tube 2.2 mm. long, pubescent on upper half; lobes 2, 5–7 × 1–1.5 mm. and perhaps larger, oblong-elliptic, pubescent. Buds capitate, the limb distinctly pubescent. Corolla colour unknown, probably white; tube 8.5–10 mm. long, very narrowly cylindrical, lobes 2.5 × 1 mm., oblong, hairy outside, the hairs longer and more setiform at apex. Filaments exserted about 1.5 mm. Style exserted 1.5 mm.; stigma capitate. Capsule 3.5 × 2.5 mm., ellipsoid, pubescent above; immature seeds 2.5 × 1.1 mm.

Zimbabwe. S: Masvingo, fl. & imm. fr. 1909, *Monro* 942 (BM; SRGH).

Known only from this specimen; ecology unknown.

It is extraordinary that no other material seems to be available. It differs too much from the previous species to be considered an abnormal variant but is clearly related to it.

3. **Spermacoce subvulgata** (K. Schum.) Garcia in Mem. Junta. Invest. Ultram. Sér. 2, No. 6: 49 (1959). —Verdc. in F.T.E.A., Rubiaceae 1: 352 (1976); in Kew Bull. **38**: 96 (1983). Types from Tanzania, Angola and Zimbabwe, near Mutare, *Schlechter* 12184 (B, syntype†; BM; K).
Borreria subvulgata K. Schum. in Engl., Bot. Jahrb. **28**: 111 (1899).

Annual or perennial erect herb 15–90 cm. tall, with ± woody base (rarely very woody up to 8 mm. wide); stems glabrous except for pubescence just below the nodes or in vertical grooves on the internodes between the ribs, arranged bifariously. Leaf blades 3–6 cm. × 1–2(8) mm., linear, ± acute at the apex, narrowed to the base, ± scabrid on the margins and midrib beneath to distinctly shortly hairy; stipule base hairy, 3 mm. long, with 6–11 fimbriae 3–6.5 mm. long, often purplish. Flowers in clusters at the nodes, the apical ± 7 running together to form a sort of cone with the leaf-like bracts protruding and the stipules overlapping or, in some specimens, the clusters remaining quite separate. Calyx tube 1.5 mm. long, ellipsoid; lobes 2 (rarely 3) or 2 large and 2 small or in one variety 4, 3–4.5 mm. long, linear. Corolla white or more rarely blue; tube 4–5 mm. long, narrowly funnel-shaped; lobes 2.5–5 mm. long, lanceolate, mostly with a few hairs at the apex outside. Filaments exserted 1.5–2.5 mm. Style exserted c. 2–5 mm. Capsule 3–4 mm. long, oblong-ellipsoid, hairy or less often ± glabrous. Seeds dark brown, 3–3.3 × 1.2 × 0.7 mm., narrowly oblong, narrowed at one end, with a ventral groove which is opened out at basal end and bears a trace of a white aril; dorsal surface rugulose.

Var. **subvulgata**. —Verdc. in F.T.E.A., Rubiaceae 1: 352, fig. 49/7 (1976). TAB. **38**, fig. 3.
Tardavel kaessneri S. Moore in Journ. Bot. Lond. **43**: 250 (1905). Type from Kenya.
Anthospermum mazzocchi-alemanii Chiov. in Bull. Soc. Bot. Ital. **1924**: 40 (1924). Type from Angola.

Calyx lobes 2, or exceptionally sometimes 3 or 2 long and 2 short or ± vestigial.

* Affinis *Spermacoce arvensis* (Hiern) Good sed habitu nec gracili nec simplicicauli autem possibiliter perenni caule basi valde ramosa, calycis lobis majoribus usque saltem 7 × 1.5 mm., verticillis florentibus superioribus plerumque separatis differt.

Zambia. B: Mongu, fl. & fr. 20.ii.1966, *Robinson* 6843 (K). N: Mbala, fl. 15.iv.1952, *Richards* 1429 (K). W: Mufulira, fl. & fr. 18.iv.1948, *Cruse* 326 (K). C: Great East Road between Undaunda and Rufunsa, fl. & fr. 6.iv.1972, *Kornaś* 1513 (K; PR). E: Lukusuzi Nat. Park, fl. 18.iii.1971, *Sayer* 1076 (K; SRGH). S: Machili, fl. & fr. 18.iii.1961, *Fanshawe* 6485 (K; NDO; SRGH). **Zimbabwe**. N: Central Umvukwes Mts., near Mutorashanga Pass, fl. & fr. 24.iv.1948, *Rodin* 4455 (K; UC; SRGH). W: Matobo, Besna Kobila Farm, fl. & fr. iv.1955, *Miller* 2766 (K; SRGH). C: Enterprise, fr. 5.vi.1928, *Eyles* 5835 (K; SRGH). E: Mutare, Commonage, Quaggas Hoek, fl. 22.iii.1953, *Chase* 4877 (K; LISC; SRGH). **Malawi**. N: Nkhata Bay Distr., N. Viphya, 3 km. NW. of Chikwina, fl. 22.v.1970, *Brummitt* 11046 (K; MAL). C: 13 km. S. of Nkhota Kota, near Sani Hill, fl. 16.vi.1970, *Brummitt* 11473 (K; MAL; SRGH). S: Zomba, Lake Chilwa, near Fisheries Research Unit, fr. 1.vii.1969, *Banda* 1114 (K; SRGH). **Mozambique**. N: 16 km. NE. of Mandimba Border Post, fl. 3.v.1960, *Leach* 9916 (K; LISC; SRGH). Z: outskirts of Maganja da Costa, fl. & fr. 31.vii.1943, *Torre* 5742 (BM; EA; J; LISC; LMA; M). MS: Báruè, Serra de Choa, about 7 km. from Choa towards R. Caneresi, fr. 25.v.1971, *Torre & Correia* 18639 (LISC; SRGH).

Also in Zaire, Burundi, Rwanda, Uganda, Kenya, Tanzania, Angola and SW. Africa. Grassland, bushland and woodland, also roadsides, dambo grassland, rock outcrops and termite mounds and as a common weed in maize and coffee plantations and other cultivations; (190)640–1650 m.

Spermacoce subvulgata is an exceptionally variable species. The typical rather coarse plant, with the upper whorls run together to form a spike-like inflorescence, is very easy to recognise, but slender forms with the whorls mostly separated are often difficult to place. These slender forms scarcely differ from *S. kirkii* (Hiern) Verdc. (*Borreria diodon* K. Schum.) known from a few specimens collected on the Mozambique coast at about 18°54'–23° S. Apart from the ecological differences the stipules are also different, the slender forms of *S. subvulgata* having much longer more slender fimbriae. There are also slight differences in the leaves and fruits but further work may show the two have to be considered variants of one species — in that case the species as a whole will have to be known by the much older name *S. kirkii* (type: Mozambique, mouth of W.R. Luabo, *Kirk* (K, holo.!)). Some specimens from Nampula and Mocimboa da Praia to Diaca, km. 34 (*Torre* 1235 (LISC) & *Torre & Paiva* 11967 (COI; LMU; SRGH; WAG)) are intermediate.

Var. **quadrisepala** Verdc. in Kew Bull. **30**: 303 (1975); in F.T.E.A., Rubiaceae 1: 353 (1976). Type from Tanzania.

Calyx lobes 4, mostly equal.

Mozambique. T: Macanga, Mt. Furancungo (Elefante) fl. & fr. iii.1966, *Pereira* et al. 1761 (BM; LMU). MS: Chemba, Chiov. C.I.C.A. Expt. Station, fl. & fr. 12.iv.1960, *Lemos & Macúacua* 80 (K; LISC; LMA; SRGH).

Also in Tanzania. *Brachystegia–Julbernardia–Uapaca* woodland; (640) 1265–1380 m.

It was at first thought this might be a narrow-leaved variant of *S. senensis*, and typical specimens are not closely similar to var. *subvulgata*, but many specimens of the latter exist, even those with the inflorescences congested into spikes, where there are 2–3 calyx lobes or 2 large and 2 small. There is a possibility that var. *quadrisepala* is of hybrid origin. Coarse specimens of *S. huillensis* are also similar. Two specimens with the facies of this variety, *Torre & Paiva* 10539 (BR; K; LMA; SRGH) (Mozambique. Z: Gúruè, Mt. Currarre, fl. 11.ii.1964) and *Torre & Paiva* 10936 (COI; LMU) (Mozambique. N: Serra do Ribáuè, 23.i.1964) have c. 2 calyx lobes up to 6 mm. long and very long broader leaves to 8 mm. wide. Broader leaves may be a better character than the number of calyx lobes.

4. **Spermacoce kirkii** (Hiern) Verdc. in Kew Bull. **30**: 303 (1975); in F.T.E.A., Rubiaceae 1: 353 adnot. (1976). Type: Mozambique, mouth of western R. Luabo, *Kirk* (K, holotype).

Diodia kirkii Hiern in F.T.A. **3**: 230 (1877).

Borreria diodon K. Schum. in Engl., Bot. Jahrb. **28**: 109 (1899). Type: Mozambique, Massinga (Machingas or Machisugu), *Schlechter* 12123 (B, holotype†; BM; K).

Sometimes annual but probably often perennial much-branched prostrate or perhaps sometimes suberect subshrubby herb 30–40(?) cm. long; stems subangular, bifariously sparsely puberulous in narrow longitudinal grooves, otherwise glabrous. Leaves 0.5–6 cm. × 1–3 mm., linear to narrowly linear-oblanceolate, acute at the apex, attenuate at the base, probably subsucculent in life, rigid, glabrous; stipule sheath 2 mm. long with 2 angular lines of pubescence and pubescent margin with 5 pubescent setae 3 mm. long. Flowers in congested clusters c. 1 cm. wide at many if not most nodes, the upper 3 clusters condensed into a head or at least more approximate, the others well spaced. Calyx tube 3 mm. long, the upper half pilose with erect hairs; lobes 2 (sometimes more?) 1.5–2.5 to 3.5 (in fruit) mm. long, narrowly oblong, ciliate. Corolla white or greenish white; tube 3.5 mm. long, funnel-shaped, hairy within; lobes 3–4 × c. 1.2 mm., triangular-lanceolate, pubescent at apex. Filaments exserted about 3 mm.; style exserted 4 mm., stigmatic club capitate-bilobed. Fruits 4.5 × 2.2 mm., oblong-ovoid, the valves opening widely. Seeds dark purplish brown, 2.5 × 1.3 mm., compressed oblong-ovoid, very slightly widest at one end, with a fairly narrow groove beneath widened at one end containing some rhaphides and with a very reduced white proximal elaiosome.

Mozambique. MS: Beira fl. vi.1911, *Dawe* 381 (K). GI: Inharrime, fl. & fr. 27.ii.1955, *Exell, Mendonça & Wild.* 666 (BM; LISC; SRGH).

Endemic. Open sandy places, sea-shore, strand-line in mangrove association (fide *Peter* 31057); 0–100 m.

Very little good recent material of the species seems to be available. It is very close to the open form of *Spermacoce subvulgata* and I at one time considered that should be sunk into it but, apart from a very different ecology, the prostrate habitat is distinctive.

5. **Spermacoce bangweolensis** (R.E. Fries) Verdc. in Kew Bull. **30**: 308 (1975). TAB. **38**, fig. 5. Type: Zambia, Lake Bangweulu, Kasoma, *Fries* 662 (UPS, holotype).

Borreria bangweolensis R.E. Fries in Wiss. Ergebn. Schwed. Rhod.-Kongo Exped. Ergänzungsheft: 17 (1921).

Prostrate or suberect clump-forming subshrub up to c. 50 cm.; stems brown, the upper parts 4-angled, bifariously pubescent with papilla-like hairs. Leaves somewhat fleshy in life, rigid when dry, 1.5–4 cm. × 1.5–2.5 mm., linear, acute at the apex, subsessile, glabrous save for short stiff marginal hairs rendering margin very scabrid; lateral nerves not visible; stipule sheath 3 mm. long, papillate-puberulous, with 1–5 setae, 0.5–6 mm. long. Flowers up to c. 20 in terminal congested clusters and at 1–3 immediately lower nodes. Calyx lobes 2, 3–6 mm. long, linear, acute. Corolla pale blue, 4–7 mm. long, 3 mm. wide at the throat, narrowly funnel-shaped, glabrous; lobes 3–4 × 1–1.5 mm., triangular-oblong, acute, with very few apical hairs outside. Anthers and style exserted c. 2 mm.; stigmatic club bifid. Capsule 3 mm. long, obovoid. Seeds ± chestnut, 2.25–2.5 mm. long, oblong.

Zambia. N: Lake Bangweulu, Samfya, fl. & fr. 30.i.1959, *Watmough* 181 (K; LISC; SRGH).

Not known elsewhere. Bare soil among grass clumps; 1150 m.

6. **Spermacoce filifolia** (Schum. & Thonn.) Lebrun & Stork in Kew Bull. **39**: 778 (1984). —Vollesen in Opera Botanica No. 59: 72 (1980) (as (Schum. & Thonn.) DC.). TAB. **38**, fig. 6. Type from Ghana.

Octodon filifolium Schum. & Thonn., Beskr. Guin. Pl.: 74 (1827). —DC. Prodr.: 540 (1830). —Hiern in F.T.A. **3**: 241 (1877). —Hutch. & Dalz., F.W.T.A. **2**: 136 (1931). —F.W. Andr., Fl. Pl. Anglo-Egypt. Sudan **2**: 448 (1952).

Borreria filifolia (Schum. & Thonn.) K. Schum. in Engl. & Prantl, Pflanzenfam. **4**, 4: 144 (1891). —Hepper, F.W.T.A. ed, 2, **2**: 221 (1963); W. Afr. Herb. Isert & Thonning: 105 (1976).

Annual erect strict herb 15–90 cm. tall with 1-several slender unbranched or sparsely branched glabrous or nearly glabrous stems. Leaves (1)3–10 cm. × 0.2–0.8 mm., filiform-subulate, pointed at the apex, sessile on to stipule sheaths which are long and narrow at sterile nodes c. 5 × 2 mm. and with 1–2 setae mostly 3 mm. long but much wider at flowering nodes 0.5–1 cm. wide, with 3 short subulate setae up to c. 1 mm. long or sometimes ± truncate. Flowers 10–50 in tight clusters up to 1.3 cm. wide, supported by the stipular involucre at the terminal and 1–2 immediately lower nodes; bracteoles filiform, 2–3.5 mm. long, broadened at the base, shortly spathulate, purple and hairy at the apex. Calyx tube 1 mm. long, hairy above, lobes 8, ± purple, minute, c. 0.5 mm. long. Corolla tube white, c. 1.5 mm. long, funnel-shaped; lobes blue or purple, c. 1 mm. Capsule with upper half purplish, c. 4 mm. long, ± obovoid, opening at the apex but valves cohering, densely covered with distinctive flattened white or purple hairs on upper half. Seeds shiny brown, 2.5–2.8 × 0.9–1.2 mm., narrowly oblong-ellipsoid, slightly broadened at the apex, very obscurely angled dorsally, ventrally with a narrow furrow containing a few rhaphides and with a basal cream aril-like cellular elaiosome.

Zambia. C: Lusaka, Chukupi Estate, fr. 12.iv.1963, *Van Rensburg* 1898 (K; SRGH). S: Mazabuka Distr., Tara Protected Forest, fl. & fr., 1.iii.1960, *White* 7502 (FHO; K; SRGH). **Zimbabwe**. N: Sengwa Res. Sta., fl. & fr. 6.iii.1984, *Mahlangu* 929 (SRGH). C: Chegutu, Poole, fr. 3.iv.1946, *Wild* 1011 (K; SRGH). **Malawi**: N: Mwanemba, fl. & fr. ii-iii.1903, *McClounie* 24 (K).

Also Senegal to N. Nigeria, Zaire, Sudan and S. Tanzania. Mostly in damp or wet places, mixed *Acacia* woodland and *Hyparrhenia* grassland, termite mounds by dambos, road drains, etc., clay dambo, dambo with wet *Oryza–Leersia* grassland and paddy fields but also drier *Brachystegia–Isoberlinia* woodland; 1120–1200 m.

The Mwanemba specimens are said to be from 8000 ft. but this must be erroneous and they might even be from near the lake. The combination attributed by the Index Kewensis and by Vollesen (loc. cit.) to DC was published in synonymy and not valid.

7. **Spermacoce chaetocephala** DC., Prodr. 4: 554 (1830). —Verdc. in F.T.E.A., Rubiaceae 1: 354, fig.

49/10 (1976). —Gonçalves in Garcia de Orta, Sér. Bot. **5**(2): 206 (1982). TAB. **38**, fig. 7. Type from Senegal.

Spermacoce kotschyana Oliv. in Trans. Linn. Soc. **29**: 88, t. 53 (1873). —Hiern in F.T.A. **3**: 239 (1877). Types from Sudan and Uganda.

Spermacoce compacta Hiern in F.T.A. **3**: 239 (1877). Type from Sudan.

Spermacoce hebecarpa var. *major* Hiern in F.T.A. **3**: 237 (1877). Type from Ethiopia.

Borreria compacta (Hiern) K. Schum. in Engl. & Prantl, Pflanzenfam. IV. **4**: 144 (1891).

Borreria chaetocephala (DC.) Hepper in Kew Bull. **14**: 256 (1960); F.W.T.A. ed. 2, **2**: 220 (1963).

A rather coarse erect mostly sparsely branched annual herb 10–60 cm. tall, the stems often reddish or pinkish, 4-angled, hairy below the nodes. Leaf blades 2.5–6 cm. × 3–8 mm., linear-lanceolate, acute at the apex, mostly broadened at the base, pubescent or ± scabrid above with white hairs and with longer white hairs beneath, particularly or sometimes only on the broadened base; stipule bases 1–2 mm. long, hairy, with 7 setae 0.2–1.5 cm. long. Flowers (3)4-merous, congested into usually quite large axillary heads 0.5–2 cm. wide, usually made up of normal axillary clusters together with those at the nodes of much reduced axillary shoots, the whole coalescing, each compound inflorescence thus supported by numerous leaves and also with leaves emerging from its centre; bracteoles very numerous, reddish-brown, 3 mm. long; pedicels 0.5 mm. long. Calyx tube 1.4 mm. long, ellipsoid; lobes 4, 1–2 mm. long, linear-lanceolate or slightly spathulate, scabrid on the margins. Corolla white or pale mauve; tube narrowly cylindrical with upper quarter funnel-shaped and c. 4 times as wide as basal part of tube, altogether 1.3 mm. long; lobes 0.8 × 0.4 mm., ovate-elliptic. Filaments exserted 0.6 mm. Style exserted 0.7 mm.; stigma with long papillae. Capsule 2.5 mm. long, oblong-ellipsoid, hairy at the apex. Seeds chestnut-brown, very shiny, 1.8–2.2 × 0.8–0.9 × 0.5 mm., narrowly oblong-ellipsoid, with a deep ventral groove but ± no sculpture.

Zambia. C: 19 km. S. of Lusaka, Mt. Makulu Research Station, fr. 2.vi.1956, *Angus* 1314 (K; LISC; SRGH). E: 14km. from Chipata on Lundazi road, 23.ii.1971, *Anton-Smith* in GHS 214851 (SRGH). S: Mapanza, W., fl. 22.ii.1954, *Robinson* 548 (K; SRGH). **Zimbabwe**. N: Lutope floodplain, Sengwa Res. Sta., fr. 8.iv.1969, *Jacobsen* 566 (SRGH). W: Insiza, Fort Rixon, fl. 8.ii.1974, *Mavi* 1518 (SRGH). C: Chegutu, Poole Farm, fr. 7.iv.1954, *Wild* 4556 (K; LISC; SRGH). E: Mutare, road to Chipondomwe Mt., 13.iii.1955, *Chase* 5502 (BM; K; LISC; SRGH). **Mozambique**. N: Between Cuamba and Mutuáli, near bridge over R. Lurio, fl. & fr. 24.iv.1961, *Balsinhas & Marrime* 431 (K; LISC; LMA; SRGH). T: Songo, right bank of R. Zambezi down river from Barragem, fl. & fr. 8.iii.1972, *Macedo* 5020 (LISC; LMA).

Also in Senegal, Mali, N. Nigeria, Sudan, Ethiopia, Uganda, Kenya and Tanzania. Bushland, woodland and mixed woodland/grassland, also by roadsides; 690–1170 m.

Hepper divides *Spermacoce chaetocephala* into two varieties. His *Borreria chaetocephala* DC. var. *minor* Hepper in Kew Bull. **14**: 256 (1960); in F.W.T.A. ed. 2, **2**: 220 (1963) is based on *Borreria hebecarpa* A. Rich., Tent. Fl. Abyss. **1**: 347 (1848). —Hiern in F.T.A. **3**: 326 (1877). —F.W. Andr., Fl. Pl. Anglo-Egypt. Sudan **2**: 427 (1952). Types: Ethiopia, Djeladjeranne, *Quartin-Dillon* (P, syntype) & *Schimper* 1712 (P, syntype, K, isosyntype!). Others consider this a variant of *Spermacoce pusilla* Wall.

I am not altogether certain of the determination of *Chase* 5502 as this species. Technically it agrees but the facies is somewhat different; *Angus* 2499, 40 km. E. of Lusaka is identical. The SRGH sheet of *Chase* 5502 appears to be the same as *Spermacoce pusilla* (fide *Drummond*).

8. **Spermacoce** sp. A

Slender strict annual 5–15 cm. tall, the unbranched stem ± 4-angled, with scattered short, downwardly directed hairs, those just below the uppermost node somewhat tubercular-based. Leaves 1.2–4 cm. × 1–3 mm., linear-oblong, the cauline ones narrowed at both ends, attenuate-acute at the apex, sessile on to the stipule sheath at the base, those supporting the inflorescence slightly widened at the base, both sorts pubescent on both surfaces; stipule sheath 2–2.5 mm. long, densely pubescent, with 3–5 glabrous yellow-brown setae up to 4.5 mm. long. Inflorescences terminal, few-flowered, consisting of a small head made up of two nodes at least, about 5 mm. across; bracteoles linear, scarious 2.5 mm. long. Calyx tube c. 1 mm. long, 0.5 mm. wide, narrowly oblong, puberulous; lobes 4, 1–1.5 mm. long, linear, puberulous. Corolla white or pale blue; tube filiform, about 1.3 mm. long; lobes c. 1 mm. long, elliptic, the stamens and style exserted and about equalling the lobes. Fruits not known.

Zimbabwe. N: Hurungwe, 1.6 km. N. of R. Mauora, 28.ii.1958, *Phipps* 954 (K; SRGH).

Presumed endemic. Gregarious in *Colophospermum–Brachystegia boehmii* ecotone, open woodland; 600–725 m.

This was thought to be a depauperate form of *Spermacoce chaetocephala* but an exactly similar group of specimens has been collected on the Zambezi Escarpment, 15.ii.1981, *Philcox et al.* 8597 (K; SRGH).

This has suggested it might be a distinct species.

9. **Spermacoce huillensis** (Hiern) Good in Journ. Bot., Lond. **65**, Suppl. 2: 41 (1927). —Verdc. in Kew Bull. **30**: 304 (1975); in F.T.E.A., Rubiaceae 1: 355 (1976). Types from Angola.
Tardavel huillensis Hiern, Cat. Afr. Pl. Welw. **1**: 503 (1898).

Annual herb, rarely putting on new growth after fruiting, (10)15–30(70) cm. tall, mostly branched near the base or quite unbranched; stems purplish-brown, usually densely covered with spreading white hairs or sometimes only slightly scabrid-pubescent or glabrescent. Leaf blades 1.3–2.6(7) cm. × 1.8–3.5(5) mm., lanceolate or lowermost elliptic, acute at the apex, slightly narrowed at the base into the stipule sheath, slightly shining and with scattered subscabrid hairs or glabrous above, hairy beneath at least on the midrib, the margins slightly thickened, revolute; stipule sheath 2.5–3 mm. long, hairy, with 7 usually reddish-brown or purple setae 3–10 mm. long. Flowers in axillary verticillate congested clusters 0.8–1.3 cm. wide, the terminal cluster made up of the true terminal and 1–2 lower nodes condensed together subtended by 4(8) leaves; below this compound terminal cluster and well separated is a further cluster subtended by 2–4 leaves and sometimes a third cluster, although sometimes only the terminal one is present; filiform bracteoles 3–6 mm. long. Calyx tube 1.5 mm. long, ellipsoid, hairy; lobes 4, 2.5–3 mm. long, linear-lanceolate, often hairy, the margins scabrid-ciliate. Corolla white to pale mauve, often with bluish or purple marks at the throat in dried material; tube almost filiform beneath, narrowly funnel-shaped above, 5(9) mm. long, glabrous outside, with an internal ring of hairs at junction of the two parts; lobes 2 mm. long, 0.6 mm. wide, elliptic-oblong, incurved at the apex and tipped with a few long hairs. Filaments exserted 2 mm. Style 7.5 mm. long; stigma small. Capsule 3 × 2.2 mm., ovoid, spreading pubescent. Seeds dark chestnut-brown, 2.5 × 1.3 × 0.8 mm., semi-ellipsoid, slightly narrowed to the base, slightly rugulose, ventrally deeply grooved.

Zambia. W: Mufulira, fl. 20.iii.1948, *Cruse* 302 (K). C: Serenje, fl. 18.ii.1955, *Fanshawe* 2073 (K; NDO; SRGH). S: Kalomo Distr., near Choma, Mochipapa, fl. 9.iii.1962, *Astle* 1448 (K; SRGH). **Zimbabwe**. N: Gokwe, fl. & fr. 19.xii.1963, *Bingham* 806B (K; SRGH). **Malawi**. N: Mzimba Distr., Lunyangwa R. descent, Mzuzu Waterworks, fl. 29.iii.1974, *Pawek* 8254 (K; MAL; MO; SRGH; UC).

Also in S. Tanzania (atypical) and Angola. *Brachystegia* woodlands and derived cultivations, pans, roadsides, etc.; 1140–1440 m.

The specimen cited from Zambia C has unusually large flowers with corolla tubes to 9 mm. and *Pawek* 8517 (Malawi, N. Mzimba, Champira Forest) has small tubes c. 2 mm. long but all other material seen has them uniformly c. 5 mm. long. Investigation of populations is necessary to see if these are distinct variants or casual variation.

10. **Spermacoce radiata** (DC.) Hiern in F.T.A. **3**: 237 (1877). —Verdc. in F.T.E.A., Rubiaceae 1: 356, fig. 49/12 (1976). TAB. **39**, fig. 10. Type from Senegal.
Borreria radiata DC., Prodr. **4**: 542 (1830). —F.W. Andr., Fl. Pl. Anglo-Egypt. Sudan **2**: 427 (1952). —Hepper, F.W.T.A. ed. 2, **2**: 219 (1963).
Tardavel andongensis Hiern, Cat. Afr. Pl. Welw. **2**: 506 (1898). Types from Angola.

Branched or unbranched rigid annual herb 9–40(100) cm. tall, with a ± simple root; stem densely covered with spreading white hairs. Leaf blades 2–5.5 cm. × 1.2–5.5 mm., very narrowly elliptic-lanceolate, acute at the apex, narrowed at the base into the stipule sheath, with scattered long hairs above and on the midnerve beneath or ± entirely glabrous save for the obscurely scabrid margins; midrib and margins distinctly thickened, white; stipules often white, base 4 mm. long, hairy, bearing 5–9 fimbriae 6.5 mm. long. Flowers in very dense terminal heads consisting of several condensed nodes, the leaves of which form radiating bracts, often without any further flowers at lower nodes but if so then these are smaller heads terminating undeveloped lateral branches; bracts of 2 sorts, one leaf-derived, 5–7 × 1.5–2 mm., ovate-lanceolate, hyaline at the base, stiff and green at the apex, the others stipule-derived of narrow thin fimbriae 3 mm. long from a narrow base. Calyx tube 2 mm. long, narrowly oblong, hairy above; lobes 4, 1.5 mm. long, filiform, hairy. Corolla greenish or white; tube, 1 mm. long, funnel-shaped; lobes 0.6 mm. long, 0.4 mm. wide, ovate-triangular. Filaments exserted 0.5 mm. Style 1.3 mm. long; stigma 0.2 mm. wide. Capsule 3 × 1.2 mm. compressed-cylindrical, with a median furrow, hairy in upper half. Seeds pale brown, 1.8 × 0.7 × 0.4 mm., oblong-ellipsoid, shiny.

Zimbabwe. S: Great Zimbabwe, fl. iv.1934, *Goldschmidt* s.n. (BM). **Mozambique**. N: Cuamba, fl. 13.v.1948, *Pedro & Pedrogão* 3358A (EA).

Senegal to Cameroon and Sudan, Uganda, W. Kenya and Angola. Ecology not known.

11. **Spermacoce pusilla** Wall. in Roxb., Fl. Indica, ed. Carey & Wall. **1**: 379 (1820). —Verdc. in F.T.E.A., Rubiaceae 1: 356, fig. 50/13 (1976). —Gonçalves in Garcia de Orta Sér Bot. **5**: 207 (1982). TAB. **39**, fig. 11. Type from Nepal.

Borreria pusilla (Wall.) DC., Prodr. **4**: 543 (1830). —F.W. Andr., Fl. Pl. Anglo-Egypt. Sudan **2**: 427 (1952). —Hepper, F.W.T.A. ed. 2, **2**: 220 (1963).

Borreria stricta sensu Hutch. & Dalz., F.W.T.A. **2**: 135 (1931). —Agnew, Upland Kenya Wild Fl.: 409, fig. (1974) et auctt. mult. non (L.f.) K. Schum., nec G.F.W. Mey.*

Borreria hebecarpa sensu Hutch. & Dalz., F.W.T.A. **2**: 135 (1931) non A. Rich.

Erect or rarely prostrate annual usually branched herb (2)7.5–60 cm. tall, with grooved ± glabrous or slightly papillate-puberulous often reddish stems. Leaf blades 1–5.3 cm. × 2–5.5 mm., linear-lanceolate to narrowly lanceolate, acute at the apex, narrowed to the base, minutely scabrid above, glabrous beneath; true petiole absent; stipule sheath 1.5–2 mm. long, glabrous or pubescent, bearing 7 setae 2–4 mm. long. Flowers in dense very compact spherical clusters at most nodes, 0.6–1(1.5) cm. in diam.; bracteoles filiform, numerous, 2 mm. long. Calyx tube 1 mm. long, ovoid, pubescent; teeth 4, equal or slightly unequal, 0.6–1.2 mm. long, subulate, scabrid. Corolla white or pink; tube 1.3 mm. long, narrowly funnel-shaped; lobes 0.8–1.1 × 0.4 mm. with some long flattened hairs at the apex. Filaments exserted c. 1 mm. Style exserted 0.5 mm. Capsule 1.5 mm. long, ellipsoid; glabrous or ± sparsely pubescent. Seeds chestnut-brown, shiny, 1.3 × 0.55 × 0.4 mm., oblong-ellipsoid or narrowly oblong, with a wide ventral groove.

Zambia. B: 30 km. W. of Kaoma (Mankoya), fr. 11.iv.1966, *Robinson* 6903 (K; SRGH). N: Mbala Distr., Kalambo Road, fl. & fr. 20.iv.1970, *Sanane* 1151 (K). W: Solwezi, fl. & fr. 9.iv.1960, *Robinson* 3474 (K; SRGH). C: 11.2 km. SE. of Lusaka, fl. & fr. 16.iii.1952, *Best* 19 (K). E: Katete, St. Francis' Hospital, fl. 27.ii.1957, *Wright* 162 (K). S: Mumbwa to Nangoma, fl. & fr. 20.iii.1963, *Van Rensburg* 1791 (K; SRGH). **Zimbabwe**. N: Makonde (Lomagundi), Trelawney, fl. & fr. 29.iii.1944, *Jack* 249 (K; SRGH). C: Harare, Cranborne, fl. & fr. 31.iii.1946, *Wild* 1035 (K; SRGH). E: Mutare, Quagga's Hoek, Commonage, fr. & fr. 22.iii.1953, *Chase* 4876 (BM; SRGH). **Malawi**. N: Nkhata Bay Distr., Chinteche Beach, fl. & fr. 23.v.1971, *Pawek* 4855 (K). C: Ntcheu Distr., Dedza Plateau, 38.4 km. S. of Dedza, fl. & fr. 29.iii.1978, *Pawek* 14156 (K; MAL; MO). S: Zomba Plateau, fl. & fr. 29.iii.1937, *Lawrence* 349 (K). **Mozambique**. N: 38.4 km. E. of Ribáuè, fl. & fr. 17.v.1961, *Leach & Rutherford-Smith* 10902 (K; LISC; SRGH). Z: Namagoa, fr. 18.iv.1948, *Faulkner* 252 (K). T: Macanga, Mt. Furancungo, fl. & fr. 15.iii.1966, *Pereira et al.* 1717 (BR; LMU). MS: Chimoio, Serra do Garuzo, fl. & fr. 2.iv.1948, *Barbosa* 1340 (LISC; LMU).

Widespread in tropical Africa also Madagascar and tropical Asia to Japan and Philippines. Seasonally damp grassland, bushland, *Brachystegia* & *Julbernardia* woodland, sandy lake shores and roadsides, stream banks and cleared ground mostly on sand; 120–2010 m.

Drummond (annot. in herb.) treats *Borreria hebecarpa* as a synonym of this species but I think Hepper is probably correct in treating it as a variety of *S. chaetocephala*.

12. **Spermacoce quadrisulcata** (Bremek.) Verdc. in Kew Bull. **30**: 305 (1975); in F.T.E.A., Rubiaceae 1: 359, fig. 50/16 (1976). TAB. **39**, fig. 12.Type from Zaire.

Spermacoce compressa sensu Hiern in F.T.A. **3**: 235 (1877) quoad syntypum *Barter* 1231.

Tardavel stricta sensu Hiern, Cat. Afr. Pl. Welw. **1**: 503 (1898) quoad *Welwitsch* 3219 & 3220 non (L.f.) Hiern.

Borreria compressa sensu Hutch. & Dalz., F.W.T.A. **2**: 135 (1931) pro parte non Hutch. & Dalz. sensu stricto.

Borreria quadrisulcata Bremek. in Bull. Jard. Bot. Brux. **22**: 102 (1952).

Borreria paludosa Hepper in Kew Bull. **14**: 259 (1960); in F.W.T.A. ed. 2, **2**: 221 (1963). Type from Sierra Leone.

Annual ± glabrous herb 0.35–1.7 m. tall from a fibrous root; stems grooved, branched or ± simple. Leaf blades 2.6–8 cm. × 1.5–9(11) mm., linear to linear-lanceolate, acute at the apex, cuneate at the base, ± scabrid or glabrous above; stipule sheath 3–5 mm. long, with (3)5–7 setae 1–3 mm. long; petiole ± obsolete. Flowers in axillary clusters c. 6 mm. in diam., 3–4-merous; bracteoles composed of fascicles of short fimbriae present. Calyx tube 1–1.5

* *Spermacoce stricta* L.f., Suppl. Pl.: 120 (1781), was based on *Crateogonum* (*amboinicum*) *minus. verum* Rumph., Herb. Amboin. **6**: 25, t. 10 (1750) and possibly also a specimen. Rumphius' plate does not look much like the plant described above and Merrill in his "Interpr. Rumph. Herb. Amboin.": 479 (1917) says that Rumphius' plant is *Hedyotis tenelliflora* Blume. A specimen from Linnaeus' Uppsala Garden labelled *S. stricta* preserved at the Linnean Society (125.5) is not the plant either so it seems best to use the name *S. pusilla*. It must be noted that if *Borreria* is kept up the name *stricta* is not usable in any case. G.F.W. Meyer stated "forte *S. stricta* L.f." and must be assumed to have described a new plant — his specimens are in any case *B. verticillata* (L.) G.F.W. Mey.; K. Schumann's combination *B. stricta* (L.f.) K. Schum. is thus illegitimate being predated by *B. stricta* G.F.W. Mey.

mm. long, ellipsoid, glabrous; lobes very small, 0.25–0.3 mm. long, triangular, with ± ciliolate margins. Corolla rose-violet or white, sometimes with mauve lobes; tube 1.5–1.7 mm. long, funnel-shaped; lobes 1mm. long, ovate-triangular. Stamens with filaments exserted 0.4–0.8 mm. Style exserted 0.8–1.2 mm.; stigma subcapitate. Capsule 2–3 × 1 mm., oblong-ellipsoid, glabrous, crowned with the minute calyx teeth. Seeds (1.8)2–2.4 × (0.5)0.75–0.8 × 0.5–0.6 mm., narrowly oblong, rugulose-pitted, grooved ventrally, the groove widened at the base and with a white granular aril.

Botswana. N: Kwara Bochai R., fl. 29.iv.1973, *Smith* 553 (K; LISC; SRGH). **Zambia**. B: Mongu, fl. & fr. 5.iii.1966, *Robinson* 6866 (K; SRGH). N: Samfya, behind shore of Lake Bangweulu, fl. & fr. 30.i.1959, *Watmough* 188 (K; LISC; SRGH). W: Mwinilunga Distr., 6.4 km. N. of Kalene Mission, Zambezi R. rapids, fl. & fr. 10.xi.1962, *Lewis* 6212 (K; MO). C: 53 km. NW. of Kabwe, Kelongwe R., Mpunde Mission, fl. & fr. 20.i.1973, *Kornaś* 3043 (K: KRA). **Malawi**. C: Bua R. below Mude R. confluence, fl. & fr. 10.ii.1959, *Robson* 1540 (BM; LISC; K; SRGH).

From Guinée to Nigeria, Central African Republic, Zaire, Burundi, Ethiopia, Uganda, Tanzania, Angola and northern Namibia. In shallow water on flood plains, and by rivers, also *Papyrus* swamps, by lakes, etc., but also in dambo grassland which is dry for some of the year and as a weed in cultivations derived from wet areas; 900–1200 m.

13. **Spermacoce natalensis** Hochst. in Flora **27**: 555 (1844). —Sond. in Harv. & Sond., F.C. **3**: 24 (1865). —Hiern in F.T.A. **3**: 236 (1877) in obs. sub. sp. 7. —Verdc. in F.T.E.A., Rubiaceae 1: 359, fig. 50/17 (1976). TAB. **39**, fig. 13. Type from S. Africa.

Spermacoce stricta sensu Hiern in F.T.A. **3**: 236 (1877) pro parte non L.f.

Borreria natalensis (Hochst.) S. Moore in Journ. Linn. Soc. Bot. **40**: 103 (1911). —Hepper, F.W.T.A. ed. 2, **2**: 220 (1963).

Diodia natalensis (Hochst.) Garcia in Mem. Junta Invest. Ultramar., sér. 2, Bot. **6**: 47 (1959).

Perennial erect, prostrate or trailing herb 15–55 cm. long or tall; stems several, branched or ± unbranched, 4-ribbed, glabrous except at nodes or scabrid with very short prickle-like hairs. Leaf blades 0.8–4.5 cm. × 1.5–8.5(16) mm., ± linear, elliptic, or oblong-elliptic to narrowly elliptic-lanceolate, subacute at the apex, narrowed to the base, scabrid with very short hairs near the margins above but otherwise glabrous; petiole obsolete; stipules with base c. 1 mm. long, pubescent, bearing 5 fimbriae 2–4.5 mm. long. Flowers in small clusters at most nodes, often appearing a little 1-sided due to the fact that they are present in 1 axil only; clusters to 8 mm. in diameter in fruiting stage; bracts stipule-like, with filiform fimbriae, 2–2.5 mm. long. Calyx tube 1.2–1.5 mm. long, obconic, glabrous; limb tube c. 0.2 mm. long; lobes 4, 1–1.8 mm. long, linear-lanceolate to oblong-triangular, sometimes reddish-brown. Corolla white tinged pink; tube 1 mm. long; lobes 1.2–1.8 × 1.2 mm., triangular, densely white hairy inside, obscurely papillate outside. Anthers borne at the sinuses. Style and stigma together very short, 0.3 mm. long. Capsule squarish with rounded sides in outline, 1.2–1.3 × 1.5–1.8 mm., compressed, 1 mm. thick, glabrous, very finely transversely wrinkled, pitted inside the valves, crowned with the persistent lobes. Seeds blackish-purple, 1.3 × 0.65 × 0.4 mm., deeply densely punctate.

Zambia. W: Mwinilunga Distr., about 1.5 km. S. of Matonchi Farm, fl. 20.xii.1937, *Milne-Redhead* 3753 (BM; K). **Zimbabwe**. E: Inyanga, fl. & fr. 20.i.1931, *Norlindh & Weimarck* 4467 (K; LD; SRGH). **Malawi**. N: Nkhata Bay Distr., Viphya Plateau, Kawandama road, about 98 km. S. of Mzuzu, fl. & fr. 8.iv.1978, *Pawek* 14354 (K; MAL; MO; SRGH; UC). C: Dedza Mt. slopes, fl. & fr. 24.xi.1967, *Salubeni* 910 (K; SRGH). S: Kirk Range, Goche, fl. 30.i.1959, *Robson* 1350 (BM; LISC; K). **Mozambique**. Z: Errego, 3 km. from Monte Ile, fl. & fr. 3.iii.1966, *Torre & Correia* 14978 (BR; LD; LISC; LMU; P; SRGH). MS: Expedition Is., fl. & fr. xi.1859, *Kirk* (K). GI: outskirts of Xai-Xai (Vila João Belo) fr. 14.i.1941, *Torre* 2594 (COL; LISC; LMU; SRGH; WAG). M: Namaacha, fl. & fr. ii.1931, *Gomes e Sousa* 433 (K).

Also in Nigeria, Cameroon, Uganda, Tanzania and S. Africa. Grassland, by roadsides and as a weed in cultivations; (50)600–1650 m.

14. **Spermacoce mauritiana** Osia Gideon in Kew Bull. **37**: 547 (1983). TAB. **39**, fig. 14. Type from Mauritius.

Borreria repens DC., Prodr. **4**: 544 (1830). Type from Mauritius.

Spermacoce ocymoides sensu auctt. mult. e.g. Verdc. in F.T.E.A., Rubiaceae 1: 361. fig. 50/19 (1976) pro parte non Burm.f.

Tardavel ocymoides sensu Hiern, Cat. Afr. Pl. Welw. **1**: 504 (1898) non (Burm.f.) Hiern sensu stricto.

A weak erect, decumbent or procumbent annual herb, usually ± well-branched, 3–40 cm. tall, with fine fibrous roots; stems with sparse to fairly dense crisped hairs on the ± wing-like prominent angles. Leaf blades 0.4–3.8 × 0.25–1.8 cm., elliptic to elliptic-lanceolate, rounded to ± acute at the apex, concavely narrowed into the petiole at the base, glabrous on both surfaces save for short marginal hairs or ± pubescent on the main nerve beneath; petiole 0–8 mm. long, with scattered hairs; stipules with base 2 mm. long, bearing c. 7 fimbriae 1–3 mm. long. Flowers in small few-flowered clusters at many of the nodes, attaining 3–6 mm. in diam. in the fruiting state; stipule-like bracteoles with fimbriae 1.8–2 mm. long. Calyx tube transversely oblong, 0.5 mm. long; lobes 2(4), 0.6–0.8 mm. long, ciliate. Corolla white; tube 0.3 mm. long; lobes 0.35 × 0.3–0.4 mm., triangular, with a few short hairs inside. Anthers situated just above the sinuses of the corolla lobes. Style c. 0.2 mm. long; stigma 0.2 mm. wide. Fruit 1 × 0.8 × 0.5–0.7 mm., oblong, compressed, finely transversely wrinkled and very shortly pubescent. Seeds chestnut-brown, 0.7–0.8 × 0.4 × 0.3–0.35 mm., oblong-ellipsoid, strongly reticulate with raised ribs, the foveae elongated in the direction of the short axis.

Zambia. W: Kitwe, by R. Kafue, Jenning's Poultry Farm, fl. & fr. 9.iii.1967, *Anton-Smith* 100 (K; SRGH). **Malawi**. N: 12.8 km. S. of main Nkhata Bay–Mzuzu Road, Chinteche Road, fl. & fr. 4.iv.1971, *Pawek* 4576 (K; MAL).

Also in Senegal to Cameroon, Fernando Po, Principe, St. Tomé, Annobon, Cabinda, Congo (Brazzaville), Uganda & Tanzania; widespread in Madagascar, Mauritius, Asia, Central America, West Indies and S. America to Argentina. Lake shores and surrounding bushland also roadsides and paths at rain-forest edges, tea-plantations, etc.; 460–1560 m.

I (F.T.E.A. Rubiaceae 1: 361) formerly followed Bremekamp (in Pulle, Fl. Suriname **4**: 285 (1935)) et auctt. in considering *Spermacoce ocymoides* Burm.f. (*Borreria ocymoides* (Burm.f.) DC.) a widespread species but Fosberg and Osia Gideon have convinced me that several distinct species are involved.

15. **Spermacoce princeae** (K. Schum.) Verdc. in Kew Bull. **30**: 307 (1975); in F.T.E.A., Rubiaceae 1: 362 (1976). Type from Tanzania.

Borreria princeae K. Schum. in Engl., Bot. Jahrb. **34**: 341 (1904). —Hepper, F.W.T.A. ed. 2, **2**: 222 (1963). —Verdc. in Kew Bull. **17**: 500 (1964). —Agnew, Upland Kenya Wild Fl.: 407 (1974).

Scrambling or decumbent perennial herb with ascending branchlets, much branched, 0.3–1.5 m. long; stems often dark crimson, square, sparsely to densely hairy on the angles. Leaf blades 1.2–7 × 0.35–3 cm., elliptic to ovate, acute at the apex, cuneate at the base, glabrous to pubescent above, with sparse hairs to pubescent beneath and always scabrid on the margins; venation impressed, giving a bullate and plicate appearance to the blade; petiole 0–1 mm. long; stipule base 4–6 mm. long, hairy, at least above, with 5–9 fimbriae 7–8.5 mm. long. Flowers in dense few- to many-flowered axillary clusters at most nodes; fimbriae of stipuliform bracts 6 mm. long, ciliate. Calyx tube 1.2–3 mm. long, fusiform or obconic; lobes (2)4(5), foliaceous, 3.5–5(9) mm. long, 0.8–1 mm. wide, lanceolate to linear-lanceolate, with pubescent margins. Corolla white or sometimes tinged pinkish; tube (5)6.5–10.5 × 0.4–2.5 mm., cylindrical or narrowly funnel-shaped; lobes 3–4(6) × 0.8–2.5 mm., oblong or elliptic. Filaments exserted 1–1.5 mm. Style exserted 2.5–4 mm.; stigma lobes 0.5–1 mm. long. Capsule 5–6 × 2–3 mm., oblong-ellipsoid, the valves opening widely, finely transversely wrinkled, glabrous or finely pubescent, crowned with the persistent calyx lobes. Seeds dark purple-brown, 4 × 1.6–2 × 0.7–0.8 mm., fusiform, deeply grooved ventrally, narrowed at one end, shiny, finely rugulose and sometimes with a shallow dorsal groove as well.

Var. **princeae**. —Verdc. in F.T.E.A., Rubiaceae 1: 362, fig. 52 (1976).

Diodia stipulosa S. Moore in Journ. Linn. Soc., Bot. **37**: 310 (1906). Types from Kenya, Tanzania and Cameroon.

Stems pubescent on the angles; leaf blades glabrescent or with sparse hairs beneath and on margins. Calyx lobes 4; corolla tube wider, 2–3 mm. wide at throat; lobes 3–4 × 0.8–2.5 mm.; capsule ± glabrescent or sometimes pubescent.

Zambia. W: Solwezi, R. Meheba, fr. 21.vii.1930, *Milne-Redhead* 742 (K). C: 7 km. from Serenje, fl. & fr. 23.xii.1963, *Symoens* 10689 (BR; K). E: Nyika Plateau, 8.8 km. SW. of Rest House, fr. 25.x.1958, *Robson* 353 (BM; K; LISC). **Malawi**. N: Nyika National Park, Chowo Rock. fl. & fr. 28.iv.1973, *Pawek* 6681 (K). S: Zomba Plateau, Mulungusi stream, fl. & fr. 26.iii.1937, *Lawrence* 300 (K).

Also in Cameroon, Zaire, Burundi, Ethiopia, Uganda, Kenya and Tanzania. Riverine evergreen forest and upland forest edges and derived disturbed vegetation; 1350–2250 m.

Var. **mwinilungae** Verdc. var. nov.* Type: Zambia, Mwinilunga, *Brummitt, Chisumpa & Polhill* 14020 (K, holotype).

Indumentum similar to var. *princeae*; calyx lobes 2–3; corolla tube very slender, scarcely 1 mm. wide at the throat; lobes longer and narrower to 6 × 1.5 mm.

Zambia. W: 8 km. N. of Mwinilunga, West Lunga R., fl. & fr. 23.i.1975, *Brummitt, Chisumpa & Polhill* 14020 (K).

Endemic. Riverine forest; 1300 m.

More gatherings of this are required to establish its status. Many of the flowers have but 2 sepals but some have 3. Study of populations might well reveal 4 as well.

Var. **pubescens** (Hepper) Verdc. in Kew Bull. **30**: 307 (1975); in F.T.E.A., Rubiaceae 1: 364 (1976). Type from Fernando-Po.

Borreria princeae var. *pubescens* Hepper in Kew Bull. **14**: 259 (1960); F.W.T.A. ed. 2, **2**: 222 (1963).

Stems, both leaf surfaces and capsule densely hairy. Flowers as in var. *princeae.*

Zambia. N: 40 km. S. of Mbala, fl. & fr. 22.vii.1930, *Hutchinson & Gillett* 4024 (BM; K). **Malawi**. N: Mzimba Distr., 4.8 km. W. of Mzuzu, Katoto, fl. 19.v.1974, *Pawek* 8642 (K; MO; SRGH; UC). C: Dedza, Chongoni Forest, fl. & fr. 31.x.1968, *Salubeni* 1180 (K; SRGH). **Mozambique**. N: Vila Cabral, S. of Serra de Massangulo, fl. 25.ii.1964, *Torre & Paiva* 10764 (C; COI; K; LISC; LMU; SRGH; WAG).

Also in Fernando Po, Cameroon, Zaire, Burundi, Uganda and Kenya. Streamsides and swampy places, also *Brachystegia* woodland; 1180–1740 m.

16. **Spermacoce schlechteri** K. Schum. ex Verdc. sp. nov.** **TAB. 41**. Type: Mozambique, Zambezia, Pebane, dunes near lighthouse, fl. & fr. 8.iii.1966, *Torre & Correia* 15081 (COI; K; LISC, holotype; LMA).

Perennial herb with straggling or prostrate stems 15–50 cm. long, woody at the base, glabrous or with sparse pubescence in the longitudinal grooves and two rows of pubescence descending from the outer margins of the stipule bases. Leaves ± subsessile, 1–6 × 0.4–2.7 cm., elliptic to oblong-elliptic or elliptic-lanceolate, broadly to narrowly acute at the apex, cuneate at the base, perhaps slightly fleshy in life, glabrous, minutely scabrid on the margins or sometimes finely puberulous beneath and often with rhaphides and resin patches rather evident in dried material; stipule bases c. 2 mm. long, pubescent at apex and with lateral lines of pubescence or sometimes pubescent all over, with c. 7 filiform setae up to 7 mm. long. Flowers 2–6 at each node, the apparent pedicel becoming short and thick, 2 mm. long in fruit. Calyx glabrous or pubescent; tube c. 2 mm. long, obconic; limb tube c. 1.2 mm. long; lobes 2–4 × 0.8–1.4 mm., lanceolate, glabrous or with short marginal setae. Corolla white; tube 3.5–4 mm. long, narrowly funnel-shaped; lobes 3.5 × 1–1.2 mm., lanceolate-elliptic. Anthers short, curved and medifixed, exserted about 1 mm.; style exserted 2–3 mm., the stigmatic head biglobose 1.2 mm. wide. Capsule 3–4.5 × 3–4 mm, subglobose or ovoid, ± faintly wrinkled in dry state and with numerous resin blotches, glabrous or shortly pubescent, the valves finally opening very widely. Seeds brown, 2.5–3.5 × 1.8 × c. 1.2 mm., ± oblong, minutely rugulose dorsally, shiny, ventrally with broad depression.

Mozambique. N: Angoche (Antonio Enes), Nantangulu Beach, fl. 24.x.1965, *Mogg* 32481 (J; LISC; SRGH). Z: Pebane, dunes near lighthouse, fl. & fr. 8.iii.1966, *Torre & Correia* 15081 (COI; K; ISC; LMA). MS: Dondo, 25 Miles Station, fr. 14.iv.1898, *Schlechter* 12294 (BM; K). GI: Inhambane, Barra Beach, fl. & fr. 12.x.1968, *Balsinhas* 1373 (LISC).

Endemic but possibly conspecific with one undescribed plant from Tanzania (Mafia I.). Coastal dunes with thicket of *Strychnos, Mimusops, Diospyros, Syzygium, Flagellaria,* etc.; 20–60 m.

Material annotated *Borreria pilosa* by K. Schumann but not published (*Schlechter* 12113 (BM; K)) has the leaves quite densely shortly pubescent on both surfaces.

* Differt a *S. princeae* var. *princeae* calycis lobis 2–3; fauce corollae tubi gracili vix 1 mm. lata, lobis longioribus angustioribusque usque 6 × 1.5 mm.

** *Spermacoce sp. D* in Flora Africae Tropicae Orientali citata (Rubiaceae 1: 369 (1976)) ob valvas capsulae tandem valde divergentes valde affinis foliis majoribus latioribus, floribus fructibusque majoribus differt sed possibiliter eadem est; *Spermacoce princeae* (K. Schum.) Verdc. affinis sed inflorescentiis paucifloris, corollae tubo breviore, setis stipulae brevioribus, habitatione sublittorali differt; cum *Diodia sarmentosa* Sw. confusa sed capsula valde dehiscenti diversa.

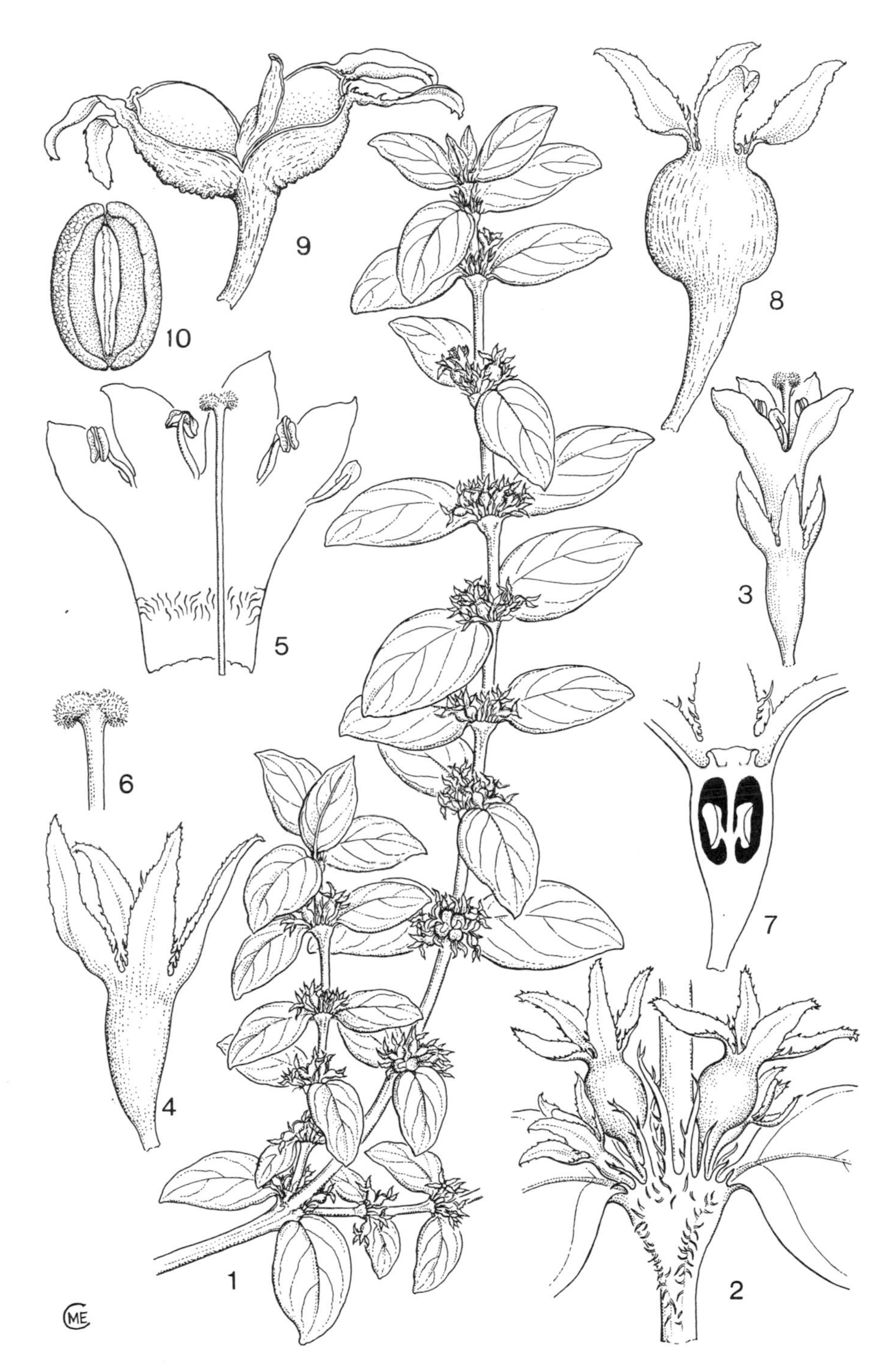

Tab. 41. SPERMACOCE SCHLECHTERI. 1, habit (× $\frac{2}{3}$); 2, node showing stipules, flower buds and young fruits (× 4); 3, flower (× 4); 4, calyx (× 6); 5, corolla opened out (× 6); 6, stigma lobes (× 10); 7, ovary, longitudinal section (× 8); 8, capsule (× 6); 9, dehisced capsule (× 6); 10, seed (× 8), all from *Torre & Correia* 15081.

17. **Spermacoce senensis** (Klotzsch) Hiern in F.T.A. **3**: 236 (1877). —Verdc. in F.T.E.A., Rubiaceae 1: 365, fig. 51/23 (1877). TAB. **39**, fig. 17. Type: Mozambique, Tete, Rios de Sena, *Peters* (B, holotype †).

Diodia senensis Klotzsch in Peters, Reise Mossamb., Bot. **1**: 289 (1861).

Mitracarpum dregeanum Sond. in Harv. & Sond., F.C. **3**: 25 (1864). Types from S. Africa (Natal).

Borreria senensis (Klotzsch) K. Schum. in Abh. Preuss. Akad. Wiss.: 23 (1894); in Engl., Pflanzenw. Ost-Afr. **C**: 394 (1895).

Diodia benguellensis Hiern, Cat. Afr. Pl. Welw. **1**: 502 (1898). Types from Angola.

Borreria stolzii K. Krause in Engl., Bot. Jahrb. **57**: 52 (1920). Type from Tanzania.

Borreria squarrosa Schinz. in Viert. Nat. Ges. Zürich **68**: 438 (1923). Type from Namibia.

Borreria rhodesica Suesseng. in Trans. Rhodes. Sci. Assoc. **43**: 130 (1951). Type: Zimbabwe, Marondera, *Dehn* 98 (K; M, holotype).

Borreria ruelliae sensu auctt. mult. non (DC.) H. Thoms. —Gonçalves in Garcia da Orta Sér. Bot. **5** (2): 208 (1982).

Borreria scabra sensu auctt. mult. non (Schumach. & Thonn.) K. Schum.

Annual unbranched or sparsely branched erect or straggling herb (3)16–60 cm. tall, with 4-ribbed stems ± densely covered with spreading white hairs. Leaf blades 2–7 × 0.35–2.2 (2.7) cm., narrowly elliptic to elliptic-lanceolate, acute at the apex, narrowly cuneate at the base, with ± long hairs on both surfaces and scabrid at the margins; true petiole absent; stipule sheath 3 mm. long, densely hairy, with 7 setae 0.2–1.1 cm. long. Flowers in clusters at the nodes, up to 1.5–2 cm. in diam. in fruiting stage. Calyx tube 2.5 mm. long, subcylindric, hairy above; lobes 4, ± equal, 2–4 × 0.2–0.8 mm., mostly lanceolate or narrowly oblong, scabrid on the margins. Corolla white, sometimes with mauve streaks reaching into the throat; tube 3–8 mm. long, funnel-shaped; lobes 1.5–4.5 × 1–3 mm., triangular, pubescent at the apex outside. Filaments exserted 1–3 mm. Style exserted 1–3.5 mm. Capsule pale, often streaked red-brown, 3 × 2 mm., ellipsoid, densely hairy. Seeds pale chestnut-coloured to deep blackish red, 2.6–2.8 × 1.2–1.7 × 0.8–1 mm., oblong-ellipsoid, with a deep narrow ventral groove, covered with a fine shallow reticulation.

Caprivi Strip. 29 km. W. of Katima Mulilo, fl. 12.ii.1969, *de Winter* 9131 (K; PRE; SRGH). **Botswana**. N: Shakawe, fr. 1.v.1975, *Biegel et al.* 5044 (K; SRGH). SE: 16 km. N. of Gaborone, fl. & fr. 19.iii.1976, *Mott* 904 (K; UB). **Zambia**. B: Mongu, fl. 29.i.1966, *Robinson* 6823 (K; SRGH). N: Lake Tanganyika, Cassava Sand Dunes, fl. 14.iv.1957, *Richards* 9226 (K). W: Ndola, Kitwe-Kapiri Mposhi road, 5 km. NW. of Ndola road junction, Katuba R., fl. & fr. 10.iii.1973, *Kornaś* 3459 (K; KRA). C: 50km. W. of Lusaka, Sanje, fl. & fr. 9.iii.1972, *Kornaś* 1351 (K; KRA). E: Katete, St. Francis' Hospital, fl. & fr. 13.ii.1957, *Wright* 152 (K). S: Mapanza Mission, fl. 27.ii.1953, *Robinson* 100 (K). **Zimbabwe**. N: Hurungwe, Rifa R., fl. 24.ii.1953, *Wild* 4084 (K; LISC; SRGH). W: Matopos Research Station, fl. 15.ii.1952, *Plowes* 1419 (K; SRGH). C: Marondera, fr. 5.iv.1950, *Wild* 3315 (K; SRGH). E: Inganga, Cheshire, fl. 3.ii.1931, *Norlindh & Weimarck* 4789 (K; LD; SRGH). S: Mwenezi, Malongwe R., SW. Mateke Hills, fr. 6.v.1958, *Drummond* 5626 (K; SRGH). **Malawi**. N: Chitipa Distr., Kaseye Mission, fl. & fr. 19.iv.1975, *Pawek* 9399 (K; MAL; MO). C: Dowa Distr., 13 km. N. of Lombadzi on road to Ntchisi, Chimwere, fl. & fr. 24.iii.1970, *Brummitt* 9340 (K; MAL). S: Blantyre Distr., 2 km. NE. of Mpatamanga, fl. 9.ii.1970, *Brummitt* 8469 (K; MAL; SRGH). **Mozambique**. N: Malema, Mutuáli, fl. & fr. 6.iv.1962, *Lemos & Marrime* 326 (K; LISC; LMA). Z: Lugela-Mocuba Distr., Namagoa Estate, fl. & fr. vi.1944, *Faulkner* PRE 300 (K; PRE; SRGH). T: Songo, right bank R. Zambezi downstream from Barragem, fl. & fr. 8.iii.1972, *Macêdo* 5021 (LISC; LMA). MS: Between Tete and the sea coast, Shamuara, fr. iii.1860, *Kirk* (K). GI: Vilanculos, fl. 24.iii.1952, *Barbosa & Balsinhas* 4998 (BM; LMA). M: Delagoa Bay, fl. 9.i.1898, *Schlechter* 12011 (K).

Also in Burundi, Kenya, Tanzania, Angola, South & South West Africa. Grassland, bushland and various types of woodland, clearings, lakeside sand dunes, fringing vegetation of pans, also as a weed of disturbed ground, roadsides, airstrip edges, etc., sometimes riverine; 25–1650 m.

The differences between this and *S. ruelliae* are slight but in general throughout most of their ranges they are fairly easily separable by the seed and calyx differences; unfortunately in Uganda the calyx character often seems to be unreliable and it is difficult to believe the two can be distinct species. I am convinced, however, that they are best kept distinct. There has been much confusion with *S. sphaerostigma* when no fruit has been available but the more terminal nature of the inflorescences and larger flowers will usually separate that species; in fruit the two are utterly distinct. Hiern saw Peter's specimen but no authentic material is extant. *Spermacoce senensis* is also closely related to *Spermacoce latituba* which is restricted to sandy shores of Lake Malawi but in some areas of the Lake both occur and very probably hybridize since some specimens show intermediate characters. *Fanshawe* 9703 (Zambia, Luano, 20.iv.1966 (K; NDO)) having a more robust habit and longer calyx lobes to about 6 mm. long, may be no more than a distinctive variant but is only in fruit; it could be a distinct species.

18. **Spermacoce deserti** N.E. Br. in Kew Bull. **1909**: 115 (1909). TAB. **39**, fig. 18. Type: Botswana, Kalahari Desert, near Bachakuru, *Lugard* 233 (K, holotype).

Herb 30–45 cm. tall with 3–6 ± 4-angled stiff unbranched or sparsely branched stems from a woody rootstock; stems scabrid with short seta-like hairs along the angles but otherwise glabrous. Leaves rigid, 2–5 × 0.25–0.8 cm., linear-oblong-lanceolate, acute at the apex, slightly narrowed at the base, sessile, with scabrid margins but otherwise glabrous; stipule base 2–3 mm. long with 5–6 setae 2.5–4 mm. long. Flowers numerous in tight nodal clusters at terminal node and up to 9 subsequent nodes beneath it. Calyx tube 1 mm. long, turbinate, compressed, pubescent on upper half; lobes 4, erect, 1.2–5 × 0.5–1.5 mm., lanceolate, acute, scabrid-ciliate at the margins. Corolla white; tube 3.5–4 mm. long, funnel-shaped, glabrous; lobes 4, 2–3 × 1.3–2 mm., triangular, ovate or ovate-oblong, sparsely pubescent outside. Stamens exserted. Style exserted, c. 5 mm. long; stigma shortly bifid. Fruit c. 3 mm. long, oblong, compressed, pubescent on upper half. Immature seeds dark brown, c. 3 mm. long, oblong; ventral furrow narrow, widened at both ends and there with packed raphides.

Botswana. SW: 10 km. NE. of Kalkfontein on Ghanzi road, fl. 11.ii.1979, *Skarpe* 303 (K; SRGH). SE: Kweneng District, Mantswabese Ranch, fl. 17.iii.1977, *Hansen* 3082 (C; GAB; K; PRE; SRGH).

Also in S. Africa. Wooded grassland with *Terminalia, Acacia, Grewia,* etc. on Kalahari Sand; 500–?800 m.

19. **Spermacoce latituba** (K. Schum.) Verdc. in Kew Bull. **30**: 304 (1975); in F.T.E.A., Rubiaceae 1: 354, fig. 49/9 (1976). TAB. **40**, fig. 19. Type: Malawi, without locality, *Buchanan* 1138* (B, holotype †; K, isotype).

Borreria latituba K. Schum. in Engl., Bot. Jahrb. **28**: 110 (1899).

Annual herb with ascending stems to over 50 cm. tall or sometimes a shrub to 2 m.; stems 4-angled, glabrous to densely pilose. Leaf blades 1–5.5 × 0.2–1.6 cm., linear-elliptic, acute at the apex, narrowed at the base into the stipule sheath, very shortly scabrid above and on the margin, or densely pilose above, glabrous to scabrid beneath; petioles ± obsolete; stipule bases glabrous or scabrid-pubescent, 2 mm. long, with 4 fimbriae 1.5–3.5 mm. long. Flowers in well-spaced clusters in the axils of the apical 3–5 nodes; leaf-like bracts 1–3.5 cm. long; stipuliform bracteoles with ciliate fimbriae 8 mm. long. Calyx tube 2 mm. long, ellipsoid, glabrous or hairy above; lobes 4, 3 × 0.8 mm., triangular-lanceolate, closely scabridulous-ciliate along the margins or sometimes quite hairy. Corolla white with faint mauve stripes; tube 8 × c. 7–8 mm., at the throat but strongly narrowed at the base to c. 1 mm.; funnel-shaped; lobes 3–4.5 × 3.2 mm., triangular, glabrous to pubescent outside. Filaments exserted 2 mm. Style exserted 4–6 mm.; stigma capitate, 0.8 mm. wide. Inflorescences up to 1.8 cm. wide in fruit. Capsule 4 mm. long and wide, oblong-ellipsoid, finely transversely wrinkled, glabrous or hairy above, crowned with the persistent calyx lobes. Seeds brown, 3 × 1.5 × 1 mm., oblong-ellipsoid, deeply grooved beneath.

Malawi. N: Karonga, fl.xi.1887, *Scott* (K). C: Mvera, fl. *Kenyon* 38 (K). S: Salima, lake shore, by Grand Beach Hotel, fl. & fr. 29.iv.1970, *Brummitt* 10275 (K; MAL). **Mozambique**. N: Lichinga (Vila Cabral), Meponda, fl. 13.ix.1958, *Rui Monteiro* 63 (LISC) (robust form).

Also in S. Tanzania. Sandy lakeshore; 420–490 m.

Various altitudes and localities from 410–2400 m. given on old labels are I think very suspect, particularly *McClounie* 84 & 136 Mwanemba 8000' and 47 Nymkowa 6500'. Apart from the sheet cited above there is no recent material save from Tanzania. This is very much more closely related to *Spermacoce senensis* than indicated in the F.T.E.A. account, in fact they may hybridize. They both occur on L. Malawi beaches.

20. **Spermacoce sphaerostigma** (A. Rich.) Vatke in Linnaea **40**: 196 (1876). —Verdc. in F.T.E.A., Rubiaceae 1: 367, fig. 53 (1976). —Gonçalves in Garcia de Orta, Sér. Bot. **5**: 268 (1982). Types from Ethiopia.

Hypodematium sphaerostigma A. Rich., Tent. Fl. Abyss. **1**: 348 (1847). —Hiern in F.T.A. **3**: 240 (1877). —Verdc. in Kew Bull. **7**: 360 (1952); in Bull. Jard. Bot. Brux. **28**: 278 (1958).

Hypodematium ampliatum A. Rich., Tent. Fl. Abyss. **1**: 349 (1847). Types from Ethiopia.

Spermacoce ampliata (A. Rich.) Oliv. in Trans. Linn. Soc. Bot. **29**: 88, t. 54 excl. fig. 4–6 (1873).

* K. Schumann gives the number as 1198 but I am convinced this is a misreading of a poorly written 3.

*Spermacoce senensis** sensu Hiern in F.T.A. **3**: 236 (1877) pro parte quoad *Kirk* Soche Hill et *Stewart* Zambesiland non Klotzsch.

Arbulocarpus sphaerostigma (A. Rich.) Tennant in Kew Bull. **12**: 386 (1958). —Cufodontis, Enum. Pl. Aethiop.: 1023 (1965). —Agnew, Upland Kenya Wild Fl: 409 (1974).

Annual erect herb 10–90 cm. tall, with unbranched to much-branched stems, sparsely to densely covered with spreading hairs. Leaf blades 1–10 × 0.4–3.5 cm., lanceolate to elliptic-lanceolate, acute at the apex, cuneate at the base, sparsely to distinctly hairy on both surfaces, sometimes very scabrid all over but nearly always so along the margin; free petiole 0–2 mm. long, joined to the stipule sheath beneath; stipules hairy, base 1–4.5 mm. long, with 5–7 fimbriae 2–8 mm. long, ciliate. Flowers in axillary clusters up to 1.8 cm. in diameter in fruiting stage, at apical and lower nodes, those at the apex supported by usually 4 bract-like leaves; stipule-derived bracteoles filiform, 2–3 mm. long. Calyx tube 2.5–3.5 mm. long, obconic, hairy at the apex; lobes 4, 2.8–5 × 1.3–1.8 mm., ovate to lanceolate, leafy, scabrid-ciliate. Corolla white, blue, purple or white with blue streaks; buds mostly tipped with erect hairs; tube funnel-shaped, often quite broadly so, 5–8.5 × 0.8 mm. at the base, 5 mm. wide at the throat; lobes 3–4 × 1.5–2 mm., triangular, usually with some long hairs outside at the apex, ± finely papillate inside. Filaments exserted 1.5–2.2 mm. Style exserted 2–4.5 mm., usually thickened upwards and ± roughened; stigma slightly bifid, 0.5 mm. wide. Capsule green, 3.5–4 × 2–3 mm., ovoid, splitting from the base upwards but remaining in one piece, the split not crossing the apical rim bearing the persistent ± spreading calyx lobes; valves shortly hairy in upper half, finely transversely rugulose; central septum a persistent white oblong left behind on the inflorescence. Seeds brown, 3–3.2 × 1.3–1.5 × 0.8 mm., narrowly oblong-ellipsoid, slightly narrowed to the base, shiny, finely obscurely reticulate, with a broad ventral groove and usually a small basal aril 0.5–1 mm. long.

Zambia. N: Mbala Distr., Chilongowelo, fl. 3.iii.1952, *Richards* 867 (K). W: Ndola, Kamboa, fl. 7.ii.1954, *Fanshawe* 792 (K; NDO). C: Luangwa Valley Game Reserve (S), Lunda Plain, young fruit 19.iii.1966, *Astle* 4677 (K; SRGH). E: Katete, St. Francis' Hospital, fl. 13.ii.1957, *Wright* 151 (K). **Malawi**. N: Mzimba Distr., Lunyangwa R., 1.6 km. SW. of M14, fl. 7.iii.1976, *Pawek* 10910 (K; MAL; MO; SRGH; UC). C: N. of Chitala on Kasache road, fl. 12.ii.1959, *Robson* 1573 (BM; K; LISC). S: Lower Kirk Range, Ntcheu, fr. 17.iii.1955, *Banda* 74 (K). **Mozambique**. N: Massangulo, fl. iii.1933, *Gomes e Sousa* 1342 (K). Z: Milange, about 4 km. from Sebalue in direction of Liefrio, fl. 17.ii.1972, *Correia & Marques* 2684 (BR; LMU). T: Mágoè 2 km. from Estima towards Merara, fl. 9.ii.1970, *Torre & Correia* 17833 (LISC; LMA; LMU; SRGH; WAG).

Also in Nigeria, Cameroon, Zaire, Burundi, Sudan, Ethiopia, Uganda, Kenya and Tanzania. Tall grassland, often in marshy places, *Brachystegia* woodland, cultivations, roadsides, etc.; 600–1650 m.

21. **Spermacoce congensis** (Bremek.) Verdc. in Kew Bull. **30**: 308 (1975); in F.T.E.A., Rubiaceae 1: 369, fig. 51/28 (1976). TAB. **40**, fig. 21. Type from Zaire.

Dichrospermum congense Bremek. in Bull. Jard. Bot. Brux. **22**: 75 (1952).

Erect annual herb 17–60 cm. tall; stems ± simple or sparingly branched, 4-angled, bifariously shortly pubescent in the grooves. Leaf blades 1.3–8.5 × 0.22–1.25 cm., linear-lanceolate to lanceolate, acute at the apex, cuneate or ± rounded at the base, sessile, ± thick, with thickened whitish margins and midrib, glabrous save for the margins and main nerve beneath or scabrid to pilose on both surfaces, the margins nearly always ciliate with short stiff hairs; stipule sheath united with petioles, 5–6 mm. long, ridged, hairy or shortly pubescent with 5–6 ciliolate setae 0.2–1.2 cm. long. Flowers sessile in axillary and pseudoterminal (1)3–6-flowered clusters at many of the upper nodes, the apical 2–3 often run together to form a short spike-like inflorescence; bracteoles 7 mm. long. Calyx tube glabrous, oblong, 1.5 mm. long; lobes 2, 0.3–2 cm. × 0.8–1.8 mm., linear-lanceolate to lanceolate, similar to the leaves in texture and indumentum, ciliate. Corolla white, pink or blue; tube 6–7(13) × 3.5 mm. at the mouth, cylindrical to narrowly funnel-shaped, with a ring of hairs inside 2.5–3 mm. above the base; lobes 3–6 × 1.2–2.2 mm. oblong-lanceolate. Filaments 1–3.5 mm. long. Style exserted 1.5–6 mm., finely papillate; stigma capitate, c. 7 mm. wide. Capsule 3.5–5 × 2.5 mm., subglobose, crowned with the calyx lobes, opening right out, the valves joined only at the base, almost circular in outline, convex, 3.5 × 4 mm., glabrous, transversely wrinkled; septum persistent, shiny.

* Hiern (loc. cit.) actually mentions the difficulty of distinguishing *Spermacoce senensis* from *Hypodematium sphaerostigma*.

Seeds dark brown to black, 3–3.6 × 1.5 × 1.3 mm., oblong-ellipsoid, shining, very finely reticulate, with a ventral groove and a conical yellowish basal appendage (elaiosome) 1.1–1.5 mm. long.

Zambia. N: Mbala, Katuka, fl. 12.iii.1950, *Bullock* 2621 (K). W: Luanshya, fr. 4.iv.1954, *Fanshawe* 1064 (K; NDO). **Malawi**. N: Mzimba Distr., 9.6 km. N. of Mzambazi and 17.6 km. S. of Mperembe, 10.iii.1978, *Pawek* 13957 (K; MAL; MO; SRGH; UC).

Also in Zaire and S. Tanzania. Tall grassland, *Brachystegia* woodland, dambos, usually on sand and often in swampy places but also by roadsides and sometimes as a weed on cleared ground; 1200–1600 m.

The possession of a seed-appendage does not seem sufficient reason for erecting a new genus; traces of such appendages are present in other species of *Spermacoce* and the facies is exactly that of *Spermacoce*. Whether or not the appendage has any connection with dispersal by ants must await observations by a field observer.

22. **Spermacoce** sp. B.

Perennial herb to 25 cm. tall with c. 15 very slender almost filiform unbranched stems from very slightly woody base, these slightly ridged in dry state and with very narrow bifarious lines of puberulence, drying dark bluish green. Leaves 1.5–3 cm. long, linear, almost filiform but rigid, scarcely more than 0.5 mm. wide on stem, but those acting as floral bracts 1 mm. wide and widening to over 2 mm. at base, mucronulate, slightly widened at base, sessile, ascending, the margins and midrib slightly thickened; stipules at flowering nodes with 1 mm. base and 3 filiform setae up to 3 mm. long, at non-flowering lower nodes with one seta up to 5 mm. long. Flowers few, at the two uppermost nodes, only one open at each at any one time; bracteoles linear, scarious, few, 3 mm. long, ciliolate. Calyx tube obovoid, 2 mm. long; lobes 2, similar in texture to the leaves, 4 × 0.8 mm., with hyaline margins and similar ± winged midrib. Corolla deep sky blue, darker at base of lobes, the tube whitish; tube 7 mm. long, cylindrical funnel-shaped, with a ring of hairs inside below the middle; lobes 4–5 × 2 mm. at the base, narrowly triangular, acute. Style well-exserted, micropapillate, the capitate stigma just overtopping the lobes; anthers just exserted. Capsule not known.

Zambia. W: Mwinilunga, between Ikelenge and Zambezi source, W. of main road, fl. 24.ii.1975, *Hooper & Townsend* 355 (K; SRGH).

Only this specimen known. Damp grassland; 1450 m.

Clearly related to *Spermacoce congensis* (Bremek.) Verdc. but with narrower leaves, shorter sepals, branched from base and presumably perennial.

23. **Spermacoce dibrachiata** Oliv. in Trans. Linn. Soc. **29**: 87, t. 52 (1873). —Hiern in F.T.A. **3**: 239 (1877). —Verdc. in F.T.E.A., Rubiaceae 1: 370, fig. 51/30 (1976). —Gonçalves in Garcia de Orta, Sér. Bot. **5**: 207 (1982). TAB. **40**, fig. 23. Type from Tanzania.
Borreria dibrachiata (Oliv.) K. Schum. in Engl. & Prantl, Pflanzenfam. IV, **4**: 144 (1891).
Pentas involucrata Bak. in Kew Bull. **1895**: 66 (1895). Type: Zambia, Fwambo, *Carson* 40 of 1894 coll. (K, holotype).

Annual or sometimes biennial or perennial very variable herb 6–90 cm. tall, with simple or sparsely branched stems covered with dense to sparse spreading often asperous hairs with tubercular bases and shorter hairs within the longitudinal grooves between the longitudinal ribs; the basal internodes are often reduced giving the appearance of a rosette of leaves; rootstock often swollen, particularly at stem junction. Leaf blades 4.5–12 × 0.35–1.5(2.8) cm., lanceolate, acute at the apex, narrowed at the base into the stipule sheath, with long hairs on both surfaces or only on the midrib or glabrous except for the ± scabrid margins; stipules hairy, base 7–9 × 4.5 mm. at normal nodes, 0.8–2 cm. wide at the apical inflorescence-supporting node; fimbriae c. 5, 3–8 mm. long at lower nodes, 8–11, 2 mm. long at apical node. Flowers in very dense mostly many-flowered apical heads, closely subtended by 2–4 leafy bracts, the heads varying greatly in size, 1–2.5 cm. wide; bracteoles linear, 5–7 mm. long, joined at the base, colourless, glabrous beneath, greenish and ciliate above; rarely an inflorescence at node below the apical one. Calyx tube 1.7–5 × 2.8 mm., obconic or ellipsoid, glabrous; free limb 0.7–1.5 mm. long; lobes 4, 3–5.5 × 0.8–1.8 mm. at base but strongly tapering, linear-lanceolate, margins and sinuses scabridulous; some small intermediate lobes may also be present. Corolla blue or violet-blue or with white tube and blue to violet limb with a white eye; tube (0.8)1–2.3 cm. long, filiform-cylindrical; lobes 2.5–6 × 1.5–3 mm., oblong-elliptic, with a few hairs at the apex outside. Filaments exserted 2–4 mm. Style exserted 2–4.5 mm.; stigma capitate, 1 mm.

wide, slightly bifid, the lobes c. 0.6 mm. long. Capsule straw-coloured with numerous red longitudinal streak-like marks, 3.5 × 3 × 1.3 mm., glabrous. Seeds dark brown, 2.6 × 1.3 × 0.95 mm., ellipsoid, finely reticulate and rugulose, with a ventral narrow groove which widens at base and apex.

Zambia. B: 35 km. WSW. of Kabompo on Zambezi (Balovale) Road, fl. 24.iii.1961, *Drummond & Rutherford-Smith* 7292 (K; LISC; SRGH). N: Mporokoso Distr., Nsama, fl. 4.iv.1957, *Richards* 9015 (K). W: Nkana/Kitwe Sewage Disposal Works, fl. 19.iii.1959, *Shepherd* 30 (K). C: Mkushi, fl. 27.iii.1961, *Angus* 2509 (FHO; K; SRGH). E: Lundazi, fl. 27.iv.1952, *White* 2482 (FHO; K) (form with very short white corolla growing in waterlogged dambo). S: Mumbwa, fl. 19.iii.1963, *Van Rensburg* 1701 (K; SRGH). **Zimbabwe**. N: Umvukwe Mts., central area near Mutorashanga Pass, fl. 24.iv.1948, *Rodin* 4458 (K; SRGH; UC). W: Bulawayo, *Cheesman* 70 (BM). C: 16 km. from Harare on Bulawayo road, fl. 18.iv.1963, *Loveridge* 654 (K; SRGH). E: Mutare, Commonage, fl. 1.iii.1950, *Chase* 2106 (BM; K; LISC; SRGH). **Malawi**. N: Chitipa Distr., Misuku Hills, fl. 26.iv.1972, *Pawek* 5246 (K). C: Ntcheu Distr., Dedza Plateau, 38.5 km. S. of Dedza, fl. 4.iv.1978, *Pawek* 14212 (K; MAL; MO). S: Zomba Plateau, fl. 4.v.1970, *Brummitt & Banda* 10380 (K; SRGH). **Mozambique**. N: Massangulo, fl. v.1933, *Gomes e Sousa* 1381 (K). Z: Mocuba, Namagoa, fl. vii.1943, *Faulkner* 95 (K; PRE). T: between Furancungo and Chipacasse, fl. 15.v.1948, *Mendonça* 4239 (LISC). MS: 15 km. NW. of Vila de Manica, Edmundio Copper Mine, fl. 29.iii.1966, *Simon* 777 (K; LISC; SRGH).

Also in Zaire, Rwanda, Burundi, Tanzania and Angola. Grassland, bushland, *Brachystegia* woodland and derived clearings, also old cultivations, dry marshland edges, occasionally by river and stream banks; 750–2100 m. (90 m. in Mozambique Z)

Field studies on the lifespan of this species would be of interest. Contrary to my statement in F.T.E.A. it can certainly be perennial. Some specimens have entirely glabrous stems and stipular involucres but the majority are densely bristly hairy; intermediates occur.

24. **Spermacoce samfya** Verdc. sp. nov.* TAB. **40**, fig. 24. Type: Zambia, Lake Bangweulu, Kalasa Mukose, *Chabwela* 029 (K, holotype; SRGH).

Perennial herb 10–20 cm. tall with 5–? stems from a slender woody root; stems densely covered with ascending grey hairs giving a rather woolly appearance. Leaves 2.5–3.5 cm. × 1.5–3 mm., linear-lanceolate, acute at the apex, sessile, glabrous, the margins thickened and similar to the very evident midrib but other venation not visible; stipule base 2–3 mm. long, densely hairy with 3–5 dark slender pubescent setae up to 7 mm. long. Flowers fairly numerous in compact heads, c. 1.5 cm. diam. (excluding leaves); bracteoles numerous, filamentous, 4–5 mm. long, scarious and ciliolate. Calyx tube c. 2 mm. long, ellipsoid, glabrous save near limb; limb tube 1 mm. long ± fimbriate; lobes 2, 4 × 0.5 mm., linear-lanceolate, margined with short almost tooth-like stiff hairs. Corolla colour unknown; tube 1.3 cm. long, cylindric, widened at throat to 2.5 mm., hairy for a short distance at middle inside; lobes 5.5 × 2 mm., oblong-ovate, with few short hairs at apex. Style exserted 2.5 mm.; stigma biglobose, 0.7 mm. wide; anthers 2 mm. long, the filaments exserted, 2.5 mm. Capsule not known.

Zambia. N: Samfya, Lake Bangweulu, Kalasa Mukose, fl.iii.1970, *Chabwela* 029 (K; SRGH).

Not known elsewhere. Seasonally flooded flats by lake; 1150 m.

Had been named *Spermacoce dibrachiata* but has only 2 calyx lobes and in general much narrower leaves.

25. **Spermacoce phyteumoides** Verdc. in Kew Bull. **30**: 308, fig. 9 (1975); in F.T.E.A., Rubiaceae 1: 371 (1976). Type: Zambia, Mbala, *Bullock* 2636 (EA; K, holotype; SRGH).

Suberect or ± prostrate annual or perennial herb 15–40(60) cm. tall; stems dark, covered with spreading white hairs. Leaf blades 2–7(14) cm. × 2.5–10(13) mm. wide, elliptic-lanceolate, acute at the apex, narrowed at the base into the stipule sheath, glabrous on both surfaces save for hairs on the midrib beneath (but even these may be absent) or densely scabrid-hairy on both surfaces; margins always scabrid with short stiff hairs; stipule bases very hairy, 2–3 mm. long, with 3–5 purple fimbriae 2.5–4.5 mm. long; stipules supporting the leaves which subtend the inflorescences with 7 fimbriae of varying thickness c. 5–6 mm. long and colleters 1.5 mm. long. Flowers in terminal heads very similar to those in *species 23*; bracteoles 3–4 mm. long, linear, ciliolate, hyaline. Calyx tube 1–3 mm. long, oblong to obconic, without a free limb, lobes 2, 3.5–5 × 0.6–1.3 mm.

* Similis *Spermacoce dibrachiatae* Oliv. sed calycis lobis 2, foliis plerumque angustioribus; *Spermacoce phyteumoidis* Verdc. affinis sed corollae tubo 1.3 cm. longo lobisque 5.5 ×2 mm., indumento caulis adscendenti sublanato diversa.

wide, linear to lanceolate, with hairy margins. Corolla white or usually deep blue; tube 6–7 mm. long; lobes 2–4.5 × 1.2–2 mm., ovate-oblong, with short to long hairs at the apex outside. Filaments exserted 2.5 mm. Style exserted 4 mm.; stigma 1.5 mm. wide. Capsule straw-coloured, streaked with brown or purplish, 2.5–3 × 1.5–2 × 1 mm., oblong-ellipsoid or obovoid, glabrous beneath, ± hairy at apex, crowned with the remains of the calyx lobes. Seeds chestnut-brown, 2.5 × 1 × 0.6 mm., narrowly oblong-ellipsoidal, dorsally convex, ventrally sulcate, rugulose.

Var. **phyteumoides**. —Verdc. in F.T.E.A., Rubiaceae 1: 371, fig. 51/31 (1976). TAB. **40**, fig. 25.
Borreria phyteuma sensu R.E. Fries, Wiss. Ergebn. Schwed. Rhod.-Kongo-Exped.: 17 (1921) non (Hiern) K. Schum.
Borreria dibrachiata sensu Hutch., Botanist in S. Afr.: 504 (1946) non (Oliv.) K. Schum.

Corolla white or very pale, rarely dark blue; mostly annual.

Zambia. N: Mbala Distr., Chilongowelo, top of escarpment, fl. 20.iii.1952, *Richards* 1069 (K). W: Luanshya, fl. 12.iii.1955, *Fanshawe* 2135 (K; NDO; SRGH). C: 8 km. E. of Lusaka, fl. 4.v.1955, *King* 3 (K).
Also in Tanzania. Long and short grassland, bushland and *Brachystegia* woodland, sometimes on marshy ground; 950–1590 m.
In F.T.E.A. I recorded var. *caerulea* from Zambia in error.

26. **Spermacoce perennis** Verdc. in Kew Bull. **30**: 314 (1975). Type: Zambia, Mporokoso, Kasinghi Dambo, *Wright* 352 (K, holotype).

Perennial herb 20–55 cm. tall with strict caespitose erect obtusely 4-angled glabrous or shortly pubescent stems from a woody many-headed rootstock. Leaves ± erect, 1.2–10 cm. × 1–3.5 mm., linear, acute at apex, narrowed at the base, somewhat coriaceous, glabrous save for the minutely scabrid-ciliolate margin; stipule sheath glabrous or densely pubescent, at flowering nodes funnel-shaped with 5–7 fimbriae 0.3–2 mm. long, at sterile nodes with 0–3 fimbriae up to 0.8 mm. long. Flowers in small terminal heads and sometimes also at penultimate node; bracts stipuliform with a triangular median lobe and one lanceolate lateral lobe on each side; bracteoles fimbriform, 6 × 0.5 mm., hyaline, brown-lined and ciliolate. Calyx tube 2 mm. long, obconic, glabrous; lobes 2–4, 2 linear 4.5 × 0.6 mm., ciliolate and 0–2 linear, hyaline, 2 mm. long or with no lobes but limb reduced to numerous fimbriae 1 mm. long. Corolla tube white, 7 mm. long, funnel-shaped, glabrous outside but with a ring of hairs inside at the middle; lobes pale blue, 3 × 1.8 mm., with a few hairs at the apex. Filaments exserted 1.8 mm. Style exserted 3 mm.; stigma biglobose, 0.9 mm. wide. Capsule not known.

Var. **perennis**. TAB. **40**, fig. 26.

Stems and stipules glabrous. Calyx lobes 2–4.

Zambia. N: Luwingu, fl. iv.1922, *Jelf* 36 (BM).
Not known elsewhere. Dambos, bushland on sandy soil; 1350 m.

Var. **fimbriolata** Verdc. in Kew Bull. **30**: 314 (1975). Type: Zambia, near Luwingu, *Astle* 530 (K, holotype; SRGH).

Stems and stipules pubescent. Calyx limb reduced to numerous fimbriae 1 mm. long.

Zambia. N: near Luwingu, Chishinga Ranch, fl. 27.iv.1961, *Astle* 530 (K; SRGH).
Not known elsewhere. Dambos, bushland; 1560 m.

27. **Spermacoce thymoidea** (Hiern) Verdc. in Kew Bull. **30**: 314 adnot. (1975). TAB. **40**, fig. 27. Types from Angola.
Tardavel thymoidea Hiern, Cat. Afr. Pl. Welw. **1**: 507 (1898).
Borreria thymoidea (Hiern) K. Schum. in Just, Jahresb. **26**: 391 (1900).
Borreria hockii De Wild. in Fedde, Repert. **11**: 511 (1913). Type from Zaire.

Perennial herb much-branched near the base from a slender ± woody rootstock; stems several–30, ± unbranched, slender, erect or radiating, 7.5–15 cm. tall, bifariously pubescent in grooves which alternate at each successive internode or typically bristly pubescent all round. Leaves 1–3.2 cm. × 1.5–3 mm., linear to linear-elliptic or linear-lanceolate, acute and minutely mucronate, cuneate, sessile, glabrous to minutely scabrid at least at the margins; margins distinctly thickened similar to midrib but other venation not evident; stipule base 1 mm. long with 1–5 microsetulose subulate teeth up to 3 mm. long. Flowers 1–3 to 20 in Angola, in apical clusters and at 1–2 successive lower nodes, the 2 upper nodes often approximate; bracts leaf-like forming a kind of involucre with the main leaves. Calyx tube 1 mm. long; lobes 2, 3–4 × 0.5 mm., linear-lanceolate, acute. Corolla violet-purple or blue; tube narrowly funnel-shaped, 5–6 mm. long, 2.5 mm. wide at the throat; lobes 4 × 2 mm., triangular to ovate, acute or obtuse. Style exserted 3–5 mm., sometimes shortly papillose; anthers just exserted. Capsule ovoid, 2.5 mm. long. Ripe seeds not seen.

Zambia. W: N. of Mwinilunga, Luakera Falls, fl. 25.i.1938, *Milne-Redhead* 4338 (K).
Also in Angola. Riverine open sandy slopes; 1500 m.
The above-cited material is consistently more slender than the type and certain other material from Angola and has bifariously pubescent stems and thinner leaves but similar plants do occur in Angola. Without more material the strong possibility that there are two subspecies cannot be confirmed.

28. **Spermacoce annua** Verdc. in Kew Bull. **30**: 316, fig. 11 (1975). Type: Zambia, Mwinilunga, 11.2 km. N. of Kalene Hill, *Robinson* 6615 (K, holotype).

Erect gregarious ephemeral herb 15–20 cm. tall with simple or sparsely branched ± 4-angled graceful chestnut stems covered with ± spreading white hairs. Leaves 1.4–2.8 cm. × 0.5–1.3 mm., linear, acute at the apex, attenuate at the base, densely covered with ± spreading white hairs; stipule-sheath narrow, 2 mm. long with 3–5 setiform pilose fimbriae (2)3–6.5 mm. long. Flowers several in a terminal head supported by an involucre of 6–8 bract-like leaves, the apical 3–4 internodes entirely suppressed. Calyx tube 1 mm. long, obconic, hairy at the apex; lobes 2, 3–5 mm. long, linear, pilose. Corolla blue; tube 4.3 mm. long, funnel-shaped, glabrous both inside and out; lobes 3 × 2.5 mm., triangular, with white hairs at apex and along middle outside. Filaments exserted 1.5 mm. Style 7.5 mm. long, the stigma 0.35 mm. wide. Capsule ellipsoid; valves 2 × 1 mm., transversely rugulose, pilose at the apex. Ripe seeds not seen.

Zambia. W: Mwinilunga, Zambezi Rapids, fl. 18.v.1969, *Mutimushi* 3356 (K; NDO; SRGH).
Not known elsewhere. Dry dambos, damp soil on rocky outcrops; 1500 m.

34. MITRACARPUS Zucc.

Mitracarpus Zucc. in Schultes & Schultes, Syst. Veg. **3**, Mant. : 210 (1827); in obs. —Anderson in Taxon **20**: 643 (1971).

Erect or prostrate annual or perennial herbs with 4-angled stems. Stipules connate with the petioles to form a fimbriated sheath. Leaves opposite; lamina linear-lanceolate to ovate or broadly elliptic. Flowers not heterostylous or only slightly so, in dense spherical sessile terminal or axillary heads. Calyx tube obconic, obovoid or subglobose; teeth 4–5, 2 often longer, sometimes with minute supplementary ones between. Corolla salver-shaped or funnel-shaped, the tube often with an internal ring of hairs; lobes 4; throat glabrous or hairy. Stamens inserted in the throat; anthers included or exserted. Disk fleshy. Style short or long, divided into 2 short linear branches. Ovary 2(3)-locular; ovules solitary in each locule, attached by the middle to the septum. Fruit a thin circumscissile capsule, the upper part splitting off together with the calyx lobes, the septum persistent. Seeds oblong or globose, the ventral face divided into 4 distinct areas; endosperm fleshy.

A genus of about 30–40 species mostly confined to tropical America, but 1 species now common throughout tropical Africa (see note at end of species) and other parts of the Old World.

Mitracarpus hirtus (L.) DC., Prodr. **4**: 572 (1830). —K. Schum. in Martius, Fl. Brasil. **6**: 84 (1888); in Engl. & Prantl., Pflanzenf. **4**: 142, fig. 46/U (1891). —Standley in Field Mus. Publ. Bot. **7**: 157 (1930), 331 (1931) & 473 (1931). —Nicolson in Taxon **26**: 573 (1977). —Tjaden in Taxon **30**: 301 (1981). TAB **42**. Type from Jamaica.
Spermacoce hirta L. Sp. Pl. ed. 2: 149 (1762).
Spermacoce villosa Sw., Prodr. Veg. Ind. Occ.: 29 (1788).
Mitracarpus villosus (Sw.) DC., Prodr. **4**: 572 (1830). —Rendle in Fl. Jam. **7**: 127, fig. 39 (1936). —Gooding, Loveless & Proctor, Fl. Barbados: 405 (1965). —Adams, Fl. Pl. Jam.: 733 (1972). —Verdc. in Kew Bull. **30**: 317-322 (1975); in F.T.E.A., Rubiaceae 1: 375 (1976). Type from Jamaica.

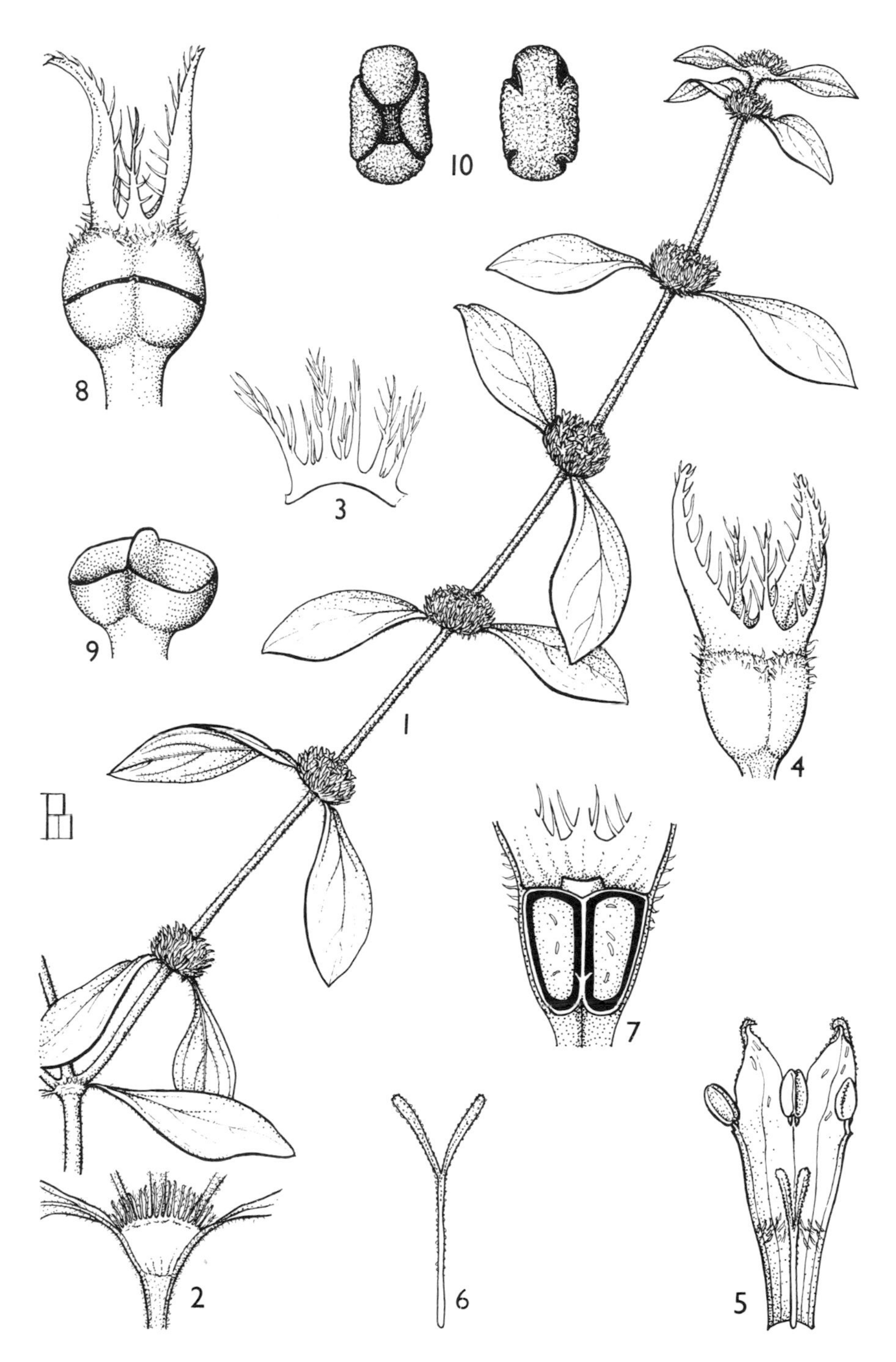

Tab. 42. MITRACARPUS HIRTUS. 1, habit (× $\frac{2}{3}$); 2, node showing stipule (× 3); 3, bracteole (× 6); 4, calyx (× 14); 5, short-styled flower, longitudinal section (× 14), 1–5 from *Chancellor* 24; 6, style from long-styled flower (× 14), from *Haarer* 2084; 7, ovary, longitudinal section (× 20); 8, capsule (× 14); 9, lower part of dehisced capsule (× 14), 7–9 from *Chancellor* 24; 10, seed, two views (× 20), from *Haarer* 2084. From F.T.E.A.

Mitracarpus scaber Zucc. in Schultes & Schultes, Mant. Syst. Veg. **3**: 210, 399 (1827). —Hiern in F.T.A. **3**: 243 (1877). —Hepper in F.W.T.A. ed. 2, **2**: 222 (1963). Type from Africa (Forte Louis —not traced).

Staurospermum verticillatum Schumach. & Thonn., Beskr. Guin. Pl.: 73 (1827). Type from Ghana.

Mitracarpus senegalensis DC., Prodr. **4**: 572 (1839). —Oliv. in Trans. Linn. Soc. **29**: 89 (1873) nom. illegit., based on last.

Mitracarpus verticillatus (Schumach. & Thonn.) Vatke in Linnaea **40**: 196 (1876). —K. Schum. in Engl., Pflanzenw. Ost-Afr. **C**: 394 (1895). —Hutch. & Dalz., F.W.T.A. **2**: 136 (1931). —Verdc. in Kew Bull. **7**: 360 (1952). —Sebastine & Ramamurthy in Bull. Bot. Surv. India **9**: 291 (1968). —Agnew, Upland Kenya Wild Fl.: 409 (1974).

Erect or spreading annual herb (5)9–40 cm. tall, with unbranched or sparsely to much-branched stems; branchlets pubescent with short curled ± appressed hairs and often with spreading ones as well, the older with epidermis eventually peeling; sometimes quite woody at the base. Leaf blades 1–6 × 0.3–2.3 cm., elliptic, subacute at the apex, cuneate at the base, glabrescent to scabrid-pubescent above, glabrescent or glabrous beneath save for hairs on the main nerves; margins often scabrid; petiole c. 1 mm. long, often densely pubescent and with ciliate margins; stipule sheath 1–3 mm. long, divided into 6–9(15) often colleter-tipped fimbriae, 1–5 mm. long, ciliate. Inflorescences numerous, present in most axils, subglobose, (0.5)0.8–1.8 cm. in diam.; flowers sessile or almost so; bracteoles filamentous, white, 1–2 mm. long. Calyx tube 1–1.4 mm. long; limb tube 0.15–0.4 mm. long; lobes 4, 2 oblong-lanceolate, green with hyaline margins, rather thick, 1.3–2.3(3) mm. long, and 2 hyaline, 0.55–1.5 mm. long, triangular-lanceolate, narrower than the others, all with usually ciliate margins and often hairy below. Corolla white, glabrous or slightly hairy outside; tube 1.4–1.9 mm. long; lobes 0.6–1 × 0.3–0.9 mm, ovate. Flowers showing very slight heterostyly, the anthers varying in their degree of exsertion; style 1.1–1.6 mm. long; stigma 0.3–0.5 mm. long. Capsule straw-coloured, c. 1 mm. long and wide. Seeds pale yellow-brown, 0.8 × 0.5 mm., compressed ellipsoid-rectangular, of very characteristic appearance (see TAB. **42**, fig. 10.), dorsally resembling a rectangle with a square portion removed from each corner, ventrally separated into 4 distinct areas by 4 impressed lines radiating from the hilum, rugulose and reticulate.

Zambia. N: Kawambwa Distr., Kafulwe, Lake Mweru, fl. 24.iv.1957, *Richards* 9429 (K). **Malawi**. N: Nkhata Bay Distr., Chinteche, fl. 7.vi.1974, *Pawek* 8684 (K; MO; UC).

Widespread in tropical Africa from Mauretania to Angola, Zaire and Central African Republic, Sudan, East Africa, Seychelles and Cape Verde Is., also in India, Burma, Selangor, New Guinea, Marianas Is., West Indies and tropical S. America. Weed in gardens and cultivations on sandy soil, also by roadsides, dambos, dry rocky hills etc; 510–1200 m.

The long standing argument about the correct name for this species has now been resolved. Details will be found in the Verdcourt (1975) and Nicolson (1977) references cited above.

35. RICHARDIA L.

Richardia L., Sp. Pl.: 330 (1753); Gen. Pl., ed. 5: 153 (1754). —Lewis & Oliver in Brittonia **26**: 271–301 (1974).

Richardsonia Kunth in Mém. Mus. Paris **4**: 43 (1818).

Annual or perennial erect or prostrate hairy herbs. Leaves opposite, sessile or shortly petiolate, mostly ovate, elliptic or oblong, the lateral nerves evident; stipule sheath connate with the petioles, bearing several fimbriae. Flowers hermaphrodite or said sometimes to be polygamo-dioecious, not heterostylous, small, in dense terminal heads enclosed in an involucre formed of (2)4 leaves which are several-nerved from the base. Calyx tube turbinate or subglobose, the limb deeply lobed; lobes 4–8, lanceolate, ovate or subulate, persistent. Corolla shortly funnel-shaped; lobes 3–5(6), ovate to lanceolate; throat glabrous but tube with a narrow area of hairs inside near the base. Stamens with anthers exserted. Style filiform; stigmas 3–4, linear or spathulate, exserted. Ovary (2)3–4(6)-locular; ovules solitary, affixed to the middle of the septum. Capsule 3–4-coccous, crowned by the persistent calyx, eventually splitting into separate cocci, sometimes leaving a very small persistent axis. Cocci mostly obovoid, smooth or more often muricate or papillose. Seeds oblong-ellipsoid or obovoid, dorsally convex, ventrally with 2 grooves; endosperm corneous.

A small genus of about 15 species in Central and S. America but several now widely spread throughout the tropics and subtropics.

Cocci with inner face having smooth depressed area [i.e. actual septa of ovary] almost as broad as the face; leaf blades with upper surface hairy - - - - - - - - - 1. *brasiliensis*
Cocci with inner face having smooth depressed area very narrow; leaf blades with upper surfaces practically glabrous save near the margins - - - - - - - - - - 2. *scabra*

1. **Richardia brasiliensis** Gomes, Mem. Ipecac.: 31, t. 2 (1801). —Rendle in Fl. Jam. **7**: 129, fig. 40 (1936). —Wild, Common Rhod. Weeds: fig. 59 (1955). —Hepper, F.W.T.A., ed. 2, **2**: 216 (1963). —Lewis & Oliver in Brittonia **26**: 276 (1974). —Agnew, Upland Kenya Wild Fl.: 407, fig. (1974). —Verdc. in F.T.E.A., Rubiaceae 1: 378, fig. 56/9 (1976). TAB. **43**, fig. B. Type from Brazil.
Richardsonia brasiliensis (Gomes) Hayne, Arzn. Gen. 8, t. 21 (1822). —K. Schum. in Martius, Fl. Bras. **6**: 94 (1888).
Richardia scabra sensu Hiern in F.T.A. **3**: 242 (1877) et auctt. mult. non L.

Perennial (or ? annual) prostrate herb, often forming a mat, from a central taproot; stems 7–40 cm. long, densely covered with spreading hairs. Leaf blades 1–6.5 × 0.4–2.7 cm., elliptic, acute or subacute at the apex, very narrowly attenuated to the base, the apparent petiole up to 1.5 cm. long, mostly with short hairs all over the upper surface and on the margins and nerves beneath; basal narrowed part with longer hairs; stipular sheath 1–3.5 mm. long, with 3–5 fimbriae, 1–4 mm. long, usually with long hairs. Inflorescences 0.7–1.2 cm. in diam.; bracts ovate-elliptic, rounded at the base, the long ones 1.5–3.5 × 0.65–2 cm., the short ones 1–1.7 cm. × 4–9.5 mm., or sometimes lacking, with a similar indumentum to that on the leaves, save that there are much longer hairs towards the base; basal part of bracts often subhyaline. Calyx tube 1.2–1.7 mm. long; lobes 5–6, 1–1.5 × 0.3–1 mm., ovate-triangular, the margins conspicuously ciliate, basal united part of the limb 0.5–1 mm. long. Corolla white often tinged pink, 2.7–3.2 mm. long; lobes 4–6, 1–1.4 × 0.5–0.8 mm. Style 3–4 mm. long, the branches 0.2–0.5 mm. long; stigmas 0.2–0.3 mm. long, spathulate. Cocci brown, 2–2.6 × 1.4–1.6 mm., oblong-obovoid, inner face with smooth depressed area (i.e. actual septum of ovary) almost as broad as the face, dorsal face covered with short flat hairs which are longer in the middle or a mixture of papilla-like hairs and longer hairs. Seeds brown, 2.5 × 1.8 mm., compressed oblong-obovoid, with ventral face broadly grooved, with 2 short basal projections 0.1 mm. long.

Zambia. N: Liunzua Agricultural School, fl. & fr. 29.iv.1952, *Richards* 1578 (K). W: Kitwe, fl. & fr. 26.i.1954, *Fanshawe* 716 (K; NDO; SRGH). C Mkushi, D. Moffat's Farm, fl. & fr. 15.iii.1981, *Drummond & Vernon* 10787 (SRGH). **Zimbabwe**. C: Harare, fl. & fr. 11.xii.1933, *Eyles* 7658 (K; SRGH). E: Haroni-Lusitu Junction, fl. & fr. 10.i.1969, *Mavi* 866 (K; SRGH). S: Mberengwa, Mt. Buchwa, lower NW. slopes, fl. & fr. 27.iv.1973, *Pope* 937 (K; LISC; SRGH). **Malawi**. N: Mzimba District, Mzuzu, Marymount, fl. & fr. 1.xii.1973, *Pawek* 7541 (K; MAL; MO; SRGH; UC). C: Dedza, Chongoni Forest Reserve, fl. & fr. 9.xi.1967, *Salubeni* 874 (K; SRGH). S: Mulanje, Mimosa Tea Research Station, fl. & fr. 27.x.1958, *G. Jackson* 2247 (K; SRGH). **Mozambique**. Z: Quelimane, 1863, *Kirk* (K). MS: Manica, Serra de Vumba, fl. & fr. 2.i.1948, *Barbosa* 781 (LISC; LMA). M: near Maputo, 8.v.1946, *Gomes e Sousa* 3449 (K).

A South American species now widely naturalized, specimens having been seen from Mauritius, Hong Kong, S. India, Sri Lanka, Java, New Guinea, Australia, Hawaii, USA, Jamaica and Mexico; in Africa it is also known from Ghana, Nigeria, Cameroon, East Africa and South Africa (Natal & Transvaal). Grassy roadsides, weed in gardens and cultivations, bare laterite, also in open places in dune forest; c. 0–2010 m.

Kirk's 1863 specimen is the earliest seen from the area. It is rapidly spreading in many areas.

2. **Richardia scabra** L., Sp. Pl.: 330 (1753). —Lewis & Oliver in Brittonia **26**: 282 (1974). —Verdc. in F.T.E.A., Rubiaceae 1: 380, fig. 56/1–8. TAB. **43**, fig. A. Type from Mexico.
Richardsonia scabra (L.) St.-Hil., Pl. Usuelles Brasiliens, t. 8 (1824) (as to name only, the description and figure refer to *R. brasiliensis* which St. Hilaire included in the synonymy of *R. scabra*).

Very similar to previous species, with erect or ascending stems 5–55 cm. long. Leaf blades (0.9)1.3–5.5 × 0.4–2.3 cm., the apparent petiole 0.2–1.2 cm. long, with short rather stiff conical hairs on the margins and marginal areas above and also on nerves beneath, but most of the upper surface is ± glabrous; stipule sheath 2–4 mm. long, with 3–7 fimbriae 1–4 mm. long. Inflorescences 0.65–1.6 cm. in diam.; long bracts 1–2.7 × 0.6–1.5 cm.; short bracts 0.5–1.5 cm. × 3–9 mm. Calyx lobes usually 6, 1.8–2.3 × 1 mm., triangular to oblong-lanceolate, basal united part of the limb c. 1 mm. long. Corolla white or pale pink or mauve or only tips of lobes tinged pink; tube 5.8–6.3 mm. long; lobes 2.5 × 1.5 mm.,

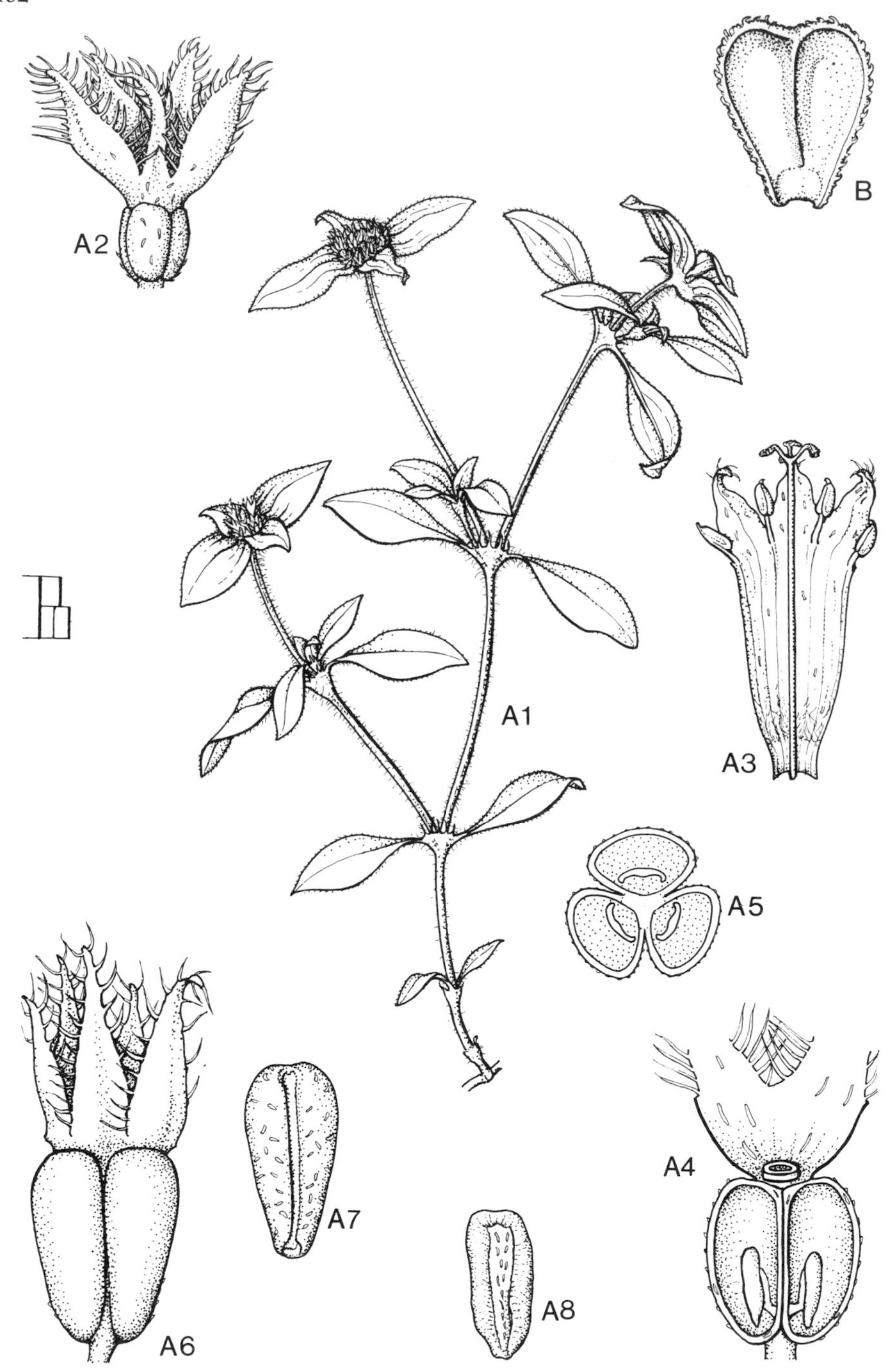

Tab. 43. A. — RICHARDIA SCABRA. A1, habit (×$\frac{2}{3}$); A2, calyx (× 8); A3, flower, longitudinal section (× 8); A4, ovary, longitudinal section (× 14); A5, transverse section of ovary (× 14); A6, capsule (× 10); A7, coccus, view of inner face (× 10); A8, seed, ventral view (× 10), A1–8 from *Drummond & Hemsley* 2888. B. — RICHARDIA BRASILIENSIS. B, coccus, view of inner face (× 10), *Drummond & Hemsley* 2070. From F.T.E.A.

ovate-triangular, sometimes with a few hairs at the apex. Style 7.5 mm. long, the 3 short branches 0.7 mm. long; stigmas spathulate, 0.7 mm. long. Cocci grey-brown, 2.5–2.8 × 1.5 mm.; inner face with smooth depressed area very narrow, other parts particularly the dorsal area densely verrucose or less often practically or entirely smooth. Seeds purplish brown, c. 2.5 × 1.3 mm., almost oblong; hilum very narrow.

Zambia. B: Kaoma, fl. & fr. 3.iv.1982, *Drummond & Vernon* 11161 (K; SRGH). C: Mkushi, fl. & fr. 15.iii.1981, *Drummond & Vernon* 10788 (K; LISC; MO; PRE; SRGH). S: Mochipapa Res. Sta., 10 km. E. of Choma, fl. & fr. 26.iii.1982, *Drummond & Vernon* 11017 (K; LISC; MO; PRE; SRGH). **Zimbabwe**. N: Mazoe, Glendale, fl. & fr. 7.iii.1944, *G.M. Taylor* in GHS 11807 (K; LISC; SRGH). C: Chegutu, Cedrela, fl. & fr. 9.iii.1969, *Hornby* 3464 (K; SRGH). E: Chipinge, map square 2032 D1, fl. & fr. 29.i.1975, *Gibbs-Russell* 2643 (K; SRGH). **Malawi**. S: Mulanje Distr., Blantyre to Mulanje via Madimba road, Phalombe Plain, fl. & fr. 2.iv.1978, *Pawek* 14182 (K; MAL; MO).

Also in Tanzania, S. Africa (Transvaal), S. & Central America, Jamaica, Cuba and Florida. Roadsides, weed on agricultural land etc.; 600–1500 m.

Tribe **12. RUBIEAE**

Herbs. Leaves and leaf-like stipules forming (pseudo)whorls. Flowers hermaphrodite, occasionally also unisexual; corolla rotate to campanulate (or salver-shaped), tube mostly very short; disk present; ovary bilocular, each locule with a single ovule affixed to septum. Fruit dry and dehiscing into 2 mericarps, or fleshy; one carpel occasionally aborted. Seed with membranous testa adhering to fruit wall, dorsal side convex, central side plane to concave.

Note. In the following descriptions the morphologically more precise but awkward term "leaves and leaf-like stipules in whorls of" is abbreviated to "leaves in whorls of". Note also that leaf descriptions and measurements strictly refer to the middle region of flowering shoots; in leaf width measurements the reflexed margins are not taken into account.

Only two genera of this ± cosmopolitan tribe occur wild in the Flora Zambesiaca area.

Leaves petiolate (in the Flora Zambesiaca area); corolla mostly 5-merous; fruit fleshy, glabrous - - - - - - - - - - - - - - - **36. Rubia**
Leaves (sub)sessile; corolla mostly 4-merous; fruits dry (in the Flora Zambesiaca), glabrous or variously hairy - - - - - - - - - - - - - - **37. Galium**

36. RUBIA

Rubia L., Sp. Pl. **1**: 109 (1753); Gen. Pl. ed. 5: 47 (1754).

Scrambling, creeping or climbing perennial herbs, somewhat woody near the base; roots quite woody, reddish (yielding a red dye). Stems branched, 4-angled, ± brittle, mostly beset with recurved prickles. Leaves in whorls of (3)4–8(12), petiolate (in the Flora Zambesiaca area), blades cordate or broadly ovate, ovate-lanceolate to linear, rounded to ± cordate at base or narrowed to base, with 1–5(7) prominent veins. Flowers in axillary and/or terminal cymes, hermaphrodite, 5 (rarely 4 or 6) -merous. Calyx lobes obsolete. Corolla rotate to subcampanulate, greenish, yellowish green to yellow; lobes ± triangular, (long) acuminate. Anthers exserted. Ovary bilocular, one ovule in each chamber, crowned by a minute disk; style branches 2, joined below; stigmas capitate. Fruit fleshy, glabrous, consisting of 2 round mericarps, each with a single seed, round on dorsal side, plane to convex on ventral side; one mericarp often aborted.

A widely distributed genus (Europe, Asia and Africa) of c. 60(?) species; only 4 species occur in Africa south of the Sahara.

Leaves in whorls of 4 (very rarely 3 or 5) - - - - - - 1. *cordifolia* subsp. *conotricha*
Leaves in whorls of 6–8(12) - - - - - - - - - - - - - 2. *horrida*

1. **Rubia cordifolia** L., Syst. Nat., ed. 12, 3 (App.): 229 (1768); Mant. Alt. 197 (1771). —DC., Prodr. **4**: 588 (1830). Type said to be from Majorca.

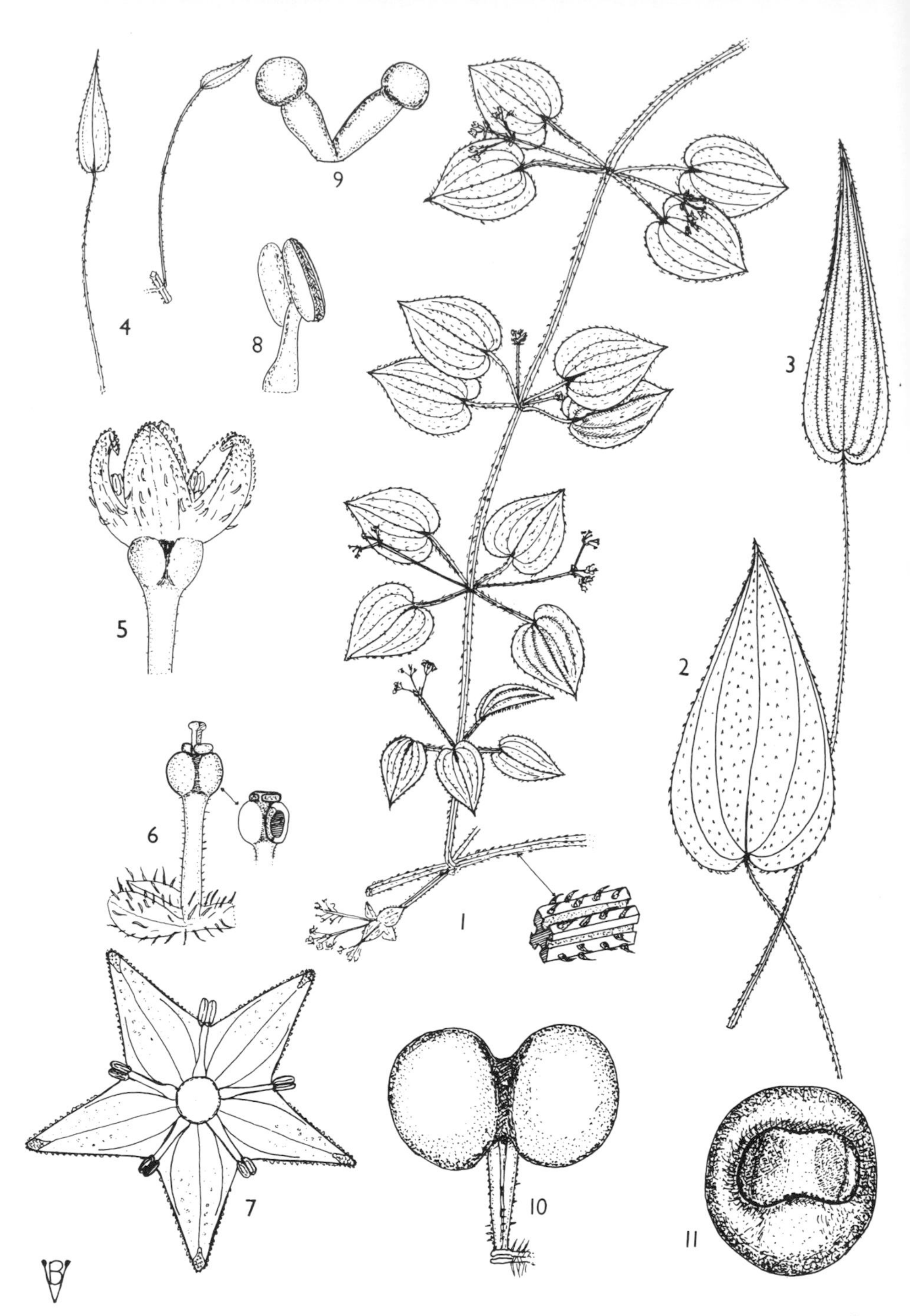

Tab. 44. RUBIA CORDIFOLIA subsp. CONOTRICHA. 1, flowering branch with enlargement of stem (× 4) & (×½), from *Greenway* 3055; 2, large leaf (×½), from *Drummond & Hemsley* 3166; 3, large leaf (×½), from *Brodhurst Hill* 358; 4, small leaves (×½), from *Heriz-Smith & Paulo* 852 and *Newbould* 3400; 5, young flower, lateral view (× 7); 6, young flower with corolla removed (× 5); 7, small flower flattened out (× 10); 8, stamen, lateral view (× 20); 9, styles and stigmas (× 20), 5–9 from *Verdcourt* 592; 10, fruit (× 2½); 11, seed (× 10), 10–11 from *Milne-Redhead & Taylor* 9013. From F.T.E.A.

Subsp. **conotricha** (Gand.) Verdc. in Kew Bull. **30**: 323 (1975); in F.T.E.A., Rubiaceae 1: 381, fig. 57 (1976). —Puff in Journ. S. Afr. Bot. **50**: 349 (1984); in Fl. Southern Afr. **31**, 1(2): 62 (1986). TAB. **44**. Type from S. Africa.

Rubia conotricha Gand. in Bull. Soc. Bot. Fr. **65**: 35 (1918). Type as above.

Rubia longipetiolata Bullock in Kew Bull. **1932**: 497 (1932). Type from Kenya.

Stems to c. 5(6) m. long, distinctly 4-ribbed, with recurved prickles on ribs, occasionally with longish white hairs between ribs, at least around nodes. Leaves in whorls of 4 (very rarely 3 or 5); blades (15)20–65(90) × (5)10–30(40) mm., (narrowly) lanceolate to broadly ovate, rounded to cordate at base, acute to acuminate at apex, with 5(7; very rarely 3) prominent veins; margins and veins below with recurved prickles, upper surface ± glabrous to scabrid, lower surface ± glabrous or sparsely to ± densely covered with longish white hairs, blades rarely glabrous; petioles (10)15–80(120) mm. long, mostly with recurved prickles and occasionally also hairy. Inflorescence several- to many-flowered, ± lax to rather dense; peduncles and pedicels 1–15(20) mm. long, glabrous, somewhat pubescent or with small prickles; ultimate bracts c. 1–2 mm. long or sometimes absent. Flowers 5 (very rarely 6) -merous; corolla 3–6 mm. in diam.; filaments to 1 mm., anthers to 0.5 mm. long; ovary 0.4–0.8 mm. long. Fruit dark purple to black, each mericarp 2.5–4 mm. in diam.

Zambia. N: Mbala Distr., banks of Chisunga R., fl. 11.ii.1952, *Richards* 686 (K). W: Solwezi, fr. 27.vii.1967, *Fanshawe* 8855 (K). **Zimbabwe**. C: Wedza Mt., 27.ii.1964, *Wild* 6358 (BM; K; SRGH). E: Chimanimani Distr., Kasipiti, fl. & fr. 1.vi.1966, *Loveridge* 1586 (K; SRGH). S: Bikita Distr., c. 8 km. SE. of Silveira Mission, fl. 7.v.1969, *Biegel* 3058 (K; SRGH). **Malawi**. N: Nyika Plateau, c. 4 km. SW. of Rest House, 24.x.1958, *Robson & Angus* 306 (BM; K; SRGH). C: Dedza Mt., fl. 2.ii.1967, *Salubeni* 531 (K; MAL; SRGH). S: Chiradzulu Mt., fl. 3.iv.1970, *Brummitt & Banda* 9594 (K; SRGH). **Mozambique**. Z: Mambane, fl. iv.1859, *Kirk* s.n. (K). MS: Manica, serra de Mavita, fl. & fr. 23.iv.1948, *Barbosa* 1534 (LISC). M: Namaacha, *Marques* 2761 (LMU).

Widely distributed in E. tropical Africa; also in S. Africa (Transvaal to E. Cape) and Angola. Mostly at forest edges, in forest clearings, or in scrub or bush-clumps.

The species (as a whole) extends from Africa to E. and SE. Asia. Subsp. *conotricha* is very variable in leaf size and shape, petiole length, indumentum and extensiveness of the inflorescence.

2. **Rubia horrida** (Thunb.) Puff in Kew Bull. **32**: 432 (1978); in Journ. S. Afr. Bot. **50**: 357, fig. 2b, c (1984); in Fl. Southern Afr., **31**, 1(2): 65 (1986). Type from S. Africa.

Galium horridum Thunb. in Hoffmann Phytograph. Blätter **1**: 16 (1803); Fl. Cap. **1**: 556 (1813). —Sond. in Harv. & Sond., F.C. **3**: 37 (1865). Type as above.

Rubia petiolaris var. *heterophylla* Sond. in Harv. & Sond., F.C. **3**: 35 (1865). Types from S. Africa.

Stems to c. 2–3 m. long, distinctly 4-ribbed, with recurved prickles on ribs, younger parts often with short hairs below nodes, occasionally also with some short whitish hairs between ribs. Leaves in whorls of 6–8(12), blades 10–50(70) × 2–8(10) mm., lanceolate, linear-lanceolate to linear, mostly narrowed to base, blades occasionally difficult to distinguish from faintly winged petioles, acute at apex, only midvein prominent; margins and midvein below with recurved prickles, seldom scabrid above; petioles 10–25(40) mm. long, ± 3-angular in section or somewhat canaliculate, mostly with (2 rows of) recurved prickles. Inflorescence several- to many-flowered, rather lax to ± dense; peduncles and pedicels 1–6 mm. long, glabrous or sometimes hairy; ultimate bracts c. 2–5 mm. long, linear. Flowers 5 (very rarely 4 or 6) -merous; corolla 3–6 mm. in diam.; filaments to 0.6 mm., anthers to 0.5 mm. long; ovary c. 0.4–0.8 mm. long. Fruit blackish or dark purple, each mericarp 2.5–3.5 mm. in diam.

Botswana. SE: Ga-Ngwaketse, Kanye, *Kelaole* A97 (SRGH). **Zimbabwe**. W: Matopos Research Station, *Kennan* 222 (SRGH). C: Harare Distr., Calgary Farm, fl. 12.i.1966, *Howard-Williams* 37 (K; SRGH).

Also in S. Africa and SW. Africa/Namibia (Windhoek Bergland). Mostly in bush-clumps in woodland areas.

37. GALIUM L.

Galium L., Sp. Pl. **1**: 105 (1753); Gen. Pl. ed. 5: 46 (1754).

Perennial herbs or annuals. Stems erect, creeping or climbing, often distinctly 4-angled, glabrous, hairy or with recurved prickles. Leaves in whorls of 4–10, (sub)sessile, blades linear, lanceolate to (ob)ovate, mostly with a prominent midvein (1-nerved). Flowers in axillary and/or terminal many-to few-flowered cymes, hermaphrodite (and protandrous), rarely unisexual, 4-merous. Calyx lobes obsolete. Corolla rotate or (sub)campanulate, greenish(-yellow), creamy white to bright yellow, lobes ovate to triangular, acute to acuminate. Anthers exserted. Ovary bilocular, one ovule in each chamber, crowned by an often 2-lobed disk; style branches 2, joined below; stigmas capitate. Fruit dry (in the Flora Zambesiaca area), glabrous or variously hairy, dehiscing into 2 (sub)globose or reniform mericarps, each with a single seed, round on dorsal side, ± excavated on ventral side; one mericarp occasionally aborted.

A cosmopolitan genus of several hundred species, most abundant in temperate regions.

1. Leaves 3-nerved, in whorls of 4 - - - - - - - - - 1. *thunbergianum*
– Leaves 1-nerved, in whorls of (5)6–10 - - - - - - - - - - - 2
2. Inflorescences very reduced, flowers in groups of 3 to 1 at nodes; leaves (linear)lanceolate to (broadly) obovate - - - - - - - - - - - - - - - 3
– Inflorescences ± extensive, several- to many-flowered, if ± few-flowered, leaves linear (no. 5) 4
3. Fruits covered with white, well spaced hooked hairs less than 0.5 mm. long; pedicels (8)10–20(35) mm. long, mostly arising singly from nodes of the main stems and branches - - - - - - - - - - - - - 2. *spurium* subsp. *africanum*
– Fruits covered with brownish, densely spaced hooked hairs c. 1 mm. long; flowers subtended by 4–1 bracts, true pedicels to 1 mm. long - - - - - - - 3. *chloroionanthum*
4. Leaf margins with ± massive, densely spaced reversed prickles; flowers yellow 4. *scabrellum*
– Leaf margins without distinct and conspicuous prickles; if inconspicuous denticles present flowers white - - - - - - - - - - - - - - - 5
5. Leaf blades with a distinct, filiform acumen at the apex; flowers yellow, corolla rotate 5. *bussei*
– Leaf blades obtuse to ± acute at the apex but without a filiform acumen; flowers white, corolla (sub)campanulate - - - - - - - - - - - - - - - 6
6. Leaves to 35 mm. long, linear to linear-elliptic, often ascending - - - 6. *stenophyllum*
– Leaves to c. 20 mm. long, linear-lanceolate to elliptic, (strongly) reflexed 7. *scioanum*

1. **Galium thunbergianum** Eckl. & Zeyh., Enum. Pl. Afr. Austr. Extratrop.: 369 (1836). —Verdc. in F.T.E.A., Rubiaceae 1: 387 (1976). —Puff in Journ. S. Afr. Bot. **44**: 221 (1978); in Fl. Southern Afr. **31**, 1(2): 66 (1986). Type from S. Africa (Cape Prov.).

Perennial with extensive rhizome. Stems climbing, suberect or procumbent, (8)10–40(60) cm. long, glabrous or with short spreading white hairs. Leaves in whorls of 4, 3-nerved, (7)10–18(22) × (3)4–8(10) mm., ovate, elliptic or ± rhombic, ± acute or with a short hyaline point at apex, narrowed to base; glabrous or both surfaces and margins with short spreading hairs. Inflorescence broadly pyramidal to ± cylindrical, cymes ± many-flowered, ultimate branches with 1–2 minute bracts; pedicels 1–3(5) mm. long, slightly elongating after anthesis, glabrous or hairy, ± divaricate in fruit. Corolla (1)1.5–2.5(3) mm. in diam., rotate, often somewhat hairy outside, greenish, greenish-white to yellowish, lobes longer than wide, acute; stamens c. half as long as lobes; ovary c. 0.3–0.5 mm. long. Fruit densely covered with white tuberculate hooked hairs; mericarps subglobose, each (0.8)1–1.5 mm. in diam.

Stems, leaves, peduncles and pedicels glabrous or nearly so (occasionally a few hairs at some upper nodes or along nerves of leaves) - - - - - - - - - - var. *thunbergianum*
Stems (mainly angles), leaves, peduncles and pedicels ± densely covered with short spreading white hairs - - - - - - - - - - - - - - - - - - var. *hirsutum*

Var. **thunbergianum**

Galium rotundifolium sensu Sond. in Harv. & Sond., F.C. **3**: 39 (1865) non L.
Galium natalense Rouy in Fl. Fr. **8**: 9 (1903) adnot. Type from S. Africa (Natal).
Galium rotundifolium var. *normale* Kuntze, Rev. Gen. **3**: 120 (1898). Type from S. Africa (Natal).

Zimbabwe. E: Inyanga Distr., Stapleford, Nuza slopes, fl. & fr. 5.iv.1962, *Wild* 5706 (K; LISC; SRGH).

Also in Ethiopia, W. Kenya, Swaziland, Lesotho and S. Africa. In forest edge vegetation, in scrub, occasionally along streams or in gulleys, often in damp to wet, ± shady places; c. 1880–2300 m.

Var. **hirsutum** (Sond.) Verdc. in Kew Bull. **30**: 326 (1975). Type from S. Africa (Cape Prov.),

Galium rotundifolium var. *hirsutum* Sond. in Harv. & Sond., F.C. **3**: 39 (1865). Type as above.
Galium rotundifolium sensu Thunb., Fl. Cap. **1**: 551 (1813) non L.
Galium dasycarpum Schweinf., Beitr. Fl. Aethiop.: 135 (1867). Type from Ethiopia.
Galium biafrae Hiern in F.T.A. **3**: 245 (1877). Syntypes from Fernando Po and Cameroon Mtn.

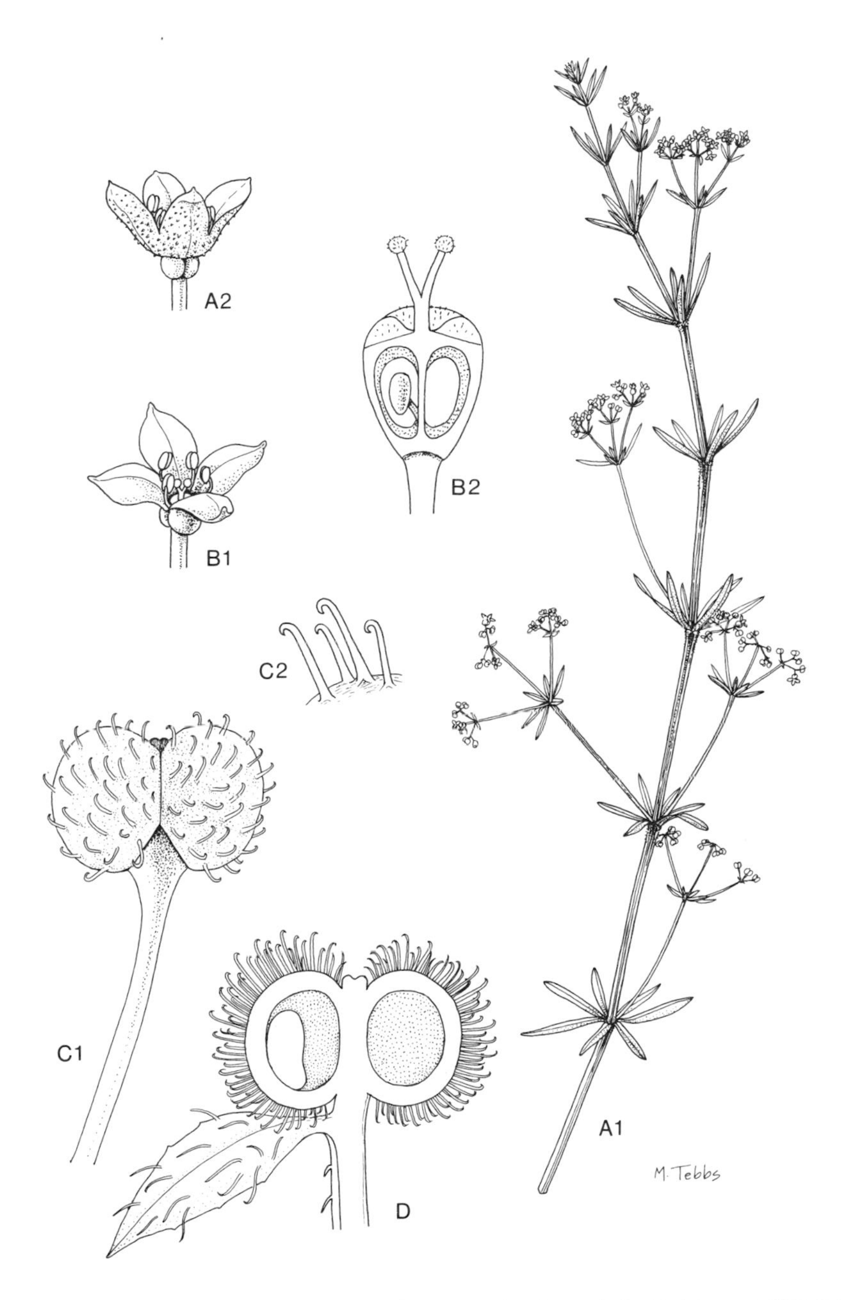

Tab. 45. A. —GALIUM STENOPHYLLUM. A1, habit (× ½); A2, flower (× 8), A1–2 from *Puff* 781230-2/1. B. —GALIUM BUSSEI var. GLABRUM. B1, flower (× 1); B2, ovary, longitudinal section (× 8), B1–2 from *Puff* 780210-1/1. C. —GALIUM SPURIUM subsp. AFRICANUM. C1, fruit (with upper part of pedicel) (× 8); C2, part of mericarp surface showing hooked hairs (× 25), C1–2 from *Puff* 780214-1/1. D. —GALIUM CHLOROIONANTHUM. Immature fruit with subtending bract, longitudinal section (× 8), from *Puff* 780210-2/1.

Zimbabwe. E: Inyanga Distr., Inyanga Downs, fl. & fr. 30.i.1931, *Norlindh & Weimarck* 4740 (BR; K; SRGH).

Widely but disjunctly distributed in afromontane areas (E., W. & S. Africa). Habitats as for var. *thunbergianum*.

2. **Galium spurium** L., Sp. Pl. **1**: 106 (1753). Type a plant grown in Uppsala Botanic Garden (no specimen in LINN).

Subsp. **africanum** Verdc. in Kew Bull. **30**: 324 (1975); in F.T.E.A., Rubiaceae 1: 390 (1976). —Puff in Journ. S. Afr. Bot. **44**: 271 (1978); in Fl. Southern Afr. **31**, 1(2): 67 (1986). TAB. **45**, fig. C. Type from Kenya.

Galium horridum sensu Eckl. & Zeyh., Enum. Pl. Afr. Austr. Extratrop.: 370 (1836) non Thunb.

Galium aparine sensu Sond. in Harv. & Sond., F.C. **3**: 38 (1865) non L.

Annual. Stems weak, prostrate to semi-erect, (0.1)0.3– c. 2 m. long, ± few-branched, with recurved prickles on angles. Leaves in whorls of 6–8, 1-nerved, (10)15–40(45) × (1.5)3–6(7) mm., (linear)lanceolate to obovate, with a distinct ± filiform acumen at apex, narrowed to base; upper surface glabrous or with a few scattered ± straight or curled hairs, midrib (beneath) and margins with coarse recurved prickles. Inflorescence extremely reduced, flowers mostly single, on ± arcuate pedicels, (8)12–20(35) mm. long when in fruit; pedicels occasionally separated from peduncle by a small ± linear bract and flowers in pairs. Corolla (1)1.2–1.8(2) mm. in diam., rotate, greenish, greenish-white or creamy white, lobes longer than wide, pointed; stamens very short; ovary c. 0.5–0.8 mm. long. Fruit covered with white non-tuberculate hooked hairs less than 0.5 mm. long; mericarps ± globose, each (2)2.5–4 mm. in diam.

Malawi. N: Misuku Hills, fr. vii.1896, *Whyte* s.n. (K). C: Dedza Distr., summit of Dedza Mt., fl. & fr. 14.ii.1978, *Puff* 780214–1/1 (J; WU). S: Zomba Plateau, Chingwe's Hole, *Exell, Mendonça & Wild* 760 (LISC; SRGH).

Widely distributed in tropical NE. and E. Africa and Lesotho, also in S. Africa. Often in disturbed or cultivated ground, also along streams and in forest edge vegetation; c. 1800–2300 m.

Typical *G. spurium* is a widely distributed weed (from Europe and N. Africa to W. Asia). It is believed to be introduced in certain parts of Africa (S. Africa, Cape Prov.!); no records are known from the Flora Zambesiaca area.

3. **Galium chloroionanthum** K. Schum. in Engl., Bot. Jahrb. **30**: 417 (1901). —Brenan in Mem. N.Y. Bot. Gard. **8**: 456 (1954). —Verdc. in F.T.E.A., Rubiaceae 1: 388 (1976). —Puff in Journ. S. Afr. Bot. **44**: 269 (1978). TAB. **45**, fig. D. Type from Tanzania.

Perennial. Stems climbing or creeping, to 2.3 m. long, few-branched, with ± large, hyaline recurved prickles on the 4 distinct, whitish angles, and often densely hairy at nodes. Leaves in whorls of 6, 1-nerved, (20)25–35(40) × (5)7–12(15) mm., (broadly) elliptic to obovate, abruptly acuminate at apex, narrowed to base; upper surface glabrous or with a few short, straight hairs, margins and midrib (beneath) with coarse, ± hyaline recurved prickles. Inflorescence very reduced, 3–1-flowered, peduncles 10–20 mm. long after anthesis, pedicels 0.5–1 mm. long subtended by 4–1 ± large to minute bracts. Corolla c. 3–4 mm. in diam., rotate, somewhat hairy outside, greenish-white to greenish-yellow; lobes about as long as wide, ± acute; stamens short; ovary c. 0.4–0.8 mm. long. Fruit densely covered with brownish, hooked hairs c. 1 mm. long; mericarps ± reniform, each 1–1.5 mm. in diam.

Zimbabwe. E: Inyanga Distr., Rhodes Inyanga National Park, Pungwe Gorge, fl. & fr. 15.vii.1955, *Chase* 5664 (BM; BR; COI; LISC; PRE; SRGH). **Malawi**. N: Nyika Plateau, below Sangule kopje, 7 km. SW. of Chelinda Camp, fr. 15.v.1970, *Brummitt* 10769 (BR; K; SRGH). S: Mulanje Mt., Luchenza Plateau, 1850 m., fl. & fr. 1.vii.1946, *Brass* 16574 (K; SRGH).

Also in the mountains of tropical E. Africa and Madagascar. Mostly in forest edge vegetation; c. 1800–2250 m.

4. **Galium scabrellum** K. Schum. in Engl., Bot. Jahrb. **28**: 113 (1899). — Brenan in Mem. N.Y. Bot. Gard. **8**: 457 (1954). — Verdc. in F.T.E.A., Rubiaceae 1: 401 (1976). — Puff in Journ. S. Afr. Bot. **44**: 255 (1978). Type: Malawi, Nyika Plateau, *Whyte* ("Carsson") 269 (B, holotype; K; NU; WU, photos).

Galium bequaertii De Wild. in Rev. Zool. Afr. **9**, Suppl. Bot.: 12 (1921); Pl. Bequaert. **2**: 307 (1923). — Robyns, Flore Spermatoph. Parc Nat. Albert **2**: 381 (1947). Type from Zaire.

Perennial. Stems scrambling or ± climbing, c. 0.4–2(3) m. long, ± much-branched, with ± straight, spreading hairs on the 4 distinct angles. Leaves in whorls of (6)8(10), 1-nerved,

(7)10–15 × (0.5)1–2(3) mm., linear to narrowly elliptic, with a distinct, whitish acumen at apex, narrowed to base; upper surface smooth or with a few scattered, ± straight hairs, lower surface with hairs at least on the midrib, the usually reflexed margins with densely spaced recurved prickles. Inflorescence ± cylindrical, cymes many-flowered, ultimate branches often ebracteate; peduncles 1–2 mm. long, hairy, pedicels 1–3 mm. long, hairy or glabrous, divaricate in fruit. Corolla 2–2.5(3) mm. in diam., rotate, often hairy outside, mostly pale or bright yellow; lobes much longer than wide, ± triangular, acuminate; stamens c. half as long as lobes; ovary c. 0.3–0.5 mm. long. Fruit mostly glabrous; mericarps ± globose, each c. 1 mm. in diam.

Zambia. N: Mbala, Itembwe Gorge, fl. & fr. 22.ii.1967, *Richards* 22138 (BR; K; LISC). E: Nyika Plateau, fl. 30.xii.1962, *Fanshawe* 7311 (BR; K). **Zimbabwe**. E: Inyanga Distr., circular road Mt. Inyangani-Troutbeck Inn, on tributary of Nyamziwa R., fl. 25.i.1979, *Puff* 790125-5/1 (WU). **Malawi**. N: Nkhata Bay Distr., c. 41.6 km. on Camp Garden road, fl. & fr. 21.ix.1969, *Pawek* 2731 (K; MAL; MO).

Also in Uganda, Tanzania, Zaire, Rwanda and Burundi. In forest edges, clearings or in bush-clumps; c. (1200)1700–2300 m.

5. **Galium bussei** K. Schum. & K. Krause in Engl., Bot. Jahrb. **39**: 571 (1907). —Brenan in Mem. N.Y. Bot. Gard. **8**: 456 (1954). —Verdc. in F.T.E.A., Rubiaceae 1: 399 (1976). —Puff in Journ. S. Afr. Bot. **44**: 246 (1978). —Gonçalves in Garcia de Orta, Sér. Bot. **5**: 192 (1982). Type from Tanzania.

Perennial with a somewhat woody rootstock. Stems erect or semi-erect, (15)25–60(70) cm. long, with usually few ± short lateral branches, the 4 distinct, often whitish angles with short ± spreading hairs or glabrous; nodes usually hairy to densely hairy. Leaves in whorls of (5)6–10, 1-nerved, (10)15–30(40) × 0.5–1.5(2.5) mm., linear to linear-oblanceolate, often apparently terete due to strongly recurved margins, with a distinct, filiform brownish acumen at apex; upper and lower surface with short, ± spreading hairs, glabrous or with minute, forwardly directed prickles above. Inflorescence broadly pyramidal to ± narrowly cylindrical, cymes many-flowered and dense to rather few-flowered and lax, ultimate branches with (0)1–4 minute bracts; pedicels 1–3(6) mm. long, ± filiform, glabrous, strongly divaricate in fruit. Corolla 2–3(4) mm. in diam., rotate, glabrous, yellow, pale yellow or greenish-yellow; lobes longer than wide, ± acuminate; stamens ± half as long as lobes; ovary c. 0.4–0.8 mm. long. Fruit glabrous, ± granulated; mericarps subglobose, each 1.5–2 mm. in diam.

1. Inflorescence dense and many-flowered - - - - - - - - - - - 2
- Inflorescence rather lax and diffuse, several- to rather few-flowered - - - - - 3
2. Stems (at least on the angles) and leaves (mainly lower surface) covered with short, ± spreading white hairs - - - - - - - - - - - - - - - - - - var. *bussei*
- Stems (except nodes) and leaves glabrous - - - - - - - - - var. *glabrum*
3. Stems (at least on the angles) and leaves (mainly lower surface) covered with short, ± spreading white hairs - - - - - - - - - - - - - - - - - - var. *strictius*
- Stems (except nodes) and leaves glabrous - - - - - - - - var. *glabrostrictius*

Var. **bussei**

Galium stenophyllum var. *flavoviride* Utzschn. & Merxm. in Transact. Rhod. Sc. Ass. **43**: 57 (1951). Type: Zimbabwe, Marondera Distr., *Dehn* 719A (BR; M, holotype SRGH).

Galium stenophyllum auct., non Bak.

Zimbabwe. C: Ruwa Distr., Tanglewood Farm, fl. xii.1960, *Miller* 7558 (K; LISC; P; SRGH). E: Inyanga Distr., Inyanga Downs near village, fl. 29.i.1931, *Norlindh & Weimarck* 4677 (K; PRE; SRGH). **Malawi**. N: Nyika Plateau, Nganda Peak, fl. 10.iv.1969, *Pawek* 2076 (K; MAL). C: Dedza Distr., Chongoni Forest Reserve, base of Chiwau Hill, fl. 15.ii.1978, *Puff* 780215-1/2 (J; WU). S: Blantyre, fl. 12.ii.1948, *Faulkner* 202 (K). **Mozambique**. N: Niassa Distr., Lichinga (Vila Cabral), *Torre* 945 (LISC; ± var. *bussei*).

Also in Tanzania and Zaire (Shaba). Mostly in grassland or grassy places in *Brachystegia* woodland; c. 1200–2250(2600) m.

Var. **glabrum** Brenan in Mem. N.Y. Bot. Gard. **8**: 456 (1954). —Verdc. in F.T.E.A., Rubiaceae 1: 399 (1976). —Puff in Journ. S. Afr. Bot. **44**: 247 (1978). —Gonçalves in Garcia de Orta, Sér. Bot. **5**: 192 (1982). TAB. **45**, fig. B. Type from Tanzania.

Zambia. W: Mufulira, fl. 11.v.1934, *Eyles* 8345 (K). E: Nyika Plateau, nr. (Zambian) Rest House, fl. 28.xi.1955, *Lees* 125 (K). **Zimbabwe**. C: Shurugwi (Selukwe) Distr., Ferny Creek, fl. 5.xii.1966, *Biegel* 1495 (SRGH). E: Inyanga, fl. 30.x.1930, *Fries, Norlindh & Weimarck* 2445 (BR; K; SRGH).

S: Masvingo Distr. (Fort Victoria), Kyle National Park, base of Mtunumashava Hill, fl. 21.v.1971, *Grosvenor* 511 (K; SRGH). **Malawi**. N: Nkhata Bay Distr., Viphya link road, c. 107 km. S. of Mzuzu, fl. 22.xi.1975, *Pawek* 10364 (K; MO). C: Lilongwe Distr., Damboleni Dambo, Senyala, *Jackson* 689 (MAL). S: Zomba Plateau, *Binns* 532 (MAL). **Mozambique**. Z: Serra do Gúruè, *Mendonça* 2251 (LISC). T: Mt. Furancungo, 1380–1420 m., fl. & fr. 15.iii.1966, *Pereira, Sarmento & Marques* 1741 (LMU). MS: Manica, Mavita, Serra Mocuta, *Pereira & Marques* 1076 (LMU).

Also in Tanzania and Zaire (Shaba). In grassy places in woodland, occasionally in disturbed areas; c. 1200–2250 m.

Var. **strictius** Brenan in Mem. N.Y. Bot. Gard. **8**: 457 (1954). —Verdc. in F.T.E.A., Rubiaceae 1: 400 (1976). Type: Malawi, Zomba, *Purves* 59 (K, holotype).

Zambia. N: Mbala Distr., Chipululu Farm, fl. & fr. 6.i.1955, *Richards* 19423 (K). W: Kitwe, fl. 15.xii.1955, *Fanshawe* 2657 (BR; K; LISC; SRGH). C: Lusili R., 5 km. W. of Serenje, fr. 3.ii.1973, *Kornaś* 3139 (K). **Malawi**. N: Nyika Plateau, Lake Kaulime, fl. & fr. 7.1.1959, *Richards* 10535 (K). C: Kasungu Distr., Chimaliro Forest, Phaso road, fl. & fr. 10.i.1975, *Pawek* 8892 (K; SRGH). S: Zomba, fl. xii.1900, *Purves* 59 (K).

Also in S. Tanzania. Mostly in wooded grassland; c. 1000–2250 m.

Var. **glabrostrictius** Brenan in Mem. N.Y. Bot. Gard. **8**: 457 (1954). —Verdc. in F.T.E.A., Rubiaceae 1: 400 (1976). —Gonçalves in Garcia de Orta, Sér. Bot. **5**: 192 (1982). Type: Zambia, Mwinilunga Distr., just N. of Matonchi farm, *Milne-Redhead* 4283 (K, holotype; BM).

Zambia. N: c. 48 km. W. of Nakonde (Tunduma), fl. 1.i.1959, *Robinson* 2975 (K; SRGH). W: Mulenga Protected Forest area, c. 20 km. NW. of Kansanshi, fr. 19.iii.1961, *Drummond & Rutherford-Smith* 7045 (BR; K; LISC; SRGH). E: near Chipata, Kapatamoya, fl. & fr. 5.i.1959, *Robson* 1038 (BR; K; LISC; SRGH). **Malawi**. N: Misuku Hills, Mughesse, fl. 31.xii.1970, *Pawek* 4276 (K). C: Dedza Distr., Chongoni Forest Reserve, near Kanjole Hill, fl. & fr. 12.i.1967, *Salubeni* 484 (K; LISC; MAL; SRGH). S: Ntcheu Distr., Lower Kirk Range, 17.iii.1955, Chipusiri, *Exell, Mendonça & Wild* 978 (LISC; SRGH). **Mozambique**. N: Lichinga (Vila Cabral), Serra de Massangulo, *Torre & Paiva* 10768 (LISC). T: Mt. Zóbuè, c. 1000 m., fr. 11.iii.1964, *Torre & Paiva* 11135 (LISC).

Also in Tanzania and Zaire (Shaba). Mostly in grassy places in woodland or open grassland; c. 1200–1800 m.

Only extreme forms of these varieties are clearly distinguishable; other forms often are difficult to place.

6. **Galium stenophyllum** Bak. in Kew Bull. **1895**: 68 (1895). —Brenan in Mem. N.Y. Bot. Gard. **8**: 456 (1954). —Verdc. in F.T.E.A., Rubiaceae 1: 397 (1976). TAB. **45**, fig. A. Syntypes: Zambia, Fwambo, *Carson* 40, 41 & 80 of 1893 collections and s.n. of 1889 collections (K).

Perennial with a slightly woody base. Stems several to numerous, suberect or straggling, c. 10–30(45) cm. long, glabrous or with whitish spreading hairs. Leaves in whorls of 6, 1-nerved, 10–35 × 0.5–1.5(3) mm., linear to linear-elliptic, obtuse to ± acute at apex but without a distinct filiform acumen, narrowed to base; glabrous to ± densely covered with spreading hairs, margins mostly recurved. Inflorescence ± cylindrical, cymes ± many-flowered, ultimate branches with 1–2 minute bracts; pedicels 1–3(4) mm. long, often recurved in fruit. Corolla 2–3(3.5) mm. in diam., ± campanulate, white, lobes longer than wide, ± acute; stamens ±. half as long as lobes; ovary c. 0.3–0.6 mm. long. Fruit glabrous; mericarps subglobose, each c. (1.8)2–3 mm. in diam.

Zambia. N: Mbala Distr., Safu marsh, Safu R. beyond Chilwa village, fl. 23.x.1967, *Richards* 22402 (K; LISC). C: 7 km. from Serenje, fl. 23.xii.1963, *Symoens* 10690 (K). **Malawi**. N: Mzimba Distr., base of Hora Mt., fl. & fr. 30.xii.1978, *Puff* 781230-2/1 (BM; J; WU). C: Dedza, 1500 m., fl. 13.ix.1946, *Brass* 17633 (BR; K; SRGH).

Also in Tanzania and Zaire. At the edge of dambos and in other wet to moist places.

7. **Galium scioanum** Chiov. in Ann. Bot. Roma **9**: 322 (1911). —Brenan in Kew Bull. **5**: 372 (1951). —Verdc. in F.T.E.A., Rubiaceae 1: 398 (1976). Type from Ethiopia.

Perennial. Stems scrambling or climbing, to c. 1 m. long, glabrous or with spreading whitish hairs, especially on the 4 distinct, whitish angles. Leaves often strongly reflexed, in whorls of 6, 1-nerved, (5)8–15(25) × 1–7(10) mm., linear-lanceolate to (narrowly) elliptic, ± acute at apex, narrowed to base; surfaces glabrous to hairy, the flat or recurved margins often hairy or with small, inconspicuous denticles. Inflorescence pyramidal to ± cylindrical, cymes many-flowered, ultimate branches with 1–2 minute bracts; pedicels 1–5 mm. long, divaricate or recurved in fruit. Corolla 2–4 mm. in diam., (sub)campanulate, white, lobes much longer than wide, ± acute; stamens ±. half as long as lobes; ovary c. 0.3–0.5 mm. long. Fruit glabrous; mericarps subglobose, each c. 1.5–2(2.5) mm. in diam.

Leaf blades 1–2.5(3) mm. wide, margins usually distinctly revolute - - - var. *glabrum*
Leaf blades (2.5)3–7(10) mm. wide, margins usually flat - - - - - var. *latum*

Var. **glabrum** Brenan in Kew Bull. **5**: 372 (1951). —Verdc. in F.T.E.A., Rubiaceae 1: 398 (1976). Type: Zambia, Mwinilunga Distr., Lunga R., just below Mudjanyama R. junction, *Milne-Redhead* 3401 (BM; K, holotype).

Zambia. W: Chifubwa, Solwezi, fl. 5.i.1962, *Holmes* 1451 (K). C: Kabwe Distr., Mufukushi R. near Chipepo, 51 km. NW. of Kabwe, fl. & fr. 20.i.1973, *Kornaś* 3051 (K).
Widely but disjunctly distributed in tropical Africa. Swampy and marshy areas.

Var. **latum** (De Wild.) Verdc. in Kew Bull. **30**: 326 (1975); in F.T.E.A., Rubiaceae 1: 399 (1976). Type from Zaire (Shaba).
Galium latum De Wild. in Fedde, Repert., Beih. **13**: 140 (1914). Type as above.

Zambia. W: Solwezi, fl. & fr. 12.x.1953, *Fanshawe* 403 (BR; K).
Also in Ethiopia, Tanzania and Zaire. Swampy and marshy areas.
Var. *scioanum* is known from Ethiopia, Tanzania and Zaire. The rather arbitary subdivision of *G. scioanum* into three varieties on the basis of hairiness and leaf width often makes it impossible to assign clearly a collection to one variety or another.

INDEX TO BOTANICAL NAMES